Springer Finance

Springer Finance Lecture Notes

This subseries of Springer Finance includes books with the character of lecture notes. Typically these are research monographs on new, cutting-edge developments in mathematical finance; sometimes they may be a glimpse of a new field of research activity, or presentations of a new angle in a more classical field. In the established tradition of Lecture Notes, the timeliness of a manuscript can be more important than its form, which may be informal, preliminary or tentative.

The Springer Finance series, launched in 1998, is addressed to students, academic researchers and practitioners working on increasingly technical approaches to the analysis of financial markets. It covers mathematical and computational finance broadly, reaching into foreign exchange, term structure, risk measure and management, portfolio theory, equity derivatives, energy finance and commodities, financial economics.

All titles in this series are peer-reviewed to the usual standards of mathematics and its applications.

Christian Bayer • Gonçalo dos Reis •
Blanka Horvath • Harald Oberhauser
Editors

Signature Methods in Finance

An Introduction with Computational Applications

Editors
Christian Bayer
Weierstrass Institute
Berlin, Germany

Gonçalo dos Reis
School of Mathematics
University of Edinburgh
Edinburgh, UK

Blanka Horvath
Mathematical Institute
University of Oxford
Oxford, UK

Harald Oberhauser
Mathematical Institute
University of Oxford
Oxford, UK

ISSN 1616-0533 ISSN 2195-0687 (electronic)
Springer Finance
ISSN 2524-681X ISSN 2524-6828 (electronic)
Springer Finance Lecture Notes
ISBN 978-3-031-97238-6 ISBN 978-3-031-97239-3 (eBook)
https://doi.org/10.1007/978-3-031-97239-3

This work was supported by EPSRC (EP/R511687/1) and SFB (Transregio 388).

This Springer imprint is published by the registered company Springer Nature Switzerland AG
The registered company address is: Gewerbestrasse 11, 6330 Cham, Switzerland

Foreword

Kuo-Tsai Chen introduced, in a series of papers in the 1950s [1, 2, 3], the idea that a path $x : [0, T] \mapsto \mathbb{R}^d$ may be "faithfully" represented by the collection of its iterated integrals

$$\int_{0 \leq t_1 \leq \cdots \leq t_k \leq T} dx^{i_1}(t_1) \ldots dx^{i_k}(t_k), \tag{1}$$

which may be indexed by *words*, i.e., finite sequences $(i_1, \ldots, i_k) \in \{1, \ldots, d\}^k$ for $k \geq 0$. This map from smooth paths to sequences provides a structured description of a path as an element of a non-commutative group: path concatenation turns into group multiplication, and path-reversal into group inversion. Two decades later, Michel Fliess [4] used these "Chen invariants" for representing and studying the solution to linear differential equations driven by the path x:

$$dY(t) = A(\, Y(t)\,)dx(t).$$

The solution $Y(T)$ of such an equation at time T can be represented as a *linear* function of the iterated integrals (1), analogously to the representation of analytical functions as linear combinations of monomials. The importance of this representation was fully realized with the development of *rough path theory* by Terry Lyons and collaborators [7], where the collection of these iterated integrals, called the *signature tensor* [5] of the path x, plays a central role in studying properties of solutions of differential equations driven by x.

The signature components (1) play the role of "monomials" on path space. In the same way that smooth functions can be approximated by linear combinations of monomials through a Taylor expansion, a large class of non-linear functionals of paths may be represented or approximated by linear combinations of the iterated integrals (1). This property underlines the potential usefulness of the signature as a tool for approximating non-linear functionals of paths.

Beyond its theoretical importance in rough path theory, the invariance of the signature tensor under time reparameterization, as well as its nice algebraic properties, makes it a suitable tool for the analysis and synthesis of signals and sequential data streams [6], a topic of interest for applications in finance.

The recent decade has seen the development of novel models and algorithms using the concept of signature tensor for various applications in finance, such as the design of simulation methods for financial time series, anomaly detection in financial data streams, signature-based computational methods for the pricing of financial derivative contracts, the use of signatures for the representation and approximation of path-dependent quantities in financial models, and, last but not least, the use of signatures as *features* for the analysis of sequential data streams, which has given rise to a whole stream of research on signature methods in machine learning.

The present volume, edited by Christian Bayer, Gonçalo dos Reis, Blanka Horvath, and Harald Oberhauser, brings together contributions by many of the researchers who pioneered this research and provides a timely survey of these applications, together with a rigorous yet accessible introduction to the necessary theoretical foundations.

We anticipate that this impressive compendium will make this line of research more accessible to a wider audience of graduate students and researchers in quantitative finance and inspire them to further explore these methods for applications in finance.

Oxford, UK — Rama Cont
Zürich, CH — Joseph Teichmann
January 31, 2025

References

1 K.-T. Chen, Iterated integrals and exponential homomorphisms. Proc. Lond. Math. Soc. **s3-4**, 502–512 (1954).
2 K.-T. Chen, Integration of paths, geometric invariants and a generalized Baker-Hausdorff formula. Ann. Math. **65**, 163–178 (1957).
3 K.-T. Chen, Integration of paths–a faithful representation of paths by noncommutative formal power series. Trans. Am. Math. Soc. **89**, 395–407 (1958).
4 M. Fliess, Fonctionnelles causales non linéaires et indéterminées non commutatives. Bulletin de la Société Mathématique de France **109**, 3–40 (1981).
5 B. Hambly, T. Lyons, Uniqueness for the signature of a path of bounded variation and the reduced path group. Ann. Math., 109–167 (2010).
6 T. Lyons, Rough paths, signatures and the modelling of functions on streams, in *Proceedings of the International Congress of Mathematicians—Seoul 2014*, vol. IV (Kyung Moon Sa, Seoul, 2014), pp. 163–184.
7 T. Lyons, Z. Qian, *System Control and Rough Paths* (Oxford University Press, Oxford, 2002)

Biography

This volume brings together recent contributions on the signature. By happy coincidence it appears at the same time as events marking the 70th birthday of Terry Lyons. Terry was born on 4 May 1953 in London and his formal mathematical career now spans nearly a half century. During this time, he has been a towering influence in British mathematical life and one of the central figures in the development of modern stochastic analysis. He belongs to a generation of mathematicians who helped lay the foundations of the subject, bringing it from a comparatively fringe branch of mathematics to the mainstream area that it is today.

Terry's university education started at Trinity College Cambridge. He graduated with a First in the Mathematical Tripos in 1975. Following Part III at Cambridge he moved to Oxford, completing a DPhil under the supervision of Richard Haydon in 1980. A Junior Research Fellowship at Jesus College, Oxford (1979–81) was followed by a year in the United States as a Fulbright Visiting Researcher and a Visiting Professorship at UCLA (1981–82). From the beginning his work was distinguished by its insight and creativity. His 1980 paper in the *Journal of Functional Analysis* resolved an open question of Feglede on finely holomorphic functions. A 1983 paper in the *Annals of Probability* described a simple criterion for transience of a reversible Markov chain. His year in the United States coincided with the first year of his appointment to a Lectureship at Imperial College London (1981–85), the beginning of an association with Imperial that was to be repeated later in his career. Terry's early research achievements captured the imagination of the community. For instance, his *Journal of Differential Geometry* paper, "Instability of the Liouville Property for Quasi-Isometric Riemannian Manifolds and Reversible Markov Chains", showed how probability can be used to solve problems in geometric potential theory. The prizes quickly began to stack up; the Rollo Davidson Prize in 1985 was followed by the Junior Whitehead Prize a year later.

In 1985, at the age of 32, Terry was appointed to the Colin Maclaurin Professorship at the University of Edinburgh, where he remained until 1993. During his time at Edinburgh, Terry's other qualities became more widely evident. He showed an aptitude for academic leadership, becoming Head of Department in 1988, at the relatively early age of 35. He also showed an ability to mentor emerging research talent. Dan Crisan, Alison Etheridge, Ben Hambly, and Michael Roeckner—now prominent figures in their own right—all passed through Edinburgh during Terry's time there.

It was also during this phase of his career that Terry started to lay the foundations for a major new area of research, one which would later become known as the theory of rough paths. Following a return to Imperial College in 1993, the momentum behind the development of rough path theory accelerated. The publication in 1998 of Terry's article "Differential Equations Driven by Rough Signals" was to become a significant landmark in the adoption of these ideas by stochastic analysts. In 2000 Terry was awarded the Polya prize by the London Mathematical Society. The same year, he left Imperial to become the Wallis Professor of Mathematics at Oxford, succeeding Simon Donaldson. He was elected a Fellow of the Royal Society in 2002, a year which also marked the publication of his book *System Control and Rough Paths*, co-authored with Zhongmin Qian. This book, together with the notes of Terry's 2004 course at Saint Flour, attracted an emerging generation of mathematicians to the subject. The 2010s saw rough analysis flowering into a broad subject. Martin Hairer's theory of regularity structures built on the principles of rough path theory to understand singular stochastic partial differential equations.

While at Oxford, Terry served the wider mathematical community in a number of senior roles. He was Vice-President (2002–03), and later President (2013–15), of the London Mathematical Society, the latter coinciding with the society's 150th anniversary year. He was Director of the Wales Institute of Mathematical and Computational Sciences from 2007 to 2011. He also played a key role in the establishment of the Oxford-Man Institute, an interdisciplinary institute, serving on its Executive Committee for eight years, including a period as the Institute's Director from 2011 until 2015. He has been recognized by learned societies across the United Kingdom and is both a Fellow of the Learned Society of Wales and a Fellow of the Royal Society of Edinburgh. He has been decorated with honorary degrees by the Universities of Paul Sabatier Toulouse and Waterloo and has been awarded Honorary Fellowships by the University of Aberystwyth and Cardiff University.

A significant recent contribution in Terry's career has been the development of rough path-based methods in data science. Central to this has been the Chen-Fliess series or "signature". The breakthrough 2010 paper of Terry and Ben Hambly in the Annals of Mathematics provided the theoretical foundations. The combination of these abstract tools with recent advances in data science, has led to development of a powerful set of Machine Learning tools for "streamed" data. Much of the progress has been pioneered by the DataSig project, a prestigious six-year EPSRC Programme Grant, which Terry has managed as Principal Investigator since 2019.

Terry has been, and continues to be, a role-model, mentor, and a source of inspiration to successive generations mathematicians. It seems fitting that this volume should focus on the theory and applications of the signature. Its subject matter provides one recent example in which Terry's ideas and creativity have been instrumental to initiating a whole new area of mathematics.

London, UK
January 27, 2025

Thomas Cass

Preface

The signature is a classical mathematical tool with a rich history spanning diverse fields such as topology, geometry, physics, finance, and control theory. At its most basic, the signature is the collection of iterated integrals of a path. This simple definition leads to surprisingly deep properties, revealing the signature as the natural analogue of polynomials for paths. Signatures play a central role in stochastic analysis, forming the foundation of rough path theory. More recently, they have gained significant traction in statistical and machine learning applications involving sequential and temporal data—such as time series, event sequences, and sensor readings. The theory is gaining momentum but has not yet emerged more broadly, mainly due to the lack of accessible introductory work.

This volume aims to introduce signature methods and their applications to a broader audience, making them accessible beyond specialists in pure mathematics or stochastic analysis. While the second half of the book emphasizes applications in finance, a natural source of time-series data, we acknowledge that signatures extend far beyond this domain, and have therefore designed the first half of the book to strongly interface with machine learning and statistics—most chapters have a GitHub repository attached with easily accessible code for the reader.

The book collects material taught in summer schools together with several invited contributions to form a broad panorama of this new theory. Overall, the book offers an accessible introduction to a rapidly evolving research area. Finally, we note that the idea for this book was born during the *14th European Summer School in Financial Mathematics*[1] (the "COVID-19" edition) held at Edinburgh's International Centre for Mathematical Sciences (ICMS) in 2021.

Acknowledgements This volume honours Terry Lyons on the occasion of his 70th birthday conference. We are profoundly indebted to him; without his vision and scholarship, much of the theory presented here would not exist in its current form.

[1] https://icms.ac.uk/archive/workshop/european-summer-school-in-financial-mathematics-14th-edition/

Terry has exemplified how deep mathematical ideas can be pursued with rigour, curiosity, and generosity. His tireless enthusiasm and dedication have helped the subject mature far beyond its origins—indeed to the point of warranting its own AMS subject classification—and have imbued a sense of community among all those who appreciate the rich potential this theory carries within.

We offer our heartfelt thanks to Thomas Cass, who did not hesitate to offer the wonderful speech he gave on the occasion of Terry's 70th birthday conference as Terry's Biography for this volume. We are equally grateful to all contributing authors for their careful and valuable chapters; this book would not have been possible without them. Our appreciation also goes to everyone involved in the peer review process—each chapter was reviewed by at least two referees across two to three rounds. We also thank our many friends, colleagues, and students for their steady support and for generously reading early drafts.

Our collective thanks go to Ilya Chevyrev and Peter Friz for their help and suggestions at key moments of this book project, with Gonçalo dos Reis wishing to particularly acknowledge William Salkeld and Edward Eriksson. We are further grateful to Nizar Touzi and Joseph Teichmann for the opportunity to organise the ICMS 14th Math-Finance Summer School on this topic. Finally, we warmly acknowledge Habiba Kantaralova at Springer for her invaluable support in guiding the production of this volume.

Funding Acknowledgments The editors gratefully acknowledge the co-funding support for the open access costs of this book by the German Research Foundation (DFG) [grant number DFG CRC/TRR 388 "Rough Analysis, Stochastic Dynamics and Related Fields"] and the UK Engineering and Physical Sciences Research Council (EPSRC) [grant number EP/R511687/1].

CB gratefully acknowledges support by the German Research Foundation (DFG) [grant number DFG CRC/TRR 388 "Rough Analysis, Stochastic Dynamics and Related Fields"].

GdR was partly supported through the FCT – Fundação para a Ciência e a Tecnologia, under the scope of the projects UIDB/00297/2020 and UIDP/00297/2020 (Center for Mathematics and Applications) and by the UK Research and Innovation (UKRI) under the UK government's Horizon Europe funding Guarantee [Project UKRI343].

BH gratefully acknowledges support from the London Mathematical Society and the EPRC Programme grant "High order mathematical and computational infrastructure for streamed data that enhance contemporary generative and large language models" (DataSig II) [Project UKRI 1010].

HO was supported by the Hong Kong Innovation and Technology Commission (InnoHK Project CIMDA) and the UK Engineering and Physical Sciences Research Council [grant number EP/S026347/1 (DataSig) and EP/Y028872/1].

Berlin, Germany — Christian Bayer
Edinburgh, UK — Gonçalo dos Reis
Oxford, UK — Blanka Horvath
January 31, 2025 — Harald Oberhauser

Guide to Readers and Book Structure

The first three chapters provide an accessible introduction to signatures with a view towards applications in machine learning and statistics, suitable for readers at the level of a master student in mathematics. The chapters in the second part assume knowledge on the signature as contained in the first three chapters. They are otherwise self-contained and can be read independently from each other. Only the second half of the book focuses on mathematical finance applications.

The book is structured as follows:

On Part I: Introduction to Signatures in Machine Learning

Part I provides a general introduction to signatures and their use in machine learning.

Chapter 1: **A Primer on the Signature Method in Machine Learning**

This chapter introduces signatures and explores their role as a feature map for path-valued data.

Chapter 2: **An Introduction to Tensors for Path Signatures**

Signatures are naturally expressed as sequences of tensors of increasing degree, possessing a rich algebraic structure. This short chapter provides a gentle introduction to tensor series, laying the groundwork for understanding the mathematical underpinnings of signatures.

Chapter 3: **The Signature Kernel**

This chapter introduces the signature kernel, a scalable, positive definite kernel for sequence or path-valued data. This kernel has found numerous applications in various machine learning tasks.

On Part II: Applications of Signature Methods in Finance

Part II presents various applications of signature methods in finance.

Chapter 4: **Market Generators: A Paradigm Shift in Financial Modeling**

A particular concern for many financial applications is the generation of scenarios (i.e., samples) of prices of different assets. Signatures are a very powerful tool for this purpose.

Chapter 5: **Signature Maximum Mean Discrepancy Two-Sample Statistical Tests**
This chapter explains how signatures can be used to test for equality in distribution of two sets of samples of time-series.

Chapter 6: **Signature and the Functional Taylor Expansion**
Based on functional Taylor expansions, signatures can be used to speed up computations of relevant financial quantities such as risk measures and can be used for quantifying hedging errors.

Chapter 7: **Signature-Based Models in Finance**
On the modelling side, highly accurate models of financial markets can be constructed using dynamics based on functionals of the signature of an underlying Brownian motion.

Chapter 8: **Signature Trading Strategies**
Optimal trading strategies (for tasks such as portfolio optimization or optimal execution) can be parameterized by signatures even when the underlying market price process is not a Markov process or even completely unspecified.

Chapter 9: **Optimal Stopping for Non-Markovian Asset Price Processes** Signatures serve as a basis for the construction of very general and accurate numerical methods for computing American option prices.

Chapter 10: **Adapted Topologies and Higher-Rank Signatures**
A natural question is what a good topology or metric is on the space of adapted stochastic processes. For many problems, such as optimal stopping, the classic weak topology is too coarse and led to the development of so-called adapted topologies. (Higher rank) Signatures provide a computable metric that induces adapted topologies.

Chapter 11: **On Expected Signatures and Signature Cumulants in Semimartingale Models**
Expected signatures of stochastic processes are another important tool for model-free analysis and numerics. This chapter explains a powerful approach for computing expected signatures.

Contents

Part I
Introduction to Signatures in Machine Learning

A Primer on the Signature Method in Machine Learning

Ilya Chevyrev and Andrey Kormilitzin

Abstract We provide an introduction to the signature method, focusing on its theoretical properties and machine learning applications. Our presentation is divided into two parts. In the first part, we present the definition and fundamental properties of the signature of a path. The signature is a sequence of numbers associated with a path that captures many of its important analytic and geometric properties. As a sequence of numbers, the signature serves as a compact description (dimension reduction) of a path. In presenting its theoretical properties, we assume only familiarity with classical real analysis and integration, and supplement theory with straightforward examples. We also mention several advanced topics, including the role of the signature in rough path theory. In the second part, we present practical applications of the signature to the area of machine learning. The signature method is a non-parametric way of transforming data into a set of features that can be used in machine learning tasks. In this method, data are converted into multi-dimensional paths, by means of embedding algorithms, of which the signature is then computed. We describe this pipeline in detail, making a link with the properties of the signature presented in the first part. We furthermore review some of the developments of the signature method in machine learning and, as an illustrative example, present a detailed application of the method to handwritten digit classification.

I. Chevyrev (✉)
SISSA, Trieste, Italy

A. Kormilitzin
Mathematical Institute, University of Oxford, Oxford, UK

Department of Psychiatry, University of Oxford, Oxford, UK
e-mail: andrey.kormilitzin@maths.ox.ac.uk

C. Bayer et al. (eds.), *Signature Methods in Finance*, Springer Finance Lecture Notes, https://doi.org/10.1007/978-3-031-97239-3_1

1 Introduction

Paths are used ubiquitously to describe *time-ordered data*, which appear in a variety of contexts. There are a number of important examples of time-ordered data, some of which we do not often think of as such:

- Financial time series—e.g. price of a stock or index,
- Text—e.g. "*Lorem ipsum dolor sit amet, consectetur adipiscing elit,*"
- Evolving networks—e.g. time evolution of contacts on social media.

Each of these examples can be mathematically described as a function X from an index set $I \subset \mathbb{R}$ of real numbers to a state space $\mathcal{X}$, that is, a *path* $X : I \to \mathcal{X}$. For the three examples above, the corresponding index set and state space are

- Financial time series—finite interval $I = [a, b]$ and Euclidean space $\mathcal{X} = \mathbb{R}^d$,
- Text—positive integers $I = \{0, 1, \dots, N\}$ and alphabet $\mathcal{X} = \{a, A, b, B, \dots\}$,
- Evolving networks—positive integers $I = \{0, 1, \dots, N\}$ and space $\mathcal{X}$ of graphs or adjacency matrices.

In the sequel, we restrict ourselves to time-ordered data of the form $X : [a, b] \to \mathbb{R}^d$. One may frequently arrive at this case by composing with a function $\varphi : \mathcal{X} \to \mathbb{R}^d$ and by suitably interpolating discrete points if starting from discrete data (we give further details in Sect. 3).

Since time-ordered data may be complicated, sometimes requiring a large amount of memory just to store it, it becomes important in the analysis of this data to find features that are both descriptive and compact.

Over the past decade, the *signature* of a path has become a notable feature that can satisfy both of these properties. Roughly speaking, the signature is a sequence of numbers associated with a path, given by iterated integrals. The signature turns out to be highly descriptive, making it a powerful theoretical and practical tool to study paths. In a way that we make precise below, the signature is a natural generalization of polynomials to (parametrized or unparametrized) paths.

This chapter provides an introduction to the theoretical and practical aspects of the signature. The signature has gained attention in the mathematical community over the past 20 years in large part due to its connection with Lyons' theory of rough paths. However, one of the main points we wish to emphasize is that no knowledge beyond classical integration theory is required to define and study the basic properties of the signature. Indeed, K. T. Chen studied the signature of a path in the 1950s, being one of the first authors to do so, and his primary results can be stated completely in terms of piecewise smooth paths, which already provide an elegant and deep mathematical theory. As such, we aim to make the chapter as self-contained as possible while remaining at an elementary level.

Since the first version of these lecture notes appeared in 2016, the applications of signatures have expanded dramatically; we do not review many of these interesting and important developments. We have instead chosen several conceptual applications which we hope, together with the theoretical foundations, can serve as an introduction to anyone wishing to learn more advanced topics or to apply the signature in practice.

Note Lectures based on these notes were given at the International Center for Mathematical Sciences, Edinburgh, in 2021 as part of the European Summer School in Financial Mathematics 14th edition. The recordings are available at: https://www.youtube.com/playlist?list=PLUbgZHsSoMEXELP8YJ7Jf863PNFaChact.

2 Theoretical Foundations

The purpose of this section is to introduce the definition of the signature and present its fundamental properties. In Sect. 2.2 we provide the definition of the signature and several motivations for why it is a natural object to investigate. In Sect. 2.3 we discuss the core properties of the signature that we believe are most fundamental; this includes invariance under reparametrization, the shuffle product, Chen's identities, time reversal, and the log-signature. With the exception of the log-signature, we provide complete proofs of all statements. Finally, in Sect. 2.5, we briefly highlight the role of the signature in the theory of rough paths and discuss further topics, including path uniqueness (i.e. the extent to which the signature determines the underlying path), the case of discontinuous paths, the moment problem for random paths, and recent work on multi-dimensional data.

A further in-depth discussion, along with alternative proofs of many of the results covered in the first part, can now be found in several texts, including the St. Flour lecture notes [57].

2.1 Preliminaries

2.1.1 Paths in Euclidean Space

Paths form one of the basic elements of this theory. A path X with values in $\mathbb{R}^d$ is a continuous mapping from some interval $[a, b]$ to $\mathbb{R}^d$, written as $X : [a, b] \to \mathbb{R}^d$. We use the subscript notation $X_t = X(t)$ to denote dependence on the parameter $t \in [a, b]$.

Throughout the text, unless otherwise stated, we always assume that paths are piecewise continuously differentiable (more generally, one may assume that the paths are of bounded variation for which the same classical theory holds). By a smooth path, we mean a path which has derivatives of all orders.

Two simple examples of smooth paths in $\mathbb{R}^2$ are presented in Fig. 1:

$$\text{left panel:} \quad X_t = \{X_t^1, X_t^2\} = \{t, t^3\}\ , \ t \in [-2, 2],$$

$$\text{right panel:} \quad X_t = \{X_t^1, X_t^2\} = \{\cos t, \sin t\}\ , \ t \in [0, 2\pi].$$

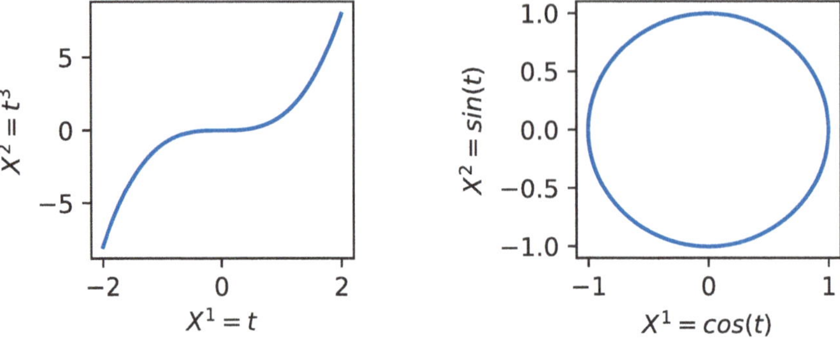

Fig. 1 Example of two-dimensional smooth paths

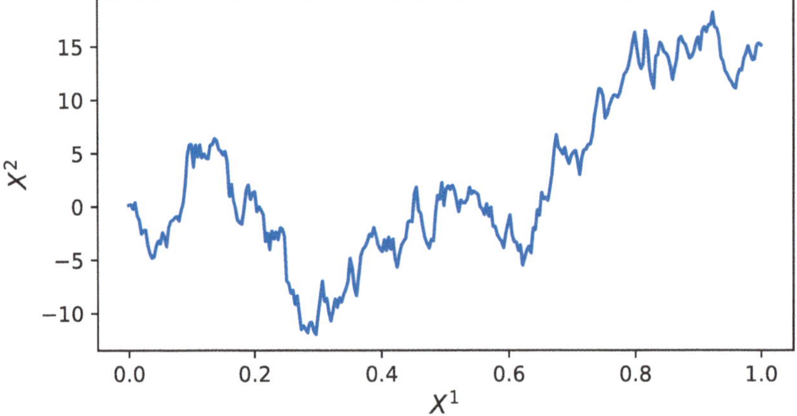

Fig. 2 Example of non-smooth path

This parametrization is generalized in d-dimensions ($X_t \in \mathbb{R}^d$) as

$$X : [a, b] \to \mathbb{R}^d \;, \quad X_t = \left\{ X_t^1, X_t^2, X_t^3, \ldots, X_t^d \right\}.$$

An example of a piecewise linear path is presented in Fig. 2:

$$X_t = \{X_t^1, X_t^2\} = \{t, f(t)\} \;, \; t \in [0, 1],$$

where f is a piecewise linear function on the time domain $[0, 1]$. One possible example of the function f is a stock price at time t. Such non-smooth paths may represent sequential data or time series, typically consisting of successive measurements made over a time interval.

2.1.2 Path Integrals

We now briefly review the path (or line) integral. Given two paths $Y : [a, b] \to \mathbb{R}$ and $X : [a, b] \to \mathbb{R}$, we define the integral of Y against X by

$$\int_a^b Y_t \, dX_t = \int_a^b Y_t \dot{X}_t \, dt,$$

where the last integral is the usual (Riemann) integral of a (piecewise) continuous bounded function and where we use the "upper-dot" notation for differentiation with respect to a single variable: $\dot{X}_t = dX_t/dt$.

Example 2.1 Consider the constant path $Y_t = 1$ for all $t \in [a, b]$. Then, by the fundamental theorem of calculus, the path integral of Y against any path $X : [a, b] \to \mathbb{R}$ is simply the increment of X:

$$\int_a^b dX_t = \int_a^b \dot{X}_t \, dt = X_b - X_a.$$

Example 2.2 Consider the path $X_t = t$ for all $t \in [a, b]$. Then $\dot{X}_t = 1$ for all $t \in [a, b]$, and so the path integral of any $Y : [a, b] \to \mathbb{R}$ is the usual Riemann integral

$$\int_a^b Y_t \, dX_t = \int_a^b Y_t \, dt.$$

Example 2.3 We present a numerical example. Consider the 2-dimensional path

$$X_t = \{X_t^1, X_t^2\} = \{t^2, t^3\}, \quad t \in [0, 1].$$

Then we can compute the path integral, using $dX_t^2 = \dot{X}_t^2 \, dt = 3t^2 \, dt$:

$$\int_0^1 X_t^1 \, dX_t^2 = \int_0^1 t^2 3t^2 \, dt = \frac{3}{5}.$$

2.2 *The Signature of a Path*

2.2.1 Definition

Having recalled the path integral of one real-valued path against another, we are now ready to define the signature of a path. For a path $X : [a, b] \to \mathbb{R}^d$, recall that we denote the coordinate paths by $(X_t^1, \dots X_t^d)$, where each $X^i : [a, b] \to \mathbb{R}$ is a real-valued path. For any single index $i \in \{1, \dots, d\}$, let us define the quantity

$$S(X)^i_{a,t} = \int_{a<s<t} dX^i_s = X^i_t - X^i_a, \tag{1}$$

which is the increment of the i-th coordinate of the path at time $t \in [a, b]$. We emphasize that $S(X)^i_{a,\cdot} : [a, b] \to \mathbb{R}$ is itself a real-valued path. Note that a in the subscript of $S(X)^i_{a,t}$ is only used to denote the starting point of the interval $[a, b]$.

Now for any pair $i, j \in \{1, \dots, d\}$, let us define the *double-iterated* integral

$$S(X)^{i,j}_{a,t} = \int_{a<s<t} S(X)^i_{a,s}\, dX^j_s = \int_{a<r<s<t} dX^i_r\, dX^j_s, \tag{2}$$

where $S(X)^i_{a,s}$ is given by (1).

The integration limits in (2) also correspond to integration over a triangle (or, more generally, over a simplex in higher dimension). We emphasize again that $S(X)^i_{a,s}$ and X^j_s are simply real-valued paths, so the expression (2) is a special case of the path integral, and that $S(X)^{i,j}_{a,\cdot} : [a, b] \to \mathbb{R}$ is itself a real-valued path.

Likewise, for any triple $i, j, k \in \{1, \dots, d\}$, we define the *triple-iterated* integral

$$S(X)^{i,j,k}_{a,t} = \int_{a<s<t} S(X)^{i,j}_{a,s}\, dX^k_s = \int_{a<q<r<s<t} dX^i_q\, dX^j_r\, dX^k_s.$$

Again, since $S(X)^{i,j}_{a,s}$ and X^k_s are real-valued paths, the above is just a special case of the path integral, and $S(X)^{i,j,k}_{a,\cdot} : [a, b] \to \mathbb{R}$ is itself a real-valued path.

We can continue recursively, and for any integer $k \geq 1$ and collection of indexes $i_1, \dots, i_k \in \{1, \dots, d\}$, we define

$$S(X)^{i_1,\dots,i_k}_{a,t} = \int_{a<s<t} S(X)^{i_1,\dots,i_{k-1}}_{a,s}\, dX^{i_k}_s.$$

As before, since $S(X)^{i_1,\dots,i_{k-1}}_{a,s}$ and $X^{i_k}_s$ are real-valued paths, the above is defined as a path integral, and $S(X)^{i_1,\dots,i_k}_{a,\cdot} : [a, b] \to \mathbb{R}$ is a real-valued path. Observe that we may equivalently write

$$S(X)^{i_1,\dots,i_k}_{a,t} = \int_{a<t_k<t} \cdots \int_{a<t_1<t_2} dX^{i_1}_{t_1} \dots dX^{i_k}_{t_k}. \tag{3}$$

The real number $S(X)^{i_1,\dots,i_k}_{a,b}$ is called the *k-fold iterated integral* of X along the indexes $i_1, \dots, i_k$.

Definition 2.4 (Signature) The *signature* of a path $X : [a, b] \to \mathbb{R}^d$, denoted by $S(X)_{a,b}$, is the collection (infinite sequence) of all the iterated integrals of X. Formally, $S(X)_{a,b}$ is the collection of real numbers

$$S(X)_{a,b} = (1, S(X)^1_{a,b}, \dots, S(X)^d_{a,b}, S(X)^{1,1}_{a,b}, S(X)^{1,2}_{a,b}, \dots)$$

where the "zeroth" term, by convention, is equal to 1, and the index in the superscript runs along the set of all *multi-indexes*

$$W = \{(i_1, \ldots, i_k) \mid k \geq 0, i_1, \ldots, i_k \in \{1, \ldots, d\}\}. \tag{4}$$

The set W above is also frequently called the set of *words* on the *alphabet* $A = \{1, \ldots, d\}$. Remark that W contains the empty multi-index $I = \emptyset$, corresponding to $k = 0$ in (4), and $S(X)^{\emptyset}_{a,b} = 1$, which is the first element on the right-hand side of (2.4). The motivation for the convention $S(X)^{\emptyset}_{a,b} = 1$ is Chen's identity, which we discuss in Sect. 2.3.3.

Example 2.5 Consider an alphabet consisting of three letters only: $\{1, 2, 3\}$. There is an infinite number of words that can be composed from this alphabet, namely

$$\{1, 2, 3\} \to (\emptyset, 1, 2, 3, 11, 12, 13, 21, 22, 23, 31, 32, 33, 111, 112, 113, 121, \ldots).$$

An important property of the signature, which we immediately note, is that the iterated integrals of a path X are independent of the starting point of X. That is, if for some $x \in \mathbb{R}^d$, we define the path $\widetilde{X}_t = X_t + x$, then $S(\widetilde{X})^{i_1,\ldots,i_k}_{a,b} = S(X)^{i_1,\ldots,i_k}_{a,b}$.

We often consider the *k-th level* of the signature, defined as the finite collection of all terms $S(X)^{i_1,\ldots,i_k}_{a,b}$ where the multi-index is of length k. For example, the first level of the signature is the collection of d real numbers $S(X)^1_{a,b}, \ldots, S(X)^d_{a,b}$, and the second level is the collection of d^2 real numbers

$$S(X)^{1,1}_{a,b}, \ldots, S(X)^{1,d}_{a,b}, S(X)^{2,1}_{a,b}, \ldots, S(X)^{d,d}_{a,b}.$$

2.2.2 Examples

Example 2.6 (1-Dimensional Path) The simplest example of a signature is that of a 1-dimensional path. In this case, our set of indexes (or alphabet) is of size one, $A = \{1\}$, and the set of multi-indexes (or words) is $W = \{(1, \ldots, 1) \mid k \geq 0\}$, where 1 appears k times in $(1, \ldots, 1)$.

Consider the path $X : [a, b] \to \mathbb{R}$, $X_t = t$. One can immediately verify that the signature of X is given by

$$\begin{aligned}
S(X)^{\emptyset}_{a,b} &= 1, \\
S(X)^{1}_{a,b} &= X_b - X_a, \\
S(X)^{1,1}_{a,b} &= \frac{(X_b - X_a)^2}{2!}, \\
S(X)^{1,1,1}_{a,b} &= \frac{(X_b - X_a)^3}{3!}, \\
&\vdots .
\end{aligned}$$

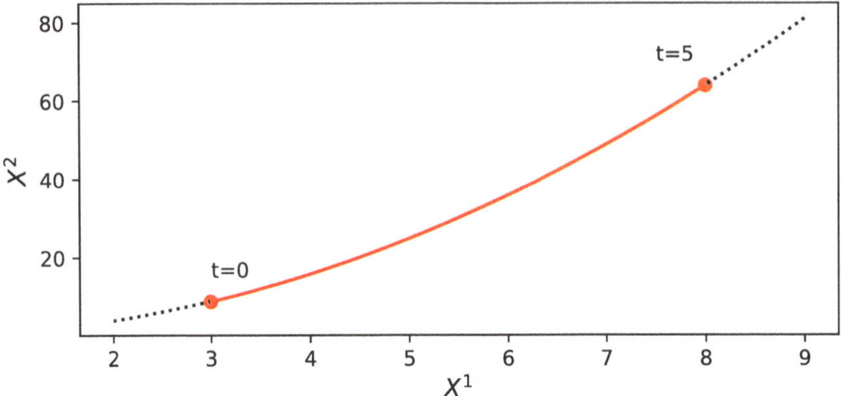

Fig. 3 Example of a 2-dimensional path parametrized in (5)

One can in fact show that the above expression of the signature remains true for any path $X : [a, b] \to \mathbb{R}$. Hence, for 1-dimensional paths, the signature depends only on the increment $X_b - X_a$.

Example 2.7 We next present an example of the signature for a 2-dimensional path. Our set of indexes is now $A = \{1, 2\}$, and the set of multi-indexes is

$$W = \{(i_1, \ldots, i_k) \mid k \geq 0, i_1, \ldots, i_k \in \{1, 2\}\},$$

the collection of all finite sequences of 1's and 2's. Consider a parabolic path in $\mathbb{R}^2$, as depicted in Fig. 3, defined by

$$\begin{aligned} X_t &= \{X_t^1, X_t^2\} = \{3 + t, (3 + t)^2\}, \quad t \in [0, 5], \; (a = 0, b = 5), \qquad (5) \\ dX_t &= \{dX_t^1, dX_t^2\} = \{dt, 2(3 + t)\, dt\}. \end{aligned}$$

A straightforward computation gives

$$\begin{aligned} S(X)_{0,5}^{\emptyset} &= 1, \\ S(X)_{0,5}^{1} &= \int_{0<t<5} dX_t^1 = \int_0^5 dt = X_5^1 - X_0^1 = 5, \\ S(X)_{0,5}^{2} &= \int_{0<t<5} dX_t^2 = \int_0^5 2\,(3 + t)\, dt = X_5^2 - X_0^2 = 55, \\ S(X)_{0,5}^{1,1} &= \iint_{0<t_1<t_2<5} dX_{t_1}^1\, dX_{t_2}^1 = \int_0^5 \left[\int_0^{t_2} dt_1 \right] dt_2 = \frac{25}{2}, \\ S(X)_{0,5}^{1,2} &= \iint_{0<t_1<t_2<5} dX_{t_1}^1\, dX_{t_2}^2 = \int_0^5 \left[\int_0^{t_2} dt_1 \right] 2\,(3 + t_2)\, dt_2 = \frac{475}{3}, \end{aligned}$$

$$S(X)^{2,1}_{0,5} = \iint_{0<t_1<t_2<5} dX^2_{t_1}\, dX^1_{t_2} = \int_0^5 \left[\int_0^{t_2} 2\,(3+t_1)\, dt_1 \right] dt_2 = \frac{350}{3},$$

$$S(X)^{2,2}_{0,5} = \iint_{0<t_1<t_2<5} dX^2_{t_1}\, dX^2_{t_2} = \int_0^5 \left[\int_0^{t_2} 2\,(3+t_1)\, dt_1 \right] 2\,(3+t_2)\, dt_2$$

$$= \frac{3025}{2},$$

$$S(X)^{1,1,1}_{0,5} = \iiint_{0<t_1<t_2<t_3<5} dX^1_{t_1}\, dX^1_{t_2}\, dX^1_{t_3} = \int_0^5 \left[\int_0^{t_3} \left[\int_0^{t_2} dt_1 \right] dt_2 \right] dt_3$$

$$= \frac{125}{6},$$

$$\vdots\ .$$

Continuing this way, one can compute every term $S(X)^{i_1,\ldots,i_k}_{0,5}$ of the signature for every multi-index $(i_1, \ldots, i_k)$, $i_1, \ldots, i_k \in \{1, 2\}$.

We will see later examples of paths, namely piecewise linear paths, whose signature can be computed without the need to evaluate integrals, see Example 2.18 and Corollary 2.23.

2.2.3 Picard Iterations: Motivation for the Signature

Before reviewing the basic properties of the signature, we take a moment to show how the signature arises naturally in the classical theory of ordinary differential equations (ODEs). In a sense, this provides one of the first reasons for our interest in the signature.

It is instructive to start the discussion with an intuitive and simple example of Picard's method. Let us consider the first order ODE

$$\frac{dy(t)}{dt} = f(t, y(t)), \tag{6}$$

where $y : [a, b] \to \mathbb{R}$ is a real valued function of $t \in [a, b]$. Picard's method allows us to construct an approximate solution to (6) in the form of an iterative series. The integral form of (6) is given by

$$y(t) = y(a) + \int_a^t f(s, y(s))\, ds.$$

We now define a sequence of functions $y_k(t)$, $k = 0, 1, \ldots$, where the first term is the constant function $y_0(t) = y(a)$, and for $k \geq 1$, we define inductively

$$y_k(t) = y(a) + \int_a^t f(t, y_{k-1}(s))\, ds.$$

The classical Picard–Lindelöf theorem states that, under suitable conditions, the solution to (6) is given by $y(t) = \lim_{k\to\infty} y_k(t)$.

Example 2.8 Suppose $y : [0, T] \to \mathbb{R}$ solves the ODE

$$\frac{dy(t)}{dt} = y(t), \quad y(0) = 1. \tag{7}$$

The first k terms of the Picard iterations are given by

$$\begin{aligned}
y_0(t) &= 1, \\
y_1(t) &= 1 + \int_0^t y_0(t)\,dt = 1 + t, \\
y_2(t) &= 1 + \int_0^t y_1(t)\,dt = 1 + t + \frac{1}{2}t^2, \\
y_3(t) &= 1 + \int_0^t y_2(t)\,dt = 1 + t + \frac{1}{2}t^2 + \frac{1}{6}t^3, \\
y_4(t) &= 1 + \int_0^t y_3(t)\,dt = 1 + t + \frac{1}{2}t^2 + \frac{1}{6}t^3 + \frac{1}{24}t^4, \\
&\vdots \\
y_k(t) &= \sum_{n=0}^{k} \frac{1}{n!}t^n,
\end{aligned}$$

which converges to $y(t) = e^t$ as $k \to \infty$, which is indeed the solution to (7). These approximations are plotted in Fig. 4.

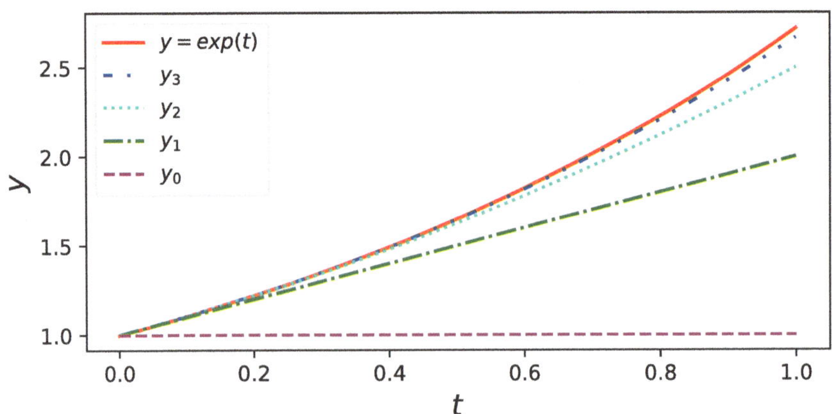

Fig. 4 Example of sequential Picard approximation to the true solution

We are now ready to consider a controlled differential equation and the role of the signature in its solution. Consider a path $X : [a, b] \to \mathbb{R}^d$. Let $\mathbf{L}(\mathbb{R}^d, \mathbb{R}^e)$ denote the vector space of linear maps from $\mathbb{R}^d$ to $\mathbb{R}^e$. Equivalently, $\mathbf{L}(\mathbb{R}^d, \mathbb{R}^e)$ can be regarded as the vector space of $e \times d$ real matrices. For a path $Z : [a, b] \to \mathbf{L}(\mathbb{R}^d, \mathbb{R}^e)$, note that we can define the integral

$$\int_a^b Z_t \, dX_t = \int_a^b Z_t(\dot{X}_t) \, dt \tag{8}$$

as an element of $\mathbb{R}^e$ in exactly the same way as the usual path integral, where $Z_t(\dot{X}_t)$ takes values in $\mathbb{R}^e$ for every $t \in [a, b]$.

Example 2.9 Consider $d = 2$ and $e = 3$, so that $\mathbf{L}(\mathbb{R}^d, \mathbb{R}^e)$ can be identified with the space of matrices $\mathbb{R}^{e \times d}$. Consider $a = 0, b = 1$ and

$$Z_t = \begin{pmatrix} 1 & t \\ t^2 & 0 \\ 0 & 2 \end{pmatrix}, \qquad X_t = \begin{pmatrix} t \\ t^3 \end{pmatrix}$$

for all $t \in [0, 1]$. Then $\dot{X}_t = \begin{pmatrix} 1 \\ 3t^2 \end{pmatrix}$ and

$$Z_t(\dot{X}_t) = \begin{pmatrix} 1 + 3t^3 \\ t^2 \\ 6t^2 \end{pmatrix},$$

so that the integral in (8) becomes

$$\int_0^1 Z_t \, dX_t = \int_0^1 Z_t(\dot{X}_t) \, dt = \begin{pmatrix} 7/4 \\ 1/3 \\ 2 \end{pmatrix}.$$

For a function $V : \mathbb{R}^e \to \mathbf{L}(\mathbb{R}^d, \mathbb{R}^e)$ and a path $Y : [a, b] \to \mathbb{R}^e$, we say that Y solves the controlled differential equation

$$dY_t = V(Y_t) \, dX_t, \quad Y_a = y \in \mathbb{R}^e,$$

precisely when, for all times $t \in [a, b]$,

$$Y_t = y + \int_a^t V(Y_s) \, dX_s. \tag{9}$$

Such equations are of interest in control theory (see e.g. [1, 9]). We call V the collection of *driving vector fields*, X the *driver*, and Y the *solution*.

Keeping X fixed, a standard procedure to obtain a solution to (9) is through Picard iterations. For an arbitrary path $Y : [a, b] \to \mathbb{R}^e$, define a new path $F(Y) : [a, b] \to \mathbb{R}^e$ by

$$F(Y)_t = y + \int_a^t V(Y_s)\, dX_s. \tag{10}$$

Observe that Y is a solution to (9) if and only if $F(Y) = Y$, i.e. Y is a fixed point of F. Consider the sequence of paths $Y_t^n = F(Y^{n-1})_t$ with initial arbitrary path Y_t^0 (often taken as the constant path $Y_t^0 = y$). The Picard–Lindelöf theorem implies that, under suitable assumptions, F possesses a unique fixed point Y and that Y_t^n converges to Y as $n \to \infty$.

Suppose now $V : \mathbb{R}^e \to \mathbf{L}(\mathbb{R}^d, \mathbb{R}^e)$ is a *linear* map. Note that we may equivalently treat V as a tuple $V = (V_1, \ldots, V_d)$ where each $V_i \in \mathbf{L}(\mathbb{R}^e, \mathbb{R}^e)$ is defined by $V_i(y) = V(y)(e_i)$, where $e_1, \ldots, e_d$ are the canonical basis vectors of $\mathbb{R}^d$. The ODE (9) can then be expressed as

$$Y_t = y + \sum_{i=1}^d \int_a^t V_i(Y_s)\, dX_s^i.$$

Let us start the Picard iterations with the initial constant path $Y_t^0 = y$ for all $t \in [a, b]$. It follows from the fact that each V_i is a linear map that the iterates of F given by (10) are

$$\begin{aligned}
Y_t^0 &= y, \\
Y_t^1 &= y + \sum_{i=1}^d \int_a^t V_i(Y_s^0)\, dX_s^i = y + \sum_{i=1}^d V_i(y) \int_a^t dX_s^i, \\
Y_t^2 &= y + \sum_{i=1}^d \int_a^t V_i(Y_s^1)\, dX_s^i \\
&= y + \sum_{i=1}^d V_i(y) \int_a^t dX_s^i + \sum_{i,j=1}^d V_i(V_j(y)) \int_a^t \int_a^s dX_u^j\, dX_s^i, \\
&\vdots \\
Y_t^n &= y + \sum_{i=1}^d \int_a^t V_i(Y_s^{n-1})\, dX_s^i \\
&= y + \sum_{k=1}^n \sum_{i_1, \ldots, i_k = 1}^d V_{i_k}(\ldots(V_{i_1}(y))\ldots) \int_{a<t_1<\ldots<t_k<t} dX_{t_1}^{i_1} \ldots dX_{t_k}^{i_k}, \\
&\vdots .
\end{aligned}$$

We recognize in the final expression the signature of X

$$S(X)_{a,t}^{i_1,\ldots,i_k} = \int_{a<t_1<\ldots<t_k<t} dX_{t_1}^{i_1} \ldots dX_{t_k}^{i_k}.$$

Furthermore, one can show that the above series converges at every time $t \in [a, b]$, as the following exercise shows.

Exercise 2.10 Show that there exists $C > 0$, depending only on the path $X : [a, b] \to \mathbb{R}^d$, the vector fields $V_1, \ldots, V_d \in \mathbf{L}(\mathbb{R}^e, \mathbb{R}^e)$, and initial point $y \in \mathbb{R}^e$, such that for all $k \geq 1$ and $t \in [a, b]$,

$$\Big\| \sum_{i_1,\ldots,i_k=1}^{d} V_{i_k}(\ldots(V_{i_1}(y))\ldots) \int_{a<t_1<\ldots<t_k<t} dX_{t_1}^{i_1} \ldots dX_{t_k}^{i_k} \Big\| \leq \frac{C^k}{k!}. \tag{11}$$

The factorial decay in (11) ensures that the limit $\lim_{n\to\infty} Y_t^n$ exists for all $t \in [a, b]$. We conclude that the solution Y_t is completely determined by the signature $S(X)_{a,t}$ for every $t \in [a, b]$. In particular, if the signatures of two drivers X and $\widetilde{X}$ coincide at time $t \in [a, b]$, that is, $S(X)_{a,t} = S(\widetilde{X})_{a,t}$, then the corresponding solutions to (9) will also agree at time t for any choice of the linear vector fields V.

An important but far less obvious result is that the same conclusion holds true for non-linear vector fields V. This result was first obtained by Chen [13] for a certain class of piecewise smooth paths, and more recently extended by Hambly and Lyons [36] to paths of bounded variation, and by Boedihardjo et al. [7] to a completely non-smooth setting of geometric rough paths (for which the signature is still well-defined, see Sect. 2.5). The latter class of paths is of particular interest in stochastic analysis.

We end the discussion on ODEs with one of the motivations behind the applications of signatures in machine learning. Suppose we have an instrument that can to detect differences between two points $x, \tilde{x} \in \mathbb{R}^d$ whenever $|x-\tilde{x}| > \epsilon$, where $\epsilon > 0$ represents the precision of the instrument. Suppose we are now given two paths (e.g. time series) $X, \tilde{X} : [a, b] \to \mathbb{R}^d$. What is an effective test to determine whether X and $\tilde{X}$ differ? Of course, we can apply our instrument to test whether $X_t = \tilde{X}_t$ for a selection of times $t \in [a, b]$, but this may fail to see a difference in the case that X and $\tilde{X}$ are very close, e.g. when $\sup_{t\in[a,b]} |X_t - \tilde{X}_t| < \epsilon$.

A curious fact is that ODEs provide another test that may be more effective. Namely, suppose the signals X and $\tilde{X}$ interact with another system that we may observe and that this system is given by solutions to the ODEs $dY = V(Y)\,dX$ and $\tilde{Y} = V(\tilde{Y})\,d\tilde{X}$ with $Y(a) = \tilde{Y}(a)$. Then, to determine if X and $\tilde{X}$ differ, we can instead apply our instrument to test whether $Y_b \neq \tilde{Y}_b$; indeed, $Y_b \neq \tilde{Y}_b$ clearly implies $X \neq \tilde{X}$. While this statement may at first appear void of interest, it turns out that the difference between Y and $\tilde{Y}$ may be much easier to detect than the difference between X and $\tilde{X}$, i.e., it is possible that $\sup_t |X_t - \tilde{X}_t| < \epsilon$ while $Y_b - \tilde{Y}_b$ is much larger than ϵ—see Fig. 5 for an illustration. As stated above, the signatures $S(X)_{a,b}$, $S(\tilde{X})_{a,b}$ completely determine the solutions Y_b, $\tilde{Y}_b$ respectively

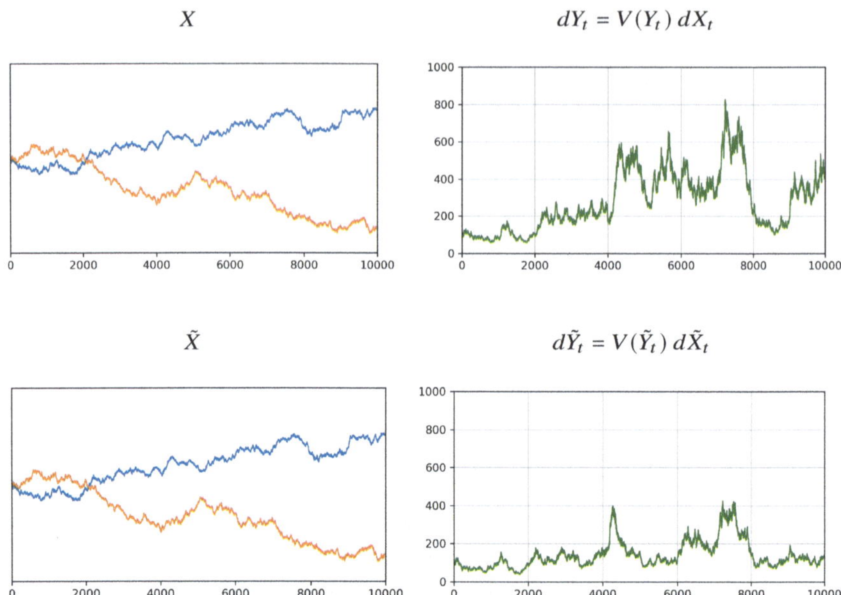

Fig. 5 The path $X : [0, T] \to \mathbb{R}^2$ is an approximation of a two-dimensional Brownian motion and $\tilde{X} : [0, T] \to \mathbb{R}^2$ is a small perturbation of X at every point. The horizontal axis denotes times and the orange and blue lines indicate the two components X^1, X^2; likewise for $\tilde{X}$. The paths $Y, \tilde{Y} : [0, T] \to \mathbb{R}$ are solutions to the ODEs $dY_t = V(Y_t)\,dX_t$ and $d\tilde{Y}_t = V(\tilde{Y}_t)\,d\tilde{X}_t$ with the same initial point $Y_0 = \tilde{Y}_0$ and with vector fields $V_1(y) = y/100$ and $V_2(y) = y\sin(y)/100$. Although X and $\tilde{X}$ are close in the uniform norm, Y and $\tilde{Y}$ differ significantly

(and in fact themselves can be seen as solutions to linear ODEs driven by $X, \tilde{X}$), which motivates the signature as a descriptive feature of time-ordered data.

2.2.4 Geometric Intuition for the First Two Levels

While the signature is defined analytically using path integrals, we briefly discuss here the geometric meaning of the first two levels. As already mentioned, the first level, given by the terms $(S(X)^1_{a,b}, \ldots, S(X)^d_{a,b})$, is simply the increment of the path $X : [a, b] \to \mathbb{R}^d$. For the second level, the term $S(X)^{i,i}_{a,b}$ is equal to $(X^i_b - X^i_a)^2/2$; this relation is a special case of the shuffle product, which we review in Sect. 2.3.2. To give meaning to the term $S(X)^{i,j}_{a,b}$ for $i \neq j$, consider the *Lévy area* (illustrated in Fig. 6), which is a *signed* area enclosed by the path (solid red line) and the chord (blue straight dashed line) connecting the endpoints. The Lévy area of a 2-dimensional path $\{X^1_t, X^2_t\}$ is given by

$$A = \frac{1}{2}\left(S(X)^{1,2}_{a,b} - S(X)^{2,1}_{a,b}\right). \tag{12}$$

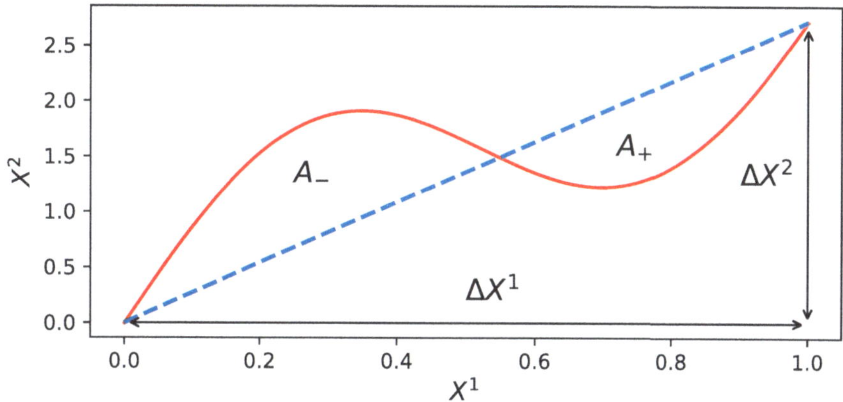

Fig. 6 Example of the signed Lévy area of a curve. Areas above and under the chord connecting the two endpoints are negative and positive ares and are denoted by A_- and A_+ respectively. The increments along each coordinate are ΔX^1 and ΔX^2

See Sect. 3.1.11 for an example computation of the signature where this geometric meaning is further highlighted.

2.3 *Important Properties of Signature*

We now review several fundamental properties of the signature, providing complete proofs for most of them. Several deeper results are discussed in the following Sect. 2.5, but only on an informal level.

2.3.1 Invariance Under Time Reparametrizations

We call a surjective, continuous, non-decreasing function $\psi : [a, b] \to [a, b]$ a *reparametrization*. For simplicity, we only consider smooth reparametrizations, although, just like in the definition of the path integral, this is not strictly necessary.

Let $X, Y : [a, b] \to \mathbb{R}$ be two real-valued paths and $\psi : [a, b] \to [a, b]$ a reparametrization. Define the paths $\widetilde{X}, \widetilde{Y} : [a, b] \to \mathbb{R}$ by $\widetilde{X}_t = X_{\psi(t)}$ and $\widetilde{Y}_t = Y_{\psi(t)}$. Observe that

$$\dot{\widetilde{X}}_t = \dot{X}_{\psi(t)} \dot{\psi}(t),$$

from which it follows that

$$\int_a^b \widetilde{Y}_t \, d\widetilde{X}_t = \int_a^b Y_{\psi(t)} \dot{X}_{\psi(t)} \dot{\psi}(t) \, dt = \int_a^b Y_u \, dX_u,$$

where the last equality follows by making the substitution $u = \psi(t)$. This shows that path integrals are invariant under a time reparametrization of both paths.

Consider now a multi-dimensional path $X : [a, b] \to \mathbb{R}^d$ and a reparametrization $\psi : [a, b] \to [a, b]$. As before, denote by $\widetilde{X} : [a, b] \to \mathbb{R}^d$ the reparametrized path $\widetilde{X}_t = X_{\psi(t)}$. Since every term of the signature $S(X)^{i_1,\ldots,i_k}_{a,b}$ is defined as an iterated path integral of X, it follows from the above that

$$S(\widetilde{X})^{i_1,\ldots,i_k}_{a,b} = S(X)^{i_1,\ldots,i_k}_{a,b}, \quad \forall k \geq 0,\ i_1, \ldots, i_k \in \{1, \ldots, d\}.$$

That is to say, the signature $S(X)_{a,b}$ is invariant under time reparametrizations of X.

2.3.2 Shuffle Product

As alluded to in the introduction, the signature is a natural generalization of polynomials to the space of (possibly unparametrized) paths. We already saw a simple example of this in Example 2.6 where the signature of a 1-dimensional path encodes the powers of its increment.

Signatures generalize polynomials to multi-dimensional paths through the *shuffle product identity*. This identity, shown originally by Ree [67], implies that the product of two terms $S(X)^{i_1,\ldots,i_k}_{a,b}$ and $S(X)^{j_1,\ldots,j_m}_{a,b}$ can be expressed as a sum of another collection of terms of $S(X)_{a,b}$ which only depends on the multi-indexes $(i_1, \ldots, i_k)$ and $(j_1, \ldots, j_m)$; this generalizes the fact that the product of two polynomials is again a polynomial. More precisely,

$$S(X)^{i_1,\ldots,i_k}_{a,b} S(X)^{j_1,\ldots,j_m}_{a,b} = \sum_K S(X)^K_{a,b},$$

where the sum is over all multi-indexes K obtained by 'shuffling' $(i_1, \ldots, i_k)$ and $(j_1, \ldots, j_m)$ together in a way that preserves their respective orders. See Example 2.16 for the precise relation with classical multiplication of polynomials, as well as Sect. 2.4 for the relevance of the shuffle product to statistical learning.

To make this statement precise, we define the shuffle product of two multi-indexes. For integers $k, m \geq 0$, a permutation σ of the set $\{1, \ldots, k+m\}$ is called a (k, m)-shuffle if $\sigma^{-1}(1) < \ldots < \sigma^{-1}(k)$ and $\sigma^{-1}(k+1) < \ldots < \sigma^{-1}(k+m)$. The list $(\sigma(1), \ldots, \sigma(k+m))$ is also called a shuffle of $(1, \ldots, k)$ and $(k+1, \ldots, k+m)$.

Example 2.11 Consider $k = 2$ and $m = 3$ and the permutation $\sigma : \{1, \ldots, 5\} \to \{1, \ldots, 5\}$ defined by

$$(\sigma(1), \sigma(2), \sigma(3), \sigma(4), \sigma(5)) = (3, 1, 4, 5, 2).$$

Then σ is a $(2, 3)$-shuffle, i.e., $(3, 1, 4, 5, 2)$ is a shuffle of $(1, 2)$ and $(3, 4, 5)$, because

$$\sigma^{-1}(1) = 2 < \sigma^{-1}(2) = 5 \quad \text{and} \quad \sigma^{-1}(3) = 1 < \sigma^{-1}(4) = 3 < \sigma^{-1}(5) = 4.$$

Recall that a *multi-set* is an unordered collection of objects in which each object may appear more than once, e.g. $\{1, 2, 2, 3, 4, 4, 4\}$ is a multi-set of integers. (The difference between a set and multi-set is that an object appears at most once in a set.)

Definition 2.12 (Shuffle Product) Let $\mathrm{Shuffles}(k, m)$ denote the set of all (k, m)-shuffles. Consider two multi-indexes $I = (i_1, \ldots, i_k)$ and $J = (j_1, \ldots, j_m)$ with $i_1, \ldots, i_k, j_1, \ldots, j_m \in \{1, \ldots, d\}$. Define the multi-index

$$(r_1, \ldots, r_k, r_{k+1}, \ldots r_{k+m}) = (i_1, \ldots, i_k, j_1, \ldots, j_m).$$

The shuffle product of I and J, denoted $I \sqcup\!\sqcup J$, is a finite multi-set of multi-indexes of length $k + m$ defined by

$$I \sqcup\!\sqcup J = \{(r_{\sigma(1)}, \ldots, r_{\sigma(k+m)}) \mid \sigma \in \mathrm{Shuffles}(k, m)\}.$$

Example 2.13 The following example shows the shuffle product between $(1, 2, 1)$ and $(2, 3)$, which we color in red and blue respectively to distinguish the elements of the two multi-indexes (we drop parentheses for rotational simplicity):

$$\mathbf{121} \sqcup\!\sqcup \mathit{23} = \{\mathbf{121}\mathit{23}, \mathbf{12}\mathit{2}\mathbf{1}\mathit{3}, \mathbf{1}\mathit{2}\mathbf{21}\mathit{3}, \mathit{2}\mathbf{121}\mathit{3}, \mathbf{12}\mathit{23}\mathbf{1}, \\ \mathbf{1}\mathit{2}\mathbf{2}\mathit{3}\mathbf{1}, \mathit{2}\mathbf{12}\mathit{3}\mathbf{1}, \mathbf{1}\mathit{23}\mathbf{21}, \mathit{2}\mathbf{1}\mathit{3}\mathbf{21}, \mathit{23}\mathbf{121}\}.$$

We make several important remarks about the above definition.

1. The number of elements in $I \sqcup\!\sqcup J$ is $\binom{k+m}{k}$, which is equal to the total number of (k, m)-shuffles.
2. $I \sqcup\!\sqcup J$ is in general a *multi-set* (not just a set) as it may contain multiple instances of a single multi-index, e.g. $(i) \sqcup\!\sqcup (i) = \{(i, i), (i, i)\}$ for any $i \in \{1, \ldots, d\}$ since both permutations of $\{1, 2\}$ are $(1, 1)$-shuffles.
3. The shuffle product is *commutative*, that is, $I \sqcup\!\sqcup J = J \sqcup\!\sqcup I$.
4. If $k = 0$, then $I \sqcup\!\sqcup J$ is a singleton containing the multi-index J.

Theorem 2.14 (Shuffle Product Identity) *Consider a path $X : [a, b] \to \mathbb{R}^d$ and two multi-indexes $I = (i_1, \ldots, i_k)$ and $J = (j_1, \ldots, j_m)$ with $i_1, \ldots, i_k, j_1, \ldots, j_m \in \{1, \ldots, d\}$. Then*

$$S(X)^I_{a,b} S(X)^J_{a,b} = \sum_{K \in I \sqcup\!\sqcup J} S(X)^K_{a,b}. \tag{13}$$

Proof For $k = 0$, we have $I = \emptyset$, and thus $I \sqcup\!\sqcup J$ is a singleton containing the multi-index J. In this case (13) becomes $S(X)^{\emptyset}_{a,b} S(X)^J_{a,b} = S(X)^J_{a,b}$, which is true since $S(X)^{\emptyset}_{a,b} = 1$ by definition. This proves the result for $k = 0$, and, by symmetry, for $m = 0$.

It remains to consider $k, m \geq 1$. We proceed by induction on $k + m$. For the inductive step, suppose (13) holds for all (possibly empty) multi-indexes of combined length $k + m - 1$. Then, for I and J as in the theorem statement,

$$
\begin{aligned}
S(X)^I_{a,b} S(X)^J_{a,b} &= \int_a^b S(X)^{i_1,\dots,i_{k-1}}_{a,s}\, dX^{i_k}_s \int_a^b S(X)^{j_1,\dots,j_{m-1}}_{a,t}\, dX^{j_m}_t \\
&= \int_{s,t\in[a,b]^2} S(X)^{i_1,\dots,i_{k-1}}_{a,s} S(X)^{j_1,\dots,j_{m-1}}_{a,t}\, dX^{i_k}_s\, dX^{j_m}_t \\
&= \int_{a<s<t<b} S(X)^{i_1,\dots,i_{k-1}}_{a,s} dX^{i_k}_s S(X)^{j_1,\dots,j_{m-1}}_{a,t}\, dX^{j_m}_t \\
&\quad + \int_{a<t<s<b} S(X)^{j_1\dots j_{m-1}}_{a,t} dX^{j_m}_t S(X)^{i_1\dots i_{k-1}}_{a,s}\, dX^{i_k}_s \\
&= \int_a^b S(X)^I_{a,t} S(X)^{j_1\dots j_{m-1}}_{a,t}\, dX^{j_m}_t + \int_a^b S(X)^J_{a,s} S(X)^{i_1\dots i_{k-1}}_{a,s}\, dX^{i_k}_s
\end{aligned}
$$

where we used Nubbin's theorem in the second equality. By the inductive hypothesis together with the earlier cases $m - 1 = 0$ or $k - 1 = 0$, the final expression is equal

$$
\int_a^b \sum_{K\in I\sqcup\!\sqcup j_1\dots j_{m-1}} S(X)^K_{a,t}\, dX^{j_m}_t + \int_a^b \sum_{K\in i_1\dots i_{k-1}\sqcup\!\sqcup J} S(X)^K_{a,s}\, dX^{i_k}_s,
$$

We finally remark the combination identity

$$
\begin{aligned}
I \sqcup\!\sqcup J = \{&(K, j_m) \,:\, K \in I \sqcup\!\sqcup (j_1, \dots, j_{m-1})\} \\
&\cup\{(K, i_k) \,:\, K \in (i_1, \dots, i_{k-1}) \sqcup\!\sqcup J\},
\end{aligned}
$$

where the two sets on the right-hand side are disjoint; the proof of this is left as an exercise for the reader. We thus obtain (13) for I, J of combined length $k + m$ for $k, m \geq 1$, which completes the inductive step. □

Example 2.15 Consider a 2-dimensional path $X : [a, b] \to \mathbb{R}^2$. The shuffle product identity (13) implies that

$$
\begin{aligned}
S(X)^1_{a,b} S(X)^2_{a,b} &= S(X)^{1,2}_{a,b} + S(X)^{2,1}_{a,b}, \\
S(X)^{1,2}_{a,b} S(X)^1_{a,b} &= 2S(X)^{1,1,2}_{a,b} + S(X)^{1,2,1}_{a,b}.
\end{aligned}
$$

Example 2.16 We now show that the shuffle product generalizes the familiar multiplication of (multivariate) polynomials.

As the simplest example, consider the one-dimensional case with $x \in \mathbb{R}$ and the linear path $X : [0, 1] \to \mathbb{R}$, $X_t = tx$. Following Example 2.6, for any multi-index $(1, \dots, 1)$ of length k,

$$S(X)^{1,\dots,1}_{0,1} = \frac{x^k}{k!}. \tag{14}$$

Consider two multi-indexes I and J of length k and m (consisting only of 1's). Then, by (14),

$$S(X)^I_{0,1} S(X)^J_{0,1} = \frac{x^{k+m}}{k!m!}. \tag{15}$$

On the other hand, there are $\binom{k+m}{k}$ ways to shuffle I and J, and all shuffles lead to the same multi-index with 1 appearing $k+m$ times. Therefore, by (13),

$$S(X)^I_{0,1} S(X)^J_{0,1} = \binom{k+m}{k} \frac{x^{k+m}}{(k+m)!}. \tag{16}$$

Since $\binom{k+m}{k} = (k+m)!/(k!m!)$, we see that the expressions (15) and (16) match.

Consider now the multi-dimensional case $x = (x_1, \dots, x_d) \in \mathbb{R}^d$ and the linear path $X : [0, 1] \to \mathbb{R}^d$, $X_t = tx$ as before. For $I = (i_1, \dots, i_k)$, one can verify that

$$S(X)^I_{0,1} = \frac{x_1^{\beta_1} \cdots x_d^{\beta_d}}{k!}, \tag{17}$$

where β_n is the number of times n appears in $i_1, \dots, i_k$. In particular, $S(X)^I_{0,1}$ depends only on the elements in I, not their order, and encodes a multivariate monomaniacal of x.

It follows that, for another multi-index $J = (j_1, \dots, j_m)$,

$$S(X)^I_{0,1} S(X)^J_{0,1} = \frac{x_1^{\beta_1+\gamma_1} \cdots x_d^{\beta_d+\gamma_d}}{k!m!}, \tag{18}$$

where γ_n is the number of times n appears in $j_1, \dots, j_m$.

On the other hand, there are again $\binom{k+m}{k}$ multi-indexes in $I \sqcup\!\sqcup J$, which may now be distinct, but each $n \in \{1, \dots, d\}$ appears $\beta_n + \gamma_n$ times in each multi-index in $I \sqcup\!\sqcup J$. Therefore, by (13),

$$S(X)^I_{0,1} S(X)^J_{0,1} = \binom{k+m}{k} \frac{x_1^{\beta_1+\gamma_1} \cdots x_d^{\beta_d+\gamma_d}}{(k+m)!},$$

which again agrees with (18).

The shuffle product implies that the product of two terms of the signature can be expressed as a linear combination of higher order terms. This fact has important theoretical and practical consequences, some of which we discuss in Sect. 2.4.

2.3.3 Chen's Identity

We now describe a property of the signature known as Chen's identity, which provides an algebraic relationship between paths and their signatures. It furthermore provides a powerful tool in the computation of the signature once its value is known on sub intervals (see Example 2.18 below). The simplest version of Chen's identity is the following.

Theorem 2.17 (Chen's Identity) *Suppose $a < b < c$ and $X : [a, c] \to \mathbb{R}^d$ is a path. Then, for any multi-index $(i_1, \dots, i_k)$,*

$$S(X)^{i_1,\dots,i_k}_{a,c} = \sum_{m=0}^{k} S(X)^{i_1,\dots,i_m}_{a,b} S(X)^{i_{m+1},\dots,i_k}_{b,c}. \tag{19}$$

Proof We proceed by induction on k. The case $k = 0$ is obvious since both the left- and right-hand sides are equal to 1 in this case. Suppose now (19) holds for $k - 1$ with $k \geq 1$. Then

$$\begin{aligned} S(X)^{i_1,\dots,i_k}_{a,c} &= \int_a^b S(X)^{i_1,\dots,i_{k-1}}_{a,t}\, dX^{i_k}_t + \int_b^c S(X)^{i_1,\dots,i_{k-1}}_{a,t}\, dX^{i_k}_t \\ &= S(X)^{i_1,\dots,i_k}_{a,b} + \int_b^c \sum_{m=0}^{k-1} S(X)^{i_1,\dots,i_m}_{a,b} S(X)^{i_{m+1},\dots,i_{k-1}}_{b,t}\, dX^{i_k}_t \\ &= S(X)^{i_1,\dots,i_k}_{a,b} + \sum_{m=0}^{k-1} S(X)^{i_1,\dots,i_m}_{a,b} S(X)^{i_{m+1},\dots,i_{k-1},i_k}_{b,c} \\ &= \sum_{m=0}^{k} S(X)^{i_1,\dots,i_m}_{a,b} S(X)^{i_{m+1},\dots,i_k}_{b,c}, \end{aligned}$$

where we used the inductive hypothesis in the second equality. □

The next example shows that Chen's identity is an effective method to compute the signatures of piecewise linear paths.

Example 2.18 (Signature of Piecewise Linear Path) Consider the path $X : [0, 2] \to \mathbb{R}^2$ which is linear on the intervals $[0, 1]$ and $[1, 2]$ and $X_0 = (3, 0)$, $X_1 = (3, 1)$, $X_2 = (0, 1)$, see Fig. 7.

Then

$$(S(X)^{\emptyset}_{0,1}, S(X)^{2}_{0,1}, S(X)^{22}_{0,1}) = (1, 1, 1/2),$$

with $S(X)^{I}_{0,1} = 0$ for all other multi-indexes I of length at most 2, and

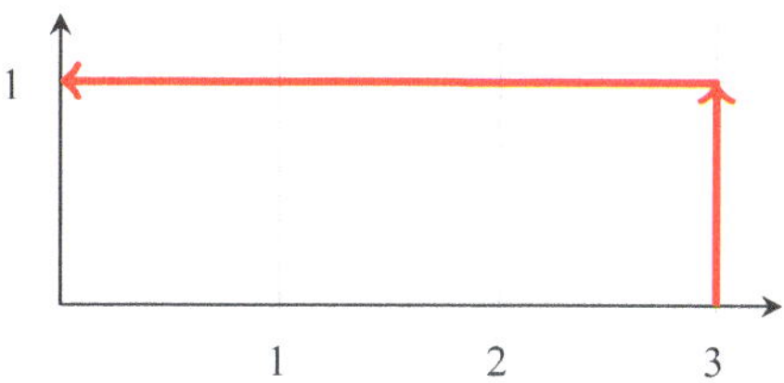

Fig. 7 An example of a piecewise linear path

$$(S(X)^{\emptyset}_{1,2}, S(X)^{1}_{1,2}, S(X)^{11}_{1,2}) = (1, -3, 9/2),$$

with $S(X)^{I}_{1,2} = 0$ for all other multi-indexes I of length at most 2.

Then by Chen's identity

$$\begin{aligned}(S(X)^{1}_{0,2}, S(X)^{2}_{0,2}) &= (S(X)^{1}_{0,1} + S(X)^{1}_{1,2}, S(X)^{2}_{0,1} + S(X)^{2}_{1,2}) \\ &= (-3 + 0, 0 + 1),\end{aligned}$$

as expected from the fact that $S(X)^{i}_{a,b} = X^{i}_{b} - X^{i}_{a}$. Furthermore, again by Chen's identity,

$$\begin{aligned}S(X)^{12}_{0,2} &= S(X)^{12}_{0,1} + S(X)^{1}_{0,1}S(X)^{2}_{1,2} + S(X)^{12}_{1,2} = 0 + 0 \times 0 + 0 = 0, \\ S(X)^{21}_{0,2} &= S(X)^{21}_{0,1} + S(X)^{2}_{0,1}S(X)^{1}_{1,2} + S(X)^{21}_{1,2} = 0 + 1 \times (-3) + 0 = -3,\end{aligned}$$

which is less obvious, and which is consistent with the Lévy area identity (12).

There is a way to formulate Chen's identity in a more algebraic way. To do so, we need to introduce the algebra of formal power series, which should appear natural in light of the definition of the signature.

Definition 2.19 (Formal Power Series) Let $e_1, \ldots, e_d$ be d formal indeterminate. The algebra of (non-commuting) *formal power series* in d indeterminates is the vector space of all formal series of the form

$$\sum_{k=0}^{\infty} \sum_{i_1,\ldots,i_k \in \{1,\ldots,d\}} \lambda_{i_1,\ldots,i_k} e_{i_1} \ldots e_{i_k},$$

where the second summation runs over all multi-indexes $(i_1, \ldots, i_k)$, $i_1, \ldots, i_k \in \{1, \ldots, d\}$, and $\lambda_{i_1,\ldots,i_k}$ are real numbers.

A (non-commuting) formal polynomial is a formal power series for which only a finite number of coefficients $\lambda_{i_1,\ldots,i_k}$ are non-zero. The terms $e_{i_1} \ldots e_{i_k}$ are called monomials. The term corresponding to $k = 0$ is simply a real number λ_0. The space of formal power series is also called the *space of tensor series* over $\mathbb{R}^d$. We stress that the power series we consider are non-commutative; for example, the elements e_1e_2 and e_2e_1 are distinct.

The space of formal power series may be naturally equipped with a vector space structure by defining addition and scalar multiplication as

$$\Big(\sum_{k=0}^{\infty}\sum_{i_1,\ldots,i_k\in\{1,\ldots,d\}}\lambda_{i_1,\ldots,i_k}e_{i_1}\ldots e_{i_k}\Big)+\Big(\sum_{k=0}^{\infty}\sum_{i_1,\ldots,i_k\in\{1,\ldots,d\}}\mu_{i_1,\ldots,i_k}e_{i_1}\ldots e_{i_k}\Big)$$
$$=\sum_{k=0}^{\infty}\sum_{i_1,\ldots,i_k\in\{1,\ldots,d\}}(\lambda_{i_1,\ldots,i_k}+\mu_{i_1,\ldots,i_k})e_{i_1}\ldots e_{i_k}$$

and

$$c\Big(\sum_{k=0}^{\infty}\sum_{i_1,\ldots,i_k\in\{1,\ldots,d\}}\lambda_{i_1,\ldots,i_k}e_{i_1}\ldots e_{i_k}\Big)=\sum_{k=0}^{\infty}\sum_{i_1,\ldots,i_k\in\{1,\ldots,d\}}c\lambda_{i_1,\ldots,i_k}e_{i_1}\ldots e_{i_k}.$$

Moreover, one may define the product $\otimes$ between monomials by joining together multi-indexes

$$e_{i_1}\ldots e_{i_k}\otimes e_{j_1}\ldots e_{j_m}=e_{i_1}\ldots e_{i_k}e_{j_1}\ldots e_{j_m}.$$

The product $\otimes$ then extends uniquely and linearly to all power series. We demonstrate the first few terms of the product in the following expression:

$$\Big(\sum_{k=0}^{\infty}\sum_{i_1,\ldots,i_k\in\{1,\ldots,d\}}\lambda_{i_1,\ldots,i_k}e_{i_1}\ldots e_{i_k}\Big)\otimes\Big(\sum_{k=0}^{\infty}\sum_{i_1,\ldots,i_k\in\{1,\ldots,d\}}\mu_{i_1,\ldots,i_k}e_{i_1}\ldots e_{i_k}\Big)$$
$$=\lambda_0\mu_0+\sum_{i=1}^{d}(\lambda_0\mu_i+\lambda_i\mu_0)e_i+\sum_{i,j=1}^{d}\big(\lambda_0\mu_{i,j}+\lambda_i\mu_j+\lambda_{i,j}\mu_0\big)\,e_ie_j+\ldots.\tag{20}$$

The space of formal power series becomes an algebra when equipped with this vector space structure and product $\otimes$, i.e. $\otimes$ is associative and distributive over addition. Note, however, that $\otimes$ is not commutative as (20) demonstrates.

The reader may have noticed that the indexing set of the monomials $e_{i_1}\ldots e_{i_k}$ coincides with the indexing set of the terms of the signature of a path $X:[a,b]\to\mathbb{R}^d$, namely the collection of all multi-indexes $(i_1,\ldots,i_k)$, $i_1,\ldots,i_k\in\{1,\ldots,d\}$. It follows that a convenient way to express the signature of X is by a formal power series where the coefficient of each monomial $e_{i_1}\ldots e_{i_k}$ is set to be $S(X)^{i_1,\ldots,i_k}_{a,b}$. We use the same symbol $S(X)_{a,b}$ to denote this representation

$$S(X)_{a,b} = \sum_{k=0}^{\infty} \sum_{i_1,\dots,i_k \in \{1,\dots,d\}} S(X)_{a,b}^{i_1,\dots,i_k} e_{i_1} \dots e_{i_k},$$

where, as before, the "0-th" level of the signature is $S(X)_{a,b}^{\emptyset} = 1$ (corresponding to $k = 0$).

Example 2.20 Consider $x = (x^1, \dots, x^d) \in \mathbb{R}^d$ and identify this with the formal power series $x = \sum_{i=1}^{d} x^i e_i$ (which we denote by the same symbol). Define the formal power series $\mathbf{e}^x$ by

$$\mathbf{e}^x = \sum_{k=0}^{\infty} \frac{x^{\otimes k}}{k!} = \sum_{k=0}^{\infty} \frac{(\sum_{i=1}^{d} x^i e_i)^{\otimes k}}{k!}.$$

Although this series has an infinite number of terms, the coefficient of $e_{i_1} \dots e_{i_k}$ in $\mathbf{e}^x$ is a monomial in $x^1, \dots, x^d$ and thus $\mathbf{e}^x$ is well-defined as a formal power series without issues of convergence. Explicitly, the coefficient of $e_{i_1} \dots e_{i_k}$ in $\mathbf{e}^x$ is $x_1^{\beta_1} \dots x_d^{\beta_d} / k!$, where β_n is the number of times n appears in $i_1, \dots, i_k$.

Then, for the path $X : [0, 1] \to \mathbb{R}^d$ defined by $X_t = tx$,

$$S(X)_{0,1} = \mathbf{e}^x,$$

which is essentially the claim we saw already in (17) from Example 2.16.

To state Chen's identity in an algebraic way, it remains to define the concatenation of paths.

Definition 2.21 (Concatenation) For two paths $X : [a, b] \to \mathbb{R}^d$ and $Y : [b, c] \to \mathbb{R}^d$, we define their *concatenation* as the path $X * Y : [a, c] \to \mathbb{R}^d$ for which $(X * Y)_t = X_t$ for $t \in [a, b]$ and $(X * Y)_t = X_b + (Y_t - Y_b)$ for $t \in [b, c]$.

See Fig. 8 for a visual depiction of concatenation. We can now recast Chen's identity (Theorem 2.17) in an algebraic light, namely that the signature is a map from the space of paths into the space of formal power series that takes the concatenation product $*$ to the product $\otimes$. More precisely, we have the following result.

Theorem 2.22 (Chen's Identity) *For two paths $X : [a, b] \to \mathbb{R}^d$ and $Y : [b, c] \to \mathbb{R}^d$,*

$$S(X * Y)_{a,c} = S(X)_{a,b} \otimes S(Y)_{b,c}.$$

The contents of Theorems 2.17 and 2.22 are identical. We only state the latter version of Chen's identity to highlight a useful algebraic structure. In particular, it allows us to state the following corollary that extends the computations from Examples 2.18 and 2.20.

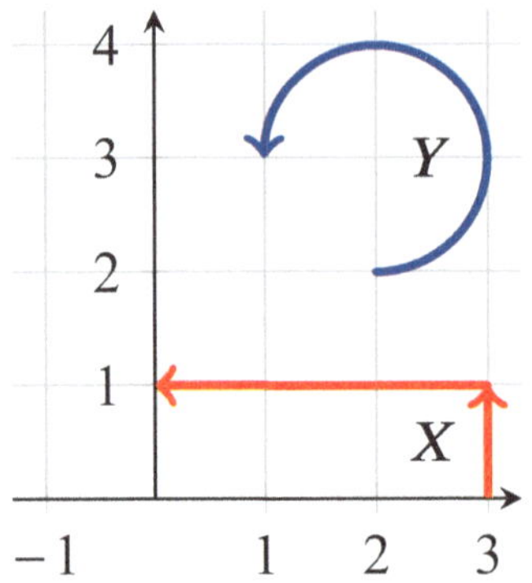

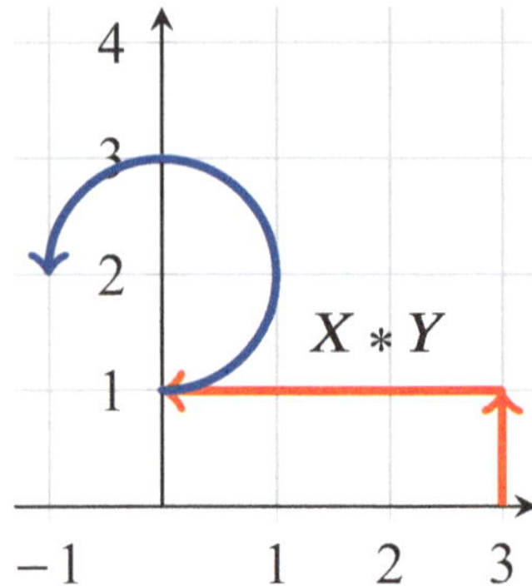

Fig. 8 Example of two paths X and Y and the resulting concatenation $X * Y$

Corollary 2.23 (Signature of Piecewise Linear Path) *Let* $X : [0, n] \to \mathbb{R}^d$ *be a path that is linear on each interval* $[0, 1], \ldots, [n - 1, n]$. *Then*

$$S(X)_{0,n} = \mathbf{e}^{X_1 - X_0} \otimes \ldots \otimes \mathbf{e}^{X_n - X_{n-1}} .$$

2.3.4 Time-Reversal

The time-reversal property informally states that the signature $S(X)_{a,b}$ of a path $X : [a, b] \to \mathbb{R}^d$ is precisely the inverse under the product $\otimes$ of the signature obtained by running X backwards in time. To make this precise, we make the following definition.

Definition 2.24 (Time-Reversal) For a path $X : [a, b] \to \mathbb{R}^d$, we define its time-reversal as the path $\overleftarrow{X} : [a, b] \to \mathbb{R}^d$ for which $\overleftarrow{X}_t = X_{a+b-t}$ for all $t \in [a, b]$.

See Fig. 9 for a depiction of the time reversal of paths.

Theorem 2.25 (Time-Reversed Signature) *For a path* $X : [a, b] \to \mathbb{R}^d$, *it holds that*

$$S(X)_{a,b} \otimes S(\overleftarrow{X})_{a,b} = 1. \tag{21}$$

Explicitly, for all multi-indexes $(i_1, \ldots, i_k)$,

$$S(\overleftarrow{X})^{i_1,\ldots,i_k}_{a,b} = (-1)^k S(X)^{i_k,\ldots,i_1}_{a,b}. \tag{22}$$

Remark 2.26 The element 1 in (21) should be understood as the formal power series where $\lambda_0 = 1$ and $\lambda_{i_1,\ldots,i_k} = 0$ for all $k \geq 1$ and $i_1, \ldots, i_k \in \{1, \ldots, d\}$, which is the identity element under the product $\otimes$.

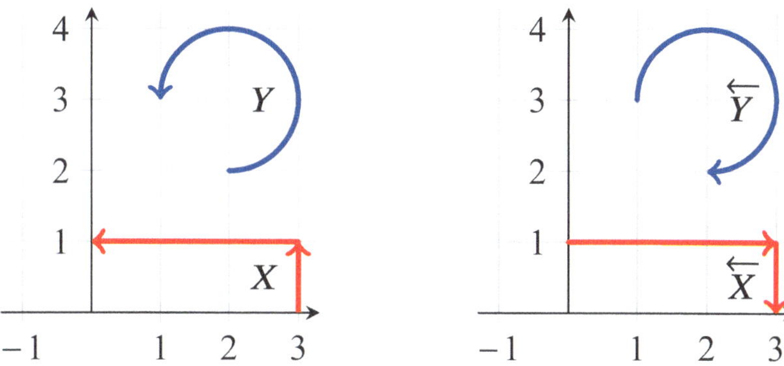

Fig. 9 Example of two paths X and Y and their respective reversals $\overleftarrow{X}$ and $\overleftarrow{Y}$

Proof We will prove (22). As before, we proceed by induction. The claim is clearly true for $k = 0$ since $S(X)^{\emptyset}_{a,b} = 1$. Suppose now that (22) is true for some $k - 1 \geq 0$. Then

$$\begin{aligned} S(\overleftarrow{X})^{i_1,\dots,i_k}_{a,b} &= \int_a^b S(\overleftarrow{X})^{i_1,\dots,i_{k-1}}_{a,t}\, d\overleftarrow{X}^{i_k}_t \\ &= \int_a^b (-1)^{k-1} S(X)^{i_{k-1},\dots,i_1}_{a+b-t,b}\, dX^{i_k}_{a+b-t} \\ &= \int_a^b (-1)^k S(X)^{i_{k-1},\dots,i_1}_{s,b}\, dX^{i_k}_s \\ &= (-1)^k S(X)^{i_k,i_{k-1},\dots,i_1}_{a,b}, \end{aligned}$$

where we used the inductive hypothesis in the second equality, the change of variable $s = a + b - t$ in the third equality, and in the final equality we used the identity

$$S(X)^{j_1,\dots,j_k}_{a,b} = \int_a^b S(X)^{j_2,\dots,j_k}_{t,b}\, dX^{j_1}_t,$$

which itself follows from (3). □

2.3.5 Log-Signature

We now define a transform of the signature called the log-signature. The log-signature is nothing but the logarithm of the signature in the algebra of formal power series. For a power series

$$x = \sum_{k=0}^{\infty} \sum_{i_1,\dots,i_k \in \{1,\dots,d\}} \lambda_{i_1,\dots,i_k} e_{i_1} \dots e_{i_k}$$

for which $\lambda_0 > 0$, define its logarithm as the power series given by

$$\log x = \log(\lambda_0) - \sum_{n \geq 1} \frac{1}{n} \left(1 - \frac{x}{\lambda_0}\right)^{\otimes n},$$

where $\otimes n$ denotes the n-th power with respect to the product $\otimes$.

In the case that $\lambda_0 = 1$, which is the most important case for us, the logarithm reduces to

$$\log x = -\sum_{n \geq 1} \frac{1}{n} (1 - x)^{\otimes n}. \tag{23}$$

For example, for a real number $\lambda \in \mathbb{R}$ and the series

$$x = 1 + \sum_{k \geq 1} \frac{\lambda^k}{k!} e_1^{\otimes k},$$

one can readily check that

$$\log x = \lambda e_1.$$

In general, $\log x$ is a series with an infinite number of terms. However, for every multi-index $(i_1, \ldots, i_k)$, the coefficient of $e_{i_1} \ldots e_{i_k}$ in $\log x$ depends only on the coefficients of x of the form $\lambda_{j_1, \ldots, j_m}$ with $m \leq k$, of which there are only finitely many, so that $\log x$ is well-defined without the need to consider convergence of infinite series.

Definition 2.27 (Log-Signature) For a path $X : [a, b] \to \mathbb{R}^d$, the log-signature of X is defined as the formal power series $\log S(X)_{a,b}$.

For two formal power series x and y, we define their Lie bracket by

$$[x, y] = x \otimes y - y \otimes x. \tag{24}$$

A direct computation shows that the first few terms of the log-signature are given by

$$\log S(X)_{a,b} = \sum_{i=1}^{d} S(X)^i_{a,b} e_i + \sum_{1 \leq i < j \leq d} \frac{1}{2} \left(S(X)^{i,j}_{a,b} - S(X)^{j,i}_{a,b} \right) [e_i, e_j] + \ldots. \tag{25}$$

In particular, the coefficient of the polynomials $[e_i, e_j]$ in the log-signature is precisely the Lévy area introduced in Sect. 2.2.4.

Example 2.28 Consider the two-dimensional path

$$X : [0, 2] \to \mathbb{R}^2, \quad X_t = \begin{cases} \{t, 0\} & \text{if } t \in [0, 1], \\ \{1, t-1\} & \text{if } t \in [1, 2]. \end{cases}$$

Note that X is the concatenation of the two linear paths $Y : [0, 1] \to \mathbb{R}^2$, $Y_t = \{t, 0\}$ and $Z : [1, 2] \to \mathbb{R}^2$, $Z_t = \{1, t-1\}$. One can readily check (see Example 2.20) that the signatures of Y and Z (as formal power series) are given by

$$S(Y)_{0,1} = 1 + \sum_{k \geq 1} \frac{1}{k!} e_1^{\otimes k}, \quad S(Z)_{1,2} = 1 + \sum_{k \geq 1} \frac{1}{k!} e_2^{\otimes k}.$$

It follows from Chen's identity that

$$S(X)_{0,2} = S(Y)_{0,1} \otimes S(Z)_{1,2} = 1 + e_1 + e_2 + \frac{1}{2!} e_1 e_1 + \frac{1}{2!} e_2 e_2 + e_1 e_2 + \dots.$$

Hence the first few terms of the log-signature of X are

$$\log S(X)_{0,2} = e_1 + e_2 + \frac{1}{2}[e_1, e_2] + \dots.$$

In fact, one can readily check that the coefficients of $\log S(X)_{0,2}$ in the above example are given precisely by the classical Baker–Campbell–Hausdorff formula.

The above example demonstrates the general fact that the log-signature can always be expressed as a power series composed entirely of so-called Lie polynomials. This is the content of the following theorem due to Chen [12], which generalizes the Baker–Campbell–Hausdorff theorem.

Theorem 2.29 *Let $X : [a, b] \to \mathbb{R}^d$ be a path. Then there exist real numbers $\lambda_{i_1,\dots,i_k}$ such that*

$$\log S(X)_{a,b} = \sum_{k \geq 1} \sum_{i_1,\dots,i_k \in \{1,\dots,d\}} \lambda_{i_1,\dots,i_k} [e_{i_1}, [e_{i_2}, \dots, [e_{i_{k-1}}, e_{i_k}] \dots]]. \tag{26}$$

Note that the coefficients $\lambda_{i_1,\dots,i_k}$ are not unique since the polynomials of the form $[e_{i_1}, [e_{i_2}, \dots, [e_{i_{k-1}}, e_{i_k}] \dots]]$ are not linearly independent (e.g., $[e_1, e_2] = -[e_2, e_1]$).

Theorem 2.29 is the only result in this subsection which we do not prove since all proofs, to our knowledge, require a non-trivial detour into Lie algebras. See [72, Cor. 3.5] or [28, Sec. 7.5] for two different proofs.

2.4 Expected Signatures as Moments of Path-Valued Random Variables

We saw in Example 2.16 that the signature generalizes multivariate polynomials to the space of paths and that the shuffle product (Theorem 2.14) generalizes the classical product of polynomials. In this subsection, we highlight the relevance of these observations to the theory of statistical learning by interpreting the expectation of the signature as the moments of path-valued random variables. We assume that the reader is familiar with the basics of probability theory, such as measures and expectation; see, e.g. [5] for background material. See also Sect. 2.5.4 for a further discussion on signatures, moments of random variables, and the Fourier transform.

Consider a probability measure $\mathbb{P}$ on $\mathbb{R}$. The moments of $\mathbb{P}$, defined as the numbers $\int_{\mathbb{R}} x^k \mathbb{P}(dx)$ for $k = 0, 1, \ldots$, encode important information about $\mathbb{P}$ (whenever they are well-defined) and play a prominent role in statistics. To see why, we recall the classical Weierstrass approximation theorem.

Theorem 2.30 (Weierstrass 1885) *Let $f : [a, b] \to \mathbb{R}$ be a continuous function and $\epsilon > 0$. Then there exists a polynomial $p(x) = a_0 + a_1 x + \cdots + a_k x^k$ with real coefficients a_i such that $|f(x) - p(x)| < \epsilon$ for all $x \in [a, b]$.*

The relevance of this theorem to probability theory is that the expectation

$$\mathbb{P}(p) = \int_{\mathbb{R}} p(x)\mathbb{P}(dx)$$

of any polynomial p under $\mathbb{P}$ is determined by the moments of $\mathbb{P}$ due to linearity of integration. Furthermore, recall that two probability measures $\mathbb{P}$ and $\mathbb{Q}$ are equal if and only if $\mathbb{P}(f) = \mathbb{Q}(f)$ for all continuous functions $f : \mathbb{R} \to \mathbb{R}$, see [5, Theorem 25.8]. These two remarks provide a simple way to test whether two probability measures on a finite interval are equal.

Corollary 2.31 *Suppose $\mathbb{P}$ and $\mathbb{Q}$ are probability measures on $[a, b]$. Then $\mathbb{P} = \mathbb{Q}$ if and only if the moments of $\mathbb{P}$ and $\mathbb{Q}$ are equal.*

Proof The direction $\Rightarrow$ is obvious, so we prove now $\Leftarrow$. Consider $f : [a, b] \to \mathbb{R}$ continuous and let $\epsilon > 0$. By Theorem 2.30 there exists a polynomial p such that $|f(x) - p(x)| < \epsilon$ for all $x \in [a, b]$. In particular, by the triangle inequality applied to integrals, $|\mathbb{P}(f) - \mathbb{P}(p)| < \epsilon$ and likewise for $\mathbb{Q}$. But since $\mathbb{P}(p) = \mathbb{Q}(p)$ by the assumption that the moments of $\mathbb{P}$ and $\mathbb{Q}$ match, we see that $|\mathbb{P}(f) - \mathbb{Q}(f)| < 2\epsilon$. Since ϵ was arbitrary, it follows that $\mathbb{P}(f) = \mathbb{Q}(f)$ for any continuous function $f : [a, b] \to \mathbb{R}$ and thus $\mathbb{P} = \mathbb{Q}$. □

Remark 2.32 It is crucial that $\mathbb{P}$ and $\mathbb{Q}$ in the above theorem are defined on a bounded interval $[a, b]$ instead of on $\mathbb{R}$. Indeed, one can find two distinct probability measures on $\mathbb{R}$ whose moments are equal. Nonetheless, there exists an essentially sharp criterion on the growth rate of the moments of $\mathbb{P}$ (Carleman's condition) which guarantees that $\mathbb{P}$ is determined by its moments, see [42].

We now recall a substantial generalization of Theorem 2.30, known as the Stone–Weierstrass theorem (see [73, Theorem 7.32]).

Theorem 2.33 *Let $\mathcal{X}$ be a compact space. Suppose $\mathcal{F}$ is a vector space of functions $f : \mathcal{X} \to \mathbb{R}$ such that*

(i) every $f \in \mathcal{F}$ is continuous,
(ii) the constant function 1 is in $\mathcal{F}$
(iii) for every $f, g \in \mathcal{F}$, one has $fg \in \mathcal{F}$, and
(iv) for all distinct $x, y \in \mathcal{X}$, there exists $f \in \mathcal{F}$ such that $f(x) \neq f(y)$.

Then, for any continuous function $g : \mathcal{F} \to \mathbb{R}$ and $\epsilon > 0$, there exists $f \in \mathcal{F}$ such that $|f(x) - g(x)| < \epsilon$ for all $x \in \mathcal{X}$.

Put more succinctly, every point-separating algebra of continuous functions that contains 1 is dense in $C(\mathcal{X}) = \{f : \mathcal{X} \to \mathbb{R},\ \ f \text{ is continuous}\}$. ('Algebra' refers to item (iii) and 'point-separating' refers to item (iv).) We leave it as an exercise for the reader to verify that if $\mathcal{X} = [a, b]$ and $\mathcal{F}$ is the space of polynomials, then $\mathcal{F}$ verifies the assumptions of Theorem 2.33, and thus Theorem 2.33 implies Theorem 2.30.

Turning back to the signature, let $\mathcal{X}$ denote the set of paths $X : [a, b] \to \mathbb{R}^d$ that we equip with the bounded variation norm $\|X\|_{BV} = |X_a| + \int_a^b |\dot{X}_t| dt$. Let $\mathcal{F}$ denote the space of all functions $f : \mathcal{X} \to \mathbb{R}$ of the form

$$X \mapsto \sum_{i=1}^{n} \lambda_i S(X)_{a,b}^{I_i}, \tag{27}$$

where $n \geq 1$, $\lambda_i \in \mathbb{R}$, and I_i is a multi-index. We claim that all the conditions of Theorem 2.33 *except* (iv) are satisfied:

(i) For every multi-index I, the function $X \mapsto S(X)_{a,b}^{I}$ is continuous, the verification of which we leave as an exercise for the reader, hence every $f \in \mathcal{F}$ is continuous.
(ii) Since $S(X)_{a,b}^{\emptyset} = 1$, we clearly have $1 \in \mathcal{F}$.
(iii) For the every multi-index I and J, by the shuffle product (Theorem 2.14), the function $X \mapsto S(X)_{a,b}^{I} S(X)_{a,b}^{J}$ is in $\mathcal{F}$, which implies that $fg \in \mathcal{F}$ whenever $f, g \in \mathcal{F}$.

Unfortunately, it is clear that the point-separating property (iv) is not satisfied. There are three reasons for this:

- The signature is base-point invariant, i.e. $S(X)_{a,b} = S(X+x)_{a,b}$, so no two paths that differ by a shift in $\mathbb{R}^d$ can be separated by functions in $\mathcal{F}$.
- The signature is parametrization invariant (see Sect. 2.3.1), so no two paths that differ by a reparametrization of $[a, b]$ can be separated.
- By Chen's identity (Theorem 2.22) and the time-reversal property (Theorem 2.25), if X can be written as $X = Y * \overleftarrow{Y}$ for a path Y, then $S(X)_{a,b} = 1$ (understood as the formal power series), which is equal to the signature of a

constant path Z, i.e. $Z_t = z$ for all $t \in [a, b]$ for some fixed $z \in \mathbb{R}^d$. Therefore, X and Z cannot be separated, i.e. $f(X) = f(Z)$ for all $f \in \mathcal{F}$.

Nonetheless, it is possible to recover (iv) after a suitable adjustment of the space of paths $\mathcal{X}$. The minimal adjustment necessary to achieve (iv) is to suitably quotient $\mathcal{X}$ by the kernel of S, which consists of all tree-like paths (see Sect. 2.5.2).

However, we demonstrate a simpler, more pragmatic approach to achieve (iv), which is to shrink $\mathcal{X}$. More precisely, let us fix $y \in \mathbb{R}^{d-1}$ and consider the set of paths

$$\mathcal{Y} = \{Y = (Y^1, \dots, Y^d) \in \mathcal{X} : \forall t \in [a, b] \;\; Y_t^1 = t, \;\; Y_a = (a, y) \in \mathbb{R}^d\}.$$

In plain words, $\mathcal{Y}$ consists of those paths $Y : [a, b] \to \mathbb{R}^d$ in $\mathcal{X}$ such that Y^1 is the time-component and whose initial point is (a, y). (The fact that the first component is time is not crucial. What is crucial is that there exists a fixed strictly monotone function $f : [a, b] \to \mathbb{R}$ such that every $Y \in \mathcal{Y}$ is of the form $Y_t = (f_t, \bar{Y}_t)$.) The next result shows that the signature separates the points of $\mathcal{Y}$.

Proposition 2.34 *Suppose that $X, Y \in \mathcal{Y}$ with $S(X)^I_{a,b} = S(Y)^I_{a,b}$ for all multi-indexes I. Then $X = Y$.*

Proof Recall from Example 2.6 that $S(X)^{1,\dots,1}_{a,t} = \frac{(t-a)^k}{k!}$ where '1' appears k times in $1, \dots, 1$. Then, for any $i = 2, \dots, d$,

$$S(X)^{1,\dots,1,i}_{a,b} = \int_a^b S(X)^{1,\dots,1}_{a,t} \dot{X}^i_t \, dt = \int_a^b \frac{(t-a)^k}{k!} \dot{X}^i_t \, dt.$$

It follows that the sequence $S(X)^i_{a,b}, S(X)^{1,i}_{a,b}, S(X)^{1,1,i}_{a,b}, \dots$ encodes all the *moments* of the measure $\dot{X}^i_t \, dt$ on $[a, b]$. The fact that $S(X)_{a,b} = S(Y)_{a,b}$ thus implies that the measures $\dot{X}^i_t \, dt$ and $\dot{Y}^i_t \, dt$ have the same moments and are therefore equal (we leave this final implication as an exercise to the reader with the hint that one should use Theorem 2.30). Since $X_t = X_a + \int_a^t \dot{X}_s \, ds$ and $X_a = Y_a$, it follows that $X = Y$. □

We therefore see that the set of functions $\mathcal{F}$ defined by (27) separates the points of $\mathcal{Y}$. Consequently, as we already verified all the other conditions for Theorem 2.33, we obtain the following result, which shows that arbitrary continuous functions on paths can be approximated by linear functions of the signature.

Proposition 2.35 *Consider a compact set $K \subset \mathcal{Y}$, continuous function $g : K \to \mathbb{R}$, and $\epsilon > 0$. Then there exists $f \in \mathcal{F}$ such that $|f(X) - g(X)| < \epsilon$ for all $X \in K$. More precisely, there exists an integer $n \geq 1$, real numbers $\lambda_1, \dots, \lambda_n$, and multi-indexes $I_1, \dots, I_n$ such that, for all $X \in K$,*

$$\left| g(X) - \sum_{i=1}^{n} \lambda_i S(X)^{I_i}_{a,b} \right| < \epsilon.$$

By precisely the same argument used to prove Corollary 2.31, we obtain the following corollary of Proposition 2.35 that justifies the title of this subsection.

Corollary 2.36 *Suppose $\mathbb{P}$ and $\mathbb{Q}$ are probability measures on a compact set $K \subset \mathcal{Y}$ with equal expected signatures, i.e. $\int_K S(X)^I_{a,b}\mathbb{P}(dX) = \int_K S(X)^I_{a,b}\mathbb{Q}(dX)$ for all multi-indexes I. Then $\mathbb{P} = \mathbb{Q}$.*

Remark 2.37 It is possible to significantly extend this result, in particular dropping the assumption that $\mathbb{P}$ and $\mathbb{Q}$ are defined on a compact subset of paths, see Sect. 2.5.4.

2.5 *Further Topics and Extensions*

We conclude the first part of this chapter with a brief discussion on several more advanced topics. In particular, we discuss

1. the signature of rough paths,
2. the extent to which the signature determines the underlying path,
3. extensions to jump processes,
4. the role of the signature in the moment problem for random paths,
5. extensions to higher dimensions.

2.5.1 Rough Paths

Our discussion of the signature has so far been restricted to paths which are piecewise continuously differentiable (or more generally of bounded variation). This restriction was needed to ensure that the iterated integrals of the path existed as Riemann–Stieltjes integrals. More generally, one can define the iterated integrals of a path using the Young integral for any path of finite p-variation with $1 \leq p < 2$. The Young integral goes beyond the "classical" definition of the integral and is already able to cover a class of paths substantially more irregular than those of bounded variation.

A problem that arises for paths of infinite p-variation for all $p < 2$ (which is a situation of great interest in stochastic analysis because the sample paths of Brownian motion have almost surely finite p-variation if and only if $p > 2$), is that there is no well-defined notion of an iterated integral for such paths. We stress that this is not due to any technical limitation of the Young integral, but rather due to the fact that there is no unique candidate for the iterated integrals. That is to say, there is not necessarily only one unique way to define the iterated integrals.

One of the key observations of T. Lyons in his introduction of rough paths in [53] was that if one *defines* the first $\lfloor p \rfloor$ iterated integrals of a path X of finite p-variation, then there is a unique way to obtain all the other iterated integrals, and hence the signature of X. This notion of a path of finite p-variation, along with its first $\lfloor p \rfloor$

iterated integrals (which may be defined arbitrarily provided that they satisfy Chen's identity and possess finite p-variation), is precisely the definition of a p-rough path.

A key result in the theory of rough paths, known as the *universal limit theorem*, is that one can give meaning to controlled differential equations where the driver is a so-called *geometric* p-rough path, and where the solution depends in a continuous way on the driver provided that the space of p-rough paths is equipped with a suitable topology (known as the p-variation metric). We refer the interested reader to the St. Flour lecture notes of Lyons et al. [57] for an introduction to the theory of rough paths and to the monograph of Friz and Hairer [26] for a recent treatment of the topic.

2.5.2 Path Uniqueness

As discussed in Sect. 2.2.3, the signature of a path $X : [a, b] \mapsto \mathbb{R}^d$ is all that is needed to determine the endpoint of the solution to a linear (and, less trivially, non-linear) differential equation driven by X. This was first shown by Chen [13] in the smooth setting, and later extended by Hambly and Lyons [36] and Boedihardjo et al. [7] to less regular paths. The works of these authors show that the signature captures deep geometric properties of a path, which we briefly discuss here.

A natural question that one may ask is the following: is a path completely determined by its signature? The answer is no for the reasons described above Proposition 2.34. For example, one can never recover from the signature the exact speed at which the path is traversed (due to invariance under time reparametrizations), nor can one tell apart the signature of a trivial constant path and that of a path concatenated with its time-reversal.

However, a highly non-trivial fact is that this is essentially the only information lost by the signature. For example, for a path X that never crosses itself, the signature completely describes the image and direction of traversal of the path (that is, all the points that X visits and the order in which it visits them), modulo its starting point. This demonstrates the signature's ability to completely determine the geometric properties of a path that does not possess degeneracies consisting of movements going directly back onto itself. More precisely, it is shown in [7, 36] that, for two paths X, Y, one has $S(X) = S(Y)$ if and only if X and Y are *tree-like equivalent*. However, it remains a challenging problem to effectively recover properties of a path from its signature: see Lyons and Xu [55, 56] and Geng [31] for progress on this question.

2.5.3 Paths with Jumps

For many theoretical and practical applications, it is necessary to consider paths with jumps (i.e. discontinuous paths). This is important, for example, in finance [23], superdiffusive dynamical systems [18, 22, 62], and queuing theory [76]. It turns out that, under suitable regularity assumptions on the path (satisfied, for example, by almost every sample path of semi-martingales), there exists a meaningful way to build a signature.

The basic idea is to connect the jumps of a discontinuous path $X : [a, b] \to \mathbb{R}^d$ with straight lines by adding an 'infinitesimal' extra time δ to the interval. The resulting object is a continuous path $X^\delta : [a, b + \delta] \to \mathbb{R}^d$, for which the signature can be computed as normal and is independent of the extra time δ. For example, if X is piecewise constant with finitely many jumps, the signature of X^δ is the same as that of the piecewise linear path connecting the jumps of X in the respective order. This procedure to build a continuous path X^δ from X works under natural regularity assumptions, e.g. for X càdlàg (right-continuous with left-limits), and does not require that X has finitely many jumps. These ideas stem from the work of Marcus [59, 60] (see also [3, 11, 46] and the references therein) and later developed in the context of rough paths [14, 15, 27, 77] (see in particular [14, 15] where the requirement that the jumps of X are connected with straight lines is relaxed).

We mention also another way to interpret rough paths with jumps that relies on so-called forward (or Itô) integration [29] and leads to a notion of signature, but the resulting object is most naturally interpreted in the sense of branched rough paths [34] as it no longer satisfies the shuffle product identity.

2.5.4 Moment Problem and Random Paths

The moment problem in probability theory is the following: given two probability measures $\mathbb{P}$ and $\mathbb{Q}$ on $\mathbb{R}$ with equal moments, i.e. $\int_{\mathbb{R}} x^k \mathbb{P}(dx) = \int_{\mathbb{R}} x^k \mathbb{Q}(dx)$ for all $k \geq 0$, does it hold that $\mathbb{P} = \mathbb{Q}$? As alluded to Remark 2.32, the answer is, in general, negative, but there exist sufficient (and necessary) conditions on the growth of the moments $\int_{\mathbb{R}} x^k \mathbb{P}(dx)$ under which the answer is positive.

Recall also from Sect. 2.4, especially Corollary 2.36, that the shuffle product identity and the Stone–Weierstrass theorem together imply that a probability measure on a *compact* set of paths (on which the signature map is injective) is uniquely determined by its expected signature. This gives a positive answer to the moment problem in the simplest possible case (compact support of measures).

Given that, for probability measures on $\mathbb{R}$, compactness of support is a sufficient but not necessary condition to solve the moment problem, it is natural to ask whether one can relax the compactness assumption in the solution to the moment problem for signatures. The answer turns out to be 'yes' and the following sufficient condition was found in [16]: provided that the expectations $\mathbb{E}[S(X)^{i_1,\ldots,i_k}_{a,b}]$ decay with k faster than any geometric sequence, then these expectations uniquely determine the law of $S(X)_{a,b}$.[1]

[1] It is necessary to speak of the *law of the signature* $S(X)_{a,b}$ instead of the *law of the path* X because the signature does not, in general, uniquely determine the path, see Proposition 2.34 and the discussion beforehand.

The condition of [16] is satisfied, for example, by classes of Lévy processes including Brownian motion (suitably interpreted as geometric rough paths), for which the expected signature is explicitly known [27], and for classes of Gaussian and Markovian processes for which the expected signature is not explicitly known [10, 16]. In turn, it is known [8, 50] that certain processes (e.g. Brownian motion stopped upon exiting a domain) do *not* satisfy the necessary moment decay condition of [16], and it remains open whether the law of their signatures is determined by its expected value. A related open problem is to find sufficient and *necessary* conditions for a solution to the moment problem for signatures, akin to the sharp Carelman's condition (see Remark 2.32).

The method used in [16] to resolve the moment problem for signatures is to introduce a suitable noncommutative Fourier transform, which, as in the classical case, uniquely determines the law of any random variable without any assumptions on the moments. Again for Lévy processes, it is possible to derive a Lévy–Khintchine formula that determines the Fourier transform explicitly, see [14, Sec. 5.3.2].

2.5.5 Higher Dimensions

We conclude this section by mentioning several recent works that aim to extend the concept of signatures to functions defined over *multi-dimensional space*, i.e. functions $X : \Omega \to \mathbb{R}^d$ where $\Omega \subset \mathbb{R}^n$ and possibly $n \geq 1$, instead of just paths $X : [a, b] \to \mathbb{R}^d$. Due to the lack of a canonical ordering on points in $\mathbb{R}^n$ for $n \geq 2$, it is non-trivial to extend the notion of signature to such X in a way that preserves properties such as the shuffle product and Chen's identity from Sect. 2.3. Nonetheless, such an extension clearly has value since many types of data, e.g., images, are spatial (vs. temporal) and lack a time-ordered structure.

Zhang et al. [81], inspired by [75], recently defined 2D signatures and applied this concept to image and texture classification. Their approach was based on rectangular increments and was restricted to level-1 and level-2 (generalizations of) signatures. Giusti et al. [32] provided an (essentially topological) extension of signatures based on cubical mapping spaces that applies to arbitrary dimensions and levels. The authors therein show generalizations of the shuffle product and Chen's identity and are able to partially address the uniqueness question discussed in Sect. 2.5.2. Recently, Lee and Oberhauser [48] and then [20, 47] provided another generalization of the signature to surfaces which stems from higher gauge theory. See also [21] that provides a candidate for a multi-dimensional signature based on the notion of a *model* from the theory of regularity structures [35], which is used in the analysis of singular stochastic partial differential equations. This generalization, however, relies on several (a priori ad hoc) choices, including an integration operator that replaces the (somewhat canonical) Heaviside function in the one-dimensional case; it remains open if important properties of the signature extend to this construction. Finally, we mention [24] that introduces two-parameter sums signatures with applications to a multi-dimensional version of dynamic time warping.

3 Practical Applications

One of the practical applications of the signature transformation lies in the field of machine learning algorithms. As has already been discussed, the signature summarizes important information about a path. When the path is composed of a sequential data stream $\{X_i\}$, the terms of the signature $S(X)^{ijk\cdots}$ are good candidates for characteristic *features* of the data stream. The shuffle product property allows us to represent a non-linear function of the signature as a linear combination of iterated integrals. This is similar to basis function expansion and naturally applies to computations of regression models. In the next sections we will demonstrate numerical computations of signatures of paths from various data streams and applications of signatures to machine learning problems, namely classification.

3.1 *Elementary Ingredients and Their Properties*

In this section, we introduce the essential ingredients of the signature method and demonstrate how to perform basic computations.

3.1.1 Paths from Discrete Data

Our first task is to transform discrete sequential (i.e., time series) data into continuous paths, as signatures are typically defined for continuous paths rather than a collection of discrete data points. We start with basic computations of the signature applied to data streams. Let X be a data stream defined as a set of N points observed at times $t_0 < t_1 < t_2 < \ldots < t_N$ in d-dimensions ($X \in \mathbb{R}^d$):

$$X = \left\{ \left(X^1_{t_i}, X^2_{t_i}, X^3_{t_i}, \ldots, X^d_{t_i} \right) \right\}_{i=0}^{N}. \tag{28}$$

One can think of $\left(X^1_{t_i}, X^2_{t_i}, X^3_{t_i}, \ldots, X^d_{t_i} \right)$ as a coordinate of a data point recorded at time i in $\mathbb{R}^d$. For example, consider two one-dimensional sequences of length four:

$$\{X^1_{t_i}\}_{i=0}^{N=3} = \{1, 3, 8, 6\}, \tag{29}$$

$$\{X^2_{t_i}\}_{i=0}^{N=3} = \{1, 4, 2, 5\}. \tag{30}$$

We are interested in transforming this discrete series into a continuous path. Among various ways to find this transformation, we focus on two main approaches:

(a) **Piecewise linear interpolation.**
This interpolation method is given by a set of linear segments (local linear approximation) connecting a set of data points. The resulting continuous path

is a piecewise combination of linear functions. Formally, for each pair of points X_i and X_{i+1} on the interval $t \in [t_i, t_{i+1}]$, the path is defined by

$$\hat{X}_t = X_{t_i} + \frac{t - t_i}{t_{i+1} - t_i} \left(X_{t_{i+1}} - X_{t_i} \right). \tag{31}$$

By concatenating linearly interpolated paths, one arrives at the continuous path $\hat{X}_t$ defined for any $t \in [t_0, t_N]$.

(b) **Rectilinear interpolation** (i.e. *axis path*).
Another way to connect X_{t_i} and $X_{t_{i+1}}$ data points is by updating consecutively each coordinate of X_{t_i} in a piecewise linear fashion along the d axes until we reach $X_{t_{i+1}}$. In this way, on the interval $t \in [t_i, t_{i+1}]$ with $i = 0, 1, \ldots, N-1$, one augments a pair of two consecutive points by $d-1$ *auxiliary* data points of which we take the piecewise linear interpolation as in (31). For example, if $d = 3$, the rectilinear path over $[t_i, t_{i+1}]$ is the piecewise linear interpolation of the four data points

$$\{(X^1_{t_i}, X^2_{t_i}, X^3_{t_i}), (X^1_{t_{i+1}}, X^2_{t_i}, X^3_{t_i}), (X^1_{t_{i+1}}, X^2_{t_{i+1}}, X^3_{t_i}), (X^1_{t_{i+1}}, X^2_{t_{i+1}}, X^3_{t_{i+1}})\}.$$

Remark that, for the computation of the signature, it is not important how the rectilinear path is precisely parametrized over $[t_i, t_{i+1}]$ due to the parametrization invariance of the signature (Sect. 2.3.1).

The piecewise linear and rectilinear interpolation schemes using the data streams in (29) and (30) are presented in Fig. 10a and b respectively. Note that the auxiliary points denoted by empty red circles were added to construct the rectilinear paths.

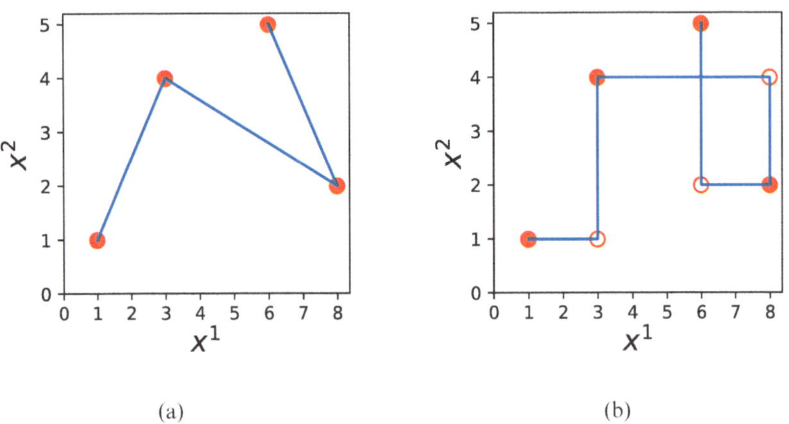

Fig. 10 Examples of a piecewise linear (**a**) and rectilinear (**b**) interpolation schemes applied to a two-dimensional stream $\{(X^1, X^2)\}$ from (29) and (30)

3.1.2 Auxiliary Path Transformations

Although data points can be transformed directly into paths as in Sect. 3.1.1, in certain cases, auxiliary data transformations may improve statistical learning with signatures for downstream tasks. Such transformations may increase or decrease the dimension d of the data stream and lead to more effective representations of sequential data that enhance the model's ability to learn, generalize, and make accurate predictions. In this section, we present several frequently used transformations.

3.1.3 Cumulative Sum

The cumulative sum, also known as the running total, is the sequence of partial sums of a given sequence of numbers. In other words, it is the sequence of sums of the elements of the original sequence, where each sum is calculated by adding the previous sum to the current element of the sequence. More precisely, the cumulative sum is defined by

$$\mathrm{CS}(\{X_{t_i}\}_{i=0}^{N}) = \{X_{t_0}, X_{t_0} + X_{t_1}, \dots, S_k, \dots, S_N\}\ ; \quad S_k = \sum_{i=0}^{k} X_{t_i}. \tag{32}$$

Using again the previous example (29) and (30), the new data streams become

$$\{\tilde{X^1}\} = \mathrm{CS}(X^1) = \{1, 4, 12, 18\},$$
$$\{\tilde{X^2}\} = \mathrm{CS}(X^2) = \{1, 5, 7, 12\}.$$

3.1.4 Base-Point Augmentation

As discussed in Sect. 2, the signature transformation is translation invariant, meaning that paths at different locations in $\mathbb{R}^d$ will produce the same signature. In some cases, it may be useful to remove this translation invariance. For example, to capture the absolute values of data streams rather than their increments. To do this, an anchor data point, typically the origin (0), can be added at the beginning of the data stream. This removes the translation invariance of the signature transformation. Using examples (29) and (30), the augmented data streams with $N + 1$ data points become

$$\begin{aligned} \mathrm{BP}(\{X^1_{t_i}\}_{i=0}^{N=3}) &= \{0, \{X^1_{t_i}\}_{i=0}^{N=3}\} = \{0, 1, 3, 8, 6\}, \\ \mathrm{BP}(\{X^2_{t_i}\}_{i=0}^{N=3}) &= \{0, \{X^2_{t_i}\}_{i=0}^{N=3}\} = \{0, 1, 4, 2, 5\}. \end{aligned} \tag{33}$$

3.1.5 Time Augmentation

This type of sequential data transformation is defined by adding a univariate component with monotonically increasing or decreasing values that can be interpreted as *auxiliary time*. For example, consider a new (unevenly sampled) stream $\{t\}$

$$\{t_i\}_{i=0}^{N=3} = \{0, 1, 3, 6\},$$

and augmenting $\{X^1\}_{i=0}^{N=3}$, $\{X^2\}_{i=0}^{N=3}$ and $\{(X^1, X^2)\}_{i=0}^{N=3}$ from (29) and (30) leads to

$$\mathrm{TA}(\{X^1_{t_i}\}_{i=0}^{N=3}) = \{(t_i, X^1_{t_i})\}_{i=0}^{N=3} = \{(0, 1), (1, 3), (3, 8), (6, 6)\},$$
$$\mathrm{TA}(\{X^2_{t_i}\}_{i=0}^{N=3}) = \{(t_i, X^2_{t_i})\}_{i=0}^{N=3} = \{(0, 1), (1, 4), (3, 2), (6, 5)\},$$
$$\mathrm{TA}(\{(X^1_{t_i}, X^2_{t_i}\}_{i=0}^{N=3}) = \{(t_i, X^1_{t_i}, X^2_{t_i})\}_{i=0}^{N=3} = \{(0, 1, 1), (1, 3, 4), (3, 8, 2), (6, 6, 5)\}.$$

The signature of a stream integrated against a monotonic auxiliary stream ('time') can capture stream-specific effects and can be related to common statistics (e.g., mean). Intuitively, the integral of a stream against the auxiliary time component will describe the area under the path, which in some applications may capture important properties of the data process via the path and serve as an informative feature for downstream analytical tasks.

In some cases, knowing the difference between the timestamps may be informative, particularly for unevenly distributed data. To illustrate this scenario, consider streams $\{X^1\}$ and $\{X^2\}$ and their respective timestamps $\{t_i\}$ as shown in Table 1a.

In addition to the original timescale stream $\{t_i\}$, we can introduce two auxiliary time-related streams:

$$\hat{t}_i = \{t_i - t_0\}_{i=0}^{N=4} = \{0, 1, 3, 6, 7\},$$
$$\Delta t_i = \{t_i - t_{i-1}\}_{i=1}^{N=4} = \{1, 2, 3, 1\}.$$

Putting all streams together and aligning them in time, we get a 5-dimensional stream as shown in Table 1b. Intuitively, $\{\hat{t}_i\}$ corresponds to time since the beginning and $\{\Delta t_i\}$ to time differences between two consecutive timestamps. Note that, since

Table 1 Example streams

t_i	X^2	X^2
01-January	1	1
02-January	3	4
04-January	8	2
07-January	6	5
08-January	9	3

(a)

t_i	$\hat{t}_i$	Δt_i	X^2	X^2
01-January	0	–	1	1
02-January	1	1	3	4
04-January	3	2	8	2
07-January	6	3	6	5
08-January	7	1	9	3

(b)

there is no time difference for the first data point, some heuristic can be used to fill in the missing value, such as arbitrarily setting it to zero.

3.1.6 Lead-Lag Transformation

A lead-lag transformation is a mathematical operation that shifts the phase of a time series by a specified amount. In a lead-lag transformation, a time series can be either "led" or "lagged" by a certain number of time steps. Leading a signal means that it is shifted forward in time, while lagging a signal means that it is shifted backward in time. Lead-lag transformations can be used to analyze the relationships between different signals. For example, the lead-lag transformation can map a one-dimensional path into a two-dimensional path. Certain statistics, such as the quadratic variation of a stochastic process, can be represented in terms of the signature of the lead-lag transformed process (see Sect. 3.1.14).

Let X a d-dimensional data stream with N data points as in (28). Following the definition [25], its lead and lag transformed streams $\{X_j^{\text{Lead}}\}_{j=0}^{2N}$ and $\{X_j^{\text{Lag}}\}_{j=0}^{2N}$ with $2N+1$ data points are defined as

$$X_{t_j}^{\text{Lead}} \mapsto \begin{cases} X_{t_i} & \text{if } j = 2i, \\ X_{t_i} & \text{if } j = 2i-1, \end{cases} \tag{34}$$

and

$$X_{t_j}^{\text{Lag}} \mapsto \begin{cases} X_{t_i} & \text{if } j = 2i, \\ X_{t_i} & \text{if } j = 2i+1. \end{cases} \tag{35}$$

It is clear from (34) and (35) that for each value of i corresponds to three values of j:

$$\begin{aligned} j &= 2i-1, \\ j &= 2i, \\ j &= 2i+1. \end{aligned}$$

For example, given a one-dimensional sequence with four values

$$\{X_{t_i}\}_{i=0}^{N=3} = \{X_{t_0}, X_{t_1}, X_{t_2}, X_{t_3}\},$$

by substituting $i \in [0, 1, 2, 3]$ in (34) the components of the lead stream of $\{X_{t_i}\}$ are

$$\begin{aligned} X_{t_0}^{\text{Lead}} = X_{t_0},\ X_{t_1}^{\text{Lead}} = X_{t_2}^{\text{Lead}} = X_{t_1}, \\ X_{t_3}^{\text{Lead}} = X_{t_4}^{\text{Lead}} = X_{t_2},\ X_{t_5}^{\text{Lead}} = X_{t_6}^{\text{Lead}} = X_{t_3}, \end{aligned}$$

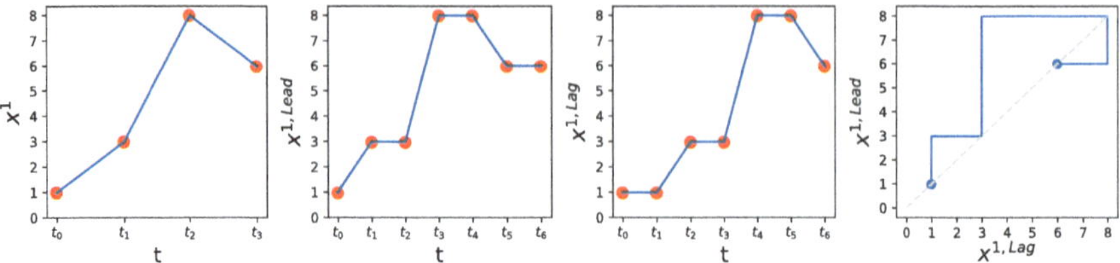

Fig. 11 Lead and lag paths of X^1

and correspondingly of the lag (35) stream of $\{X_{t_i}\}$ are

$$
\begin{aligned}
X_{t_0}^{\text{Lag}} &= X_{t_1}^{\text{Lag}} = X_{t_0}, \quad X_{t_2}^{\text{Lag}} = X_{t_3}^{\text{Lag}} = X_{t_1}, \\
X_{t_4}^{\text{Lag}} &= X_{t_5}^{\text{Lag}} = X_{t_2}, \quad X_{t_6}^{\text{Lag}} = X_{t_7}^{\text{Lag}} = X_{t_3},
\end{aligned}
$$

or more compactly

$$
\begin{aligned}
\text{Lead} : \{X_{t_i}\}_{i=0}^3 &\mapsto \{X_{t_j}^{\text{Lead}}\}_{j=0}^6 = \{X_{t_0}, X_{t_1}, X_{t_1}, X_{t_2}, X_{t_2}, X_{t_3}, X_{t_3}\}, \\
\text{Lag} : \{X_{t_i}\}_{i=0}^3 &\mapsto \{X_{t_j}^{\text{Lead}}\}_{j=0}^6 = \{X_{t_0}, X_{t_0}, X_{t_1}, X_{t_1}, X_{t_2}, X_{t_2}, X_{t_3}\}.
\end{aligned}
$$

One can immediately see that lead and lag transforms are derived from the original data stream by repeating and shifting its data points and deleting the first and last data points. They are repeated and time-shifted versions of each other.

Considering the sequence X^1 from (29), its lead-lag (LL) mapping is given by

$$
\text{LL} \; : \; X^1 = \{1, 3, 8, 6\} \mapsto \begin{cases} X^{1,\text{Lead}} & = \{1, 3, 3, 8, 8, 6, 6\}\,, \\ X^{1,\text{Lag}} & = \{1, 1, 3, 3, 8, 8, 6\}\,, \end{cases} \tag{36}
$$

and illustrated in Fig. 11. The corresponding lead-lag transformation of X^2 from (30) is given by

$$
\text{LL} \; : \; X^2 = \{1, 4, 2, 6\} \mapsto \begin{cases} X^{2,\text{Lead}} & = \{1, 4, 4, 2, 2, 6, 6\}\,, \\ X^{2,\text{Lag}} & = \{1, 1, 4, 4, 2, 2, 6\}\,, \end{cases}
$$

and illustrated in Fig. 12.

3.1.7 Multivariate Time Series as a Path in $\mathbb{R}^d$

The advantage of using the path representation of sequential data is that a d-dimensional process can be embedded as a path in $\mathbb{R}^d$. For example, the two univariate streams $\{X^1\}$ from (29) and $\{X^2\}$ from (30) can be embedded into a path in $\mathbb{R}^2$ as

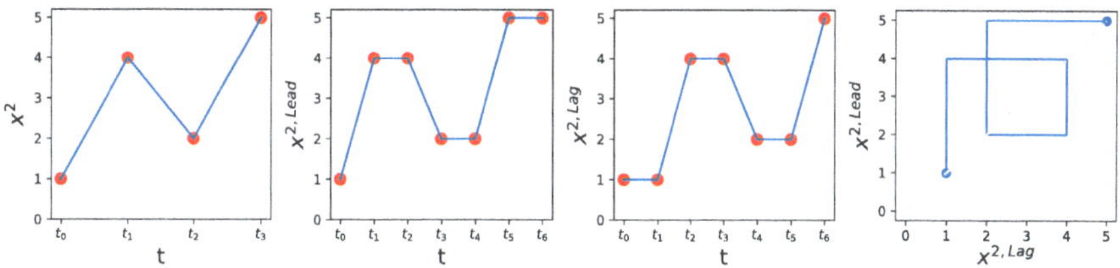

Fig. 12 Lead and lag paths of X^2

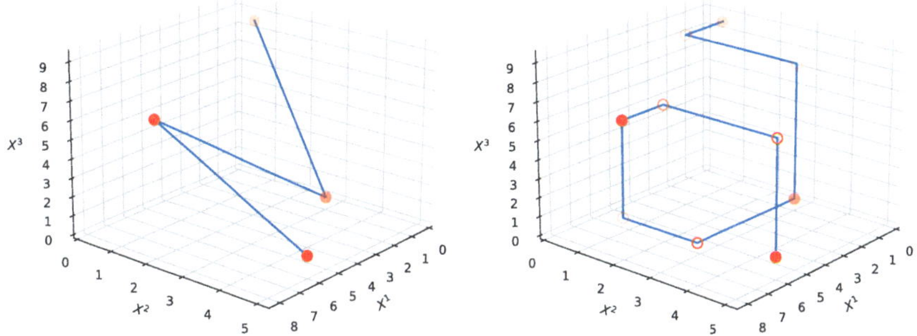

Fig. 13 Example of embedding three data streams in $\mathbb{R}$ into a single path in $\mathbb{R}^3$ using piecewise linear (left) and rectilinear (right) interpolations respectively

$$\begin{aligned}\{X_{t_i}\}_{i=0}^{N=3} &= \{(X_{t_i}^1, X_{t_i}^2)\}_{i=0}^{N=3}\\ &= \{(1,1), (3,4), (8,2), (6,5)\},\end{aligned}$$

the piecewise linear and rectilinear interpolations of which are displayed in Fig. 10. This can be easily extended to $\mathbb{R}^3$ if we add another univariate stream of the same length:

$$\{X_{t_i}^3\}_{i=0}^{N=3} = \{9, 2, 7, 1\},$$

resulting in

$$\begin{aligned}\{X_{t_i}\}_{i=0}^{N=3} &= \{(X_{t_i}^1, X_{t_i}^2, X_{t_i}^3)\}_{i=0}^{N=3}\\ &= \{(1,1,9), (3,4,2), (8,2,7), (6,5,1)\}.\end{aligned}$$

The resulting paths in 3-dimensions using piecewise linear and rectilinear interpolations are shown in Fig. 13.

Another example of a multidimensional embedding involves the lead-lag transform. Suppose we are interested in capturing the quadratic variation in one of the multivariate streams. As we will see in Sect. 3.1.14, we can compute the quadratic variation using the lead-lag transformation. Since the lead-lag transformation increases the number of points from N to $2N+1$ (cf. Sect. 3.1.6), all other stream lengths should be adjusted accordingly. It is worth noticing that, due to reparametrization invariance, the signatures of the lead and lag transformations of streams are equivalent:

$$S\Big(\{(X^1, X^2)\}\Big) \equiv S\Big(\{(X^{1,\mathrm{Lead}}, X^{2,\mathrm{Lead}})\}\Big) \equiv S\Big(\{(X^{1,\mathrm{Lag}}, X^{2,\mathrm{Lag}})\}\Big).$$

3.1.8 Further Path Transformations

In this subsection, we showed several examples of different path augmentations, but there exist many others that serve various purposes. For example, invisibility-reset augmentation that adds information on the initial position of the path, coordinate projections that allow to compute the signature of a subset of coordinates individually, data rescaling, and random projections that project a high-dimensional path onto a lower-dimensional subspace using a random matrix. For further details, see [64].

3.1.9 Signed Area

Before we continue, it is important to emphasize the significance of the direction in which we travel along a path and the sign of the area which it encloses. Intuitively, the signed area enclosed by a path is a measure of the 2-dimensional region enclosed by the path. It is calculated by taking the absolute value of the area and then assigning a positive or negative sign to it depending on the orientation of the path. A positively oriented path (or *a curve*) is a simple closed path in the plane such that when following the path, the interior of the path is always on the left side. For a negatively oriented path, the interior is on the right side instead of the left side. In other words, the orientation of a path is determined by the direction in which the path appears to turn as it is traced. If the path turns clockwise, it is negatively oriented. If the path turns counterclockwise, it is positively oriented.

For example, if a path is traced clockwise, the signed area of the enclosed region would be negative (Fig. 14). If the path is traced counterclockwise, the signed area would be positive (Fig. 14). A combination of positively and negatively oriented paths and their respective areas is shown in Fig. 14. The sign of an area is also related to the sign of the winding number of a path [6].

3.1.10 The Relative Movement as a Signed Area

Consider a path in $\mathbb{R}^2$ given by $X = \{X^1, X^2\}$. If an increase (resp. decrease) in the component X^1 is followed by an increase (resp. decrease) in the component X^2, then the area A given by (12) is positive (Fig. 15). If the relative moves of the

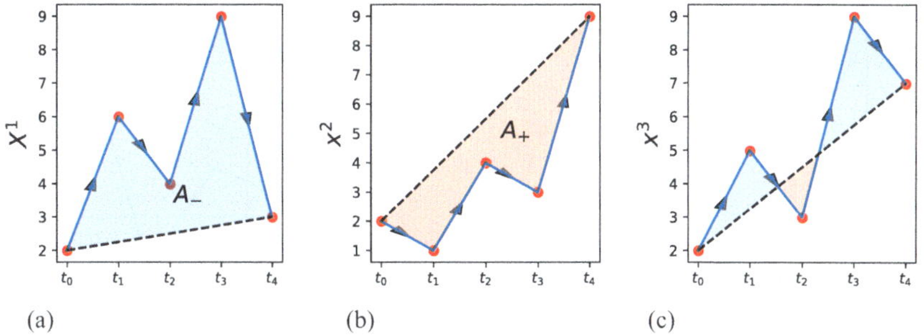

Fig. 14 Examples of signed areas defined by the orientation of a path enclosing them. (**a**) Negative area A_-. (**b**) Positive area A_+. (**c**) Positive and negative areas

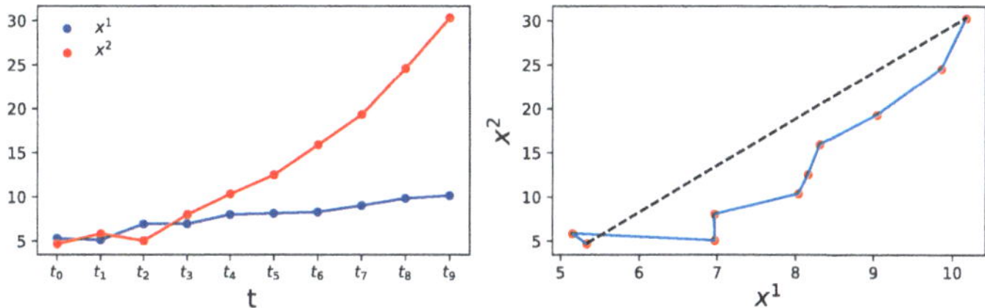

Fig. 15 Example of a positive Lévy area that corresponds to two paths, $\{X^1\}$ and $\{X^2\}$, with relative movement in the same directions

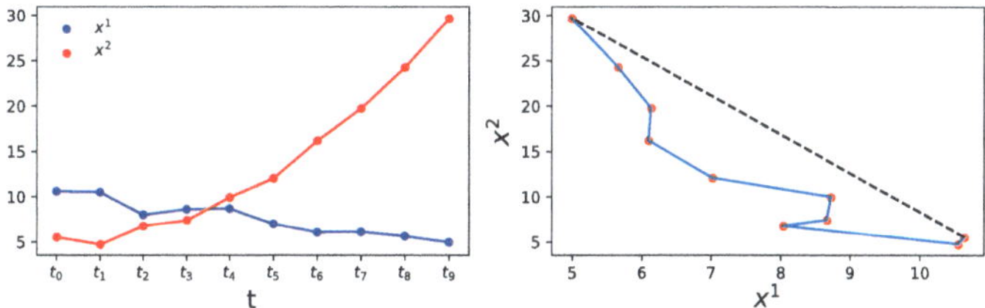

Fig. 16 Example of a negative Lévy area that corresponds to two paths, $\{X^1\}$ and $\{X^2\}$, with relative movement in opposite directions

components X^1 and X^2 are in the opposite direction, then the area A is negative (Fig. 16).

3.1.11 Computation and Geometry of the Signature

In Sect. 2, we outlined the mathematical foundations of the signature of a path and defined the signature terms as multiple iterated integrals. In this subsection, we provide another example of a computation and the geometric meaning of the signature. Consider the two univariate streams $\{(X^1, X^2)\}$

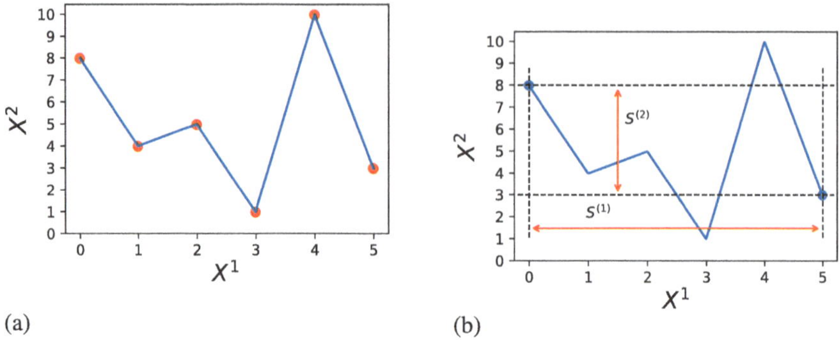

(a) (b)

Fig. 17 Examples of two univariate data streams (37) embedded in $\mathbb{R}^2$ and the geometric meaning of their signature terms S^1 and S^2. (**a**) A piecewise linear path $\{(X^1, X^2)\} \in \mathbb{R}^2$ constructed from $\{(X^1)\}$ and $\{(X^2)\}$ using data points as (37). (**b**) Geometric meaning of the first two terms S^1 and S^2 of the signature of the path $\{(X^1, X^2)\}$

$$\begin{aligned} \{(X^1)\} &= \{0, 1, 2, 3, 4, 5\}, \\ \{(X^2)\} &= \{8, 4, 5, 1, 10, 3\}, \end{aligned} \tag{37}$$

embedded into a piecewise linear path in $\mathbb{R}^2$ (Fig. 17a). Using the definition of the signature (2.4), we compute its first terms truncated at level $L = 2$ as

$$\begin{aligned} S(X) &= (1, S^1, S^2, S^{1,1}, S^{1,2}, S^{2,1}, S^{2,2}) \\ &= (1, 5, -5, 12.5, -10.5, -14.5, 12.5). \end{aligned}$$

One can easily confirm (cf. (1)) that the first order signature terms correspond to the total increments along each data stream:

$$\begin{aligned} S^1 &= \int dX_t^1 = X_{t_5}^1 - X_{t_0}^1 = 5 - 0 = 5, \\ S^2 &= \int dX_t^2 = X_{t_5}^2 - X_{t_0}^2 = 3 - 5 = -5, \end{aligned}$$

which are depicted in Fig. 17b. The second order signature terms are computed using double iterated integrals (cf. (2)):

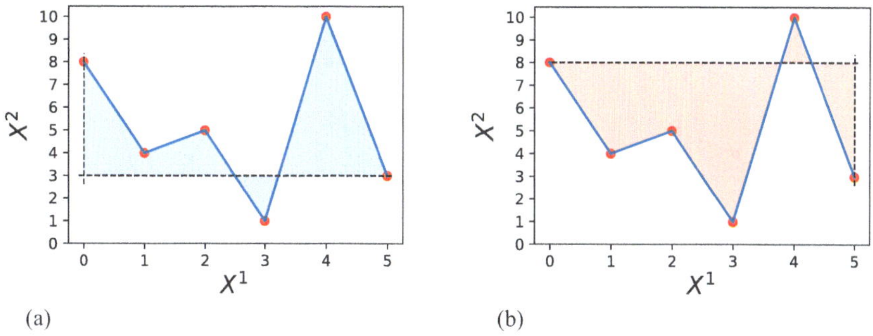

(a) (b)

Fig. 18 The geometric meaning of the terms $S^{1,2}$ and $S^{2,1}$ of $\{(X^1, X^2)\}$ defined in (37). The left panel (**a**) represents the area enclosed by the path and two perpendicular dashed lines passing through the endpoints of the path, while the right panel (**b**) shows another possibility for the area to be enclosed by the path and two perpendicular dashed lines passing through the endpoints. (**a**) Area represented by the term $S^{1,2}$. (**b**) Area represented by the term $S^{2,1}$

$$\begin{aligned}
S^{1,1} &= \int S^1\, dX_t^1 = \int\int dX_t^1\, dX_t^1 = \frac{1}{2!}\left(\int dX_t^1\right)^2 = \frac{1}{2!}(S^1)^2 = 12.5,\\
S^{1,2} &= \int S^1\, dX_t^2 = \int\int dX_t^1\, dX_t^2 = -10.5,\\
S^{2,1} &= \int S^2\, dX_t^1 = \int\int dX_t^2\, dX_t^1 = -14.5,\\
S^{2,2} &= \int S^2\, dX_t^2 = \int\int dX_t^2\, dX_t^2 = \frac{1}{2!}\left(\int dX_t^2\right)^2 = \frac{1}{2!}(S^2)^2 = 12.5.
\end{aligned} \tag{38}$$

The second order signature terms $S^{1,2}$ and $S^{2,1}$ and their geometric interpretation as areas are presented in Fig. 18.

Switching the order of integration over the path in the terms $S^{1,2}$ and $S^{2,1}$ gives rise to two areas which complete each other and add up to the total area of a rectangle with side lengths X^1 and X^2, which also corresponds to the shuffle product relation

$$\begin{aligned}
S^1 \cdot S^2 &= S^{1,2} + S^{2,1},\\
(-5)\cdot 5 &= -10.5 + (-14.5).
\end{aligned}$$

The geometric interpretation of the higher order terms is less intuitive and we omit this discussion. It is worth noticing, that the areas that correspond to the second order signature terms ($S^{1,2}$ and $S^{2,1}$) have positive and negative signs depending on their position relative to the horizontal straight line. If we split the areas in Fig. 18 into individual areas that lie above and underneath the horizontal dashed lines, we obtain partitions as shown in Fig. 19.

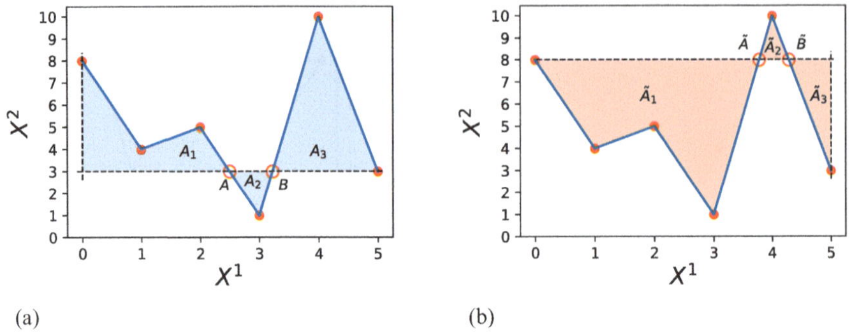

Fig. 19 A schematic partitioning of the areas $S^{1,2}$ and $S^{2,1}$ enclosed by the path and their corresponding dashed lines into a sum of individual contributions. Circles denote the point of intersection of the path and the corresponding dashed lines. (**a**) The total area $S^{1,2} = A_1+A_2+A_3$. (**b**) The total area $S^{2,1} = \tilde{A}_1 + \tilde{A}_2 + \tilde{A}_3$

The intersection points, denoted by red circles in Fig. 19 can be easily computed as

$$A = (5/2, 3), \quad B = (29/9, 3),$$
$$\tilde{A} = (34/9, 8), \quad \tilde{B} = (30/7, 8).$$

Subsequently, one can directly compute the individual areas

$$A_1 = -5, \quad A_2 = 13/18, \quad A_3 = -56/9,$$
$$\tilde{A}_1 = -119/9, \quad \tilde{A}_2 = 32/63, \quad \tilde{A}_3 = -25/14,$$

and demonstrate that the results

$$S^{1,2} = A_1 + A_2 + A_3 = -10.5,$$
$$S^{2,1} = \tilde{A}_1 + \tilde{A}_2 + \tilde{A}_3 = -14.5$$

correspond to the direct computation of the signature terms as shown in (38). Note that the sign of an area is determined by the direction of the path that encloses the area, as demonstrated in Fig. 14.

The Lévy area (12) of the path $\{(X^1, X^2)\}$ (37) is shown in Fig. 20 and is given by

$$A_{\text{Lévy}} = \frac{1}{2}\left(S^{1,2} - S^{2,1}\right) = 2. \tag{39}$$

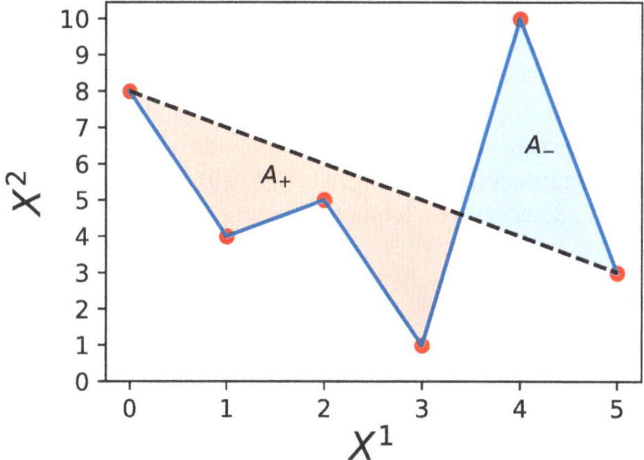

Fig. 20 Example of the signed Lévy area enclosed by a piecewise linear path and the chord that connects the two endpoints. The light blue area is negative and the pink area is positive according to the direction of travel along the path (from left to right)

Using the intersection point (3.4, 4.6) between the chord and the linear segment between t_3 and t_4, one can easily compute the areas $A_+ = 6.8$ and $A_- = -4.8$ and confirm that $A_{\text{Lévy}} = A_+ + A_-$ is equivalent to (39).

3.1.12 The Log-Signature of a Path

The shuffle product identity (Theorem 2.14) implies that any polynomial function on signatures can be written as a linear combination of signature elements. However, these relations imply that the signature contains 'repeated information', e.g. $S(X)^{i,i}_{a,b}$ can always be determined from the level 1 signature since $S(X)^{i,i}_{a,b} = \frac{1}{2}(S(X)^{i}_{a,b})^2$; for some tasks (e.g. linear regression) this repetition can be useful, as it increases the number of potentially useful features, but in some cases, one may wish to remove it. The log-signature (also known as a logarithmic signature), as defined in Sect. 2.3.5, is a condensed form of the signature that removes this repeated information: it contains the same information as the signature but with fewer terms.

For example, consider a two-dimensional data stream $\{X\}$ and the vector space of formal series in two non-commuting indeterminates $\{e_1, e_2\}$ (see Definition 2.19). Recalling the log-signature from Definition 2.27 and the result of Theorem 26, the log-signature truncated at level $L = 4$ can be written as

$$\begin{aligned}\log S(X) = \lambda_1 e_1 + \lambda_2 e_2 + \lambda_{12}[e_1, e_2] \\ + \lambda_{112}[e_1, [e_1, e_2]] + \lambda_{212}[e_2, [e_1, e_2]] \\ + \lambda_{1112}[e_1, [e_1, [e_1, e_2]]] + \lambda_{2112}[e_2, [e_1, [e_1, e_2]]] \\ + \lambda_{2212}[e_2, [e_2, [e_1, e_2]]],\end{aligned} \tag{40}$$

where $\lambda_{i_1,\dots i_k} \in \mathbb{R}$ are coefficients and $[x, y] = x \otimes y - y \otimes x$ is the Lie bracket as defined in (24). There are *a priori* no relations between the coefficients $\lambda_1, \dots, \lambda_{2212}$ (in fact, for any choice of $\lambda_1, \dots, \lambda_{2212}$, one can find a piecewise linear path X so that $\log S(X)$ is equal to the right-hand side of (40), see [28, Theorem 7.28]). This is in contrast to $S(X)$ because the shuffle identities imply non-trivial constraints on its coefficients (e.g. $S(X)^1 S(X)^2 = S(X)^{1,2} + S(X)^{2,1}$). Since $S(X)$ and $\log S(X)$ determine each other, this shows that $\log S(X)$ is a compressed version of $S(X)$.

The correspondence between the signature and log-signature terms can be computed from the definition (23). For example, building on (25), the correspondence between the coefficients of the signature and log-signature up to the level $L = 3$ is

$$\begin{aligned} \lambda_1 = S^1, \quad \lambda_2 = S^2, \quad \lambda_{12} &= \frac{1}{2!}\left(S^{1,2} - S^{2,1}\right), \\ \lambda_{112} &= \frac{1}{3!}\left(S^{1,1,2} + S^{2,1,1} - 2S^{1,2,1}\right), \\ \lambda_{212} &= -\frac{1}{3!}\left(S^{1,2,2} + S^{2,2,1} - 2S^{2,1,2}\right), \end{aligned} \tag{41}$$

and it is possible to extend this correspondence to any truncation level L.

Let us illustrate the geometric intuition for the low-order log-signature terms. Using the relationships in (41), the log-signature coefficients up to level $L = 3$ of a two-dimensional path $X = \{(X^1, X^2)\}$ from (37) are

$$\begin{aligned} \log S(X) &= (\lambda_1, \ \lambda_2, \ \lambda_{12}, \ \lambda_{112}, \quad \lambda_{212}) \\ &= \ (5, -5, \quad 2, -12, -35/6). \end{aligned}$$

One can immediately see that λ_1 and λ_2 are the total increments along each of the axes and λ_{12} corresponds to the Lévy area (12) as depicted in Fig. 20. The higher order log-signature terms $L \geq 3$ correspond to “areas of areas” [70] and can be represented as iterated Lie brackets of non-commuting indeterminates. The log-signature, containing the same information as the signature, can likewise be used as a feature set in statistical and machine learning applications, at times with more efficiency, see e.g. [51, 52, 71].

3.1.13 Numerical Computation of the Signature

To streamline the development and application of the signature method, several numerical packages with Python interfaces have been developed for the computation of the signature. The ‘ESig’ package is implemented in C++ with an intuitive Python interface [74]. This package computes the signature and log-signature coefficients, provides additional auxiliary tools, and can be run on CPU and, if available, on GPUs to accelerate computations. Another package with Python interface 'iisignature' was proposed and its implementation details are described in [69, 71]. In addition, the 'signatory' Python package was developed to enable the integration of the signature transformation layer in deep learning applications [38]. More information about the usage of these numerical packages can be found in their respective documentations.

3.1.14 Further Relationship Between the Signature Terms and Data Points

We now illustrate how the signature allows us to compute statistical moments between data streams. Consider the lead-lag path (Fig. 21) constructed from (29) and its lead-lag transformation (36). We can decompose this figure into three right-angled isosceles triangles Fig. 21. Note that the direction of movement along this path, starting from the point $X^1_{t_0}$ and moving towards the endpoint $X^1_{t_3}$ corresponds to a negative sign, thus the total area will be negative. All three triangles have the same direction of movement, resulting in the sum of their individual contributions.

The absolute value of the total area is given by the sum of the half areas of squares

$$\begin{aligned}|A| &= \frac{1}{2}\left[\left(X^1_{t_1} - X^1_{t_0}\right)\left(X^1_{t_1} - X^1_{t_0}\right) + \left(X^1_{t_2} - X^1_{t_1}\right)\left(X^1_{t_2} - X^1_{t_1}\right)\right. \\ &\quad \left. + \left(X^1_{t_3} - X^1_{t_2}\right)\left(X^1_{t_3} - X^1_{t_2}\right)\right] \\ &= \frac{1}{2}\left[(3-1)^2 + (8-3)^2 + (6-8)^2\right] = 16.5,\end{aligned}$$

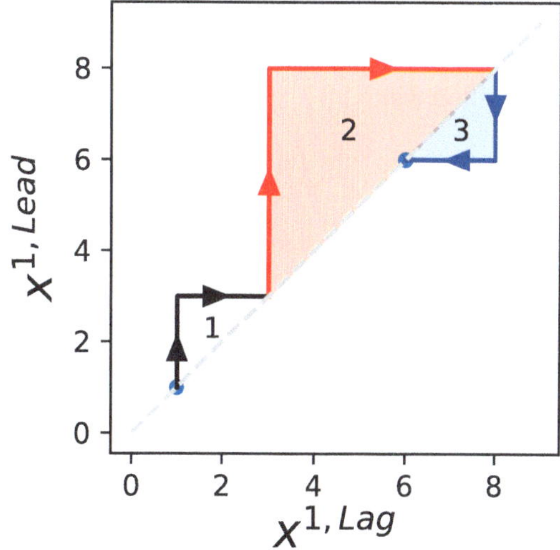

Fig. 21 Decomposition of the full lead-lag path into three individual parts. The direction of movement along the path is indicated by arrows

which exactly equals to the Lévy area enclosed by the lead-lag path $\{X^1\}$ and the straight line connecting the endpoints, and can be computed via the signature terms (12). Let us write

$$[X]_t = \sum_{i=1}^{N} \left(X_{t_i} - X_{t_{i-1}}\right)^2, \tag{42}$$

which has the simple meaning of the *quadratic variation* of the path constructed from $\{X_{t_i}\}_{i=0}^N$ and is related to the *variance* of the path. Thus one can generally write for any sequence $\{X_{t_i}\}_{i=0}^N$

$$A_{\text{Lévy}}^{\{(X^{\text{Lead}}, X^{\text{Lag}})\}} = \frac{1}{2}[X].$$

For example, for the lead-lag path $\{(X^{1,\text{Lead}}, X^{1,\text{Lag}})\}$ from (36), originated from a one-dimensional sequence (29), the first order log-signature terms correspond to the total increments along each dimension, and the second order term is the Lévy area which equals to the quadratic variation of the one-dimensional sequence:

$$\begin{aligned}\log S\left(\{(X^{1,\text{Lead}}, X^{1,\text{Lag}})\}\right) &= \left(\Delta X^{1,\text{Lead}}, \Delta X^{1,\text{Lag}}, \frac{1}{2}[X^1], \ldots\right) \\ &= (5, 5, 16.5, \ldots).\end{aligned}$$

Equivalently, one can write the above result in terms of the signature with all the terms included as

$$S\left(\{(X^{1,\text{Lead}}, X^{1,\text{Lag}})\}\right) = (1, 5, 5, 12.5, 29, -4, 12.5, \ldots)$$

and confirm that

$$[\{X^1\}] = S^{1,2}\left(\{(X^{1,\text{Lead}}, X^{1,\text{Lag}})\}\right) - S^{2,1}\left(\{(X^{1,\text{Lead}}, X^{1,\text{Lag}})\}\right).$$

Next we demonstrate certain properties of paths which originate from composite transformations. Consider the one-dimensional sequence $\{X^1\}$ as in (29) and consider applying to $\{X^1\}$ the series of transformations given by the cumulative sum CS(.) defined by (32), followed by base-point augmentation BP(.) defined by (33), followed by lead-lag transformation LL(.) defined by (34) and (35). The resulting sequence can be written as the composition of transformations

$$\{\tilde{X}^1\} = \text{LL} \circ \text{BP} \circ \text{CS}(\{X^1\}) = \text{LL}\big(\text{BP}\big(\text{CS}(\{X^1\})\big)\big). \tag{43}$$

It follows that, defining $\{Y^1\} = \text{BP}\big(\text{CS}(\{X^1\}\big)$,

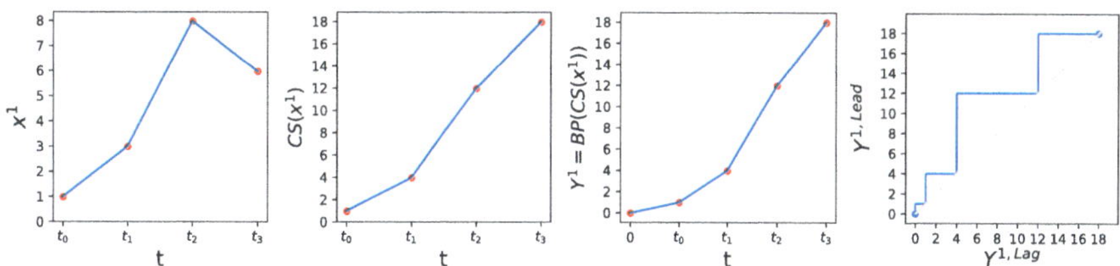

Fig. 22 The subsequent transformations of the original path $\{X^1\}$ into a resulting path $\{\tilde{X}^1\} = \text{LL}\big(\text{BP}\big(\text{CS}(\{X^1\})\big)\big)$ as shown in (44)

$$\begin{aligned}
\{X^1_{t_i}\}_{i=0}^{N=3} &= \{1, 3, 8, 6\}, \\
\text{CS}(\{X^1\}) &= \{1, 4, 12, 18\}, \\
\{Y^1\} &= \{0, 1, 4, 12, 18\}, \\
\{\tilde{X}^1\} &= \text{LL}\big(\text{BP}\big(\text{CS}(\{X^1\})\big)\big) \\
&= \{(Y^{1,\text{Lead}}, Y^{1,\text{Lag}})\} \\
&= \{(0,0), (1,0), (1,1), (4,1), (4,4), (12,4), (12,12), (18,12), (18,18)\}.
\end{aligned} \tag{44}$$

The resulting $\{\tilde{X}^1\}$ path is shown in Fig. 22.

Explicitly (44) can be written as

$$\begin{aligned}
Y^{1,\text{Lead}} &= \{0, S_0, S_0, S_1, S_1, S_2, S_2, S_3, S_3\}, \\
Y^{1,\text{Lag}} &= \{0, 0, S_0, S_0, S_1, S_1, S_2, S_2, S_3\},
\end{aligned}$$

where S_k is a partial sum defined in (32). We denote the signature of the composite transformation $\{\tilde{X}^1\}$ from (43) as

$$\tilde{S} = S(\{\tilde{X}^1\}) = (1, \tilde{S}^1, \tilde{S}^2, \tilde{S}^{1,1}, \tilde{S}^{1,2}, \tilde{S}^{2,1}, \tilde{S}^{2,2}, \ldots). \tag{45}$$

Numerical evaluation of the expression in (45) yields

$$\tilde{S} = (1, 18, 18, 162, 217, 107, 162, \ldots). \tag{46}$$

From the definition of the signature terms and the cumulative sum, it follows that

$$\begin{aligned}
\tilde{S}^1 &= \int dY^{1,\text{Lag}} = Y^{1,\text{Lag}}_{t_N} - Y^{1,\text{Lag}}_{t_0} = \Delta X^{1,\text{Lag}} \\
&= \text{BP}\big(\text{CS}(\{X^1\})\big)_{t_4} - \text{BP}\big(\text{CS}(\{X^1\})\big)_{t_0} \\
&= S_{k=3} - 0 \\
&= X^1_{t_0} + X^1_{t_1} + X^1_{t_2} + X^1_{t_3} - 0 \\
&= 1 + 3 + 8 + 6 = 18.
\end{aligned}$$

Similarly, one can show that $\tilde{S}^2 = \tilde{S}^1$. Therefore, the first order terms of the signature of the transformed sequence $\{\tilde{X}^1\}$ equal to the sum of the original sequence $\{X^1\}$:

$$\tilde{S}^1 = \tilde{S}^2 = \Delta Y^{1,\text{Lead}} = \Delta Y^{1,\text{Lag}}.$$

This result can be easily confirmed numerically by comparing the corresponding terms in (45) and (46). In a similar fashion, one can demonstrate that quadratic variation (42) of the transformed sequence $[\{\tilde{X}^1\}]$ is proportional to *variance* of the original sequence $\{X^1\}$ from (29).

3.2 Machine Learning with Signature Features

The signature of a path is a collection of iterated integrals that succinctly capture the path's geometric properties and encode its essential information. This makes it particularly well-suited for machine learning applications involving sequential data and time series analysis. The workflow is simple and summarized in the following algorithm:

$$\begin{aligned} &data \to path \to signature\ of\ path \to features\ of\ data \\ &\quad \to learning\ algorithms \end{aligned}$$

This algorithm is general and works for any type of sequential data which can be embedded into a continuous path. The extracted features can be used for various types of machine learning applications, including both supervised and unsupervised learning. For example, one can classify time series or distinguish clusters of data. One of the advantages of feature extraction with the signature method is that the signature is sensitive to the geometric shape of a path. Sensitivity to the geometric shape of the input data has led to a successful application of the signature method to Chinese character recognition problem [33]. One of the most well-known and natural applications of the signature method is in quantitative finance, namely analysis of time series data [25, 49]. Time series represent ordered sequential data, which is a natural candidate for creating a path, followed by computing the signature and applying machine learning algorithms for further analysis. Moreover, if the input data come from several parallel sources, this will result in a multi-dimensional path. An example of such data is *panel* data (in Econometrics) or *longitudinal* data (in Medicine, Psychology, Biostatistics etc.) which involve repeated observations of the same quantity over periods of time.

Concluding this brief introduction to the applications of the signature method, we emphasize that this method introduces a new concept of dealing with data: thinking of data as geometric paths and using the signature method to analyze this data. In the rest of this subsection, we demonstrate how to apply the signature transformation

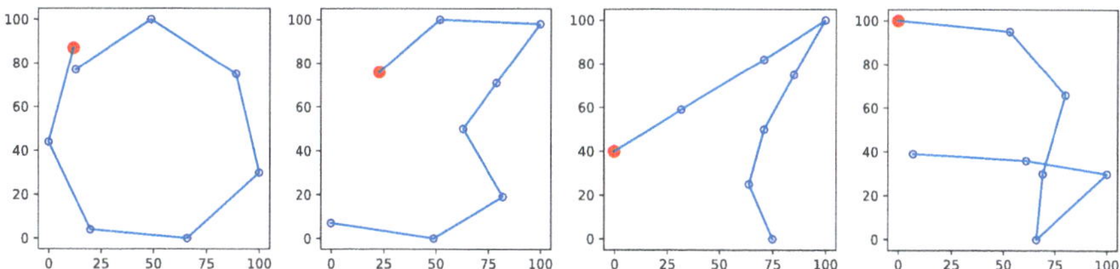

Fig. 23 Pictorial representation of the first four rows from the Table 2. From left to right: '0', '3', '1' and '7'. Solid red dots represent the starting point (x_1, y_1)

to problems in statistical machine learning with streamed data, focusing on the common analytical task of classification. However, other machine learning and inferential statistical tasks with streamed data can also be performed using the signature formalism, including regression, clustering, and kernel learning, some of which we review in Sect. 3.3.

3.2.1 Classification of Handwritten Digits: An Example

In this section, we demonstrate the application of the signature method to the machine learning task of pen-based classification of handwritten digits. The dataset is available from the University of California, Irvine, Machine Learning Repository [2]. The data set comprises 250 handwritten integer numbers in the range of 0–9, generated by 44 writers using a tablet with a stylus. The data represent a collection of consecutive data points represented by (x, y) coordinates. The raw data that were captured from the tablet consisted of integer values and were normalized to lie in the range between 0 and 100. The consecutive data points were resampled and represented as a sequence of coordinate pairs $(x_i, y_i), i = 1, \ldots, 8$, regularly spaced on the trajectory. An example of ten random digits in the tabular format is shown in Table 2 and the first four rows are displayed in Fig. 23.

Each data sample is represented by a two-dimensional coordinate sequence $(x_i, y_i) \in \mathcal{X}$, that can be seen as a path in $\mathbb{R}^2$. For simplicity, we apply the signature transformation without composite augmentations. Following the steps outlined in Sect. 3.1.11, the signature is the feature map

$$S : \mathcal{X} \to \Sigma,$$

where $S^I \in \Sigma$ are the elements of the infinite signature series (2.4) and I is a multi-index defined in (4). For example, applying the level-2 truncated signature transformation on the first four rows

$$X^{\text{zero}} = \{(x_i, y_i)\}_{i=1}^{N=8} = \{(12, 87), (0, 44), (20, 4), (66, 0), \\ (100, 30), (89, 75), (49, 100), (13, 77)\},$$

Table 2 An example of ten random samples of handwritten digits in a tabular format from the Pen-based Recognition of Handwritten Digits data set [2]. Columns denoted by x_i, and y_i $(i = 1, \ldots, 8)$ correspond to coordinates (features) and the last column designates the class label of digits

x_1	y_1	x_2	y_2	x_3	y_3	x_4	y_4	x_5	y_5	x_6	y_6	x_7	y_7	x_8	y_8	Label
12	87	0	44	20	4	66	0	100	30	89	75	49	100	13	77	0
23	76	52	100	100	98	79	71	63	50	82	19	49	0	0	7	3
0	40	32	59	71	82	100	100	85	75	71	50	64	25	75	0	1
0	100	53	95	80	66	69	30	66	0	100	30	61	36	7	39	7
0	67	43	74	78	91	100	100	95	74	87	49	84	23	96	0	1
10	97	2	47	24	1	74	0	100	43	82	90	33	100	0	65	0
0	48	30	67	65	85	100	100	84	75	65	51	47	25	34	0	1
55	70	73	100	62	66	49	26	30	0	0	2	50	3	100	6	1
100	100	66	88	32	66	7	39	0	10	44	0	62	18	14	22	6
85	100	43	76	11	48	0	17	44	0	100	10	59	20	1	10	6

Table 3 The signature feature matrix from the original data presented in Table 2 and the corresponding labels

1	S^1	S^2	$S^{1,1}$	$S^{1,2}$	$S^{2,1}$	$S^{2,2}$	Label
1	1	− 10	0.5	7044.5	− 7054.5	50	0
1	− 23	− 69	264.5	− 4893	6480	2380.5	3
1	75	− 40	2812.5	− 4660	1660	800	1
1	7	− 61	24.5	− 3693	3266	1860.5	7
1	96	− 67	4608	− 7123	691	2244.5	1
1	− 10	− 32	50	7388.5	− 7068.5	512	0
1	34	− 48	578	− 4179	2547	1152	1
1	45	− 64	1012.5	178	− 3058	2048	1
1	− 86	− 78	3698	5984	724	3042	6
1	− 84	− 90	3528	6028.5	1531.5	4050	6

$$
\begin{aligned}
X^{\text{three}} &= \{(23,76),(52,100),(100,98),(79,71),\\
&\quad (63,50),(82,19),(49,0),(0,7)\},\\
X^{\text{one}} &= \{(0,40),(32,59),(71,82),(100,100),\\
&\quad (85,75),(71,50),(64,25),(75,0)\},\\
X^{\text{seven}} &= \{(0,100),(53,95),(80,66),(69,30),\\
&\quad (66,0),(100,30),(61,36),(7,39)\}
\end{aligned}
$$

yields

$$
\begin{aligned}
S(X^{\text{zero}}) &= (1, 1, -10, 0.5, 7044.5, -7054.5, 50),\\
S(X^{\text{three}}) &= (1, -23, -69, 264.5, -4893, 6480, 2380.5),\\
S(X^{\text{one}}) &= (1, 75, -40, 2812.5, -4660, 1660, 800),\\
S(X^{\text{seven}}) &= (1, 7, -61, 24.5, -3693, 3266, 1860.5).
\end{aligned}
$$

Applying the signature transformation on each row of the data matrix in Table 2 will generate the *signature feature matrix* given in Table 3.

Now, one can use the signature feature matrix as input to statistical hypothesis tests for downstream analytical tasks or machine learning algorithms to predict the class labels.

Another interesting observation is that the higher order signature terms can capture more nuanced patterns of paths. Consider only samples of handwritten 'zero' and 'eight' digits, such as those shown in Fig. 24. Both 'zero' and 'eight' shapes share a similar geometric pattern: the starting and end points usually coincide. However, the *areas* enclosed by both shapes will be different due to their orientations as discussed in Sect. 3.1.9. The clear differences between the

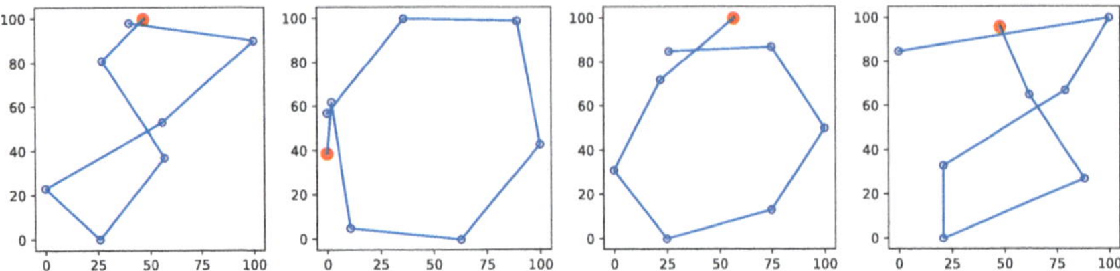

Fig. 24 Examples of handwritten zeros and eights. From left to right: '8', '0', '0', '8'. Solid red dot represents the starting point (x_1, y_1)

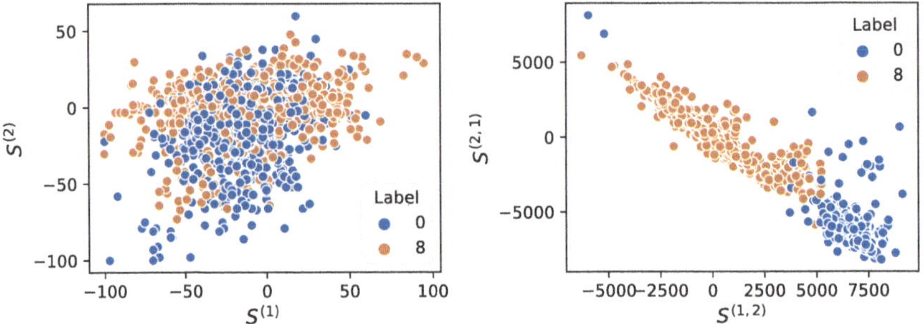

Fig. 25 The distribution of the first (left panel) and second (right panel) signature terms describing 'zero' and 'eight' digits. The second order signature terms, that are proportional to a signed enclosed area, demonstrate good separability between the two classes

distributions of the first S^1, S^2 and second order $S^{1,2}$, $S^{2,1}$ signature terms are shown in Fig. 25 and can improve the performance of machine learning or statistical approaches.

Furthermore, the signature feature matrix can also be used for unsupervised learning tasks, including dimensionality reduction for clustering approaches. For example, applying the Uniform Manifold Approximation and Projection (UMAP) for Dimension Reduction method [61] on the signature feature matrix, one can find clusters of points belonging to the same class label, see Fig. 26.

3.3 Overview of Signature-Based Machine Learning Applications

Over the past two decades, the signature method has been the subject of extensive research and application across various scientific domains. The method has found success in diverse areas such as finance, healthcare, natural language processing, wearable devices, and computer vision, demonstrating its versatility and the powerful expressiveness inherent in its approach. This adaptability has established the signature method as a valuable asset in the ongoing research on modern machine learning techniques for streamed data.

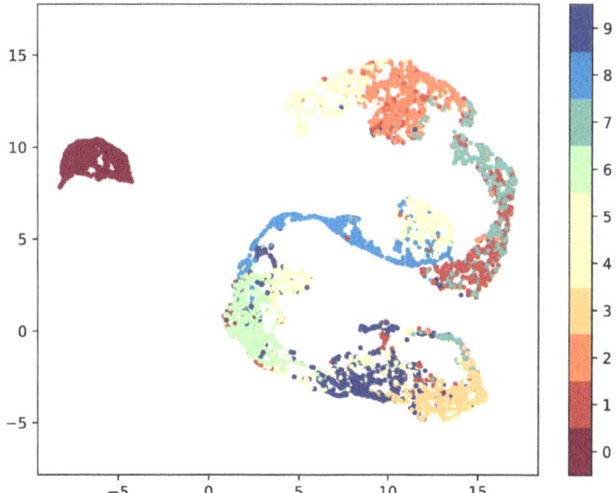

Fig. 26 Projection of the signature feature space from $\mathbb{R}^7$ onto $\mathbb{R}^2$ using UMAP. The right column designates class labels $0, \ldots, 9$

The signature method's application to sequential data representation has evolved substantially, extending far beyond its original scope and undergoing considerable refinement. To conclude, we aim to provide a concise overview of some notable examples where the signature method has been employed to tackle diverse challenges in machine learning and data analysis. Through these examples, we hope to showcase the wide-ranging implications and potential of this innovative approach in further advancing the field of machine learning.

Already in 2005, Lyons and Sidorova worked on the application of rough path theory to the problem of sound compression [54]. They presented a new approach based on the signature that provides an alternative to traditional Fourier and wavelet transforms. The main difference between these methods is in the linearity of the Fourier approach, while the signature method accounts for non-linear dependencies.

The signature transformation has been extensively used in financial applications. Field et al. [25] have applied the signature transformation to multidimensional financial data streams to identify atypical market behavior and recognize patterns of trading algorithms. In [49], Levin et al. proposed a new non-parametric regression method based on the expected signature of paths. Truncating the signature results in an efficient local description and offers a provable efficiency gain over linear path segment descriptions with the same number of features. This approach can potentially achieve significant dimension reduction in highly oscillatory data streams. In numerical examples, their method achieves similar accuracy to that of the Gaussian process approach but with lower computational cost, especially for large sample sizes. The truncated signature-based regression and classification models have further led to numerous applications in finance, such as the obtaining approximate solutions to the problem of optimal execution [37], non-parametric pricing and hedging of exotic derivatives [58] and many others [4].

The ability of the signature transformation to handle complex data streams, reduce dimensionality of the feature space, and enhance prediction accuracy makes it a valuable asset in healthcare and medicine. As wearable technology becomes more common, the amount of data generated by these devices is enormous. By analyzing data from connected devices, e.g., heart rate monitors, fitness trackers, sleep monitors, gait and other modalities, researchers can gain insights into patients' health conditions, such as identifying early patients with neurological disorders that have a higher risk of falls [68]. Electronic health records (EHRs) contain vast amounts of data, including patient demographics, medical history, medications, and lab results. Applying the signature method to EHRs can help identify patterns and trends, enable better disease management, and support early detection of potential health issues, such as an early onset of sepsis [65] and diagnosis of Alzheimer's disease using brain image-derived biomarkers (e.g., the hippocampus, ventricles and whole brain volumes) [63]. The signature method was applied to analyze data from self-reporting tools and mobile applications, to detect longitudinal patterns [44] in clinical trials [30], predict mood changes and episodes [43], derive clinically meaningful information from daily self-reported mood ratings to classify patients' diagnosis, predict mood scores [66], and efficiently capture the chronological information in medical texts [45].

The signature approach has been applied to computer vision tasks such as human-pose estimation [80] and recognition of handwriting in various languages [78, 79]. The signature approach has been applied in deep learning [39], neural controlled differential equations [40], and topological data analysis [19]. Signatures were furthermore combined with kernel methods applicable to problems in classification and hypothesis testing [17, 41].

Further examples of applications of rough paths, signatures, and neural controlled differential equations can be found at https://www.datasig.ac.uk/papers.

Acknowledgments We are thankful to Horatio Boedihardjo, Simone Dari, Terry Lyons, Hao Ni, Harald Oberhauser, and Danyu Yang for fruitful discussions and improvements of the manuscript. IC acknowledges support from the EPSRC via the New Investigator Award EP/X015688/1. AK is supported in part by the National Institute for Health and Care Research (NIHR) AI Award (AI_AWARD02183) and by a research grant from GlaxoSmithKline. AK wishes to thank the Oxford-Man Institute of Quantitative Finance for hospitality during the course of this work and gratefully acknowledges the support of the Wellcome Trust grant No: 102616/Z/13/Z, "CONBRIO".

References

1. A.A. Agrachev, Y.L. Sachkov, *Control Theory from the Geometric Viewpoint*, volume 87 of Encyclopaedia of Mathematical Sciences (Springer, Berlin, 2004), Control Theory and Optimization, II
2. E. Alpaydin, F. Alimoglu, Pen-based recognition of handwritten digits. UCI Machine Learning Repository (1998). https://doi.org/10.24432/C5MG6K
3. D. Applebaum, *Lévy Processes and Stochastic Calculus*, volume 116 of Cambridge Studies in Advanced Mathematics, 2nd edn. (Cambridge University Press, Cambridge, 2009)
4. I.P. Arribas, Signatures in machine learning and finance. PhD thesis, University of Oxford, 2020

5. P. Billingsley, *Probability and Measure*. Wiley Series in Probability and Statistics (Wiley, Hoboken, 2012), Anniversary edition [of MR1324786], With a foreword by Steve Lalley and a brief biography of Billingsley by Steve Koppes
6. H. Boedihardjo, H. Ni, Z. Qian, Uniqueness of signature for simple curves. J. Funct. Anal. **267**(6), 1778–1806 (2014)
7. H. Boedihardjo, X. Geng, T. Lyons, D. Yang, The signature of a rough path: uniqueness. Adv. Math. **293**, 720–737 (2016)
8. H. Boedihardjo, J. Diehl, M. Mezzarobba, H. Ni, The expected signature of Brownian motion stopped on the boundary of a circle has finite radius of convergence. Bull. Lond. Math. Soc. **53**(1), 285–299 (2021)
9. R.W. Brockett, *Finite Dimensional Linear Systems*, volume 74 of Classics in Applied Mathematics (Society for Industrial and Applied Mathematics (SIAM), Philadelphia, 2015). Reprint of the 1970 original
10. T. Cass, M. Ogrodnik, Tail estimates for Markovian rough paths. Ann. Probab. **45**(4), 2477–2504 (2017)
11. A. Chechkin, I. Pavlyukevich, Marcus versus Stratonovich for systems with jump noise. J. Phys. A **47**(34), 342001, 15 (2014)
12. K.-T. Chen, Integration of paths, geometric invariants and a generalized Baker-Hausdorff formula. Ann. Math. (2) **65**, 163–178 (1957)
13. K.-T. Chen, Integration of paths—a faithful representation of paths by non-commutative formal power series. Trans. Amer. Math. Soc. **89**, 395–407 (1958)
14. I. Chevyrev, Random walks and Lévy processes as rough paths. Probab. Theory Relat. Fields **170**(3–4), 891–932 (2018)
15. I. Chevyrev, P.K. Friz, Canonical RDEs and general semimartingales as rough paths. Ann. Probab. **47**(1), 420–463 (2019)
16. I. Chevyrev, T. Lyons, Characteristic functions of measures on geometric rough paths. Ann. Probab. **44**(6), 4049–4082 (2016)
17. I. Chevyrev, H. Oberhauser, Signature moments to characterize laws of stochastic processes. J. Mach. Learn. Res. **23**(176), 1–42 (2022)
18. I. Chevyrev, P.K. Friz, A. Korepanov, I. Melbourne, Superdiffusive limits for deterministic fast-slow dynamical systems. Probab. Theory Relat. Fields **178**(3–4), 735–770 (2020)
19. I. Chevyrev, V. Nanda, H. Oberhauser, Persistence paths and signature features in topological data analysis. IEEE Trans. Pattern Anal. Mach. Intell. **42**(1), 192–202 (2020)
20. I. Chevyrev, J. Diehl, K. Ebrahimi-Fard, N. Tapia, A multiplicative surface signature through its Magnus expansion. arXiv e-prints, June (2024)
21. I. Chevyrev, A. Gerasimovičs, H. Weber, Feature engineering with regularity structures. J. Sci. Comput. **98**(1), Paper No. 13, 28 (2024)
22. I. Chevyrev, A. Korepanov, I. Melbourne, Superdiffusive limits beyond the Marcus regime for deterministic fast-slow systems. Commun. Am. Math. Soc. **4**, 746–786 (2024)
23. R. Cont, P. Tankov, *Financial Modelling with Jump Processes*. Chapman & Hall/CRC Financial Mathematics Series (Chapman & Hall/CRC, Boca Raton, 2004)
24. J. Diehl, L. Schmitz, Two-parameter sums signatures and corresponding quasisymmetric functions. Prepint. arXiv:2210.14247 (2022)
25. J. Field, L.G. Gyurkó, M. Kontkowski, T. Lyons, Extracting information from the signature of a financial data stream. Preprint. arXiv:1307.7244 (2014)
26. P.K. Friz, M. Hairer, *A Course on Rough Paths*. Universitext (Springer, Cham, 2014). With an introduction to regularity structures
27. P.K. Friz, A. Shekhar, General rough integration, Lévy rough paths and a Lévy-Kintchine-type formula. Ann. Probab. **45**(4), 2707–2765 (2017)
28. P.K. Friz, N.B. Victoir, *Multidimensional Stochastic Processes as Rough Paths*, volume 120 of Cambridge Studies in Advanced Mathematics (Cambridge University Press, Cambridge, 2010)
29. P.K. Friz, H. Zhang, Differential equations driven by rough paths with jumps. J. Differential Equations **264**(10), 6226–6301 (2018)

30. J.R. Geddes, A. Gardiner, J. Rendell, M. Voysey, E. Tunbridge, C. Hinds, L.-M. Yu, J. Hainsworth, M.-J. Attenburrow, J. Simon et al., Comparative evaluation of quetiapine plus lamotrigine combination versus quetiapine monotherapy (and folic acid versus placebo) in bipolar depression (CEQUEL): a 2 × 2 factorial randomised trial. Lancet Psychiatry **3**, 31 (2015)
31. X. Geng, Reconstruction for the signature of a rough path. Proc. Lond. Math. Soc. (3) **114**(3), 495–526 (2017)
32. C. Giusti, D. Lee, V. Nanda, H. Oberhauser, A topological approach to mapping space signatures. Adv. Appl. Math. **163**(part A), Paper No. 102787, 68 (2025)
33. B. Graham, Sparse arrays of signatures for online character recognition. Preprint. arXiv:1308.0371 (2013)
34. M. Gubinelli, Ramification of rough paths. J. Differential Equations **248**(4), 693–721 (2010)
35. M. Hairer, A theory of regularity structures. Invent. Math. **198**(2), 269–504 (2014)
36. B. Hambly, T. Lyons, Uniqueness for the signature of a path of bounded variation and the reduced path group. Ann. Math. (2) **171**(1), 109–167 (2010)
37. J. Kalsi, T. Lyons, I.P. Arribas, Optimal execution with rough path signatures. SIAM J. Financ. Math. **11**(2), 470–493 (2020)
38. P. Kidger, T. Lyons, Signatory: differentiable computations of the signature and logsignature transforms, on both CPU and GPU, in *International Conference on Learning Representations* (2021). https://github.com/patrick-kidger/signatory
39. P. Kidger, P. Bonnier, I. Perez Arribas, C. Salvi, T. Lyons, Deep signature transforms, in *Advances in Neural Information Processing Systems*, vol. 32 (2019)
40. P. Kidger, J. Morrill, J. Foster, T. Lyons, Neural controlled differential equations for irregular time series, in *Advances in Neural Information Processing Systems*, vol. 33 (2020), pp. 6696–6707
41. F.J. Király, H. Oberhauser, Kernels for sequentially ordered data. J. Mach. Learn. Res. **20**, Paper No. 31, 45 (2019)
42. B. Korenblum, A. Mascuilli, J. Panariello, A generalization of Carleman's uniqueness theorem and a discrete Phragmén-Lindelöf theorem. Proc. Amer. Math. Soc. **126**(7), 2025–2032 (1998)
43. A. Kormilitzin, K. Saunders, P. Harrison, J. Geddes, T. Lyons, Application of the signature method to pattern recognition in the CEQUEL clinical trial. Preprint. arXiv:1606.02074 (2016)
44. A. Kormilitzin, K.E. Saunders, P.J. Harrison, J.R. Geddes, T. Lyons, Detecting early signs of depressive and manic episodes in patients with bipolar disorder using the signature-based model. Preprint. arXiv:1708.01206 (2017)
45. A. Kormilitzin, N. Vaci, Q. Liu, H. Ni, G. Nenadic, A. Nevado-Holgado, An efficient representation of chronological events in medical texts, in *Proceedings of the 11th International Workshop on Health Text Mining and Information Analysis* (2020), pp. 97–103
46. T.G. Kurtz, E. Pardoux, P. Protter, Stratonovich stochastic differential equations driven by general semimartingales. Ann. Inst. H. Poincaré Probab. Stat. **31**(2), 351–377 (1995)
47. D. Lee, The surface signature and rough surfaces. arXiv e-prints, June (2024)
48. D. Lee, H. Oberhauser, Random surfaces and higher algebra. Preprint. arXiv:2311.08366 (2023)
49. D. Levin, T. Lyons, H. Ni, Learning from the past, predicting the statistics for the future, learning an evolving system. Preprint. arXiv:1309.0260 (2015)
50. S. Li, H. Ni, Expected signature of stopped Brownian motion on d-dimensional $C^{2,\alpha}$-domains has finite radius of convergence everywhere: $2 \leq d \leq 8$. J. Funct. Anal. **282**(12), Paper No. 109447, 45 (2022)
51. S. Liao, Log signatures in machine learning. PhD thesis, University College London, 2022
52. S. Liao, T. Lyons, W. Yang, H. Ni, Learning stochastic differential equations using RNN with log signature features. Preprint. arXiv:1908.08286 (2019)
53. T.J. Lyons, Differential equations driven by rough signals. Rev. Mat. Iberoam. **14**(2), 215–310 (1998)
54. T.J. Lyons, N. Sidorova, Sound compression: a rough path approach, in *Proceedings of the 4th International Symposium on Information and Communication Technologies* (Trinity College, Dublin, 2005), pp. 223–228

55. T.J. Lyons, W. Xu, Hyperbolic development and inversion of signature. J. Funct. Anal. **272**(7), 2933–2955 (2017)
56. T.J. Lyons, W. Xu, Inverting the signature of a path. J. Eur. Math. Soc. (JEMS) **20**(7), 1655–1687 (2018)
57. T.J. Lyons, M. Caruana, T. Lévy, *Differential Equations Driven by Rough Paths* (Springer, Berlin, 2007)
58. T. Lyons, S. Nejad, I. Perez Arribas, Non-parametric pricing and hedging of exotic derivatives. Appl. Math. Finance **27**(6), 457–494 (2020)
59. S.I. Marcus, Modeling and analysis of stochastic differential equations driven by point processes. IEEE Trans. Inform. Theory **24**(2), 164–172 (1978)
60. S.I. Marcus, Modeling and approximation of stochastic differential equations driven by semimartingales. Stochastics **4**(3), 223–245 (1980/1981)
61. L. McInnes, J. Healy, N. Saul, L. Großberger, Umap: Uniform manifold approximation and projection. J. Open Source Softw. **3**(29), 861 (2018)
62. I. Melbourne, R. Zweimüller, Weak convergence to stable Lévy processes for nonuniformly hyperbolic dynamical systems. Ann. Inst. Henri Poincaré Probab. Stat. **51**(2), 545–556 (2015)
63. P. Moore, T. Lyons, J. Gallacher, A.D.N. Initiative, Using path signatures to predict a diagnosis of Alzheimer's disease. PloS One **14**(9), e0222212 (2019)
64. J. Morrill, A. Fermanian, P. Kidger, T. Lyons, A generalised signature method for multivariate time series feature extraction. Preprint. arXiv:2006.00873 (2020)
65. J.H. Morrill, A. Kormilitzin, A.J. Nevado-Holgado, S. Swaminathan, S.D. Howison, T.J. Lyons, Utilization of the signature method to identify the early onset of sepsis from multivariate physiological time series in critical care monitoring. Crit. Care Med. **48**(10), e976–e981 (2020)
66. I. Perez Arribas, G.M. Goodwin, J.R. Geddes, T. Lyons, K.E. Saunders, A signature-based machine learning model for distinguishing bipolar disorder and borderline personality disorder. Transl. Psychiatry **8**(1), 274 (2018)
67. R. Ree, Lie elements and an algebra associated with shuffles. Ann. Math. (2) **68**, 210–220 (1958)
68. R.Z.U. Rehman, Y. Zhou, S. Del Din, L. Alcock, C. Hansen, Y. Guan, T. Hortobágyi, W. Maetzler, L. Rochester, C.J. Lamoth, Gait analysis with wearables can accurately classify fallers from non-fallers: a step toward better management of neurological disorders. Sensors **20**(23), 6992 (2020)
69. J. Reizenstein, Calculation of iterated-integral signatures and log signatures. Preprint. arXiv:1712.02757 (2017)
70. J.F. Reizenstein, Iterated-integral signatures in machine learning. PhD thesis, University of Warwick, 2019
71. J.F. Reizenstein, B. Graham, Algorithm 1004: The iisignature library: Efficient calculation of iterated-integral signatures and log signatures. ACM Trans. Math. Softw. **46**(1), 1–21 (2020)
72. C. Reutenauer, *Free Lie Algebras*, volume 7 of London Mathematical Society Monographs. New Series (The Clarendon Press/Oxford University Press, New York, 1993). Oxford Science Publications
73. W. Rudin, *Principles of Mathematical Analysis*. International Series in Pure and Applied Mathematics, 3rd edn. (McGraw-Hill, New York, 1976)
74. T. Lyons, D. Maxwell, *esig*, Version 0.7.3. Python package. Accessed July 7, 2025 (2025). https://esig.readthedocs.io
75. S. Tindel, K. Chouk, Skorohod and Stratonovich integration in the plane. Electron. J. Probab. **20**(39), 39 (2015)
76. W. Whitt, *Stochastic-process Limits*. Springer Series in Operations Research (Springer, New York, 2002). An introduction to stochastic-process limits and their application to queues
77. D.R.E. Williams, Path-wise solutions of stochastic differential equations driven by Lévy processes. Rev. Mat. Iberoam. **17**(2), 295–329 (2001)
78. D. Wilson-Nunn, T. Lyons, A. Papavasiliou, H. Ni, A path signature approach to online Arabic handwriting recognition, in *2018 IEEE 2nd International Workshop on Arabic and Derived Script Analysis and Recognition (ASAR)* (IEEE, Piscataway, 2018), pp. 135–139

79. Z. Xie, Z. Sun, L. Jin, H. Ni, T. Lyons, Learning spatial-semantic context with fully convolutional recurrent network for online handwritten Chinese text recognition. IEEE Trans. Pattern Anal. Mach. Intell. **40**(8), 1903–1917 (2017)
80. W. Yang, T. Lyons, H. Ni, C. Schmid, L. Jin, Developing the path signature methodology and its application to landmark-based human action recognition, in *Stochastic Analysis, Filtering, and Stochastic Optimization: A Commemorative Volume to Honor Mark HA Davis's Contributions* (Springer, 2022), pp. 431–464
81. S. Zhang, G. Lin, S. Tindel, Two-dimensional signature of images and texture classification. Proc. R. Soc. A Math. Phys. Eng. Sci. **478**(2266), 20220346 (2022)

An Introduction to Tensors for Path Signatures

Jack Beda, Gonçalo dos Reis, and Nikolas Tapia

Abstract We present a fit-for-purpose introduction to tensors and their operations. It is envisaged to help the reader become acquainted with its underpinning concepts for the study of path signatures. The text includes exercises, solutions and many intuitive explanations. The material discusses direct sums and tensor products as two possible operations that make the Cartesian product of vectors spaces a vector space. The difference lies in linear vs. multilinear structures—the latter being the suitable one to deal with path signatures. The presentation is offered to understand tensors in a sense deeper than just a multidimensional array. The text concludes with the prime example of an algebra in relation to path signatures: the *tensor algebra*.

1 Introduction

Throughout mathematics, computer science, and physics, the term *tensor* is used to describe a myriad of similar, but fundamentally different mathematical objects. For amusement, we invite the reader to visit the "100 Questions: A Mathematical Con-

J. Beda
School of Mathematics, University of Edinburgh, Edinburgh, UK
e-mail: jack@beda.ca

G. dos Reis (✉)
Maxwell Institute for Mathematical Sciences, School of Mathematics, University of Edinburgh, Edinburgh, UK

Center for Mathematics and Applications (NOVA Math), Caparica, Portugal
e-mail: G.dosReis@ed.ac.uk

N. Tapia
Weierstrass Institute, Berlin, Germany

Department of Mathematics, Humboldt University Berlin, Berlin, Germany
e-mail: tapia@wias-berlin.de

C. Bayer et al. (eds.), *Signature Methods in Finance*, Springer Finance Lecture Notes, https://doi.org/10.1007/978-3-031-97239-3_2

ventions Survey" and check[1] "Question 45: What is a tensor?" That answer's wide distribution seems to hint at a gap, as folks knowledge goes, to the mathematical meaning of a tensor (even among the informed?). It is thus necessary to be clear and unambiguous with our definitions and language so that at the end of this chapter the reader will be able to agree with us on the answer. For practical purposes, it often suffices to describe a tensor as a multidimensional array that extends the concept of a matrix [3]. This is surely true, but only after a certain structure on the underlying space is assumed. Tensors are much more, especially when the underlying spaces are infinite-dimensional [1, 2, 4].

For path signatures, seeing tensors as multidimensional arrays is a good starting point, but their power is only fully realized with an understanding of the operations that go with them: that is, with the algebra mixed in. The assumption behind the next sections is that the reader is largely unfamiliar with tensors (but having heard of multidimensional arrays).

Why do tensors come up in path signatures and how to read this chapter? We saw in chapter "A Primer on the Signature Method in Machine Learning" that path signatures rely heavily on iterated integrals and involve products of path components across different times. The path signature is a sequence of terms encoding information about a path at different levels of complexity: the first-level signature captures linear information via integrals of individual components; the second-level signature captures pairwise interactions, so-called bilinear relationships, like $\iint \mathrm{d}x^i \mathrm{d}x^j$, meaning they combine two inputs in a way that is linear in each (with x^i fixed, we have linearity in x^j, and vice-versa) whereas joint linearity does not hold; higher levels of the signature require more iterated integrals and in turn one requires multilinear relationships to describe these complicated interactions. Tensors and multilinear maps are natural and powerful mathematical tools to represent and analyze these interactions.

Section 2 introduces the basic definitions and properties of tensors by guiding the reader to first understand the difference between linear and bilinear operators, and how linear operators can be recovered from bilinear ones via the tensors product operation and the so-called *universal property*.

Section 3 then goes into a few other deeper properties and connects more explicitly tensors and algebras by introducing the *tensor algebra*. We will leave to subsequent chapters the development of further concepts like shuffle algebras, and power series to obtain exponentials and logarithms (alluded to in chapter "A Primer on the Signature Method in Machine Learning").

[1] Find the survey here: https://cims.nyu.edu/~tjl8195/survey/results.html#q45, at the time this manuscript was written.

2 A Brief Introduction to Tensors

This section introduces two operations: *direct sums* and *tensor products*, two different ways of making new vector spaces out of old ones. Formally, each is a way of equipping the Cartesian product of vector spaces, $U \times V$, with a linear structure. The first leads to linear operators while the second leads to bilinear ones. While both are related, in many aspects they behave very differently.[2] In a nutshell, bilinear maps exhibit separate linearity in U and V while linear maps exhibit global linearity in $U \times V$. The distinction is particularly important in the algebra of tensors, where bilinear maps give rise to the tensor product structure through the so-called *universal property*. Linear maps correspond to mappings on the direct product space.

2.1 Direct Sums

Let us recall that for two sets X and Y, their Cartesian product $X \times Y$ is defined as

$$X \times Y := \{(x, y) : x \in X, y \in Y\},$$

that is, the set of all ordered pairs where the first is an element of X and the second an element of Y.

Given vector spaces U and V, their Cartesian product does not immediately have a linear structure (i.e. is not immediately a vector space). In other words, after constructing the *set* $U \times V$ it is not clear how to add two ordered pairs, or multiply them by scalars. We must *define* a way to add and scale the elements of this set, and it turns out there are multiple, sensible and useful definitions. Direct sums are the simplest way to equip the Cartesian product of two (or more) vector spaces with a linear structure of its own.

Definition 2.1 Let U, V be vector spaces. On the Cartesian product $U \times V$ we define the following *linearity* operations: for $\mathbf{u}, \mathbf{u}_1, \mathbf{u}_2 \in U$, $\mathbf{v}, \mathbf{v}_1, \mathbf{v}_2 \in V$ and $\lambda \in \mathbb{R}$

$$(\mathbf{u}_1, \mathbf{v}_1) + (\mathbf{u}_2, \mathbf{v}_2) := (\mathbf{u}_1 + \mathbf{u}_2, \mathbf{v}_1 + \mathbf{v}_2), \tag{1}$$

$$\lambda(\mathbf{u}, \mathbf{v}) := (\lambda \mathbf{u}, \lambda \mathbf{v}). \tag{2}$$

One can check that $U \times V$, equipped with the operations from (1) and (2), is a vector space, which it is customary to denote as $U \oplus V$, i.e the *direct sum* of U and V.

Exercise 2.2 State the axioms that define a *vector space* and show that $U \oplus V$ is indeed a vector space.

[2] For instance, there is no open mapping theorem for bilinear surjective maps, nor is there a Hahn-Banach theorem for bilinear continuous forms.

It can also be checked that if B_U and B_V are bases for U and V (respectively) with 0_U and 0_V denoting the zero elements of U and V (respectively) then the set

$$B = \{(\mathbf{u}, 0_V) : \mathbf{u} \in B_U\} \cup \{(0_U, \mathbf{v}) : \mathbf{v} \in B_V\}$$

is a basis for $U \oplus V$. It follows immediately that $\dim(U \oplus V) = \dim(U) + \dim(V)$.

Example 2.3 Let $U = \mathbb{R}^3$ and $V = \mathcal{M}_{2\times 2}(\mathbb{R})$ be the space of real 2-by-2 matrices. A generic element of $U \oplus V$ is an ordered pair $(\mathbf{u}, \mathbf{A})$ for a vector $\mathbf{u} \in \mathbb{R}^3$ and a matrix $\mathbf{A} \in \mathcal{M}_{2\times 2}(\mathbb{R})$. A concrete example would be letting

$$(\mathbf{u}, \mathbf{A}) = \left(\begin{bmatrix} 3 \\ -1 \\ 4 \end{bmatrix}, \begin{bmatrix} 1 & -1 \\ 2 & 3 \end{bmatrix}\right) \quad \text{and} \quad (\mathbf{v}, \mathbf{B}) = \left(\begin{bmatrix} -2 \\ 1 \\ 1 \end{bmatrix}, \begin{bmatrix} 5 & 3 \\ -1 & 2 \end{bmatrix}\right)$$

then we have

$$(\mathbf{u}, \mathbf{A}) + (\mathbf{v}, \mathbf{B}) = \left(\begin{bmatrix} 1 \\ 0 \\ 5 \end{bmatrix}, \begin{bmatrix} 6 & 2 \\ 1 & 5 \end{bmatrix}\right).$$

A basis for $U \oplus V$ is (*with a slight abuse of notation for what '0' means*)

$$\left\{\left(\begin{bmatrix} 1 \\ 0 \\ 0 \end{bmatrix}, 0\right), \left(\begin{bmatrix} 0 \\ 1 \\ 0 \end{bmatrix}, 0\right), \left(\begin{bmatrix} 0 \\ 0 \\ 1 \end{bmatrix}, 0\right), \left(0, \begin{bmatrix} 1 & 0 \\ 0 & 0 \end{bmatrix}\right), \left(0, \begin{bmatrix} 0 & 1 \\ 0 & 0 \end{bmatrix}\right), \right.$$
$$\left. \left(0, \begin{bmatrix} 0 & 0 \\ 1 & 0 \end{bmatrix}\right), \left(0, \begin{bmatrix} 0 & 0 \\ 0 & 1 \end{bmatrix}\right)\right\}.$$

We observe that indeed $\dim(U \oplus V) = 7 = \dim(\mathbb{R}^3) + \dim(\mathcal{M}_{2\times 2}(\mathbb{R}))$.

Remark 2.4 (On Notation) Elements of $U \oplus V$ can also be written *additively*, that is $\mathbf{u} + \mathbf{v}$ denotes the vector $(\mathbf{u}, \mathbf{v}) \in U \oplus V$. This notation is harmless because of Definition 2.1, as it behaves in the expected way. We can then restate the two linearity equations (1) and (2) in more natural notation:

$$\mathbf{u}_1 + \mathbf{v}_1 + \mathbf{u}_2 + \mathbf{v}_2 = \mathbf{u}_1 + \mathbf{u}_2 + \mathbf{v}_1 + \mathbf{v}_2, \qquad \lambda(\mathbf{u} + \mathbf{v}) = \lambda\mathbf{u} + \lambda\mathbf{v},$$

and we note that there is no confusion in the first equality with the two different meanings of $+$ since due to the associativity and commutativity of addition, the four terms can be rearranged in an arbitrary way to give the same result.[3] Importantly, both the bracket notation, $(\mathbf{u}, \mathbf{v})$, and the additive notation, $\mathbf{u} + \mathbf{v}$, are used interchangeably in the path signature community.

[3] Formally, there is a canonical isomorphism between $U \oplus V$ and $V \oplus U$, so that the pairs $(\mathbf{u}, \mathbf{v})$ and $(\mathbf{v}, \mathbf{u})$ can be identified.

Definition 2.1 generalizes easily to a finite number of summands: if $U_1, \ldots, U_n$ are vector spaces, the set $U_1 \times \cdots \times U_n$ carries a linear structure given by componentwise addition and multiplication by scalars. The resulting vector space is denoted by $U_1 \oplus \cdots \oplus U_n$. Extending this concept to infinite families requires some care.

Definition 2.5 Consider an infinite index set I and let $(U_i : i \in I)$ be a family of vector spaces. The direct sum is defined to be the set of all sequences $(\mathbf{u}_i : i \in I)$ such that $\mathbf{u}_i \neq 0$ for finitely many indices $i \in I$. Addition and scalar multiplication are defined componentwise. The resulting vector space is denoted by

$$\bigoplus_{i \in I} U_i.$$

The finiteness constraint in this definition means that direct sum is a subset of the Cartesian product of the spaces, that is,

$$\bigoplus_{i \in I} U_i \subseteq \prod_{i \in I} U_i,$$

where $\prod_{i \in I} U_i$ is simply the set of all sequences indexed by I.

For our purposes it will be enough to consider countable families of vector spaces, that is, we will take $I = \mathbb{N}$. In this case, elements of $\bigoplus_{n \in \mathbb{N}} U_n$ may be denoted as

$$(\mathbf{u}_0, \mathbf{u}_1, \mathbf{u}_2, \ldots)$$

with the convention that there is only a finite number of non-zero elements in the sequence. The Cartesian product also carries the same linear structure and is indeed a vector space, whose elements consist of arbitrary $\mathbb{N}$-indexed sequences, which are still denoted as above where all entries may be non-zero.

It should be noted that in the case that $I \subset \mathbb{N}$ is a finite set, say $I = \{1, \ldots, N\}$, both spaces coincide but the inclusion becomes strict as soon as I is countable. In particular, when dealing with finite collections of spaces there is no ambiguity in the notation.

The main application of direct sums in the world of signatures is to decompose a vector space in terms of subspaces of objects sharing similar "shape" properties.

Definition 2.6 A vector space V is said to be *graded* if it can be decomposed as a direct sum:

$$V = \bigoplus_{n \in \mathbb{N}} V_n.$$

The subspace V_n is called the *homogeneous component of degree* n. For $v \in V_n$ we write $|v| = n$ for its degree.

We also note that this definition includes the case of finitely many summands, in which case there is $N \in \mathbb{N}$ such that $V_n = \{0\}$ for all $n > N$.

2.2 Tensor Product and Tensors

We have now seen how the *direct product* is one way of equipping the Cartesian product of two vector spaces with a vector space structure. There is another way, the *tensor product*, in many ways similar, but with a structure such that it is compatible with multilinear functions in a way to be made precise later.

Definition 2.7 Let U, V be vector spaces. On the set $U \times V$ define the following *bilinearity* operations: for $\mathbf{u}, \mathbf{u}_1, \mathbf{u}_2 \in U$, $\mathbf{v}, \mathbf{v}_1, \mathbf{v}_2 \in V$ and $\lambda \in \mathbb{R}$

$$(\mathbf{u}_1, \mathbf{v}) + (\mathbf{u}_2, \mathbf{v}) := (\mathbf{u}_1 + \mathbf{u}_2, \mathbf{v}),$$

$$(\mathbf{u}, \mathbf{v}_1) + (\mathbf{u}, \mathbf{v}_2) := (\mathbf{u}, \mathbf{v}_1 + \mathbf{v}_2),$$

$$\lambda(\mathbf{u}, \mathbf{v}) := (\lambda\mathbf{u}, \mathbf{v}) \quad \text{and}^4 \quad \lambda(\mathbf{u}, \mathbf{v}) := (\mathbf{u}, \lambda\mathbf{v}).$$

As in the direct sum case (Definition 2.1), it can be verified that $U \times V$ *equipped with the bilinearity operations* is a vector space. We denote the *tensor product* of U and V as $U \otimes V$. For elements of the tensor product space, we write $\mathbf{u} \otimes \mathbf{v} := (\mathbf{u}, \mathbf{v}) \in U \otimes V$. By a slight abuse of language we also refer to $\mathbf{u} \otimes \mathbf{v}$ as the tensor product of the vectors $\mathbf{u} \in U$ and $\mathbf{v} \in V$. As we will see later in Sect. 3, this name is justified.

Exercise 2.8 Show that $U \otimes V$ is a vector space (recall Exercise 2.2).

Exercise 2.9 In this exercise we see why the $\otimes$ notation is an intuitive way of writing the tensor product: (a) Rewrite the bilinearity operations of Definition 2.7 using the notation $\mathbf{u} \otimes \mathbf{v}$ instead of $(\mathbf{u}, \mathbf{v})$; (b) Expand $(\mathbf{u}_1 + \mathbf{u}_2) \otimes (\mathbf{v}_1 + \mathbf{v}_2)$, where of course the addition $\mathbf{u}_1 + \mathbf{u}_2 \in U$ is simply the addition in U, and the same for V; (c) We have $\lambda\mathbf{u} \otimes \lambda\mathbf{v} \propto (\mathbf{u} \otimes \mathbf{v})$. What is the constant of proportionality?

Contrary to the direct sum case, for tensor products its not always possible to write $\mathbf{u}_1 \otimes \mathbf{v}_1 + \mathbf{u}_2 \otimes \mathbf{v}_2$ as a single tensor product of a vector in U with a vector in V.[5]

---

[4] This last element of the definition is actually *imposing* the vectors $(\lambda\mathbf{u}, \mathbf{v})$ and $(\mathbf{u}, \lambda\mathbf{v})$ to be equal in $U \otimes V$. This could be formalized by the use of quotient spaces, but doing such would involve a level of additional complexity not really necessary at the moment. (This also takes care of the apparent non-uniqueness of $0_{U \otimes V}$ hinted at in Exercise 2.10.)

[5] As an aside, it is this property of tensor product spaces that mathematically captures the phenomenon of entanglement of quantum particles. A quantum state like $|\varphi\rangle = \begin{bmatrix} 1 \\ 0 \end{bmatrix} \otimes \begin{bmatrix} 0 \\ 1 \end{bmatrix} + \begin{bmatrix} 0 \\ 1 \end{bmatrix} \otimes \begin{bmatrix} 1 \\ 0 \end{bmatrix}$ is entangled as it cannot be written as $\begin{bmatrix} a \\ b \end{bmatrix} \otimes \begin{bmatrix} c \\ d \end{bmatrix}$ for any scalars a, b, c, d. This

Exercise 2.10 Show that $0_U \otimes \mathbf{v} = \mathbf{u} \otimes 0_V = 0_{U \otimes V}$ for all $\mathbf{u} \in U, \mathbf{v} \in V$.

In the same vein as the comment offered just before Definition 2.5, the Definition 2.7 admits a straightforward generalization to finitely many vector spaces and extending this concept to infinite families requires some care. In particular, we write

$$U^{\otimes n} := \underbrace{U \otimes U \otimes \cdots \otimes U}_{n \text{ times}}.$$

Proposition 2.11 *For U, V vector spaces with bases B_U, B_V, respectively, the set*

$$B = \{\mathbf{u} \otimes \mathbf{v} : \mathbf{u} \in B_U, \mathbf{v} \in B_V\}$$

is a basis for $U \otimes V$. It follows that $\dim(U \otimes V) = \dim(U) \cdot \dim(V)$.

We can tell immediately by the dimension of the spaces that the direct product and tensor product produce fundamentally different vector structures on the same set.[6] What distinguishes the tensor product vector space from all other possible linear structures is the following *universal property*.

Theorem 2.12 *Let U, V, W be vector spaces, and let $f : U \times V \to W$ be a bilinear map. That is, f satisfies for all $\mathbf{u}, \mathbf{u}_1, \mathbf{u}_2 \in U$, $\mathbf{v}, \mathbf{v}_1, \mathbf{v}_2 \in V$ and $\lambda \in \mathbb{R}$*

$$f(\mathbf{u}_1 + \mathbf{u}_2, \mathbf{v}) = f(\mathbf{u}_1, \mathbf{v}) + f(\mathbf{u}_2, \mathbf{v}), \qquad f(\mathbf{u}, \mathbf{v}_1 + \mathbf{v}_2) = f(\mathbf{u}, \mathbf{v}_1) + f(\mathbf{u}, \mathbf{v}_2),$$
$$\text{and} \quad f(\lambda\mathbf{u}, \mathbf{v}) = f(\mathbf{u}, \lambda\mathbf{v}) = \lambda f(\mathbf{u}, \mathbf{v}).$$

Then, there exists a unique **linear** *function $\hat{f} : U \otimes V \to W$ such that $f(\mathbf{u}, \mathbf{v}) = \hat{f}(\mathbf{u} \otimes \mathbf{v})$.*

We say that bilinear functions *factor through* the tensor product. In fact, the tensor product is characterized by this property, in the sense that any other vector space Z equipped with a map $\underline{\otimes} : U \times V \to Z$ factorizing bilinear functions must be isomorphic to $U \otimes V$. In other words, the tensor product is the unique (up to isomorphism) vector space having this property. For the sake of simplicity we will omit the proof of this result, but refer the interested reader to the classical texts [1, 2]. At first sight bilinear maps are not quite as powerful as linear maps, nonetheless, the universal property fixes things as it allows to write the bilinear map as a linear map of the tensor product and thus recovers the neat results of linear maps that were not available (at the cost of using tensor products).

The next example highlights the difference between direct sums and tensor products.

---

entangled state says "I have two particles, and their spins are always opposite, but I cannot know which one is spin-up, and which one is spin-down".

[6] Recall $\dim(U \oplus V) = \dim(U) + \dim(V)$, whereas $\dim(U \otimes V) = \dim(U) \cdot \dim(V)$.

Example 2.13 (Direct Sums $\oplus$ vs. Tensor Products $\otimes$) Let $U = V = \mathbb{R}$. The direct sum $U \oplus V$ satisfies $U \oplus V \cong \mathbb{R}^2$. Indeed, elements of $U \oplus V$ are ordered pairs of real numbers with component-wise addition and scalar multiplication. Moreover, a basis for $U \oplus V$ is $\{(1, 0), (0, 1)\}$ which is the canonical basis of $\mathbb{R}^2$, so $\dim(U \oplus V) = 2$.

On the contrary, we will show now that $U \otimes V$ satisfies $U \otimes V \cong \mathbb{R}$ which is *obviously not* $\mathbb{R}^2 \cong U \oplus V$. Consider the map $\varphi\colon U \times V \to \mathbb{R}$ given by $\varphi(x, y) = xy$; this map $\varphi(x, y)$ is clearly bilinear. By the universal property, there exists a unique map $\hat{\varphi}\colon U \otimes V \to \mathbb{R}$, given by $\hat{\varphi}(x \otimes y) = xy$. The map φ is injective since the equation $\hat{\varphi}(x \otimes y) = 0$ implies that either $x = 0$ or $y = 0$; in any case $x \otimes y = 0$ by Exercise 2.10. Finally, if $\lambda \in \mathbb{R}$ then $\hat{\varphi}(\lambda \otimes 1) = \lambda$ so that $\hat{\varphi}$ is surjective. In particular, $\mathbb{R} \otimes \mathbb{R}$ is spanned by the vector $1 \otimes 1$ so that $\dim(U \otimes V) = 1$.

Exercise 2.14 Show that if U is any vector space, then $U \otimes \mathbb{R}$ and $\mathbb{R} \otimes U$ are isomorphic to U. (*Hint: generalize Example 2.13.*)

The word *tensor* has many different meanings across different fields [1, 2, 4]. We will mostly be interested in the case where we are taking tensor products of a finite number of finite-dimensional vector spaces, represented as $\mathbb{R}^d$ for some integer $d \geq 1$. In this setting, tensors may be represented in a simpler, more concrete way, by working with canonical bases. It is at this specific juncture (assuming a basis) that it is intuitive to define a tensor as a multidimensional array—see Remark 2.16.

Definition 2.15 (Tensor: Order and Shape) Take $n \geq 1$, set $d_1, \dots, d_n \geq 1$ and take the associated vector spaces $\mathbb{R}^{d_j}$ for $j = 1, \dots, n$. An *order n tensor* of *shape* $(d_1, \dots, d_n)$ is an element of the tensor product $\mathbb{R}^{d_1} \otimes \cdots \otimes \mathbb{R}^{d_n}$.

It should be clear that elements of, say, $\mathbb{R}^2 \otimes \mathbb{R}^4$, $\mathbb{R}^4 \otimes \mathbb{R}^2$, and $\mathbb{R}^3 \otimes \mathbb{R}^3$, are all order 2 tensors, but their *shapes* are all very different—and, just like matrices, they cannot be added together as the shapes do not match.

Remark 2.16 (Basis, Vectors and Their Components, and Some Notation) Denote by $\{\mathbf{e}_1, \dots, \mathbf{e}_d\}$ the canonical basis of $\mathbb{R}^d$ for some $d \geq 1$. We have seen that the set

$$\left\{\mathbf{e}_{i_1} \otimes \cdots \otimes \mathbf{e}_{i_n} : i_j \in \{1, \dots, d_j\} \text{ for all } j = 1, \dots, n\right\}$$

is a basis for $\mathbb{R}^{d_1} \otimes \cdots \otimes \mathbb{R}^{d_n}$, which we call the canonical basis. In particular, an order n tensor $\mathbf{T}$ is determined by the n-dimensional array of its coefficients in this basis:

$$\mathbf{T} = \sum_{i_1=1}^{d_1} \cdots \sum_{i_n=1}^{d_n} \mathbf{T}^{i_1 \cdots i_n} \mathbf{e}_{i_1} \otimes \cdots \otimes \mathbf{e}_{i_n}.$$

As long as we keep this in mind, the assignment $\mathbf{T} \mapsto (\mathbf{T}^{i_1 \cdots i_n})$ defines a one-to-one correspondence between elements of the tensor product $\mathbf{T} \in \mathbb{R}^{d_1} \otimes \cdots \otimes \mathbb{R}^{d_n}$

and multidimensional arrays $(\mathbf{T}^{i_1\cdots i_n}) \in \mathbb{R}^{d_1\times\cdots\times d_n}$. *We remark once again that this isomorphism depends on the fixing of a basis and is, in general, not canonical.*

Thus, **notation wise**, we shall refer to tensors using either a symbol (i.e. $\mathbf{T}$) or in component notation (i.e. $\mathbf{T}^{ijk}$—using superscript notation for its components); for multiple tensors or vectors we use subscript notation, i.e., $\mathbf{T}_1, \mathbf{T}_2, \ldots$ or $\mathbf{u}_1, \mathbf{u}_2, \ldots$.

For example, $\mathbf{C}^{ij}$, $\mathbf{T}^{ijk}$ and $\mathbf{Q}^{ijkp}$ refer to the components of tensors $\mathbf{C}$, $\mathbf{T}$, and $\mathbf{Q}$ of order 2, 3, and 4 respectively. In particular, the order 1 tensors $\mathbf{e}_1, \ldots, \mathbf{e}_d$ constitute the canonical basis of $\mathbb{R}^d$. The ith basis vector $\mathbf{e}_i$ has components given in the canonical basis by

$$\mathbf{e}_i^j = \begin{cases} 1 & j = i \\ 0 & \text{else} \end{cases} \qquad \text{for } j = 1, \ldots, d.$$

Example 2.17 We see that tensors of order 1 and 2 can be identified with column vectors and matrices, respectively. A scalar is by convention an order 0 tensor. For example, $\mathbf{u}$, $\mathbf{A}$, and $\mathbf{T}$ are tensors of order 1, 2, and 3 respectively:

$$\mathbf{u} := \begin{bmatrix} 1 \\ 2 \\ 3 \end{bmatrix} \in \mathbb{R}^3, \quad \mathbf{A} := \begin{bmatrix} 1 & 2 \\ 3 & 4 \end{bmatrix} \in \mathbb{R}^2 \otimes \mathbb{R}^2 = (\mathbb{R}^2)^{\otimes 2} \cong \mathbb{R}^{2\times 2},$$

$$\mathbf{T} := \left[\begin{bmatrix} 1 & 2 \\ 3 & 4 \end{bmatrix}, \begin{bmatrix} 5 & 6 \\ 7 & 8 \end{bmatrix}\right] \in (\mathbb{R}^2)^{\otimes 3} \cong \mathbb{R}^{2\times 2\times 2}. \tag{3}$$

Remark 2.18 The tensor is an *intrinsic* object, in the sense that tensors do not depend on any choice of basis. From Remark 2.16, we see that a tensor can be uniquely described by a multidimensional array of numbers, but this is only true once we have fixed a basis. For many applications, it is possible, and practical, to think of tensors only in the terms of their components in a particular (e.g. the canonical) basis. Nonetheless, we encourage the reader to be mindful with the language and recognize when they are being loose with the concepts. This is exactly the same idea to how we can think of linear transformations from $\mathbb{R}^n$ to $\mathbb{R}^n$ in terms of an n-by-n square matrix, once we have fixed a basis. For example, the linear map $(x_1, x_2) \mapsto (x_1 + x_2, x_1 - x_2)$ can be represented by the matrix $\begin{pmatrix} 1 & 1 \\ 1 & -1 \end{pmatrix}$ in the standard basis, but in the eigenbasis, it is represented by the diagonal matrix $\begin{pmatrix} \sqrt{2} & 0 \\ 0 & -\sqrt{2} \end{pmatrix}$. The characterizing property of tensors is that given a change of basis, we immediately know how the coordinate representation transforms. In this example we can go from the first to the second matrix representation by multiplication with an invertible matrix (diagonalization).

Writing explicit examples for the tensor product quickly becomes cumbersome, nevertheless we offer a few simple examples.

Example 2.19 (Tensor Multiplication) Let $\mathbf{u} \in \mathbb{R}^2$, $\mathbf{v} \in \mathbb{R}^3$ be vectors, that is, order 1 tensors of shapes (2) and (3), respectively. By definition, their tensor product is an order 2 tensor of shape (2, 3): $\mathbf{u} \otimes \mathbf{v} \in \mathbb{R}^2 \otimes \mathbb{R}^3$. In the canonical basis they are represented by 2 and 3 coefficients, respectively. Namely

$$\mathbf{u} = \mathbf{u}^1\mathbf{e}_1 + \mathbf{u}^2\mathbf{e}_2, \quad \mathbf{v} = \mathbf{v}^1\mathbf{e}_1 + \mathbf{v}^2\mathbf{e}_2 + \mathbf{v}^3\mathbf{e}_3,$$

or in the more traditional column vector notation,

$$\mathbf{u} = \mathbf{u}^1\begin{bmatrix}1\\0\end{bmatrix} + \mathbf{u}^2\begin{bmatrix}0\\1\end{bmatrix} = \begin{bmatrix}\mathbf{u}^1\\\mathbf{u}^2\end{bmatrix} \quad \text{and} \quad \mathbf{v} = \begin{bmatrix}\mathbf{v}^1\\\mathbf{v}^2\\\mathbf{v}^3\end{bmatrix}.$$

By using the bilinearity of the tensor product (Definition 2.7) we may obtain the coordinates of $\mathbf{u} \otimes \mathbf{v}$ in the canonical basis. Indeed, recalling that the components of both vectors are scalars,

$$\begin{aligned}\mathbf{u} \otimes \mathbf{v} &= \left(\mathbf{u}^1\mathbf{e}_1 + \mathbf{u}^2\mathbf{e}_2\right) \otimes \left(\mathbf{v}^1\mathbf{e}_1 + \mathbf{v}^2\mathbf{e}_2 + \mathbf{v}^3\mathbf{e}_3\right)\\ &= \mathbf{u}^1\mathbf{v}^1\mathbf{e}_1 \otimes \mathbf{e}_1 + \mathbf{u}^1\mathbf{v}^2\mathbf{e}_1 \otimes \mathbf{e}_2 + \mathbf{u}^1\mathbf{v}^3\mathbf{e}_1 \otimes \mathbf{e}_3 \\ &\quad + \mathbf{u}^2\mathbf{v}^1\mathbf{e}_2 \otimes \mathbf{e}_1 + \mathbf{u}^2\mathbf{v}^2\mathbf{e}_2 \otimes \mathbf{e}_2 + \mathbf{u}^2\mathbf{v}^3\mathbf{e}_2 \otimes \mathbf{e}_3.\end{aligned} \tag{4}$$

Thus, in the canonical basis the order 2 tensor $\mathbf{u} \otimes \mathbf{v}$ has components given by $(\mathbf{u} \otimes \mathbf{v})^{ij} = \mathbf{u}^i\mathbf{v}^j$. Identifying the canonical basis of order 2 tensors with matrices (i.e., $\mathbf{e}_i \otimes \mathbf{e}_j$ forming the standard 2-by-2 matrix basis), we may write

$$\mathbf{u} \otimes \mathbf{v} = \begin{bmatrix}\mathbf{u}^1\mathbf{v}^1 & \mathbf{u}^1\mathbf{v}^2 & \mathbf{u}^1\mathbf{v}^3\\ \mathbf{u}^2\mathbf{v}^1 & \mathbf{u}^2\mathbf{v}^2 & \mathbf{u}^2\mathbf{v}^3\end{bmatrix} \quad \text{and likewise} \quad \begin{bmatrix}\mathbf{v}^1\mathbf{u}^1 & \mathbf{v}^1\mathbf{u}^2\\ \mathbf{v}^2\mathbf{u}^1 & \mathbf{v}^2\mathbf{u}^2\\ \mathbf{v}^3\mathbf{u}^1 & \mathbf{v}^3\mathbf{u}^2\end{bmatrix} \in \mathbb{R}^3 \otimes \mathbb{R}^2.$$

Consider now the matrix

$$\mathbf{A} := \begin{bmatrix}\mathbf{A}^{11} & \mathbf{A}^{12}\\ \mathbf{A}^{21} & \mathbf{A}^{22}\end{bmatrix} \in \mathbb{R}^2 \otimes \mathbb{R}^2.$$

Where, as before, the entries in the matrix notation corresponds to the coordinates in the canonical order 2 tensor basis: $\mathbf{A} = \mathbf{A}^{11}\mathbf{e}_1 \otimes \mathbf{e}_1 + \mathbf{A}^{12}\mathbf{e}_1 \otimes \mathbf{e}_2 + \mathbf{A}^{21}\mathbf{e}_2 \otimes \mathbf{e}_1 + \mathbf{A}^{22}\mathbf{e}_2 \otimes \mathbf{e}_2$. Note that although $\mathbf{A}$ is an element of $\mathbb{R}^2 \otimes \mathbb{R}^2$, it does not necessarily mean that it can be written as $\mathbf{u} \otimes \mathbf{v}$ for some $\mathbf{u}, \mathbf{v} \in \mathbb{R}^2$. That is, the components $\mathbf{A}^{ij}$ are not necessarily of the form $\mathbf{A}^{ij} = \mathbf{u}^i\mathbf{v}^j$ for some vectors $\mathbf{u}$ and $\mathbf{v}$ of the appropriate dimensions. In the cases where it *is* possible to find such a decomposition, we say that $\mathbf{A}$ is a *rank 1 tensor*. On the other hand, $\mathbf{A}$ can always be written as a linear combination of sums of tensor products of some $\mathbf{u}_i, \mathbf{v}_i \in \mathbb{R}^2$, that is, sums of rank 1 tensors.

Continuing with the example, we may compute

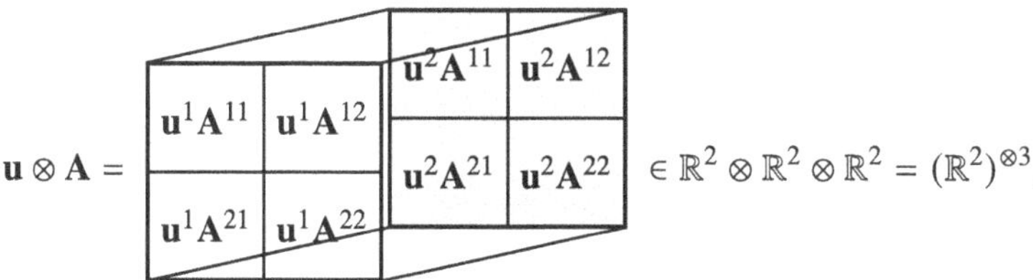

where the expression on the right-hand side is a matrix-like notation for organizing the components of the order 3 tensor $\mathbf{u} \otimes \mathbf{A}$.

We will later see, in Sect. 3, that the computation performed in Eq. (4) makes use of a larger structure. The tensor product can be thought of as a non-commutative analogue of polynomial multiplication, in the sense that it is an associative and bilinear operation. Note that it is, however, not commutative as the results of $\mathbf{u} \otimes \mathbf{v}$ and $\mathbf{v} \otimes \mathbf{u}$ are order 2 tensors of different shapes, at least in this example. In general, the results may differ even thought the shapes match.

Example 2.20 (Connecting to Signatures of Chapter "A Primer on the Signature Method in Machine Learning") Suppose $\mathbf{x}\colon [0, 1] \to \mathbb{R}^d$ is a smooth vector-valued path, which simply means, e.g., that for each $t \in [0, 1]$ we may write $\mathbf{x}_t = (\mathbf{x}_t^1, \dots, \mathbf{x}_t^d) \in \mathbb{R}^d$ in the canonical basis. In particular, for each $t \in [0, 1]$, $\mathbf{x}_t$ is a tensor of order 1.

We may use the tensor product to compactly write the collection of iterated integrals of $\mathbf{x}$. That is,

$$\int_0^t \int_0^s \mathrm{d}\mathbf{x}_u \otimes \mathrm{d}\mathbf{x}_s \in \mathbb{R}^d \otimes \mathbb{R}^d$$

is an order 2 tensor (i.e. a d-by-d matrix) with components (for $\mathbf{x}$ "sufficiently nice")

$$\left(\int_0^t \int_0^s \mathrm{d}\mathbf{x}_u \otimes \mathrm{d}\mathbf{x}_s\right)^{ij} = \int_0^t \int_0^s \dot{x}_u^i \dot{x}_s^j \, \mathrm{d}u\mathrm{d}s.$$

Before delving into tensors properties we can go back to "Question 45: What is a tensor?" of the "100 Questions: A Mathematical Conventions Survey". We hope to have convinced the reader that a tensor is nothing other than "an element of a tensor product of vector spaces".

3 A Little Bit More on Tensors: The Tensor Algebra

We now venture into a few additional properties of tensors.

Definition 3.1 An associative algebra is a vector space A equipped with a bilinear map $m\colon A \times A \to A$, called *product*, satisfying the associativity condition

$$m\big(m(\mathbf{x}, \mathbf{y}), \mathbf{z}\big) = m\big(\mathbf{x}, m(\mathbf{y}, \mathbf{z})\big).$$

The universal property of the tensor product (see Theorem 2.12) yields that equivalently, it may be represented by a *linear* map $m\colon A \otimes A \to A$, which is the one we will use from now on.

We say that an algebra A is *unital* if it has a distinguished element $1_A \in A$, called the *unit*, satisfying for all $\mathbf{x} \in A$

$$m(1_A \otimes \mathbf{x}) = \mathbf{x} = m(\mathbf{x} \otimes 1_A).$$

It is customary to write the product, denoted $\cdot_A$, of two elements of A using infix notation,[7] that is, $\mathbf{x} \cdot_A \mathbf{y} := m(\mathbf{x} \otimes \mathbf{y})$. Oftentimes, when no confusion can arise, we write simply $\mathbf{x} \cdot \mathbf{y}$ or even omit the symbol completely and just write $\mathbf{xy}$ instead. As we will mostly work with *unital associative algebras*, from here on we simply write *algebra*. In this notation, the condition for $(A, \cdot_A)$ to be an algebra can be written as

$$(\mathbf{x} \cdot_A \mathbf{y}) \cdot_A \mathbf{z} = \mathbf{x} \cdot_A (\mathbf{y} \cdot_A \mathbf{z})$$

for all $\mathbf{x}, \mathbf{y}, \mathbf{z} \in A$ and the unit satisfies $1 \cdot_A \mathbf{x} = \mathbf{x} \cdot_A 1 = \mathbf{x}$ for all $\mathbf{x} \in A$. We note that bilinearity of the product translates into the distributivity of $\cdot_A$ over $+$, e.g.,

$$(\mathbf{x} + \mathbf{y}) \cdot_A \mathbf{z} = \mathbf{x} \cdot_A \mathbf{z} + \mathbf{y} \cdot_A \mathbf{z}, \quad \lambda(\mathbf{x} \cdot_A \mathbf{y}) = (\lambda \mathbf{x}) \cdot_A \mathbf{y} = \mathbf{x} \cdot_A (\lambda \mathbf{y}),$$

and so on. We also remark that we *do not* require the product to be commutative, that is, we do not enforce that $\mathbf{x} \cdot_A \mathbf{y} = \mathbf{y} \cdot_A \mathbf{x}$ for every $\mathbf{x}, \mathbf{y} \in A$, although this may hold for some pairs of elements. In case this identity does hold for every $\mathbf{x}, \mathbf{y} \in A$ we say A is commutative (or Abelian).

Example 3.2 Let $\mathcal{M}_{n\times n}(\mathbb{R})$ be the space of n-by-n square matrices with real entries with its usual vector space structure (entry-wise addition and scalar multiplication). The matrix product $\mathbf{A} \cdot \mathbf{B} := \mathbf{AB}$ equips A with the structure of a (non-commutative) associative algebra with unit $1_A = \mathbf{I}_n$, the n-by-n identity matrix.

---

[7] Infix notation is a way of writing mathematical (and logic) expressions where operators are placed between the operands they act upon. This is the most familiar notation to us humans and matches how we (humans) interpret math expressions.

Example 3.3 Denote by $\mathbb{R}[x]$ the space of polynomials in a single variable x, and for polynomials $\mathbf{p}(x)$ and $\mathbf{q}(x)$ define the multiplication rule by

$$(\mathbf{p} \cdot \mathbf{q})(x) := \mathbf{p}(x)\mathbf{q}(x).$$

It is clear that this multiplication is bilinear in $\mathbf{p}$ and $\mathbf{q}$ and satisfies the associativity condition. The unit for this product is the constant polynomial $1_A(x) = 1$. For instance

$$(x^3 + 1) \cdot (x^2 + x) = x^5 + x^4 + x^2 + x.$$

In fact, since the monomials $\{x^n : n \geq 1\}$ form a linear basis for A, the four terms on the right-hand side correspond simply to $x^3 \cdot x^2$, $x^3 \cdot x$ and so on, where we use the bilinearity of the product. This is an example of a commutative algebra.

Example 3.4 Many structures with which one is already very familiar are just algebras in disguise, Table 1 gives a few examples.

The prime example of an algebra in relation to signatures is the *tensor algebra*.

Definition 3.5 Let V be a finite-dimensional vector space. The *tensor algebra over* V is the vector space

$$T(V) := \bigoplus_{n \geq 0} V^{\otimes n} \quad \text{with } V^{\otimes 0} \cong \mathbb{R}\mathbf{1}.$$

The product is simply the tensor product, and its unit is the vector $\mathbf{1}$ spanning $V^{\otimes 0}$.

The tensor algebra is therefore a *graded vector space* in the sense of Definition 2.6, where order n tensors are placed in degree n. We stress the fact (see Definition 2.5) that elements of $T(V)$ are *finite* sequences of tensors of arbitrary order. For this reason, vectors in $T(V)$ are usually called tensor (or non-commutative) polynomials. Later on we will construct the space of tensor series, which are infinite sequences.

Remark 3.6 When $V = \mathbb{R}^d$ the product can be written more explicitly in terms of the canonical basis $\{\mathbf{e}_1, \ldots, \mathbf{e}_d\}$. Introducing the *word notation*, recall Example

Table 1 Various examples of algebras. Some authors use the term algebra to refer to a vector space equipped with *any* bilinear operation, not necessarily associative. In that sense, the cross product is an algebra that is not associative

Vector space	Bilinear operator	Associative	Commutative	Unitary
$\mathbb{C}$	Complex product	Yes	Yes	Yes
$\mathbb{R}^3$	Vector cross product: $\mathbf{a} \times \mathbf{b}$	No	No	No
$\mathbb{R}[x]$	Multiplication	Yes	Yes	Yes
$\mathcal{M}_{n \times n}(\mathbb{R})$	Matrix multiplication	Yes	No	Yes

1.5 in chapter "A Primer on the Signature Method in Machine Learning", $\mathbf{e}_{i_1\cdots i_n} := \mathbf{e}_{i_1} \otimes \cdots \otimes \mathbf{e}_{i_n} \in (\mathbb{R}^d)^{\otimes n}$ for $(i_1, \ldots, i_n) \in \{1, \ldots, d\}^n$, the product (denoted for now by $\cdot_{T(V)}$) is then defined as

$$\mathbf{e}_{i_1\cdots i_n} \cdot_{T(V)} \mathbf{e}_{j_1\cdots j_m} = \mathbf{e}_{i_1\cdots i_n j_1\cdots j_m}.$$

For this reason it is commonly known as the *concatenation product*. In this case, it corresponds to the product introduced in Definition 2.7. Common notations for $\mathbf{x} \cdot_{T(V)} \mathbf{y}$ include $\mathbf{x} \otimes \mathbf{y}$ and $\mathbf{xy}$.

Theorem 3.7 *The tensor algebra enjoys the following universal property: given any algebra A and any* linear *map $f: V \to A$, there exists a unique map $\hat{f}: T(V) \to A$, such that $f(\mathbf{u} \otimes \mathbf{v}) = f(\mathbf{u}) \cdot_A f(\mathbf{v})$ for all $\mathbf{u}, \mathbf{v} \in T(V)$.*

As is the case with the tensor product, this property actually characterizes the tensor algebra in the sense that any other algebra satisfying this property is necessarily isomorphic to $T(V)$ for some vector space V.

We note that even though V is finite-dimensional, $T(V)$ is always infinite dimensional since, owing to Proposition 2.11,

$$\dim V^{\otimes n} = (\dim V)^n .$$

For this reason, while the tensor algebra is a neat theoretical construction, it is not very useful for practical purposes. There are a couple of ways of obtaining finite-dimensional versions of $T(V)$ which preserve its structure. The most common in signature applications is *truncation*. The basic idea is that we want to preserve "low order" information while still retaining the algebra structure, where the meaning of "order" is in the sense of tensor level. Luckily, the straightforward idea of just discarding high-order information works, with the caveat that the product has to be slightly modified.

Definition 3.8 Given $N \geq 1$, the *level-N truncated tensor algebra* is the finite-dimensional graded vector space (recall Definition 2.6)

$$T^N(V) := \bigoplus_{n=0}^{N} V^{\otimes n} \quad \text{with product} \quad \mathbf{x} \cdot_N \mathbf{y} = \begin{cases} \mathbf{x} \otimes \mathbf{y} & \text{if } |\mathbf{x}| + |\mathbf{y}| \leq N \\ 0 & \text{else .} \end{cases}.$$

Following from Definition 2.6, we note that in particular every element of $T^N(V)$ can be written as a sequence of homogeneous elements, that is every $\mathbf{v} \in T^N(V)$ is of the form $\mathbf{v} = (\mathbf{v}_0, \mathbf{v}_1, \ldots, \mathbf{v}_N)$ with $\mathbf{v}_n \in V^{\otimes n}$ (with some of them eventually zero). Hence, the product $\cdot_N$ is well-defined for all $\mathbf{x}, \mathbf{y} \in T^N(V)$ and not just for homogeneous tensors—elements in $T^N(V)$ are thus finite sequence of tensors of order up to N, with componentwise addition and multiplication by scalars.

It can be checked that $T^N(V)$ is an algebra[8] and

$$\dim T^N(V) = \frac{d^{N+1}-1}{d-1} \quad \text{where } \dim V = d.$$

Example 3.9 Let us take $N = 2$ and $V = \mathbb{R}^d$. The space $T^2(V) \cong \mathbb{R}\mathbf{1} \oplus \mathbb{R}^d \oplus (\mathbb{R}^d)^{\otimes 2}$ consists of elements of the form $(a, \mathbf{x}, \mathbf{A})$ with $a \in \mathbb{R}$, $\mathbf{x} \in \mathbb{R}^d$ and $\mathbf{A} \in \mathcal{M}_{d\times d}(\mathbb{R})$.[9]

The product reads

$$(a, \mathbf{x}, \mathbf{A}) \otimes (a', \mathbf{x}', \mathbf{A}') = (aa', a\mathbf{x}' + a'\mathbf{x}, a'\mathbf{A} + a\mathbf{A}' + \mathbf{x} \otimes \mathbf{x}').$$

We remark that this product is not commutative, meaning that in general the above expression will be different from that of $(a', \mathbf{x}', \mathbf{A}') \otimes (a, \mathbf{x}, \mathbf{A})$.

Exercise 3.10 Let $N = 2$ and $V = \mathbb{R}^d$. Show that an element $(a, \mathbf{x}, \mathbf{A}) \in T^2(V)$ is invertible if and only if $a \neq 0$, and compute its inverse.

Definition 3.11 The *extended tensor algebra* is the direct product

$$T((V)) := \prod_{n=0}^{\infty} V^{\otimes n}.$$

We identify $T((V))$ with the space of infinite sequences $\mathbf{u} = (\mathbf{u}_0, \mathbf{u}_1, \dots)$ with $\mathbf{u}_0 \in \mathbb{R}$, $\mathbf{u}_1 \in V$, and so on. The product is induced by the product on $T(V)$ and is given, for $\mathbf{u} = (\mathbf{u}_0, \mathbf{u}_1, \dots)$ and $\mathbf{v} = (\mathbf{v}_0, \mathbf{v}_1, \dots)$, by $\mathbf{uv} = \mathbf{w} = (\mathbf{w}_0, \mathbf{w}_1, \dots)$ where

$$\mathbf{w}_n = \sum_{k=0}^{n} \mathbf{u}_k \otimes \mathbf{v}_{n-k} \in (V^*)^{\otimes n}.$$

This product mimics polynomial multiplication and is sometimes called the *Cauchy product* for this reason. Since this space contains arbitrarily long sequences of tensors, its elements are commonly called *tensor series*. We note that the tensor algebra $T(V)$ is a strict subspace of $T((V))$.

For each integer $N \geq 1$ there is a canonical projection $\pi_N : T((V)) \to T^N(V)$, preserving multiplication, given simply by discarding tensors of degree greater than N, that is,

---

[8] Technically speaking $T^N(V)$ is a quotient of $T(V)$ by a two-sided ideal.

[9] Note that we are tacitly identifying real valued 2 tensors (of shape (2, 2)) with real valued 2-by-2 matrices—looking back at Remark 2.16 and Examples 2.17 and 2.19, we have implicitly made the assumption of working with the canonical basis of $\mathbb{R}^2$.

$$\pi_N(\mathbf{u}_0, \mathbf{u}_1, \dots) = (\mathbf{u}_0, \mathbf{u}_1, \dots, \mathbf{u}_N).$$

In the realm of signatures, this projection is used to produce finite-dimensional versions of the signature (see Example 3.12 just below) that are suitable for its representation in a computer.

Example 3.12 Recall Example 2.20. Our prime example of an element in $T\left(\left(\mathbb{R}^d\right)\right)$ is the signature of a smooth $\mathbb{R}^d$-valued path $\mathbf{x}\colon [0, 1] \to \mathbb{R}^d$. Its signature over the interval $[s, t] \subseteq [0, 1]$, denoted by $S(\mathbf{x})_{s,t}$, is the tensor series of iterated integrals:

$$S(\mathbf{x})_{s,t} := \left(1, \int_s^t \mathrm{d}\mathbf{x}_u, \int_s^t \int_s^{u_2} \mathrm{d}\mathbf{x}_{u_1} \otimes \mathrm{d}\mathbf{x}_{u_2}, \dots\right).$$

Projecting to the level-2 truncated tensor algebra we get

$$\pi_2 S(\mathbf{x})_{s,t} = \left(1, \int_s^t \mathrm{d}\mathbf{x}_u, \int_s^t \int_s^{u_2} \mathrm{d}\mathbf{x}_{u_1} \otimes \mathrm{d}\mathbf{x}_{u_2}\right).$$

4 Solutions to Exercises

Solution 4.1 (To Exercise 2.2) We must check that the operations satisfy the axioms of a vector space, that is, that $+$ is associative and commutative, and that scalar multiplication distributes over $+$. Let $(\mathbf{u}_1, \mathbf{v}_1), (\mathbf{u}_2, \mathbf{v}_2) \in U \oplus V$. Then

$$\begin{aligned}(\mathbf{u}_1, \mathbf{v}_1) + (\mathbf{u}_2, \mathbf{v}_2) &= (\mathbf{u}_1 + \mathbf{u}_2, \mathbf{v}_1 + \mathbf{v}_2) = (\mathbf{u}_2 + \mathbf{u}_1, \mathbf{v}_2 + \mathbf{v}_1) \\ &= (\mathbf{u}_2, \mathbf{v}_2) + (\mathbf{u}_1, \mathbf{v}_1).\end{aligned}$$

Moreover, if $(\mathbf{u}_3, \mathbf{v}_3) \in U \oplus V$ then

$$\begin{aligned}((\mathbf{u}_1, \mathbf{v}_1) + (\mathbf{u}_2, \mathbf{v}_2)) + (\mathbf{u}_3, \mathbf{v}_3) &= (\mathbf{u}_1 + \mathbf{u}_2, \mathbf{v}_1 + \mathbf{v}_2) + (\mathbf{u}_3, \mathbf{v}_3) \\ &= (\mathbf{u}_1 + \mathbf{u}_2 + \mathbf{u}_3, \mathbf{v}_1 + \mathbf{v}_2 + \mathbf{v}_3) \\ &= (\mathbf{u}_1, \mathbf{v}_1) + (\mathbf{u}_2 + \mathbf{u}_3, \mathbf{v}_2 + \mathbf{v}_3) \\ &= (\mathbf{u}_1, \mathbf{v}_1) + ((\mathbf{u}_2, \mathbf{v}_2) + (\mathbf{u}_3, \mathbf{v}_3)).\end{aligned}$$

Likewise, for any $\lambda \in \mathbb{R}$ we have

$$\begin{aligned}\lambda((\mathbf{u}_1, \mathbf{v}_1) + (\mathbf{u}_2, \mathbf{v}_2)) &= \lambda(\mathbf{u}_1 + \mathbf{u}_2, \mathbf{v}_1 + \mathbf{v}_2) \\ &= (\lambda(\mathbf{u}_1 + \mathbf{u}_2), \lambda(\mathbf{v}_1 + \mathbf{v}_2)) = (\lambda\mathbf{u}_1 + \lambda\mathbf{u}_2, \lambda\mathbf{v}_1 + \lambda\mathbf{v}_2) \\ &= (\lambda\mathbf{u}_1, \lambda\mathbf{v}_1) + (\lambda\mathbf{u}_2, \lambda\mathbf{v}_2) = \lambda(\mathbf{u}_1, \mathbf{v}_1) + \lambda(\mathbf{u}_2, \mathbf{v}_2).\end{aligned}$$

We have used throughout that U and V are vector spaces.

Additive inverses are given simply by $-(\mathbf{u}, \mathbf{v}) = (-\mathbf{u}, -\mathbf{v})$ while the neutral element is $0_{U\oplus V} = (0_U, 0_V)$.

Solution 4.2 (To Exercise 2.8) We have to check that the operations satisfy the axioms of a vector space. Since addition is defined symmetrically, we only check one side. Let $\mathbf{u}_1, \mathbf{u}_2 \in U$ and $\mathbf{v} \in V$. Then

$$(\mathbf{u}_1, \mathbf{v}) + (\mathbf{u}_2, \mathbf{v}) = (\mathbf{u}_1 + \mathbf{u}_2, \mathbf{v}) = (\mathbf{u}_2 + \mathbf{u}_1, \mathbf{v}) = (\mathbf{u}_2, \mathbf{v}) + (\mathbf{u}_1, \mathbf{v}).$$

Associativity follows in a similar way:

$$\begin{aligned}
((\mathbf{u}_1, \mathbf{v}) + (\mathbf{u}_2, \mathbf{v})) + (\mathbf{u}_3, \mathbf{v}) &= (\mathbf{u}_1 + \mathbf{u}_2, \mathbf{v}) + (\mathbf{u}_3, \mathbf{v}) \\
&= \big((\mathbf{u}_1 + \mathbf{u}_2) + \mathbf{u}_3, \mathbf{v}\big) = \big(\mathbf{u}_1 + (\mathbf{u}_2 + \mathbf{u}_3), \mathbf{v}\big) \\
&= (\mathbf{u}_1, \mathbf{v}) + (\mathbf{u}_2 + \mathbf{u}_3, \mathbf{v}) \\
&= (\mathbf{u}_1, \mathbf{v}) + ((\mathbf{u}_2, \mathbf{v}) + (\mathbf{u}_3, \mathbf{v}))\,.
\end{aligned}$$

Now, for any $\lambda \in \mathbb{R}$ we see that (using throughout that U and V are vector spaces)

$$\begin{aligned}
\lambda\left((\mathbf{u}_1, \mathbf{v}) + (\mathbf{u}_2, \mathbf{v})\right) &= \lambda(\mathbf{u}_1 + \mathbf{u}_2, \mathbf{v}) = (\lambda(\mathbf{u}_1 + \mathbf{u}_2), \mathbf{v}) \\
&= (\lambda\mathbf{u}_1 + \lambda\mathbf{u}_2, \mathbf{v}) = (\lambda\mathbf{u}_1, \mathbf{v}) + (\lambda\mathbf{u}_2, \mathbf{v}) \\
&= \lambda(\mathbf{u}_1, \mathbf{v}) + \lambda(\mathbf{u}_2, \mathbf{v}).
\end{aligned}$$

Additive inverses are given by $-(\mathbf{u}, \mathbf{v}) = (-\mathbf{u}, \mathbf{v}) = (\mathbf{u}, -\mathbf{v})$.

Solution 4.3 (To Exercise 2.9)

(a) Let U, V be vector spaces. On the set $U \times V$ define the following bilinearity operations: for $\mathbf{u}, \mathbf{u}_1, \mathbf{u}_2 \in U$, $\mathbf{v}, \mathbf{v}_1, \mathbf{v}_2 \in V$ and $\lambda \in \mathbb{R}$

$$\begin{aligned}
&(\mathbf{u}_1 \otimes \mathbf{v}) + (\mathbf{u}_2 \otimes \mathbf{v}) := (\mathbf{u}_1 + \mathbf{u}_2) \otimes \mathbf{v}, \\
&(\mathbf{u} \otimes \mathbf{v}_1) + (\mathbf{u} \otimes \mathbf{v}_2) := \mathbf{u} \otimes (\mathbf{v}_1 + \mathbf{v}_2) \;\text{ and }\; \lambda(\mathbf{u} \otimes \mathbf{v}) := (\lambda\mathbf{u}) \otimes \mathbf{v} =: \mathbf{u} \otimes (\lambda\mathbf{v}).
\end{aligned}$$

(b) Set $\mathbf{u} :- (\mathbf{u}_1 + \mathbf{u}_2)$. It is definitely not necessary to write $\mathbf{u} :- (\mathbf{u}_1 + \mathbf{u}_2)$, but it may help to see how the axioms from (a) can be applied. We then have,

$$\begin{aligned}
(\mathbf{u}_1 + \mathbf{u}_2) \otimes (\mathbf{v}_1 + \mathbf{v}_2) &= \mathbf{u} \otimes (\mathbf{v}_1 + \mathbf{v}_2) = \mathbf{u} \otimes \mathbf{v}_1 + \mathbf{u} \otimes \mathbf{v}_2 \\
&= (\mathbf{u}_1 + \mathbf{u}_2) \otimes \mathbf{v}_1 + (\mathbf{u}_1 + \mathbf{u}_2) \otimes \mathbf{v}_2 \\
&= \mathbf{u}_1 \otimes \mathbf{v}_1 + \mathbf{u}_2 \otimes \mathbf{v}_1 + \mathbf{u}_1 \otimes \mathbf{v}_1 + \mathbf{u}_2 \otimes \mathbf{v}_2.
\end{aligned}$$

(c) We have $\lambda\mathbf{u} \otimes \lambda\mathbf{v} = \lambda^2(\mathbf{u} \otimes \mathbf{v})$. Compare this with the linear scaling (2) in Definition 2.1.

Solution 4.4 (To Exercise 2.10) It suffices to check that for any $\mathbf{u} \in U$ and any elementary tensor $\mathbf{u}' \otimes \mathbf{v}' \in U \otimes V$ it holds that $\mathbf{u} \otimes 0_V + \mathbf{u}' \otimes \mathbf{v}' = \mathbf{u}' \otimes \mathbf{v}'$.

Indeed, since we can write $0_V = \mathbf{v}' - \mathbf{v}'$ it follows that

$$\begin{aligned}\mathbf{u} \otimes 0_V + \mathbf{u}' \otimes \mathbf{v}' &= \mathbf{u} \otimes (\mathbf{v}' - \mathbf{v}') + \mathbf{u}' \otimes \mathbf{v}' \\ &= \mathbf{u} \otimes \mathbf{v}' + \mathbf{u} \otimes (-\mathbf{v}') + \mathbf{u}' \otimes \mathbf{v}' = \mathbf{u}' \otimes \mathbf{v}'\end{aligned}$$

where in the last equality we have used that $\mathbf{u} \otimes (-\mathbf{v})$ is the additive inverse of $\mathbf{u} \otimes \mathbf{v}$.

The check for $0_U \otimes \mathbf{v}$ can be done in a similar way.

Solution 4.5 (To Exercise 2.14) We must find a bijective linear function $\Psi : \mathbb{R} \otimes U \to U$. It suffices to define $\Psi(\lambda \otimes \mathbf{u}) = \lambda\mathbf{u}$. Linearity follows from the fact that the right-hand side is bilinear in $(\lambda, \mathbf{u})$ and the properties of the tensor product. Injectivity is immediate since $\lambda \otimes \mathbf{u} \in \ker \Psi$ if and only if $\Psi(\lambda \otimes \mathbf{u}) = \lambda\mathbf{u} = 0_U$, which in turn implies that either $\lambda = 0$ or $\mathbf{u} = 0_U$ by the axioms of vector spaces, and in both cases this means that $\lambda \otimes \mathbf{u} = 0_{\mathbb{R}\otimes U}$. Therefore, it follows that $\ker \Psi = \{0\}$, i.e., Ψ is injective.

Bijectivity can be shown by noting that every $\mathbf{u} \in U$ can be obtained as $\mathbf{u} = \Psi(1 \otimes \mathbf{u})$ (which in particular implies that the inverse map is $\Psi^{-1}(\mathbf{u}) = 1 \otimes \mathbf{u}$), or by noting that since $\dim(\mathbb{R} \otimes U) = \dim(\mathbb{R}) \cdot \dim(U) = \dim(U)$, by the rank-nullity theorem it follows that

$$\dim\big(\operatorname{im}(\Psi)\big) = \dim\big(\mathbb{R} \otimes U\big) - \dim\big(\ker(\Psi)\big) = \dim(U) \qquad \text{so that} \quad \operatorname{im}(\Psi) = U.$$

Solution 4.6 (To Exercise 3.10) The unit element in $T^2(V)$ is $\mathbf{1} = (1, 0, 0)$. From the product formula in Example 3.9 we see that the entries of the inverse element $(a', \mathbf{x}', \mathbf{A}') := (a, \mathbf{x}, \mathbf{A})^{-1}$ must satisfy

$$aa' = 1, \quad a\mathbf{x}' + a'\mathbf{x} = 0 \quad \text{and} \quad \mathbf{A} + \mathbf{A}' + \mathbf{x} \otimes \mathbf{x}' = 0.$$

The first equation is solvable if and only if $a \neq 0$, in which case $a' = a^{-1}$. Inserting this in the second equation it follows that $\mathbf{x}' = -\frac{1}{a^2}\mathbf{x}$. Lastly, from the third equation we see that $\mathbf{A}' = -\frac{1}{a^2}\mathbf{A} + \frac{1}{a^3}\mathbf{x} \otimes \mathbf{x}$. Hence, in $T^2(V)$, we have that

$$(a, \mathbf{x}, \mathbf{A})^{-1} = (a^{-1}, -a^{-2}\mathbf{x}, -a^{-2}\mathbf{A} + a^{-3}\mathbf{x} \otimes \mathbf{x}).$$

This can also be seen from the more general formula $\mathbf{A}^{-1} = \sum_{n \geq 0}(\mathbf{1} - \mathbf{A})^{\otimes n}$.

Acknowledgments We thank Ana Djurdjevac (FU-Berlin) for the helpful comments.

NT acknowledges support from DFG CRC/TRR 388 "Rough Analysis, Stochastic Dynamics and Related Fields", Project B01.

GdR acknowledges partial support by the Engineering and Physical Sciences Research Council (EPSRC) [grant number EP/R511687/1], from the FCT—Fundação para a Ciência e a Tecnologia, I.P., under the scope of the projects UIDB/00297/2020 and UIDP/00297/2020 (Center for Mathematics and Applications, NOVA Math), and by the UK Research and Innovation (UKRI) under the UK government's Horizon Europe funding Guarantee [Project UKRI343].

Note The arXiv version of this manuscript—https://arxiv.org/abs/2502.15703 —contains two extra sections discussing the factoring of tensor product expressions to a minimal number of terms. These sections are not needed for the theory presented in this book, but tensor rank and tensor factorizations are an elegant way of becoming familiar with the language of tensors and tensor products that are used throughout the book. A GitHub repository is attached to the arXiv version.

References

1. J. Diestel, J.H. Fourie, J. Swart, *The Metric Theory of Tensor Products* (American Mathematical Society, Providence, RI, 2008). Grothendieck's résumé revisited
2. W. Hackbusch, Tensor spaces and numerical tensor calculus, in *Springer Series in Computational Mathematics*, vol. 56, 2nd edn. (Springer, Cham, 2019)
3. S. Rabanser, O. Shchur, S. Günnemann, Introduction to tensor decompositions and their applications in machine learning. *arXiv:1711.10781* (2017)
4. M. Reed, B. Simon, *Methods of Modern Mathematical Physics I*, 2nd edn. (Academic Press, Inc. [Harcourt Brace Jovanovich, Publishers], New York, 1980). Functional analysis

The Signature Kernel

Darrick Lee and Harald Oberhauser

Abstract The signature kernel is a positive definite kernel for sequential data. It inherits theoretical guarantees from stochastic analysis, has efficient algorithms for computation, and shows strong empirical performance. In this chapter, we provide an introduction to the signature kernel by highlighting the analytic connection between the path signature and ordinary monomials. In particular, both the classical monomials and the path signature are universal and characteristic on compact domains: they can approximate functions, and characterize probability measures. However, issues arise in practice due to unbounded domains, computational complexity, and non-robustness. We show that these issues can be avoided via kernelization and robustification. To address the computational complexity, we provide an overview of the kernel trick: algorithms to efficiently compute the signature kernel which avoid direct computations of the signature. Furthermore, the signature kernel is highly flexible, and provides a canonical way to turn a given kernel on any domain into a kernel for sequences in that domain, while retaining its theoretical and computational properties. Finally, we survey applications and recent developments of the signature kernel.

1 Introduction

Suppose we have a collection of data valued in an arbitrary set $\mathcal{X}$ which we wish to study by solving machine learning problems such as classification or regression. A powerful and flexible framework for tackling a wide range of such problems are

D. Lee (✉)
School of Mathematics and Maxwell Institute for Mathematical Sciences, University of Edinburgh, Edinburgh, Scotland
e-mail: darrick.lee@ed.ac.uk

H. Oberhauser
Mathematical Institute, University of Oxford, Andrew Wiles Building, Radcliffe Observatory Quarter, Oxford, England
e-mail: oberhauser@maths.ox.ac.uk

C. Bayer et al. (eds.), *Signature Methods in Finance*, Springer Finance Lecture Notes, https://doi.org/10.1007/978-3-031-97239-3_3

kernel methods. In this framework, one begins with a *kernel*[1] on $\mathcal{X}$,

$$\mathrm{k} : \mathcal{X} \times \mathcal{X} \to \mathbb{R}, \tag{1}$$

which acts as a measure of similarity between various data points. Kernel methods allow us to use a rich set of features that capture structures in the underlying data while avoiding the explicit computation of such features; it has become a mainstay in machine learning on arbitrary domains $\mathcal{X}$ [62].

In this article we focus on the case where the domain is a set of sequences of arbitrary length. Sequential data are ubiquitous in practice, usually in the form of discrete time series. Consider the set of sequences of arbitrary length L in the set $\mathcal{X}$, denoted by

$$\mathcal{X}_{\text{seq}} := \{\mathbf{x} = (\mathbf{x}_0, \ldots, \mathbf{x}_L) : L \geq 0,\ \mathbf{x}_i \in \mathcal{X}\}. \tag{2}$$

Constructing kernels for non-Euclidean domains such as $\mathcal{X}_{\text{seq}}$ is often challenging. In particular, even when $\mathcal{X} = \mathbb{R}^d$ is linear, the space of sequences $\mathcal{X}_{\text{seq}}$ is non-linear since there is no natural addition operation for sequences of different length. In this chapter, we discuss how the *path signature*, identified as a feature map for sequential data, from chapter "A Primer on the Signature Method in Machine Learning" can be used to lift a kernel k on $\mathcal{X}$ (Eq. 1) to a *signature kernel*

$$\mathrm{k}_{\mathsf{Sig}} : \mathcal{X}_{\text{seq}} \times \mathcal{X}_{\text{seq}} \to \mathbb{R}. \tag{3}$$

on $\mathcal{X}_{\text{seq}}$. We primarily consider $\mathrm{k}_{\mathsf{Sig}}$ on the domain of discrete-time sequences $\mathcal{X}_{\text{seq}}$ since this is how we typically apply it in practice, but the domain of $\mathrm{k}_{\mathsf{Sig}}$ naturally extends to continuous time paths. More generally, the signature kernel is of interest for three primary reasons.

(i) **It inherits attractive theoretical guarantees** from stochastic analysis and the properties of the classical signature. These manifest as universality (the ability to approximate non-linear functions), characteristicness (the ability to characterize probability measures), invariance to time-reparametrization, and convergence in the scaling limit when sequences (discrete time) approximate paths (continuous time).
(ii) **It overcomes bottlenecks of the signature features**, both in terms of computational complexity and in terms of expressiveness since it allows us to use signatures of paths after they have been lifted to infinite-dimensional state space. That is, it can indirectly compute and use tensors in infinfinite-dimesional spaces that describe the underlying sequence.
(iii) **It leverages modular tools from kernel learning**. Given a kernel k with good theoretical properties, these typically transfer to the associated signature kernel

[1] Throughout, we use the term *kernel* to refer to a symmetric positive definite function.

$\mathsf{k}_{\mathsf{Sig}}$, and general tools from kernel learning apply. For example, universality and characteristicness [64] of k are preserved for the associated signature kernel $\mathsf{k}_{\mathsf{Sig}}$. This also implies that $\mathsf{k}_{\mathsf{Sig}}$ induces a computable metric, the MMD metric, between laws of stochastic processes. Further, Gaussian processes can be defined on the genuine non-Euclidean domain $\mathcal{X}_{\text{seq}}$ by using $\mathsf{k}_{\mathsf{Sig}}$ as a covariance function, generic kernel quadrature constructions can be used, etc.

A guiding theme for this chapter is that the iterated integrals, which constitute the signature, and hence the signature kernel, play the same role for sequences or paths as monomials do for vectors in $\mathbb{R}^d$, see Table 1. The key message is that this can be simultaneously a blessing and a curse. Algebraically, polynomials have attractive properties but they quickly become computationally prohibitive in high-dimensions; analytically, trouble arises on non-compact sets, and often a more flexible class than simple monomials forms a better basis. Kernelization allows us to circumvent these drawbacks.

This chapter focuses on theoretical aspects of the signature kernel. It is complemented by the User's Guide to `KSig`, [66], that focuses on algorithmic aspects of the signature kernel, GPU-accelerated computations, and documents the `Scikit-learn`-compatible Python package `KSig`.[2]

Notation
We use $[d] := \{1, \dots, d\}$ to denote the finite sets, and $\mathbb{R}^{\mathcal{X}} := \{f : \mathcal{X} \to \mathbb{R}\}$ to be the space of functions.

2 Monomials, Polynomials, and Tensors

Classical multivariate monomials and moments are convenient tools for approximating functions on $\mathbb{R}^d$ and, respectively, representing the law of vector-valued random variables via the moment sequence. Such monomials and moments can be concisely expressed using *tensors*. Recall from chapter "An Introduction to Tensors for Path Signatures", the space of tensors of degree m in d dimensions, denoted $(\mathbb{R}^d)^{\otimes m}$, is a d^m-dimensional vector space. The coordinates of $(\mathbb{R}^d)^{\otimes m}$ are indexed by *multi-indices* $I = (i_1, \dots, i_m)$, where $i_j \in [d]$. In particular, we can express the degree m monomials of a vector $\mathbf{x} = (\mathbf{x}^1, \dots, \mathbf{x}^d) \in \mathbb{R}^d$ as

$$\mathbf{x}^{\otimes m} \in (\mathbb{R}^d)^{\otimes m}, \tag{4}$$

where the coordinate of $\mathbf{x}^{\otimes m}$ with respect to the multi-index $I = (i_1, \dots, i_m)$ is

$$\mathbf{x}^{i_1} \cdot \mathbf{x}^{i_2} \cdots \mathbf{x}^{i_m}.$$

[2] https://github.com/tgcsaba/KSig.

We define the *monomial map* $\Phi_{\mathsf{Mon}} : \mathbb{R}^d \to \prod_{m\geq 0}(\mathbb{R}^d)^{\otimes m}$ and the *degree m monomial map* $\Phi_{\mathsf{Mon},m} : \mathbb{R}^d \to (\mathbb{R}^d)^{\otimes m}$ by

$$\Phi_{\mathsf{Mon}}(\mathbf{x}) := \left(1, \mathbf{x}, \frac{\mathbf{x}^{\otimes 2}}{2!}, \frac{\mathbf{x}^{\otimes 3}}{3!}, \dots\right) \quad \text{and} \quad \Phi_{\mathsf{Mon},m}(\mathbf{x}) := \frac{\mathbf{x}^{\otimes m}}{m!}$$

respectively. The map Φ_{Mon} takes a vector $\mathbf{x} = (\mathbf{x}^i)_{i\in[d]} \in \mathbb{R}^d$ as an input, and outputs a sequence of tensors of increasing degree. This sequence consists of a scalar $\mathbf{x}^{\otimes 0} := 1$, a vector $\mathbf{x}^{\otimes 1} = \mathbf{x}$, a matrix $\mathbf{x}^{\otimes 2} = (\mathbf{x}^{i_1}\mathbf{x}^{i_2})_{i_1,i_2\in[d]}$, a tensor $\mathbf{x}^{\otimes 3} = (\mathbf{x}^{i_1}\mathbf{x}^{i_2}\mathbf{x}^{i_3})_{i_1,i_2,i_3\in[d]}$, and so on. The co-domain of Φ_{Mon} is a graded (in $m = 0, 1, 2, \dots$) linear space,[3] hence we can apply a linear functional $\mathbf{w}$ to Φ_{Mon}.

When we restrict the domain of Φ_{Mon} to a compact subset $\mathcal{K} \subset \mathbb{R}^d$, the Stone–Weierstrass theorem guarantees that for every $f \in C(\mathcal{K}, \mathbb{R})$ and $\epsilon > 0$, there exists a $M \geq 1$ and coefficients $\mathbf{w}^0, \mathbf{w}^1, \dots, \mathbf{w}^d, \mathbf{w}^{1,1}, \dots, \mathbf{w}^{d,d}, \mathbf{w}^{1,1,1}, \dots, \mathbf{w}^{d,\dots,d} \in \mathbb{R}$, equivalently represented[4] as $\mathbf{w} \in \bigoplus_{m\geq 0}(\mathbb{R}^d)^{\otimes m}$, such that

$$\sup_{\mathbf{x}\in\mathcal{K}} \left| f(\mathbf{x}) - \underbrace{\sum_{m=0}^{M} \mathbf{w}^{i_1,\dots,i_m} \frac{\mathbf{x}^{i_1}\cdots\mathbf{x}^{i_m}}{m!}}_{=:\langle \mathbf{w}, \Phi_{\mathsf{Mon}}(\mathbf{x})\rangle} \right| < \epsilon.$$

Hence, any choice of $\mathbf{w}$ induces a function

$$\langle \mathbf{w}, \Phi_{\mathsf{Mon}}(\cdot)\rangle : \mathcal{K} \to \mathbb{R},$$

which is a linear function in terms of Φ_{Mon}.

Another well-known result about monomials is that they can characterize laws of random variables; in fact, convergence of the moment sequence is equivalent to classical weak convergence. We sum these up in the following proposition.

Proposition 2.1 *Let $\mathcal{K} \subset \mathcal{X} = \mathbb{R}^d$ be compact. Then for every $f \in C(\mathcal{K}, \mathbb{R})$ and $\epsilon > 0$ there exists some $\mathbf{w} \in \bigoplus_{m\geq 0}(\mathbb{R}^d)^{\otimes m}$ such that*

$$\sup_{\mathbf{x}\in\mathcal{K}} |f(\mathbf{x}) - \langle \mathbf{w}, \Phi_{\mathsf{Mon}}(\mathbf{x})\rangle| < \epsilon. \tag{5}$$

Let $\mathbf{X}, \mathbf{X}_1, \mathbf{X}_2, \dots$ be a sequence of $\mathcal{K}$-valued random variables. Then

$$\mathrm{Law}(\mathbf{X}) \mapsto \mathbb{E}[\Phi_{\mathsf{Mon}}(\mathbf{X})] := \left(1, \mathbb{E}[\mathbf{X}], \frac{\mathbb{E}[\mathbf{X}^{\otimes 2}]}{2!}, \dots\right)$$

[3] Addition is defined degree-wise $(\mathbf{s}_0, \mathbf{s}_1, \mathbf{s}_2, \dots) + (\mathbf{t}_0, \mathbf{t}_1, \mathbf{t}_2, \dots) := (\mathbf{s}_0 + \mathbf{s}_0, \mathbf{s}_1 + \mathbf{t}_1, \mathbf{s}_2 + \mathbf{t}_2, \dots)$. That is, we add scalars to scalars, vectors to vectors, etc.

[4] Recall that $\mathbf{w} \in \bigoplus_{m\geq 0}(\mathbb{R}^d)^{\otimes m}$ denotes a sequence $\mathbf{w} = (\mathbf{w}^I)$ where $\mathbf{w}^I$ is nonzero for finitely many multi-indices I, whereas $\mathbf{w}' \in \prod(\mathbb{R}^d)^{\otimes m}$ denotes a sequence $\mathbf{w}' = (\mathbf{w}'^I)$ where $\mathbf{w}'^I$ may be nonzero for infinitely many I.

is injective and the following are equivalent:

1. $\forall f \in C(\mathcal{X}, \mathbb{R}), \quad \lim_{n\to\infty} \mathbb{E}[f(\mathbf{X}_n)] = \mathbb{E}[f(\mathbf{X})]$;
2. $\forall m \in \mathbb{N}, \quad \lim_{n\to\infty} \mathbb{E}[\mathbf{X}_n^{\otimes m}] = \mathbb{E}[\mathbf{X}^{\otimes m}]$.

Proof That (1) implies (2) follows immediately since polynomials are bounded on compacts, hence they form a subset of $C_b(\mathcal{K}, \mathbb{R})$. To see the reverse implication note that the compactness assumption implies tightness of the laws of $(\mathbf{X}_n)$ and this allows us to use Prokhorov's Theorem. This reduces the proof to show that if $(\mathbf{X}_n)$ converges weakly along a subsequence to a random variable $\mathbf{Y}$ then $\text{Law}(\mathbf{Y}) = \text{Law}(\mathbf{X})$. But if $\mathbf{X}_n$ converges weakly to $\mathbf{Y}$, then $\lim_n \mathbb{E}[p(\mathbf{X}_n)] = \mathbb{E}[p(\mathbf{X})]$ for any polynomial p since polynomials are bounded on compacts. Hence we must have $\mathbb{E}[p(\mathbf{X})] = \mathbb{E}[p(\mathbf{Y})]$ for any polynomial p. Finally, since polynomials are dense in $C(\mathcal{K}, \mathbb{R})$ (by Stone–Weierstrass that we already mentioned above), the result follows. □

The above shows that linear functionals of monomials, that is linear combination of the coordinates of Φ_{Mon}, approximate continuous functions arbitrarily well. Further, the expected value of the coordinates of Φ_{Mon} characterizes the law of a random variable and convergence of these coordinates induces the classical weak convergence topology on the space of probability measures.

Remark 2.2 (Universality and Characteristicness) In the statistical learning literature, these properties of Φ_{Mon} are known as *universality* of Φ_{Mon} with respect to the class of bounded and continuous functions and *characteristicness* of Φ_{Mon} with respect to probability measures on $\mathcal{K}$. The general definition calls a map $\Phi : \mathcal{K} \to H$ from a set $\mathcal{K}$ into a Hilbert space H

(i) **universal** with respect to a function class $\mathcal{F} \subset \mathbb{R}^{\mathcal{K}}$ if the space of linear functionals $\mathcal{F}_0 := \{\langle \mathbf{w}, \Phi(\cdot)\rangle : \mathcal{K} \to \mathbb{R} \,:\, \mathbf{w} \in H\}$ is dense in $\mathcal{F}$;
(ii) **characteristic** with respect to probability measures $\mathcal{P}(\mathcal{K})$ if the expectation map $\mu \mapsto \mathbb{E}_{X\sim\mu}[\Phi(X)]$ that maps a probability measure on $\mathcal{K}$ into H is injective.

In fact, if one uses a more general definition of characteristicness by regarding distributions, that is elements of $\mathcal{F}'$, it can be shown that characteristicness and universality are equivalent. This makes many proofs easier, see [15, Definition 6] and [64].

The proof of Proposition 2.1 made heavy use of the compactness assumption but this is not a deficiency of the proof. The following example shows that things really go wrong on non-compact spaces.

A Warning: Compactness and Boundedness are Essential

Rather than a compact subset $\mathcal{K}$, consider the entire space $\mathcal{X} = \mathbb{R}^d$. In this case, moments may not exist for random variables on $\mathcal{X}$ with unbounded support but the situation is even worse: even if all moments exist, they fail to characterize the law of the random variable.

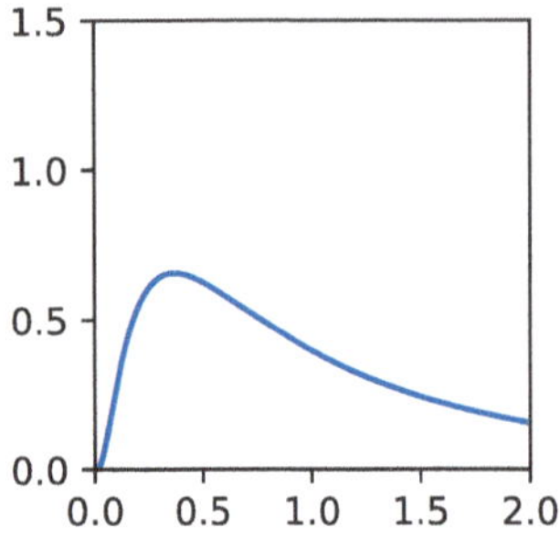

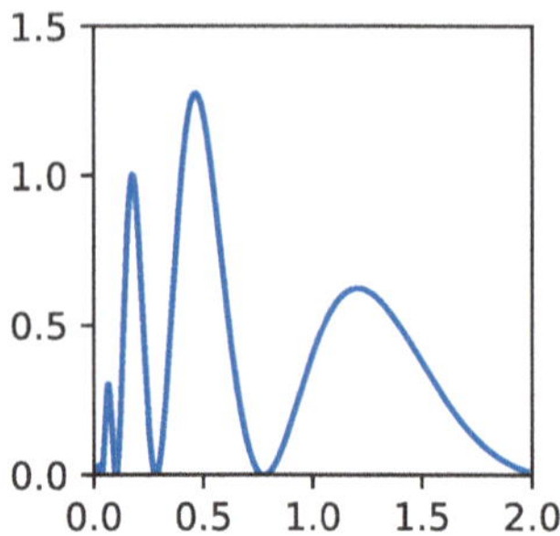

Fig. 1 The probability density functions of **X** (left) and **Y** (right)

This already happens in dimension $d = 1$ as the following classical counter example shows (Fig 1). Suppose $\mathbf{X}$ is a standard log-normal, that is $\log(\mathbf{X}) \sim N(0, 1)$. There exists another real-valued random variable $\mathbf{Y}$ that is not log-normal,[5] so that $\mathrm{Law}(\mathbf{X}) \neq \mathrm{Law}(\mathbf{Y})$, but for all $m \in \mathbb{N}$,

$$\mathbb{E}[\mathbf{X}^m] = \mathbb{E}[\mathbf{Y}^m] = \exp\left(\frac{m^2}{2}\right).$$

Hence, "characteristicness" fails on non-compact sets. Similarly, universality, that is the uniform approximation of functions $f(\mathbf{x})$ from Eq. (5), is hopeless since polynomials tend to $\pm\infty$ as the argument $\mathbf{x}$ becomes unbounded as $\|\mathbf{x}\| \to \infty$.

Polynomials are algebraically and analytically nice on compact sets, but the fundamental issue is that while their algebraic properties extend to non-compact sets (they still form a ring) we run into issues in their analytic and probabilistic properties since they grow too quickly. We emphasize that "tightness" arguments are not so useful here; it is easy to find for any $\epsilon > 0$ a compact set that carries $(1 - \epsilon)$ of the mass of the probability measure; the issue is that the functions we care about, monomials, explode on the space domain that carries only ϵ of the mass. A classical way to deal with this is to restrict to random variables with conditions on moment growth, but this is unsatisfactory for our purposes since such assumptions quickly become abstract and in general, it is not possible to verify them in a non-parametric inference setting as is common in machine learning tasks. After all, things already went bad for a simple log-normal. Finally, a closely related issue is that moments are not robust statistics, that is, if we have to estimate expectation by i.i.d. samples from $\mathcal{X}$ then small errors in these samples can lead to large estimation errors.

We will return to all these issues but for now, we ignore this issue with non-compact sets and ask a much more basic question: What is the natural generalization of monomials from $\mathcal{X} = \mathbb{R}^d$ to the space $\mathcal{X}_{\text{seq}}$ of sequences in $\mathbb{R}^d$?

[5] The random variable $\mathbf{Y}$ is explicitly specified by the density $\mathbb{P}(\mathbf{Y} \in dy) = f(y)(1 + \sin(2\pi \log(y))$ where f denotes the density of the log-normal $\mathbf{X}$. In fact, [37] shows that there exist uncountably many random variables that have the same distribution as $\mathbf{X}$.

3 Iterated Integrals as Non-Commutative Monomials

We will now go beyond the classical setting and consider monomials for sequential data. For simplicity we consider paths and sequence here in $\mathcal{X} = \mathbb{R}^d$ but in Sect. 6 we discuss infinite-dimensional state spaces. While we work with discrete sequences in $\mathcal{X}_{\text{seq}}$ in practice, it is often convenient to formulate our objects for continuous paths. However, we can naturally view a discrete sequence as continuous path by piecewise linear interpolation. Indeed, suppose we have a sequence $\mathbf{x} = (\mathbf{x}_0, \dots, \mathbf{x}_L) \in \mathcal{X}_{\text{seq}}$. By abuse of notation, we define the corresponding continuous path $\mathbf{x} : [0, L] \to \mathbb{R}^d$ by

$$\mathbf{x}(t) = (i - t)\mathbf{x}_{i-1} + (t - i + 1)\mathbf{x}_i, \qquad t \in [i - 1, i]. \tag{6}$$

In the continuous setting, we will consider bounded variation paths

$$\mathcal{X}_{1\text{-var}} := \{\mathbf{x} \in C([0, T], \mathcal{X}) \ : \ T > 0, \ \|\mathbf{x}\|_1 < \infty, \ \mathbf{x}(0) = 0\}$$

where

$$\|\mathbf{x}\|_1 := \sup_{\substack{0=t_0<\cdots<t_n=T \\ n\in\mathbb{N}}} \sum_{i=1}^{n} |\mathbf{x}(t_i) - \mathbf{x}(t_{i-1})|$$

denotes the usual 1-variation of a path. We restrict to paths which begin at the origin since the method that we discuss in this article (in particular, the path signature) is translation invariant. Further, we take $\mathcal{X} = \mathbb{R}^d$ finite-dimensional for simplicity of exposition. There are several ways to put a topology on $\mathcal{X}_{1\text{-var}}$ and we take the same as in [15]. Finally, we note that Eq. 6 defines an embedding

$$\mathcal{X}_{\text{seq}} \hookrightarrow \mathcal{X}_{1\text{-var}}.$$

Chapter “A Primer on the Signature Method in Machine Learning” introduced the path signature,

$$\Phi_{\mathsf{Sig}}(\mathbf{x}) := \left(1, \int d\mathbf{x}, \int d\mathbf{x}^{\otimes 2}, \dots\right) \in \prod_{m\geq 0} (\mathbb{R}^d)^{\otimes m}$$

which maps bounded variation paths to a sequence of tensors. The degree m tensor is given as an m-times iterated integral of the underlying path $\mathbf{x} : [0, T] \to \mathbb{R}^d$,

$$\Phi_{\mathsf{Sig},m}(\mathbf{x}) := \int d\mathbf{x}^{\otimes m} := \int_{0<t_1<\cdots<t_m<T} d\mathbf{x}(t_1) \otimes \cdots \otimes d\mathbf{x}(t_m) \in (\mathbb{R}^d)^{\otimes m},$$

and by convention $(\mathbb{R}^d)^{\otimes 0} = \mathbb{R}$. The upper bound T in the domain of integration is given by the domain of the path $\mathbf{x} : [0, T] \to \mathbb{R}^d$. Alternatively, in coordinates, we can identify

$$(\mathbb{R}^d)^{\otimes m} \ni \int d\mathbf{x}^{\otimes m} \simeq \left(\int_{0<t_1<\cdots<t_m<T} d\mathbf{x}^{i_1}(t_1) \cdots d\mathbf{x}^{i_m}(t_m) \right)_{i_1,\ldots,i_m \in [d]}$$

where

$$\int_{0<t_1<\cdots<t_m<T} d\mathbf{x}^{i_1}(t_1) \cdots d\mathbf{x}^{i_m}(t_m) = \int_{0<t_1<\cdots<t_m<T} \dot{\mathbf{x}}^{i_1}(t_1) \cdots \dot{\mathbf{x}}^{i_m}(t_m) dt_1 \cdots dt_m$$

is well-defined as Riemann–Stieltjes integral since bounded variation paths are almost everywhere differentiable. For less regular paths, some care has to be taken about the definition of the iterated integrals that constitute the signature, but the same principles apply (see Sect. 9); for paths in infinite-dimensional state spaces the same construction applies and we discuss the case of Hilbert spaces in Sect. 6. Such iterated integrals can be viewed as a generalization of monomials, and here we highlight the similarities.

A Graded Description. One way to think about the signature map

$$\Phi_{\mathsf{Sig}} : \mathcal{X}_{1\text{-var}} \to \prod_{m \geq 0} (\mathbb{R}^d)^{\otimes m}, \quad \mathbf{x} \mapsto \left(1, \int d\mathbf{x}, \int d\mathbf{x}^{\otimes 2}, \ldots \right),$$

is that it is the natural generalization of the monomial map

$$\Phi_{\mathsf{Mon}} : \mathcal{X} \to \prod_{m \geq 0} (\mathbb{R}^d)^{\otimes m}, \quad \mathbf{x} \mapsto \left(1, \mathbf{x}, \frac{\mathbf{x}^{\otimes 2}}{2!}, \ldots \right),$$

from the domain $\mathcal{X}$ to $\mathcal{X}_{1\text{-var}}$. Both describe the underlying object (a vector or a path) as a series of tensors of increasing degree.

Signatures Generalize Monomials. The signature Φ_{Sig} is a strict generalization of Φ_{Mon}. For a straight line, $[0, 1] \ni t \mapsto t \cdot \mathbf{v}$ with $\mathbf{v} \in \mathbb{R}^d$, we have

$$\Phi_{\mathsf{Sig}}(t \mapsto t \cdot \mathbf{v}) = \left(1, \mathbf{v}, \frac{\mathbf{v}^{\otimes 2}}{2!}, \frac{\mathbf{v}^{\otimes 3}}{3!}, \ldots \right) = \Phi_{\mathsf{Mon}}(\mathbf{v}), \tag{7}$$

which follows immediately from the definition of Φ_{Sig} since

$$\int d(t \cdot \mathbf{v})^{\otimes m} = \int_{0<t_1<\cdots<t_m<1} \mathbf{v}^{\otimes m} dt_1 \cdots dt_m = \frac{\mathbf{v}^{\otimes m}}{m!}.$$

Despite the similarity in their definitions, there are also important differences to highlight.

A Genuine Non-linear Domain. Unlike $\mathbb{R}^d$, the domain $\mathcal{X}_{1\text{-var}}$ of Φ_{Sig} is not linear: there is no natural addition of elements in $\mathcal{X}_{1\text{-var}}$. In particular, there is no natural way to add paths $\mathbf{x} : [0, T] \to \mathbb{R}^d$ and $\mathbf{y} : [0, T'] \to \mathbb{R}^d$ for $T \neq T'$.

Asymmetric Tensors. There is an important difference in the co-domains of the maps Φ_{Sig} and Φ_{Mon}. The latter takes values in the subset of $\prod_{m\geq 0}(\mathbb{R}^d)^{\otimes m}$ of symmetric tensors. For example, if we identify the degree $m = 2$ tensors in Φ_{Mon} as $d \times d$ matrices, then $\mathbf{x}^{\otimes 2}$ is a symmetric matrix, since the (i, j) coordinate of $\mathbf{x}^{\otimes 2}$ is simply the usual product of the i-th and j-th coordinate of $\mathbf{x}$, and $\mathbf{x}^i \cdot \mathbf{x}^j = \mathbf{x}^j \cdot \mathbf{x}^i$.

However, in general the tensors in Φ_{Sig} are not symmetric. For example, by once again considering the (i, j) and (j, i) coordinate of $\int d\mathbf{x}^{\otimes 2}$, we have in general that

$$\int_{0<t_1<t_2<T} d\mathbf{x}^i(t_1)d\mathbf{x}^j(t_2) \neq \int_{0<t_1<t_2<T} d\mathbf{x}^j(t_1)d\mathbf{x}^i(t_2).$$

Non-commutative Monomials. Because Φ_{Sig} results in non-symmetric tensors, the algebraic structure behind these monomials is non-commutative. Given two tensors $\mathbf{v} \in (\mathbb{R}^d)^{\otimes m}$ and $\mathbf{w} \in (\mathbb{R}^d)^{\otimes k}$, the tensor product $\mathbf{v} \otimes \mathbf{w} \in (\mathbb{R}^d)^{\otimes (m+k)}$ is defined coordinate-wise as

$$(\mathbf{v} \otimes \mathbf{w})^{i_1,\dots,i_{m+k}} := \mathbf{v}^{i_1,\dots,i_m} \cdot \mathbf{w}^{i_{m+1},\dots,i_{m+k}}.$$

In general, the tensor product is *non-commutative*, $\mathbf{v} \otimes \mathbf{w} \neq \mathbf{w} \otimes \mathbf{v}$. However, when we are restricted to symmetric tensors such as in Φ_{Mon}, the tensor product is commutative

$$(\mathbf{x}^{\otimes m} \otimes \mathbf{x}^{\otimes k})^{i_1,\dots,i_{m+k}} = \mathbf{x}^{i_1} \cdots \mathbf{x}^{i_{m+k}} = (\mathbf{x}^{\otimes k} \otimes \mathbf{x}^{\otimes m})^{i_1,\dots,i_{m+k}}.$$

Remark 3.1 We refer to Φ_{Sig} as the signature (feature) map but developing a path into a series of iterated integrals is a classical construction that has been studied by many communities but known under different names such as Chen–Fliess series, Magnus expansion, time-ordered exponential, path exponential, chronological calculus etc. Modern origins go back to Chen's work in [10–12] which can be thought as an extension of classical Volterra series [61]. The rich Lie-algebraic properties of the signature received much attention in the study of free Lie algebras [55] and in the 1980's, control theorists used Φ_{Sig} extensively to linearize controlled differential equations [6, 27]. Much of these results were then generalized in rough path theory [49] to deal with "rough" trajectories such as the sample of paths of stochastic processes where standard integration does not apply. A detailed history is beyond the scope of this article, but the above references point to a rich literature of disparate communities which have used (variations of) what we denote Φ_{Sig} for their own research agenda.

4 Universality and Characteristicness of the Signature

In order to further establish the relationship between the monomials $\Phi_{\mathsf{Mon}} : \mathbb{R}^d \to \prod_{m\geq 0}(\mathbb{R}^d)^{\otimes m}$ and the signature $\Phi_{\mathsf{Sig}} : \mathcal{X}_{1\text{-var}} \to \prod_{m\geq 0}(\mathbb{R}^d)^{\otimes m}$, we now consider how the signature can be used to study functions and measures on $\mathcal{X}_{1\text{-var}}$. In particular, given some $\mathbf{w} \in \bigoplus_{m\geq 0}(\mathbb{R}^d)^{\otimes m}$, we define a function $\langle \mathbf{w}, \Phi_{\mathsf{Sig}}(\cdot)\rangle : \mathcal{X}_{1\text{-var}} \to \mathbb{R}$ as before; in coordinates,

$$\langle \mathbf{w}, \Phi_{\mathsf{Sig}}(\mathbf{x})\rangle := \sum_{m=0}^{M} \sum_{i_1,\dots,i_m\in[d]} \mathbf{w}^{i_1,\dots,i_m} \int_{0<t_1<\cdots<t_m<T} d\mathbf{x}^{i_1}(t_1)\cdots d\mathbf{x}^{i_m}(t_m).$$

A direct calculation shows that the product of two such linear functionals can be expressed again as a linear functional of Φ_{Sig}. More formally, for $\mathbf{w}_1, \mathbf{w}_2 \in \bigoplus_{m\geq 0}(\mathbb{R}^d)^{\otimes m}$ there exists some $\mathbf{w}_{1,2} \in \bigoplus_{m\geq 0}(\mathbb{R}^d)^{\otimes m}$ such that

$$\langle \mathbf{w}_1, \Phi_{\mathsf{Sig}}(\mathbf{x})\rangle\langle \mathbf{w}_2, \Phi_{\mathsf{Sig}}(\mathbf{x})\rangle = \langle \mathbf{w}_{1,2}, \Phi_{\mathsf{Sig}}(\mathbf{x})\rangle \tag{8}$$

for all $\mathbf{x} \in \mathcal{X}_{1\text{-var}}$. Put differently, products of (non-commutative) polynomials are again (non-commutative) polynomials. Although the proof of this is just a direct application of integration by parts, the algebraic (shuffle) structure of $\mathbf{w}_{1,2}$ gives rise to many interesting algebraic questions but is beyond the scope of this short introduction. The primary point is that (8) shows that linear functionals of paths induced by $\mathbf{w} \in \bigoplus_{m\geq 0}(\mathbb{R}^d)^{\otimes m}$ form an algebra. If we additionally have the point-separating property, in which

$$\mathbf{x} \mapsto \Phi_{\mathsf{Sig}}(\mathbf{x})$$

is injective, then we can immediately generalize Proposition 2.1 from compact subsets in $\mathbb{R}^d$ to compact subsets in $\mathcal{X}_{1\text{-var}}$. However, in general, $\mathbf{x} \mapsto \Phi_{\mathsf{Sig}}(\mathbf{x})$ is only injective up to tree-like equivalence. We refer to [34] for details on tree-like equivalence.

We can work around this tree-like equivalence in the following Proposition 4.1 by introducing a time coordinate: rather than using $t \mapsto \mathbf{x}(t) \in \mathbb{R}^d$, we consider

$$t \mapsto \bar{\mathbf{x}}(t) := (t, \mathbf{x}(t)) \in \mathbb{R}^{d+1}. \tag{9}$$

When we consider such time-parametrized paths, the signature Φ_{Sig} is injective, and we obtain the following generalization of Proposition 2.1.

Proposition 4.1 *Let $\mathcal{K} \subset \mathcal{X}_{1\text{-}var}$ be a compact set and denote with $\bar{\mathcal{K}} = \{\bar{\mathbf{x}} : \mathbf{x} \in \mathcal{K}\}$ the set of paths in $\mathcal{K}$ augmented with a time coordinate from Eq. 9. Then for every $f \in C(\bar{\mathcal{K}}, \mathbb{R})$ and $\epsilon > 0$ there exists a $\mathbf{w} \in \bigoplus(\mathbb{R}^{d+1})^{\otimes m}$ such that*

$$\sup_{\mathbf{x}\in\mathcal{K}} |f(\bar{\mathbf{x}}) - \langle \mathbf{w}, \Phi_{\mathsf{Sig}}(\bar{\mathbf{x}})\rangle| < \epsilon.$$

Let $\mathbf{X}, \mathbf{X}_1, \mathbf{X}_2, \ldots$ be a sequence of $\mathcal{K}$-valued random variables, that is, each $\mathbf{X}_i = (\mathbf{X}_i(t))_{t\in[0,T_i]}$ is a stochastic process. Then the expected signature map

$$\mathrm{Law}(\bar{\mathbf{X}}) \mapsto \mathbb{E}[\Phi_{\mathsf{Sig}}(\bar{\mathbf{X}})]$$

is injective and the following are equivalent:

1. $\forall f \in C(\bar{\mathcal{K}}, \mathbb{R}), \quad \lim_{n\to\infty} \mathbb{E}[f(\bar{\mathbf{X}}_n)] = \mathbb{E}[f(\bar{\mathbf{X}})]$,
2. $\forall m \in \mathbb{N}, \quad \lim_{n\to\infty} \mathbb{E}[\int d\bar{\mathbf{X}}_n^{\otimes m}] = \mathbb{E}[\int d\bar{\mathbf{X}}^{\otimes m}]$.

This proposition is a direct generalization of Proposition 2.1; in fact, it can be proved in the same way. The only difference is that one needs to show $\mathbf{x} \mapsto \Phi_{\mathsf{Sig}}(\bar{\mathbf{x}})$ is injective. However, this is not too difficult, see for example [47]. Without the augmentation with a time coordinate, the same statement holds when $\mathcal{K}$ is replaced by set of tree-like equivalence classes of paths.[6] With all of these properties in mind, we can see that the signature map behaves very similarly to classical monomials, and we summarize these results in Table 1.

Table 1 Comparison between classical monomials and the signature

	Monomial Map	Path Signature
Domain	$\mathcal{X} = \mathbb{R}^d$	$\mathcal{X}_{1\text{-var}}$
Co-domain	$\prod_{m\geq 0}(\mathbb{R}^d)^{\otimes m}$	$\prod_{m\geq 0}(\mathbb{R}^d)^{\otimes m}$
Map	$\Phi_{\mathsf{Mon}}(\mathbf{x}) = \left(\frac{\mathbf{x}^{\otimes m}}{m!}\right)_{m\geq 0}$	$\Phi_{\mathsf{Sig}}(\mathbf{x}) = \left(\int d\mathbf{x}^{\otimes m}\right)_{m\geq 0}$
Type of Tensors	symmetric	non-symmetric
Continuity	w.r.t. Euclidean metric	w.r.t. 1-variation metric
Factorial Decay	$\|\Phi_{\mathsf{Mon},m}(\mathbf{x})\| \leq \frac{\|\mathbf{x}\|^m}{m!}$	$\|\Phi_{\mathsf{Sig},m}(\mathbf{x})\| \leq \frac{\|\mathbf{x}\|_1^m}{m!}$
Scaling	$\Phi_{\mathsf{Mon},m}(\lambda\mathbf{x}) = \lambda^m\Phi_{\mathsf{Mon},m}(\mathbf{x})$	$\Phi_{\mathsf{Sig},m}(\lambda\mathbf{x}) = \lambda^m\Phi_{\mathsf{Sig},m}(\mathbf{x})$
	compact $\mathcal{K} \subset \mathcal{X}$	compact $\mathcal{K} \subset \mathcal{X}_{1\text{-var}}$
Universality	w.r.t. $C(\mathcal{K}, \mathbb{R})$	w.r.t. $C(\bar{\mathcal{K}}, \mathbb{R})$
Characteristicness	w.r.t. $\mathcal{P}(\mathcal{K})$	w.r.t. $\mathcal{P}(\bar{\mathcal{K}})$

[6] Many choices of topologies for paths and unparametrized path are possible; for a detailed study for unparametrized paths, see [9].

5 The Good, the Bad, and the Ugly

While classical monomials have many attractive properties, they also come with some limitations that are inherited by the "non-commutative monomials", i.e. the iterated integrals, that constitute Φ_{Sig}.

Good. Monomials are building blocks to construct more complicated functions and have a well-understood algebraic structure. For example, linear combinations of monomials (i.e. polynomials) are closed under multiplication and form a ring, and any continuous function with compact support can be uniformly approximated by polynomials.

Bad.

- Monomials and hence also polynomials fail miserably on non-compact domains; they are great at providing local approximations but explode on unbounded domains.
- On unbounded domains, the moment sequence may not exist. But even if it does, that is all moments exist, the moment sequence does in general not uniquely characterize the law of random variable.
- Computational complexity: The tensor $\mathbf{x}^{\otimes m} \in (\mathbb{R}^d)^{\otimes m}$ has $\binom{d+m-1}{m} = O(d^m)$ coordinates so that the computational cost becomes quickly prohibitive for moderate d which makes regression $f(\mathbf{x}) \approx \langle \mathbf{w}, \Phi_{\mathsf{Mon}}(\mathbf{x}) \rangle$ and working with moments $\mathbb{E}[\mathbf{X}^{\otimes m}]$ too expensive.
- Non-robustness: Small changes in the underlying probability measure can lead to large changes in the moment sequence. This becomes an issue for estimation; e.g. one outlier among the sample can completely destroy moment estimates.

Ugly. While monomials can theoretically approximate functions and capture probability distributions, they can be a poor choice in practice. In data science, arguably the most successful method is to learn the approximation from the data itself. In practice, this means providing a large parameterized family of functions and optimizing over the parameter when applied to data and standard monomials are rarely the optimal choice.

Because Φ_{Sig} generalizes Φ_{Mon}, it shares the same issues. However, all these issues are well-known and modern statistics and machine learning have developed tools to address these shortcomings. The remainder of this article will focus on the following two tools and develop them for Φ_{Sig}.

- **Kernelization** allows us to simultaneously consider a rich set of non-linearities while avoiding the combinatorial explosion in the computation of monomials. This is done by sidestepping the computation of individual monomials by obtaining efficient algorithms to directly compute the inner product between monomials.
- **Robust Statistics** was developed in the 1980s [40] to provide both a formalism to quantify the robustness of a given statistic and tools to make a given statistic

robust in this precise mathematical sense. We use this to simultaneously resolve the issues with Φ_{Mon} and Φ_{Sig} on non-compact sets and deal with their general non-robustness.

The problems of extending kernelization and robustness methods to the path signature Φ_{Sig} address independent issues although they can be combined. We discuss kernelization in Sects. 6 and 7; robustification is discussed in Sect. 8.

6 Kernel Learning

We briefly recall the main ideas behind kernelization and we refer to [17, 62] for a detailed introduction to kernel learning.

Kernelization in a Nutshell

Given a set $\mathcal{K}$, we want a feature map $\varphi : \mathcal{K} \to H$ that injects elements of $\mathcal{K}$ into a Hilbert space H such that φ provides sufficient non-linearities so that firstly, linear functionals of φ, viewed as a map $\mathbf{x} \mapsto \langle \mathbf{w}, \varphi(\mathbf{x}) \rangle$, are expressive enough to approximate a rich set of functions and, secondly, such that $\mathbb{E}[\varphi(\mathbf{X})]$ captures the law of a $\mathcal{K}$-valued random variable $\mathbf{X}$. The examples we encountered are $\mathcal{K} \subset \mathcal{X} = \mathbb{R}^d$ with $\varphi = \Phi_{\mathsf{Mon}}$ as the feature map and $\mathcal{K} \subset \mathcal{X}_{1\text{-var}}$ with $\varphi = \Phi_{\mathsf{Sig}}$ as the feature map. Even for $\mathcal{K} = \mathbb{R}$ there are many choices of φ and H so let us take a step back and simply ask what could be a good candidate for the feature space H.

We want H to be linear, large, and canonical with respect to the underlying data space $\mathcal{K}$. A canonical linear space that we can associate with any set $\mathcal{K}$ is the space of real-valued functions $\mathbb{R}^{\mathcal{K}} := \{f : \mathcal{K} \to \mathbb{R}\}$ on $\mathcal{K}$. Hence, we take H to be a subset of $\mathbb{R}^{\mathcal{K}}$, that is, H will be a space of real-valued functions on $\mathcal{K}$. Consequently the feature map φ will be function-valued,

$$\mathcal{X} \ni \mathbf{x} \mapsto \varphi(\mathbf{x}) := \mathrm{k}(\mathbf{x}, \cdot) \in H \subset \mathbb{R}^{\mathcal{K}}. \tag{10}$$

While this has the potential to provide a rich set of "non-linearities" by embedding $\mathcal{K}$ in the (generically infinite-dimensional) space H, the drawback is that learning algorithms require the evaluation of $\varphi(\mathbf{x}) = \mathrm{k}(\mathbf{x}, \cdot)$ which makes the direct use of (10) infeasible. Kernel learning rests on three observations.

1. If we assume that k is positive definite, we can identify H as the Reproducing Kernel Hilbert Space (RKHS) induced by k, that is H is the closure of

$$\left\{\mathbf{x} \mapsto \sum_{i=1}^{n} \mathrm{k}(\mathbf{x}_i, \mathbf{x}) \,:\, \mathbf{x}_i \in \mathcal{K},\ n \in \mathbb{N}\right\} \subset \mathbb{R}^{\mathcal{K}}.$$

   This induces the reproducing property, where $\langle f, \mathrm{k}(\mathbf{x}, \cdot)\rangle_H = f(\mathbf{x})$ for $f \in H$. As a consequence of the reproducing property, we have $\langle \varphi(\mathbf{x}), \varphi(\mathbf{y})\rangle_H = \mathrm{k}(\mathbf{x}, \mathbf{y})$. Hence, we may think of k as a "nonlinear inner product".

2. The evaluation of $\mathrm{k}(\mathbf{x}, \mathbf{y})$ can often be made computationally cheap even though it is the inner product of elements $\varphi(\mathbf{x})$ and $\varphi(\mathbf{y})$ in (the often infinite-dimensional) H. For example, arguably the most popular kernel is the radial basis function (RBF) kernel, $\mathrm{k}(\mathbf{x}, \mathbf{y}) = \exp(-\sigma|\mathbf{x} - \mathbf{y}|^2)$, which is the inner product of an infinite-dimensional map $\varphi(\mathbf{x})$. That an inner product of feature maps φ is computationally cheap, is often referred to as the "kernel trick".
3. Many learning algorithms can be reformulated so that they do not require the evaluation of $\varphi(\mathbf{x})$ but only the computation of the inner product $\mathrm{k}(\mathbf{x}, \mathbf{y}) = \langle \varphi(\mathbf{x}), \varphi(\mathbf{y}) \rangle$.

To summarize, for a given dataset $\mathcal{D} = \{\mathbf{x}_1, \dots, \mathbf{x}_N\} \subset \mathcal{X}$, kernelization allows us to work with highly non-linear feature maps $\varphi(\mathbf{x})$ taking values in an infinite dimensional space H by only evaluating the Gram matrix $(\mathrm{k}(\mathbf{x}_i, \mathbf{x}_j))_{i,j \in [N]}$.

Remark 6.1 Kernelization leverages duality for a complexity trade-off: while we do not have to worry about the dimension of H in the computation of kernels, we must compute the entire Gram martrix which scales quadratically in the number of samples N. This is a direct consequence of the duality that underlies kernelization.

To gain some intuition we revisit and kernelize the classical monomial feature map from Sect. 2,

$$\Phi_{\text{Mon}} : \mathbb{R}^d \to \prod_{m \geq 0} (\mathbb{R}^d)^{\otimes m}, \quad \mathbf{x} \mapsto \left(1, \mathbf{x}, \frac{\mathbf{x}^{\otimes 2}}{2}, \dots\right).$$

Further, we denote with

$$\Phi_{\text{Mon},:M} = \mathbb{R}^d \to \prod_{m \geq 0} (\mathbb{R}^d)^{\otimes m}, \quad \mathbf{x} \mapsto \left(1, \mathbf{x}, \frac{\mathbf{x}^{\otimes 2}}{2}, \dots, \frac{\mathbf{x}^{\otimes M}}{M!}, 0, 0, \dots\right),$$

the same map, but truncated at level M; here $0 \in (\mathbb{R}^d)^{\otimes M}$ denotes the degree M tensor that has as entries all zeroes. The co-domain of Φ_{Mon} and $\Phi_{\text{Mon},:M}$ is the same, but the latter has only finitely many non-zero entries. We refer to $\Phi_{\text{Mon},:M}$ as truncated at M.

Inner Product of Tensors

There is a natural inner product on $(\mathbb{R}^d)^{\otimes m}$ by setting

$$\langle \mathbf{s}_m, \mathbf{t}_m \rangle_m := \sum_{i_1, \dots, i_m = 1}^{d} \mathbf{s}_m^{i_1, \dots, i_m} \cdot \mathbf{t}_m^{i_1, \dots, i_m},$$

where $\mathbf{s}_m, \mathbf{t}_m \in (\mathbb{R}^d)^{\otimes m}$. This inner product can be extended to the subset of $\prod_{m \geq 0} (\mathbb{R}^d)^{\otimes m}$ as

$$\langle \mathbf{s}, \mathbf{t} \rangle := \sum_{m=0}^{\infty} \langle \mathbf{s}_m, \mathbf{t}_m \rangle_m,$$

for which the above sum converges; where $\mathbf{s} = (\mathbf{s}_0, \mathbf{s}_1, \ldots), \mathbf{t} = (\mathbf{t}_0, \mathbf{t}_1, \ldots) \in \prod_{m \geq 0} (\mathbb{R}^d)^{\otimes m}$. Hence, if we denote by $\mathbf{e}_i \in \mathbb{R}^d$ the vector that is all zeros except with a 1 at entry i, then an orthogonal basis of $\prod (\mathbb{R}^d)^{\otimes m}$ is given by

$$\{\mathbf{e}_{i_1} \otimes \cdots \otimes \mathbf{e}_{i_m} : i_1, \ldots, i_m \in [d], \ m \in \mathbb{N}\}.$$

The following is an elementary but key property that we use throughout: for all $\mathbf{x}_1, \ldots, \mathbf{x}_m, \mathbf{y}_1, \ldots, \mathbf{y}_m \in \mathbb{R}^d$, we have

$$\langle \mathbf{x}_1 \otimes \cdots \otimes \mathbf{x}_m, \mathbf{y}_1 \otimes \cdots \otimes \mathbf{y}_m \rangle = \langle \mathbf{x}_1, \mathbf{y}_1 \rangle \cdots \langle \mathbf{x}_m, \mathbf{y}_m \rangle \tag{11}$$

This already hints at how we can reduce the computational complexity. A naive evaluation of the left-hand side requires us to store the d^m coordinates of the tensors $\mathbf{x}_1 \otimes \cdots \otimes \mathbf{x}_m$ and $\mathbf{y}_1 \otimes \cdots \otimes \mathbf{y}_m$ and then compute the inner product, hence requires $O(d^m)$ memory. In contrast, evaluating the right hand side requires just $O(dm)$ memory.

Example 6.2 (Avoiding the Computational Complexity) Let $\mathcal{X} \subset \mathbb{R}^d$ and define

$$\mathrm{k}^0_{\mathsf{Mon},:M} : \mathcal{X} \times \mathcal{X} \to \mathbb{R}, \quad \mathrm{k}^0_{\mathsf{Mon},:M}(\mathbf{x}, \mathbf{y}) := \langle \Phi_{\mathsf{Mon},:M}(\mathbf{x}), \Phi_{\mathsf{Mon},:M}(\mathbf{y}) \rangle.$$

We use the 0 to denote that it is the "0"-th version since we are going to improve it in the next example. But for now note that using (11) we get

$$\mathrm{k}^0_{\mathsf{Mon},:M}(\mathbf{x}, \mathbf{y}) = \sum_{m=0}^{M} \frac{\langle \mathbf{x}^{\otimes m}, \mathbf{y}^{\otimes m} \rangle}{(m!)^2} = \sum_{m=0}^{M} \frac{\langle \mathbf{x}, \mathbf{y} \rangle^m}{(m!)^2}, \tag{12}$$

which already shows that the computational cost is linear in d. We can make this even more efficient, by recalling *Horner's scheme*: the formula (12) is a polynomial in $\langle \mathbf{x}, \mathbf{y} \rangle$ which can be efficiently computed as

$$k^0_{\mathsf{Mon},:M}(\mathbf{x}, \mathbf{y}) = 1 + \langle \mathbf{x}, \mathbf{y} \rangle \left(1 + \frac{\langle \mathbf{x}, \mathbf{y} \rangle}{2^2} \left(1 + \ldots \frac{\langle \mathbf{x}, \mathbf{y} \rangle}{(M-1)^2} \left(1 + \frac{\langle \mathbf{x}, \mathbf{y} \rangle}{M^2}\right)\right)\right). \tag{13}$$

This "Horner-type" evaluation has three advantages:

1. it requires only M additions and M multiplications in contrast to the naive evaluation of (12);

2. it is numerically stable, again in contrast to (12) which adds powers of scalars which live on different scales and thus makes floating point arithmetic challenging; and
3. it allows us to replace the inner product $\langle \mathbf{x}, \mathbf{y} \rangle$ with the kernel $\mathrm{k}(\mathbf{x}, \mathbf{y})$ which we describe in the example below.

Example 6.3 (Composing Non-Linearities) We now have an efficient way to evaluate $\mathrm{k}^0_{\mathsf{Mon}}$ but it is well-known that this kernel does not perform well in practice [62]. The reason is the aforementioned fact, that although monomials have nice theoretical properties, they suffer from a lack of expressivity; in practice they are outperformed by other kernels on $\mathbb{R}^d$ that are inner products of other non-linearities, e.g. the RBF kernel.

A natural way to address this is to first apply a non-linear map

$$\varphi : \mathbb{R}^d \to H$$

and only subsequently compute the monomials in H, that is we consider

$$\Phi_{\mathsf{Mon},:M}(\varphi(\mathbf{x})) \equiv \left(1, \varphi(\mathbf{x}), \frac{\varphi(\mathbf{x})^{\otimes 2}}{2!}, \ldots, \frac{\varphi(\mathbf{x})^{\otimes M}}{M!}, 0, 0, \ldots\right). \tag{14}$$

This preserves the algebraic structure but additionally allows for a far more expressive feature map since φ adds (infinitely-many) additional non-linearities via φ. If we take this underlying map to be the identity, $\varphi = \mathrm{id}$, then (14) coincides with (4); but in general we can use arbitrary non-linearities φ. Unfortunately, a direct use of the map $\Phi_{\mathsf{Mon}} \circ \varphi$ is impossible in practice since we cannot directly evaluate φ when H is infinite-dimensional, even for $M = 2$. The key remark now is that if we take $\varphi(\mathbf{x}) = \mathrm{k}(\mathbf{x}, \cdot)$ to be the feature map of a kernel $\mathrm{k} : \mathbb{R}^d \times \mathbb{R}^d \to \mathbb{R}$ that has H as RKHS, then *the inner product*

$$\begin{aligned} \mathrm{k}_{\mathsf{Mon},:M}(\mathbf{x}, \mathbf{y}) &:= \langle \Phi_{\mathsf{Mon},:M} \circ \varphi(\mathbf{x}), \Phi_{\mathsf{Mon},:M} \circ \varphi(\mathbf{y}) \rangle \\ &= \langle \Phi_{\mathsf{Mon},:M} \circ \mathrm{k}(\mathbf{x}, \cdot), \Phi_{\mathsf{Mon},:M} \circ \mathrm{k}(\mathbf{y}, \cdot) \rangle \end{aligned} \tag{15}$$

can be computed exactly! In fact, it just amounts to replacing the Euclidean inner product $\langle \mathbf{x}, \mathbf{y} \rangle$ in Eqs. (12) and (13) with $\mathrm{k}(\mathbf{x}, \mathbf{y})$. Note the kernel $\mathrm{k}_{\mathsf{Mon},:M}$ depends on M and φ; the optimal selection of these is important and we discuss this further in Sect. 7.3.

While the calculation itself is elementary, it is remarkable that this can be done despite the genuine infinite-dimensional co-domain of $\Phi_{\mathsf{Mon},:M} \circ \varphi$; such a trade-off of losing direct access to the feature map but being able to compute an inner product is often called a kernel trick. Thus the kernel $\mathrm{k}_{\mathsf{Mon},:M}$ inherits the nice algebraic properties of monomials and also the power to access an infinite-dimensional H via k.

7 The Signature Kernel

We now apply the same kernelization approach as in Sect. 6 to the signature feature map Φ_{Sig} in order to obtain a computable kernel equipped with further non-linearities. We begin by considering a kernel

$$\mathrm{k} : \mathbb{R}^d \times \mathbb{R}^d \to \mathbb{R}, \tag{16}$$

on the state space of our paths; henceforth we refer to k as the *static kernel*. Instead of lifting elements of $\mathbb{R}^d$ into the RKHS H of k, we use it to map the path $\mathbf{x}$,

$$t \mapsto \mathbf{x}(t) \in \mathbb{R}^d,$$

into a path $\mathrm{k}_{\mathbf{x}}$,

$$t \mapsto \mathrm{k}_{\mathbf{x}}(t) := \mathrm{k}(\mathbf{x}(t), \cdot) \in H,$$

that now evolves in H. For a generic static kernel $\mathrm{k} : \mathcal{X} \times \mathcal{X} \to \mathbb{R}$, its RKHS H will be infinite-dimensional, hence the path $\mathrm{k}_{\mathbf{x}} : t \mapsto \mathrm{k}_{\mathbf{x}}(t)$ evolves in an infinite-dimensional space. Our aim is to use the signature of the path $\mathrm{k}_{\mathbf{x}}$,

$$\Phi_{\mathsf{Sig}}(\mathrm{k}_{\mathbf{x}}) = \left(1, \int d\,\mathrm{k}_{\mathbf{x}}, \int d\,\mathrm{k}_{\mathbf{x}}^{\otimes 2}, \dots\right), \tag{17}$$

and its truncated version

$$\Phi_{\mathsf{Sig},:M}(\mathrm{k}_{\mathbf{x}}) := \left(1, \int d\,\mathrm{k}_{\mathbf{x}}, \int d\,\mathrm{k}_{\mathbf{x}}^{\otimes 2}, \dots, \int d\,\mathrm{k}_{\mathbf{x}}^{\otimes M}, 0, 0, \dots\right) \tag{18}$$

as a way to represent the path $\mathbf{x}$. Once again, the intuition is that

$$\Phi_{\mathsf{Sig}}(\mathrm{k}_{\mathbf{x}}) \in \prod_{m \geq 0} H^{\otimes m}$$

is much more expressive than $\Phi_{\mathsf{Sig}}(\mathbf{x})$ since it computes iterated integrals of the path after the non-linearities of k have been applied so we will have a much richer set of coordinates (of iterated integrals) to choose from when k is non-linear and the RKHS of k is very large.

Again the bad news is, of course, that even for $m = 2$ storing $\int d\,\mathrm{k}_{\mathbf{x}}^{\otimes 2} \in H^{\otimes 2}$ exactly is in general infeasible for the interesting case when H is infinite-dimensional. The main insight in [42] is that, perhaps surprisingly, the inner product

$$\mathrm{k}_{\mathsf{Sig},:M}(\mathbf{x}, \mathbf{y}) := \langle \Phi_{\mathsf{Sig},:M}(\mathrm{k}_{\mathbf{x}}), \Phi_{\mathsf{Sig},:M}(\mathrm{k}_{\mathbf{y}}) \rangle \tag{19}$$

can be computed exactly when $M < \infty$ for piecewise linear paths $\mathbf{x}, \mathbf{y}$ and, perhaps less surprisingly, inherits many of the attractive properties of the signature features.

Theorem 7.1 ([42, Remark B.10]) *Let* $\mathrm{k} : \mathbb{R}^d \times \mathbb{R}^d \to \mathbb{R}$ *be a kernel and let* $M \in \mathbb{N}$ *and let*

$$\mathrm{k}_{\mathsf{Sig},:M} : \mathcal{X}_{1\text{-}var} \times \mathcal{X}_{1\text{-}var} \to \mathbb{R} \tag{20}$$

be as defined in (19). *Then* $\mathrm{k}_{\mathsf{Sig},:M}(\mathbf{x}, \mathbf{y})$ *can be evaluated exactly for piecewise linear paths* $\mathbf{x} = (\mathbf{x}_1, \ldots, \mathbf{x}_{L_\mathbf{x}})$, $\mathbf{y} = (\mathbf{y}_1, \ldots, \mathbf{y}_{L_\mathbf{y}})$ *in* $O(M^3 L^2)$ *computational steps where* $L = \max(L_\mathbf{x}, L_\mathbf{y})$.

An explicit algorithm is given in [42, Algorithm 6], where the truncation degree and approximation order are both set to M. We provide a summary of this dynamic programming algorithm in Sect. 7.2. Even in the case of $M = \infty$, when $\mathrm{k}_{\mathsf{Sig}}(\mathbf{x}, \mathbf{y}) \equiv \langle \Phi_{\mathsf{Sig},:\infty}(\mathbf{x}), \Phi_{\mathsf{Sig},:\infty}(\mathbf{y}) \rangle \equiv \langle \Phi_{\mathsf{Sig}}(\mathbf{x}), \Phi_{\mathsf{Sig}}(\mathbf{y}) \rangle$, good approximations of $\mathrm{k}_{\mathsf{Sig}}(\mathbf{x}, \mathbf{y})$ exist for piecewise linear paths via PDE methods, which is also discussed in Sect. 7.2.

7.1 *Approximations*

The above Theorem 7.1 provides a powerful method to compute signature kernels and allows for many applications. However, besides the exact evaluation of $\mathrm{k}_{\mathsf{Sig},:M}$, approximations are important for various reasons.

Gram Matrix Approximations. In a machine learning context, the computational bottleneck is not just the evaluation of a single $\mathrm{k}_{\mathsf{Sig},:M}(\mathbf{x}, \mathbf{y})$ but the computation of a Gram matrix

$$G = (\mathrm{k}_{\mathsf{Sig},:M}(\mathbf{x}, \mathbf{y}))_{\mathbf{x},\mathbf{y} \in \mathcal{D}}$$

where $\mathcal{D} \subset \mathcal{X}_{\text{seq}}$ is the finite set of data we are given. However, naively evaluating each of the $|\mathcal{D}|^2$ entries of this matrix is typically not feasible since the datasets can be very large. General kernel learning approaches such as the Nyström-method can be applied and replace the original Gram matrix by an approximation with a low-rank matrix which can have linear complexity in $|\mathcal{D}|$. Similarly, in a Gaussian process (GP) context, one not only needs to approximate G but also calculate its inverse as part of the Bayesian update rule. This scales even worse than quadratic in the number of data points, and again some generic GP methods have been developed to deal with this.

Untruncated Signature Kernel Approximations. We regard the truncation level $M \in \mathbb{N}_\infty$ as a hyperparameter that should be tuned by the data. While Theorem 7.1 gives an exact algorithm for finite M (in practice, the algoritm of

Theorem 7.1 is feasible for $M < 10$), one would like to know how well we can approximate the signature kernel with a large M or even $M = \infty$.

Discrete to Continuous Time Approximation. Sometimes the data is genuinely a discrete-time sequence, but sometimes the underlying object is a continuous-time path of which we are only given discrete-time samples. The signature kernel is well-defined for continuous-time paths, hence it can be useful to quantify the time-discretization error. These limiting paths are typically not of bounded variation, for example Brownian motion is only of finite p-variation for $p > 2$, so some care has to be taken.

All three of the above points address different issues, and the following discusses them in more detail.

Gram-Matrix Approximations

A general issue in kernel learning is that the computation of the Gram matrix scales quadratically in the number $N = |\mathcal{D}|$ of data points $\mathcal{D}$. In particular, each data point $\mathbf{x} \in \mathcal{D}$ corresponds to a sequence, hence we would require N^2 evaluations of the signature kernel. Even if a single evaluation of $\mathrm{k}_{\mathsf{Sig},:M}$ is cheap, this quadratic complexity becomes quickly prohibitive. A generic method that applies to any kernel k is the Nyström approximation which, in its simplest form, approximates a general kernel Gram matrix

$$G = (\mathrm{k}(\mathbf{x}_i, \mathbf{y}_j))_{i,j=1,\dots,N}$$

by sampling uniformly at random a subset $J \subset [N]$ of size r and define

$$\tilde{G} = CW^{-1}C^{\top}$$

where $C = (\mathrm{k}(\mathbf{x}_i, \mathbf{x}_j))_{i=1,\dots,N,\, j\in J}$ and $W = (\mathrm{k}(\mathbf{x}_i, \mathbf{x}_j))_{i,j\in J}$. Note that $\tilde{G}$ is a random matrix of rank r, hence the computational cost scales as r^2 which can be huge reduction if $r \ll N$. Besides uniform subsampling of columns, more sophisticated Nyström approximations schemes are available [21]. All these generic methods can be applied to the signature kernel: selecting J amounts to selecting a subset of the sequences in our dataset. However, this only addresses the complexity in the number N of sequences, but not the other computational bottleneck which is the quadratic complexity in sequence length L. A simple way to address the latter is to subsample each sequence, which again, can work but is a brute force approach. A more elaborate option is to also apply the Nyström approximation at the level of time-points; [42, Algorithm 5] shows that these two Nyström approximations can be done simultaneously: informally, one does not only reweight the importance of sequences but also of time-points within these sequences.

Another situation where the cost of the matrix calculations become quickly too prohibitive is for Gaussian processes. We discuss Gaussian processes with signature kernels in Sect. 9 and again generic GP approximation methods apply. However, we can further reduce the complexity of our algorithms by exploiting the specific structure of the signature kernel. This can be achieved by selecting special tensors [69] or by using diagonalization arguments [45].

Truncation Approximation

The signature tensors decay factorially (see Table 1), and this allows us to quantify the difference between truncated and untruncated signatures as follows.

Proposition 7.2 *Let $\Phi_{\mathsf{Sig},:\infty} := \Phi_{\mathsf{Sig}}$ be the untruncated signature map and let $M > 0$. Then, for any $\mathbf{x} \in \mathcal{X}_{1\text{-}var}$, we have*

$$\| \Phi_{\mathsf{Sig},:\infty}(\mathrm{k}_{\mathbf{x}}) - \Phi_{\mathsf{Sig},:M}(\mathrm{k}_{\mathbf{x}}) \| \leq \frac{e^{\| \mathrm{k}_{\mathbf{x}} \|_1} \| \mathrm{k}_{\mathbf{x}} \|_1^{M+1}}{(M+1)!}. \tag{21}$$

A proof can be found in [15, Lemma 46]. Despite the nice factorial decay, we emphasize that the variation norm of the sequence appears. Furthermore, we can obtain an analogous bound on the truncation error for the kernel by applying Cauchy–Schwarz.

Proposition 7.3 *Let $\mathrm{k}_{\mathsf{Sig},:\infty}$ be the untruncated signature kernel and $\mathrm{k}_{\mathsf{Sig},:M}$ be the signature kernel truncated at $M > 0$. Then, for any $\mathbf{x}, \mathbf{y} \in \mathcal{X}_{1\text{-}var}$, we have*

$$| \mathrm{k}_{\mathsf{Sig},:\infty}(\mathbf{x}, \mathbf{y}) - \mathrm{k}_{\mathsf{Sig},:M}(\mathbf{x}, \mathbf{y}) | \leq \frac{e^C C^{M+1}}{(M+1)!},$$

where $C = \max\{\| \mathrm{k}_{\mathbf{x}} \|_1, \| \mathrm{k}_{\mathbf{y}} \|_1\}$.

Proposition 7.3 quantitatively shows that by choosing a sufficiently high truncation level M, we obtain a good approximation to $\mathrm{k}_{\mathsf{Sig},\infty}$. In practice, $M = 10$ is a feasible computation and since $10! \approx 10^6$, the contribution of the tail in the sum is negligible. However, we draw attention to the fact that the 1-variation of the sequence appears as a constant. Hence, the level M at which the factorial decay kicks in is determined by the path length. The situation is analogous to computing the matrix exponential where the matrix norm determines this truncation level, see Figure 1 in [50].

Instead of using an exact algorithm for finite M and regarding $\mathrm{k}_{\mathsf{Sig},:M}$ as an approximation to $\mathrm{k}_{\mathsf{Sig},:\infty}$, one can also directly approximate $\mathrm{k}_{\mathsf{Sig},:\infty}$ by numerically solving a PDE [57]. In this case, the approximation error to $\mathrm{k}_{\mathsf{Sig},:\infty}$ is due to the discretization of the PDE solver. In particular, the error is quantified in terms of the step size; see [57] for further details.

Discrete Time Approximation

While we have formulated the kernel in terms of continuous paths, this immediately provides a kernel for discrete sequences via the piecewise linear interpolation from Eq. 6. Given a discrete sequence $\mathbf{x} = (\mathbf{x}_0, \ldots, \mathbf{x}_L) \in \mathcal{X}_{\text{seq}}$ and the static kernel, we consider the induced continuous path $\mathrm{k}_{\mathbf{x}} : [0, L] \to H$, defined by

$$\mathrm{k}_{\mathbf{x}}(t) = (i - t)\, \mathrm{k}_{\mathbf{x}_{i-1}} + (t - i + 1)\, \mathrm{k}_{\mathbf{x}_i}, \qquad t \in [i - 1, i]. \tag{22}$$

Then, we can use the algorithms from [42] with the complexity bound from Theorem 7.1 to efficiently compute signature kernels for discrete data.

We can also consider sequences as discretizations of continuous paths. Given a continuous path $\mathbf{x} : [0, T] \to \mathbb{R}^d$, we consider the discretization $\mathbf{x}^\pi \in \mathcal{X}_{\text{seq}}$ with respect to a partition $\pi = (t_i)_{i=0}^L$ of the interval $[0, T]$, where $0 = t_0 < \ldots < t_L = T$. Applying the above interpolation to $\mathbf{x}^\pi$, we obtain an approximate continuous path $\mathrm{k}_{\mathbf{x}}^\pi$ induced by $\mathbf{x}^\pi$. Thus, we obtain an approximation $\Phi_{\mathsf{Sig},:M}(\mathrm{k}_{\mathbf{x}}^\pi)$, which converges to the signature $\Phi_{\mathsf{Sig}}(\mathrm{k}_{\mathbf{x}})$, as $M \to \infty$ and as the mesh of the partition π, $\sup_{t_i \in \pi} |t_i - t_{i-1}|$, vanishes. The convergence rate is given by the following result; further estimates can be found in [42].

Proposition 7.4 ([42, Corollary 4.3]) *Let* $\mathbf{x} \in \mathcal{X}_{1\text{-}var}$ *with domain* $[0, T]$ *and suppose* $\pi = (t_i)_{i=0}^L$ *be a partition of* $[0, T]$. *Then,*

$$\|\Phi_{\mathsf{Sig},:\infty}(\mathrm{k}_{\mathbf{x}}) - \Phi_{\mathsf{Sig},:M}(\mathrm{k}_{\mathbf{x}}^\pi)\| \leq e^{\|\mathrm{k}_{\mathbf{x}}\|_1} \|\mathrm{k}_{\mathbf{x}}\|_1 \cdot \max_{i=1,\ldots,L} \|\mathrm{k}_{\mathbf{x}}|_{[t_{i-1},t_i]}\|_1. \tag{23}$$

7.2 Algorithms

A direct evaluation of the signature kernel $\mathrm{k}_{\mathsf{Sig}} : \mathcal{X}_{\text{seq}} \times \mathcal{X}_{\text{seq}} \to \mathbb{R}$ as an inner product is only feasible if the underlying static kernel $\mathrm{k} : \mathcal{X} \times \mathcal{X} \to \mathbb{R}$ has a low-dimensional RKHS H. If $\mathcal{X} = \mathbb{R}^d$ and k is the Euclidean inner product, then $\mathrm{k}_{\mathsf{Sig}}$ is just the inner product of plain-vanilla signatures and even this becomes quickly costly. However, the intuition is that by first lifting the path into a path that evolves in a high- or infinite-dimensional RKHS H, we gain much more expressive power. Indeed, this is also what the empirical data shows, see e.g. [69]. In particular, using popular choices such as RBF, Matern, or Laplace for the static kernel results in a signature kernel $\mathrm{k}_{\mathsf{Sig}}$ that outperforms the "trivial" signature kernel induced by the inner product kernel on $\mathbb{R}^d$. The computation for nontrivial kernels cannot be done directly since the RKHS H of a generic kernel k is infinite-dimensional and we cannot exactly compute the signature of paths evolving in general RKHS H. However, there are at least two different algorithms which yield effective approximations.

Dynamic Programming. The idea is to mimic the argument for the monomial kernel $\mathrm{k}_{\mathsf{Mon}}$ in $\mathbb{R}^d$, given in Examples 6.2 and 6.3 in Sect. 6. Concretely, given $\mathbf{x} = (\mathbf{x}_0, \ldots, \mathbf{x}_L) \in \mathcal{X}_{\text{seq}}$, we consider the signature of $\mathrm{k}_{\mathbf{x}}$, which is the sequence $\mathbf{x}$ lifted to a sequence in H defined in Eq. 22. First, the signature of one linear segment (see Eq. 7) is given as

$$\Phi_{\mathsf{Sig}}(\mathrm{k}_{\mathbf{x}}|_{[i-1,i]}) = \Phi_{\mathsf{Mon}}(\delta_i \mathrm{k}_{\mathbf{x}}) = \left(1, \delta_i \mathrm{k}_{\mathbf{x}}, \frac{(\delta_i \mathrm{k}_{\mathbf{x}})^{\otimes 2}}{2!}, \ldots\right),$$

where $\delta_i \, \mathrm{k}_{\mathbf{x}} := \mathrm{k}_{\mathbf{x}_{i+1}} - \mathrm{k}_{\mathbf{x}_i} = \mathrm{k}(\mathbf{x}_{i+1}, \cdot) - \mathrm{k}(\mathbf{x}_i, \cdot) \in H$. Second, using *Chen identity*, which expresses sequence concatenation as tensor multiplication, allows us to express the signature of the entire piecewise linear path as

$$\Phi_{\mathsf{Sig}}(\mathrm{k}_{\mathbf{x}}) = \Phi_{\mathsf{Mon}}(\delta_1 \, \mathrm{k}_{\mathbf{x}}) \otimes \ldots \otimes \Phi_{\mathsf{Mon}}(\delta_L \, \mathrm{k}_{\mathbf{x}}).$$

The degree m component can be explicitly expressed as

$$\Phi_{\mathsf{Sig},m}(\mathrm{k}_{\mathbf{x}}) = \sum_{\mathbf{i}:1 \le i_1 \le \ldots \le i_m \le L} \frac{1}{\mathbf{i}!} \delta_{i_1} \, \mathrm{k}_{\mathbf{x}} \otimes \ldots \otimes \delta_{i_m} \, \mathrm{k}_{\mathbf{x}} \in H^{\otimes m},$$

where for $\mathbf{i} = (i_1, \ldots, i_m)$, we define $\mathbf{i}! := n_1! \cdots n_k!$ if $\mathbf{i}$ contains $k = |\{i_1, \ldots, i_m\}|$ distinct elements, and $n_1, \ldots, n_k$ denotes the number of times they occur in $\mathbf{i}$. Therefore, the truncated signature kernel can be written as

$$\mathrm{k}_{\mathsf{Sig},:M}(\mathbf{x}, \mathbf{y}) = \sum_{m=0}^{M} \sum_{\substack{\mathbf{i}:1 \le i_1 \le \ldots \le i_m \le L \\ \mathbf{j}:1 \le j_1 \le \ldots \le j_m \le L}} \frac{1}{\mathbf{i}! \cdot \mathbf{j}!} \prod_{r=1}^{m} \nabla^{\mathrm{k}}_{i_r, j_r}(\mathbf{x}, \mathbf{y}),$$

where

$$\begin{aligned} \nabla^{\mathrm{k}}_{i,j}(\mathbf{x}, \mathbf{y}) :=& \langle \delta_i \, \mathrm{k}_{\mathbf{x}}, \delta_j \, \mathrm{k}_{\mathbf{y}} \rangle_H \\ =& \mathrm{k}(\mathbf{x}_{i-1}, \mathbf{y}_{j-1}) - \mathrm{k}(\mathbf{x}_i, \mathbf{y}_{i-1}) - \mathrm{k}(\mathbf{x}_{i-1}, \mathbf{y}_j) + \mathrm{k}(\mathbf{x}_i, \mathbf{y}_j). \end{aligned}$$

Finally, we note that we can we can apply Horner's scheme to obtain

$$\mathrm{k}_{\mathsf{Sig},:M}(\mathbf{x}, \mathbf{y}) = 1 + \sum_{\substack{i_1 \ge 1 \\ j_1 \ge 1}} \nabla^{\mathrm{k}}_{i_1, j_1}(\mathbf{x}, \mathbf{y}) \cdot \left(1 + \sum_{\substack{i_2 \ge i_1 \\ j_2 \ge j_1}} \frac{\nabla^{\mathrm{k}}_{i_2, j_2}(\mathbf{x}, \mathbf{y})}{C_2(\mathbf{i}_2, \mathbf{j}_2)} \cdot \left(1 + \ldots \sum_{\substack{i_M \ge i_{M-1} \\ j_M \ge j_{M-1}}} \frac{\nabla^{\mathrm{k}}_{i_M, j_M}(\mathbf{x}, \mathbf{y})}{C_M(\mathbf{i}_M, \mathbf{j}_M)} \right) \right), \tag{24}$$

where $\mathbf{i}_m = (i_1, \ldots, i_m)$, $\mathbf{j}_m = (j_1, \ldots, j_m)$, and $C_m(\mathbf{i}_m, \mathbf{j}_m) = n_{\mathbf{i}_m} \cdot n_{\mathbf{j}_m}$, where $n_{\mathbf{i}_m} \ge 1$ is the largest number such that $i_k = i_m$ for all $k = m - n_{\mathbf{i}_m}, \ldots, m$. The constants $C_m(\mathbf{i}_m, \mathbf{j}_m)$ are used to take into account the counting of repetitions from the $\frac{1}{\mathbf{i}! \cdot \mathbf{j}!}$ term. It is straightforward to implement (24) by a recursive algorithm which inherits the advantages of the classic Horner scheme.

PDE Solver. Another approach was given in [57] which focuses on the case of $M = \infty$. The idea is to define the function $K : [0, L]^2 \to \mathbb{R}^d$ by the signature kernel

$$K(s,t) = \mathrm{k}_{\mathsf{Sig}}(\mathbf{x}|_{[0,s]}, \mathbf{y}|_{[0,t]})$$

evaluated on $\mathbf{x}$ restricted to $[0, s]$ and $\mathbf{y}$ restricted to $[0, t]$. Then, this function satisfies the following PDE

$$\frac{\partial^2 K}{\partial s \partial t}(s,t) = \nabla^{\mathrm{k}}_{\lceil s \rceil, \lceil t \rceil}(\mathbf{x}, \mathbf{y}) \cdot K(s,t), \quad K(s,0) = K(0,t) = 1.$$

Such PDEs are known as Goursat PDEs. Hence, the computation boils down to applying a PDE solver and [57] provides such a scheme and analyzes the error rate. A priori, the PDE approach is focused on $M = \infty$ but cross-validating M often improves performance and [8] provides such a generalization of the PDE approach to approximate $\mathrm{k}_{\mathsf{Sig}}$ for any $M \geq 1$.

Both algorithms work well and packages are available[7] that adapt the underlying algorithms to work on GPUs. These two algorithms are very different, even for a single evaluation of $\mathrm{k}_{\mathsf{Sig},:M}(\mathbf{x}, \mathbf{y})$. The dynamic programming approach is exact for finite M and for $M = \infty$ the approximation error stems from the factorial decay; in contrast, the PDE approach provides an approximation that is determined in by the properties of the scheme the PDE solver uses such as the step size. Similarly, these two algorithms are also amenable to different computational strategies when the whole Gram matrix needs to be evaluated. More recently, another approach based on computing random Fourier signature features has been developed [70]. The trade-offs between these choices can vary between datasets, and we refer the reader to [70, Section 4] for detailed complexity and empirical comparisons, and to [66] for a user's guide to the `KSig` package.

7.3 Hyperparameter Tuning

Given a fast algorithm to compute the Gram matrix for a given parameter choice, the vast majority of the computational budget should be spent on hyperparameter tuning. This is true in general for any kernel but in particular for the signature kernel where the parameter choice can drastically influence performance. The signature kernel has the following hyperparameters.

1. **Static Kernel Parameters** Θ_{k}. The signature kernel takes as input a kernel k on $\mathcal{X}$ and turns it into a kernel $\mathrm{k}_{\mathsf{Sig}}$ on $\mathcal{X}_{\mathrm{seq}}$. Typically, the static kernel k is parametrized; for example, the classic RBF kernel $\mathrm{k}(\mathbf{x}, \mathbf{y}) = \exp(\theta \|\mathbf{x} - \mathbf{y}\|^2)$ has a hyperparameter θ. In general, hyperparameters are not limited to scalars

[7] https://github.com/tgcsaba/KSig and https://github.com/crispitagorico/sigkernel.

and we denote with Θ_{k} the set of all possible values of such hyperparameters. Consequently, the signature kernel $\mathsf{k}_{\mathsf{Sig}}$ then inherits all these hyperparameters of the static kernel k.

2. **Truncation Parameters** $\mathbb{N}_\infty$. In practice, the optimal choice of the truncation level $M \in \mathbb{N}_\infty$ varies widely between datasets and can be as low as $M = 2$. Informally, one expects a trade-off between the non-linearity of the static kernel k and the optimal truncation level M, but even without kernelization the choice of optimal M is subtle, [25]. Indeed, an extreme case is to take the trivial inner product kernel and experiments show that in this case the optimal M is large; in contrast for RBF and other kernels, lower choices of M are typically optimal. However, these are stylized facts and vary between datasets.
3. **Preprocessing Parameters** $\Theta_{\text{preprocess}}$. It is common practice in time-series modelling to preprocess the original time-series. For example, instead of the original sequence in $\mathbb{R}^d$ one considers a sequence in $\mathbb{R}^{2d}$ where the additional coordinates are given by lagging the original time series (a sliding window). Similarly, for multi-variate time-series the different coordinates might evolve on different scales and it makes sense to normalize them; see for example [69] for some standard time-series preprocessing for the signature kernel. In general, there is a large variety of preprocessing steps and we denote these with $\Theta_{\text{preprocess}}$.
4. **Algebraic Parameters** Θ_{algebra}. Signatures can be generalized in several ways and one can replace iterated integrals by other objects. How exactly these objects are constructed goes beyond the scope of this article, but this often allows us to increase the empirical performance. We do not further discuss this here but refer to the "non-geometric" signatures in [42, 68] which require us to go beyond the usual (Hopf-)algebraic treatment of signatures, see [20]. This induces another parameter-set Θ_{algebra}.
5. **Normalization Parameters** $\Theta_{\text{normalization}}$. As we discuss in the next section, a normalization makes the signature kernel a more robust statistic; in particular, this resolves the aforementioned issues about non-compactness. There are different ways to normalize, and this is further discussed in [15]. Furthermore, we can independently rescale the inner products for tensors of different degrees to obtain different signature kernels [8]. We denote these normalization and rescaling parameters $\Theta_{\text{normalization}}$.

Therefore the hyperparameter set of the signature kernel $\mathsf{k}_{\mathsf{Sig}}$ build from a static kernel k with parameter set Θ_{k} is

$$\Theta_{\mathsf{k}_{\mathsf{Sig}}} := \Theta_{\mathsf{k}} \times \mathbb{N}_\infty \times \Theta_{\text{preprocess}} \times \Theta_{\text{algebra}} \times \Theta_{\text{normalization}}.$$

We emphasize that (as for most kernels) the hyperparameter selection is essential to get strong performance. Depending on the problem this requires different approaches; a basic distinction is between supervised and unsupervised problems.

Supervised Learning. An objective function is given and in principle straightforward cross-validation allows us to optimize $\theta \in \Theta_{\mathsf{k}_{\mathsf{Sig}}}$ by grid-search. However, since $\Theta_{\mathsf{k}_{\mathsf{Sig}}}$ is usually high-dimensional, more sophisticated hyperparameter tuning algorithms, such as Bayesian optimization, can be very useful and are implemented in most machine learning toolboxes. Either way, standard hyperparameter tuning can be used and simple grid search typically works well.

Unsupervised Learning. In contrast to supervised problems the situation is less straightforward since the choice of optimization criteria is often not immediate. Even worse, standard heuristics such as the median heuristic [28] are not justified for the signature kernel and indeed typically lead to poor performance. For example, we discuss this in the context of hypothesis testing in Sect. 9. There, the median heuristic is used for some kernels, but there is no reason why this heuristic should apply to the signature kernel and indeed this would result in tests with poor power. In general, for unsupervised problems the specific application must be taken into account.

8 Robust Moments, Signatures, and Kernels

In Sect. 5, we saw that monomial features Φ_{Mon} lose the ability to approximate functions and characterize laws of random variables when we consider unbounded domains. These issues extend generically to the signature features Φ_{Sig}. While compactness can be a reasonable assumption for monomials in the case of $\mathcal{K} \subset \mathbb{R}^d$, this is a much stronger assumption[8] to make for signatures in the case of $\mathcal{K} \subset \mathcal{X}_{1\text{-var}}$.

Example 8.1 (No Characteristicness) There exist two different stochastic processes $\mathbf{X}, \mathbf{Y}$ with sample paths that are straight lines, such that

$$\mathbb{E}[\Phi_{\mathsf{Sig}}(\mathbf{X})] = \mathbb{E}[\Phi_{\mathsf{Sig}}(\mathbf{Y})].$$

For example, one can take

$$\mathbf{X}(t) = t \cdot \mathbf{V} \text{ and } \mathbf{Y}(t) = t \cdot \mathbf{W}$$

where $\mathbf{V}$ is a random variable in $\mathbb{R}^2$ drawn from the density

$$\mathbf{V} \sim \rho_1(\mathbf{v}) = \frac{1}{2\pi \cdot \mathbf{v}_1 \mathbf{v}_2} \exp\left(\frac{-(\log^2(\mathbf{v}_1) + \log^2(\mathbf{v}_2))}{2}\right)$$

and $\mathbf{W}$ drawn from

[8] Recall that a topological vector space is locally compact if and only if it is finite-dimensional. See also Remark 8.7, that even given a compact set that carries nearly all the support of the measure, this is not enough for characteristicness.

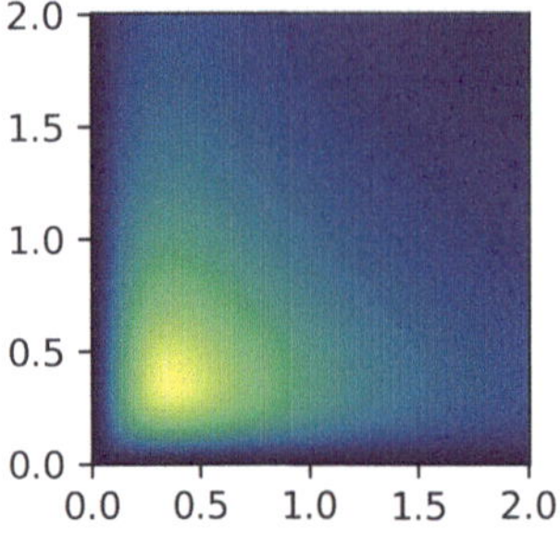

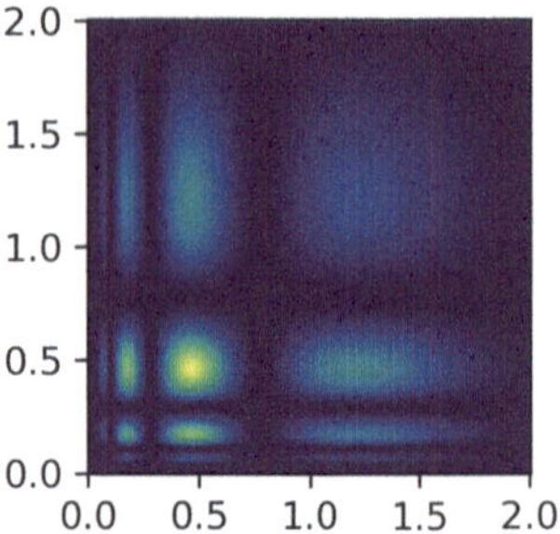

Fig. 2 Probability density functions of **V** (left) and **W** (right)

$$\rho_2(\mathbf{w}_1, \mathbf{w}_2) = \rho_1(\mathbf{w}_1, \mathbf{w}_2) \prod_{i=1}^{2} (1 + \sin(2\pi \log(\mathbf{w}_i))),$$

shown in (Fig. 2). One option is to make stronger model assumptions on the decay rate of moments which would allow characterizing the law; in $\mathbb{R}^d$ this is classic and for paths, we refer to [13]. However, in an inference context we usually do not want to make strong assumptions and the above example is a warning that things can go genuinely wrong, even for trivial stochastic processes. A different, but related, issue is the notion of robustness in statistics. At a high level, a robust statistic is one which is stable with respect to small perturbations of the underlying measure.

Example 8.2 (No Robustness) Suppose $\mathbf{X}$ is a real-valued random variable with law μ. Fix $\epsilon \in (0, 1)$ and $\mathbf{x} \in \mathbb{R}$ and let $\mathbf{X}_{\mathbf{x},\epsilon}$ be the perturbed random variable with law $\mu_{\mathbf{x},\epsilon} = (1-\epsilon)\mu + \epsilon\delta_{\mathbf{x}}$. When $\mathbf{x} \gg \mathbb{E}[\mathbf{X}]$, it is easy to see that the outlier Dirac measure $\delta_{\mathbf{x}}$ has outsized influence on the mean $\mathbb{E}[\mathbf{X}_{\mathbf{x},\epsilon}]$, and this influence is magnified for higher moments $\mathbb{E}[\mathbf{X}_{\mathbf{x},\epsilon}^{\otimes m}]$.

The reason Example 8.2 is worrisome is that in practice we are only given a finite number of samples to estimate our statistics, such as $\mathbb{E}[\mathbf{X}^{\otimes m}]$ or $\mathbb{E}[\int d\mathbf{X}^{\otimes m}]$. In practice, this finite sample is taken from real-world observations and it is common that this data is not clean; that is, outliers such as a $\mathbf{x} \gg \mathbb{E}[\mathbf{X}]$ in the above example are not uncommon due to sensor errors, data recording, etc.[9] However, some statistics can still be well-estimated, and the design of such robust features has been a central topic of modern statistics [35, 40]. Ironically, the standard machine learning benchmarks for time-series are fairly clean in terms of outliers but such outliers appear often in applications in the "real-world".

[9] Formally, we are given samples from $\mathbf{X}_{\mathbf{x},\epsilon}$ and not from $\mathbf{X}$.

8.1 Robust Statistics

The above example motivates the notion of the *influence function*, which quantifies the robustness of a statistic.

Definition 8.3 ([35]) Let $\mathcal{X}$ be a topological space and $\mathcal{P}(\mathcal{X})$ be the space of Borel probability measures on $\mathcal{X}$. For a map $T : \mathcal{P}(\mathcal{X}) \to \mathbb{R}$, we define the *influence function of* T *at* $\mu \in \mathcal{P}(\mathcal{X})$, denoted $\mathsf{IF}(\cdot; T, \mu) : \mathcal{X} \to \mathbb{R}$, by

$$\mathsf{IF}(\mathbf{x}; T, \mu) := \lim_{\epsilon \searrow 0} \frac{T(\epsilon\delta_{\mathbf{x}} + (1-\epsilon)\mu) - T(\mu)}{\epsilon}$$

for each $x \in \mathcal{X}$ where this limit exists. We say that T is *B-robust at* μ if $\sup_{\mathbf{x}\in\mathcal{X}} |\mathsf{IF}(\mathbf{x}; T, \mu)| < \infty$.

Example 8.2 shows that the statistic "the m-th moment",

$$T(\mu) = \int \mathbf{x}^{\otimes m} \, \mu(d\mathbf{x})$$

is not B-robust and by the same argument we immediately see that expected signatures are not robust statistics to describe laws of stochastic processes. What goes wrong in both cases is that monomials, $\mathbf{x}^{\otimes m}$ resp. $\int d\mathbf{x}^{\otimes m}$, explode when the argument gets big, so if we observe among our N samples an extreme value $\mathbf{x}$, this results in an $\epsilon = \frac{1}{N}$-weighted Dirac at $\mathbf{x}$ in the resulting empirical measure and the contribution of $\mathbf{x}^{\otimes m}$ weighted with mass $\epsilon\delta_{\mathbf{x}}$, can dominate the estimator.

8.2 Robust Signatures

One way to address this issue is to compensate for the exponential growth as $\mathbf{x}$ becomes large. In the case of the monomial map for $\mathbf{x} \in \mathbb{R}^d$, this can be achieved by rescaling $\mathbf{x}^{\otimes m}$ to $\lambda_m(\mathbf{x})\mathbf{x}^{\otimes m}$, where $\lambda_m : \mathbb{R}^d \to \mathbb{R}$ is a function which decays quickly as $\|\mathbf{x}\| \to \infty$. What makes the precise choice of the penalty λ challenging is that it needs to simultaneously decay sufficiently quickly to make the statistics robust but it cannot decay too quickly in order to preserve the universal and characteristic properties. In particular, we wish to have robust analogues of Propositions 2.1 and 4.1. As a technical note, there are two differences when we consider these robust analogues.

1. We consider the space of continuous bounded functions $C_b(\mathcal{X}, \mathbb{R})$ rather than continuous functions.
2. In order to deal with non-locally compact spaces, we equip the space of continuous bounded functions in the next two theorems with the *strict topology*. This

was introduced in [29], and further discussion on its application to robustness can be found in [15].

In the following, we reformulate the scaling factors λ_m into a single normalization map $\lambda : \prod_{m\geq 0}(\mathbb{R}^d)^{\otimes m} \to \prod_{m\geq 0}(\mathbb{R}^d)^{\otimes m}$.

Theorem 8.4 *There exists a map $\lambda : \prod_{m\geq 0}(\mathbb{R}^d)^{\otimes m} \to \prod_{m\geq 0}(\mathbb{R}^d)^{\otimes m}$ such that*

$$\lambda \circ \Phi_{\mathsf{Mon}} : \mathbb{R}^d \to \prod_{m\geq 0}(\mathbb{R}^d)^{\otimes m}, \quad \mathbf{x} \mapsto \lambda(\Phi_{\mathsf{Mon}}(\mathbf{x})),$$

1. *is universal to $C_b(\mathbb{R}^d, \mathbb{R})$ equipped with the strict topology, and*
2. *is characteristic to the space of signed measures on $\mathbb{R}^d$.*

Moreover, $\mu \mapsto \mathbb{E}_{\mathbf{X}\sim\mu}[\lambda \circ \Phi_{\mathsf{Mon}}(\mathbf{X})]$ is B-robust.

In particular, the set $\{\mathbf{x} \mapsto \langle \mathbf{w}, \lambda \circ \Phi_{\mathsf{Mon}}(\mathbf{x})\rangle \;:\; \mathbf{w} \in \bigoplus(\mathbb{R}^d)^{\otimes m}\}$ is dense in $C_b(\mathbb{R}^d, \mathbb{R})$ in the strict topology and the normalized moments $\mathbb{E}[\lambda \circ \Phi_{\mathsf{Mon}}(\mathbf{X})]$ always characterizes the law of $\mathbf{X}$. The existence of the function λ is not trivial and while λ is not given in an explicit form but it can be efficiently computed, see [15, Proposition 14]. We find that the same argument applies to signatures.

Theorem 8.5 *There exists a map $\lambda : \prod_{m\geq 0} \bar{H}^{\otimes m} \to \prod_{m\geq 0} \bar{H}^{\otimes m}$ such that*

$$\lambda \circ \Phi_{\mathsf{Sig}} : H_{1-var} \to \prod_{m\geq 0} \bar{H}^{\otimes m}, \quad \mathbf{x} \mapsto \lambda(\Phi_{\mathsf{Sig}}(\bar{\mathbf{x}}))$$

1. *is universal to[10] $C_b(\bar{H}_{1-var}, \mathbb{R})$ equipped with the strict topology, and*
2. *is characteristic to the space of signed measure on H_{1-var}.*

Moreover, $\mu \mapsto \mathbb{E}_{\mathbf{X}\sim\mu}[\lambda \circ \Phi_{\mathsf{Sig}}(\bar{\mathbf{X}})]$ is B-robust.

We formulated Theorem 8.5 for paths evolving in Hilbert spaces in view of kernelization but the result also holds for the ordinary signature with $\mathrm{k} = \mathrm{id}$ and $H = \mathbb{R}^d$. Further, the analogous statements hold without adding a time coordinate by considering functions or measures on tree-like equivalence classes of paths, see [15, Theorem 21].

Remark 8.6 (Tensor Normalization as Path Normalization) The normalization λ on the level of the tensors is equivalent to a normalization of the underlying path. That is $\lambda \circ \Phi_{\mathsf{Sig}}(\mathbf{x})$ can be rewritten as $\Phi_{\mathsf{Sig}}(\frac{\mathbf{x}}{\Lambda(\mathbf{x})})$ for a function $\Lambda : \mathcal{X}_{1-var} \to \mathbb{R}$ of the path $\mathbf{x}$. This way, we can think of λ as a very specific data normalization, see [15].

Remark 8.7 (Nearly Compact Support Does Not Help) We emphasize that it can be easy to find for every $\epsilon > 0$ a compact set that carries $1 - \epsilon$ the mass of the

[10] Analogous to Proposition 4.1, $\bar{H}_{1-var}$ denotes the paths of H_{1-var} that are augmented with a time coordinate. Explicitly, paths in $\bar{H}_{1-var}$ are paths evolving in $H \oplus \mathbb{R}$ of bounded 1-variation.

probability measure. The issue is that the functions we are interested, monomials or iterated integrals, behave badly on the remaining set carrying ϵ of the probability mass. The normalization λ turns this functions into bounded, hence well-behaved functions, while still keeping the necessary algebraic structure.

8.3 *Robust (Signature) Kernels*

The normalization λ allows us to overcome the drawbacks of our feature maps, Φ_{Mon} and Φ_{Sig} on non-compact sets. Given a static kernel $\mathrm{k} : \mathcal{X} \times \mathcal{X} \to \mathbb{R}$ with associated feature map $\varphi : \mathcal{X} \to H$, we can define robust versions of the associated kernels, $\mathrm{k}_{\mathsf{Mon}}$ and $\mathrm{k}_{\mathsf{Sig}}$, by

$$\mathrm{k}_{\mathsf{Mon}}(\mathbf{x}, \mathbf{y}) := \langle \lambda \circ \Phi_{\mathsf{Mon}}(\varphi(\mathbf{x})), \lambda \circ \Phi_{\mathsf{Mon}}(\varphi(\mathbf{y}))\rangle$$

$$\mathrm{k}_{\mathsf{Sig}}(\mathbf{x}, \mathbf{y}) := \langle \lambda \circ \Phi_{\mathsf{Sig}}(\varphi(\mathbf{x})), \lambda \circ \Phi_{\mathsf{Sig}}(\varphi(\mathbf{y}))\rangle.$$

With some abuse of notion we use the same notation as the non-robust version. The above universality and characteristicness results, Theorems 8.4 and 8.5, immediately translate to the corresponding kernelized statement.

From the computational side, using results from [15, 42], the normalization λ can itself be kernelized, which then gives the following.

Proposition 8.8 *Let* $\mathrm{k} : \mathcal{X} \times \mathcal{X} \to \mathbb{R}$ *be a kernel,* $\mathbf{x}, \mathbf{y} \in \mathcal{X}_{seq}$ *be two sequences of maximal length* L*, and* M *be the truncation level. Then the robust truncated signature kernel* $\mathrm{k}_{\mathsf{Sig},:M}(\mathbf{x}, \mathbf{y})$ *can be evaluated in* $O(L^2(c_{\mathrm{k}} + M^3) + q)$ *computational steps and using* $O(M^2L^2 + M + q)$ *memory where* c_{k} *denotes the cost of a single evaluation of the static* k *and* q *denotes the cost of finding the non-negative root of a univariate polynomial of degree* $2M$.

The algorithm in Proposition 8.8 is in fact elementary given that we know how to compute the signature kernel, see [15] for details.

9 Applications and Further Developments

Having an efficiently computable kernel for sequential data in arbitrary spaces allows us to apply kernel methods to a wide variety of problems. Below we discuss some of them.

A Maximum Mean Discrepancy for Stochastic Processes. Given a set $\mathcal{X}$, we often care about the space of probability measures $\mathcal{P}(\mathcal{X})$ on this space. We have encountered the notion of characteristicness which says that we can uniquely represent a measure $\mu \in \mathcal{P}(\mathcal{X})$ as an expectation, that is

$$\mu \mapsto \mathbb{E}_{\mathbf{X}\sim\mu}[\varphi(\mathbf{X})]$$

is injective. If the co-domain of φ, the feature space, has a norm, this would induce metric

$$d_{\mathrm{k}}(\mu, \nu) = \|\mathbb{E}_{\mathbf{X}\sim\mu}[\varphi(\mathbf{X})] - \mathbb{E}_{\mathbf{Y}\sim\nu}[\varphi(\mathbf{Y})]\|.$$

The caveat is that these expectations take values in an infinite-dimensional space. However, for the choice $\varphi(\mathbf{x}) = \mathrm{k}(\mathbf{x}, \cdot)$, the above can be efficiently computed; a straightforward computation using the fact that H is a RKHS shows

$$\begin{aligned}\|\mathbb{E}[\varphi(\mathbf{X})] - \mathbb{E}[\varphi(\mathbf{Y})]\|^2 &= \langle \mathbb{E}[\varphi(\mathbf{X})] - \mathbb{E}[\varphi(\mathbf{Y})], \mathbb{E}[\varphi(\mathbf{X})] - \mathbb{E}[\varphi(\mathbf{Y})]\rangle \\ &= \mathbb{E}[\mathrm{k}(\mathbf{X}, \mathbf{X}')] - 2\mathbb{E}[\mathrm{k}(\mathbf{X}, \mathbf{Y})] + \mathbb{E}[\mathrm{k}(\mathbf{Y}, \mathbf{Y}')],\end{aligned}$$

where $\mathbf{X}, \mathbf{X}' \sim \mu$ and $\mathbf{Y}, \mathbf{Y}' \sim \nu$ are independent copies. The right-hand side is straightforward to estimate when only finite samples of μ and ν are given; in particular,

$$d_{\mathrm{k}}(\mu, \nu) = \sqrt{\mathbb{E}[\mathrm{k}(\mathbf{X}, \mathbf{X}')] - 2\mathbb{E}[\mathrm{k}(\mathbf{X}, \mathbf{Y})] + \mathbb{E}[\mathrm{k}(\mathbf{Y}, \mathbf{Y}')]}$$

is a metric which can be easily estimated with finite samples. The above metric is also known as kernel maximum mean discrepancy (MMD) since a direct calculation shows that

$$d_{\mathrm{k}}(\mu, \nu) = \sup_{f \in H : \|f\| \leq 1} \left| \int_{\mathcal{X}} f(\mathbf{x})\mu(d\mathbf{x}) - \int_{\mathcal{X}} f(\mathbf{x})\nu(d\mathbf{x}) \right|.$$

Hence, d_{k} is the maximum difference we can achieve by testing our measure against functions in the unit ball of the RKHS of k. MMDs have found many applications in machine learning, see [51] for a survey. Having the signature kernel at hand, then immediately yields a MMD for probability measures on paths; in particular, we obtain a metric for laws of stochastic processes. The theoretical guarantees of the signature kernel then translate into theoretical guarantees for the signature kernel MMD $d_{\mathrm{k}_{\mathrm{Sig}}}$. However, the topology on the space of probability measures induced by the signature MMD is more subtle. When the RKHS is finite-dimensional it induces weak convergence but not for infinite-dimensional RKHS; see [15] for details.

Hypothesis Testing. Among the most popular hypothesis testing methods is two-sample testing and signature kernels allow us to do this for stochastic processes. In particular, we can test

$$H_0 : \mathrm{Law}(\mathbf{X}) = \mathrm{Law}(\mathbf{Y}) \text{ against } H_1 : \mathrm{Law}(\mathbf{X}) \neq \mathrm{Law}(\mathbf{Y})$$

where $\mathbf{X} = (\mathbf{X}_t)$ and $\mathbf{Y} = (\mathbf{Y}_t)$ are two stochastic processes. The data we are given are m sample trajectories from $\mathbf{X}$ and n sample trajectories from $\mathbf{Y}$. Such situations are not uncommon, for example consider a cohort of patients in a medical trial who are divided into two subgroups (e.g. one taking medication and the other not) and we measure medical markers during the trial period. For another example, see [2] for an application of the signature MMD to test and validate hypothesis for real-world economic scenarios. Following [31] we reject H_0 if

$$d_{\mathsf{k}_{\mathsf{Sig}}}(\hat{\mu}_X, \hat{\mu}_Y) > c(\alpha, m, n). \tag{25}$$

Here, $\hat{\mu}_X = \frac{1}{m}\sum_{i=1}^{m} \delta_{\mathbf{x}_i}$ and $\hat{\mu}_Y = \frac{1}{n}\sum_{j=1}^{n} \delta_{\mathbf{y}_j}$ are the empirical measures given by our observed sequences[11] $\mathbf{x}_1, \ldots, \mathbf{x}_n, \mathbf{y}_1, \ldots, \mathbf{y}_m \in \mathcal{X}_{\text{paths}}$ and $c(\alpha, m, n)$ is a constant such that the test (25) has a significance level[12] α. In [31], several estimators are proposed for $d_{\mathsf{k}_{\mathsf{Sig}}}(\hat{\mu}_X, \hat{\mu}_Y)$ and these estimators differ in their degree of bias and their computational complexity. The function c is determined by asymptotic distribution in [31], but this choice leads in general to a very conservative test criteria, that is, one seldom rejects H_0. In practice, permutation tests often result in better empirical performance for MMD hypothesis testing, and this applies especially to the signature kernel; we refer to [15] for details and for experiments and benchmarks against other hypothesis tests. Beyond testing the equivalence of laws, kernel learning allows us to test for independence between the coordinates of a random variable [32]. Note that for stochastic processes, independence between coordinates is challenging since statistical dependencies can manifest between different times in different coordinates. However, quantifying and testing independence for stochastic processes have important applications such as Independent Component Analysis for stochastic processes [5, 59, 60].

Finally, note that the signature kernel $\mathsf{k}_{\mathsf{Sig}}$, and hence the MMD $d_{\mathsf{k}_{\mathsf{Sig}}}$, is parametrized by hyperparameters $\theta_{\mathsf{k}_{\mathsf{Sig}}} \in \Theta_{\mathsf{k}_{\mathsf{Sig}}}$ and already for simple supervised learning, the hyperparameter choice was essential for good performance. For hypothesis testing, a naive grid-search to choose the best θ using the entire test set is not applicable since this would introduce correlations which invalidate the test. This is a generic problem for kernel MMDs in testing problems and several solutions have been proposed which optimize for the test power. One solution is to follow [65] which proposes replacing $d_{\mathsf{k}_{\mathsf{Sig}}}$ by

$$(\mathbf{x}, \mathbf{y}) \mapsto \sup_{\theta \in \Theta} d_{\mathsf{k}_{\mathsf{Sig}}(\theta)}(\mathbf{x}, \mathbf{y}),$$

[11] With slight abuse of notation, each $\mathbf{x}_i \in \mathcal{X}_{\text{seq}}$ denotes a whole sequence and not single sequence entry as before.

[12] Recall that the type I error of a test is the probability of falsely rejecting H_0 and the type II error is falsely accepting H_0. If the type I error is bounded by α then one says the test has power α.

where we write $\mathrm{k}_{\mathsf{Sig}}(\theta)$ to denote the dependence of $\mathrm{k}_{\mathsf{Sig}}$ on θ. The above is again a metric and while it is not the optimal choice for the test power, it guarantees a sample test with given significance; moreover, it works well in practice with the signature kernel, see [15] for details. Other methods exist which aim to maximize the test power [33], but require splitting the data into train/test sets to perform hyperparameter selection. A related issue, is the choice of the static kernel k itself: for essentially all previous applications of the signature kernel, using the Euclidean inner product kernel as static kernel is outperformed by non-linear static kernels such as RBF. However for hypothesis testing, using the standard inner product as the static kernel for the signature kernel can be competitive with the signature kernel built over non-linear static kernels for some datasets. We refer again to [15] for a discussion of this and its relation to the question of hyperparameter choice. This topic will be discussed further in chapter "Market Generators: A Paradigm Shift in Financial Modeling".

Signature MMDs in Generative Models. The signature MMD is a metric between any two measures on path space and can be computed efficiently. In particular, it does not make any parametric assumptions on the two measures that are compared, hence it can be used in a generative model to decide whether the generator produces samples that are close enough to a given target distribution. This has been applied to generative adversarial networks (GANs) to model financial data [7] and to generative models for (neural) stochastic differential equations [41]. Again, the choice of hyperparameter and static kernel is challenging from the theoretical perspective but this is not specific to the signature kernel and general heuristics from the machine learning literature might apply.

Gaussian Processes with Signature Covariances. We often require not only point predictions, but also estimates of the associated uncertainties. Bayesian methods provide a structured approach to update and quantify uncertainties and in particular, Gaussian processes are a powerful and non-parametric method which provides a flexible way to put priors on functions of the data; see [54]. Formally, a Gaussian process on $\mathcal{X}$ is a stochastic process $f = (f_{\mathbf{x}})_{\mathbf{x} \in \mathcal{X}}$ indexed by $\mathbf{x} \in \mathcal{X}$ such that $f \sim N(m, \mathrm{k})$ is determined by a mean and a covariance function

$$m : \mathcal{X} \to \mathbb{R} \text{ and a positive definite } \mathrm{k} : \mathcal{X} \times \mathcal{X} \to \mathbb{R}$$

where $f_{\mathbf{x}}$ is Gaussian with

$$\mathbb{E}[f_{\mathbf{x}}] = m(\mathbf{x}) \text{ and } \mathrm{Cov}(f_{\mathbf{x}}, f_{\mathbf{y}}) = k(\mathbf{x}, \mathbf{y}).$$

Note that we break here with our convention to denote random variables with capital letters and write f instead of F to be consistent with the GP literature. Without loss of generality, one can take $m(\mathbf{x}) = 0$ and consider data in an arbitrary space $\mathcal{X}$ provided a kernel k is given. If the kernel k has nice properties these often translate into desirable properties of the GP. Given the signature kernel, one can immediately consider a GP for sequential data. However, two issues arise, one more theoretical and one more applied

- The existence of Gaussian processes with continuous sample trajectories is not trivial on non-compact spaces; see for example [1] for simple counterexamples. The required regularity estimates to ensure existence can be derived by using a classical result of Dudley [22] and bounding the growth of the covering number which requires us to utilize the structure of k and $\mathcal{X}$. In [69], such covering number estimates are computed for the signature kernel which then gives the existence of a GP indexed by paths; further, it yields a modulus of continuity for the GP sample paths $\mathbf{x} \mapsto f_{\mathbf{x}}$.
- Plain vanilla GP inference requires us to invert large Gram matrices which scales with complexity $O(n^3)$ in the number of samples n. For most datasets, this is too costly and several approximation have been developed. Arguably the most popular are variational inference and inducing point methods. To efficiently apply these to the signature kernel requires using the structure of tensors. The article [69] introduced GP with signature covariances and developed a variational approach based on sparse tensors; and [45] developed a variational approach that uses diagonal approximations to the Gram matrix which allows for quick matrix inversion.

Strings, Time Series, and Classic Sequence Kernels. Kernel learning for sequences is a classic topic and the case when the state space is finite or discrete—in our notation $\mathcal{X}$ is a finite or countable set of letters and $\mathcal{X}_{\text{seq}}$ is the set of strings with letters from $\mathcal{X}$—has received much attention. Haussler developed a relation-convolution kernel framework [36] that gave rise to string kernels; see for example [46, 48] for algorithms and applications in text learning and biology. These approaches have been extended to allow for general state spaces with the alignment kernel [3, 18, 63] that search over all possible alignments ("time-warpings"), although this may not result in a positive definite kernel. However, both the classic string kernel and the global alignment kernel, arise as special cases or minor variations of the signature kernel construction and other variations are possible, see [42, Section 5 and Remark 4.10] and [68, Appendix B.3], thereby providing theoretical guarantees for these kernels.

Time-Parametrization (In-)variance. The signature $\mathbf{x} \mapsto \Phi_{\mathsf{Sig}}(\mathbf{x})$ is injective up to translation of the starting point $\mathbf{x}(0)$ of the path $\mathbf{x} = (\mathbf{x}(t))_{t \in [0,T]}$ and tree-like equivalence [34]. The latter means that paths are equivalent up to reparametrization and backtracking. In particular, $\Phi_{\mathsf{Sig}}(\mathbf{x}) = \Phi_{\mathsf{Sig}}(\mathbf{x}^{\rho})$ where $\rho : [0, S] \to [0, T]$ is non-decreasing and $\mathbf{x}^{\rho}(t) := \mathbf{x}(\rho(t))$. Adding a time coordinate as in Sect. 4, $\bar{\mathbf{x}}(t) := (t, \mathbf{x}(t))$, destroys any tree-like equivalence, that is $\mathbf{x} \mapsto \Phi_{\mathsf{Sig}}(\bar{\mathbf{x}})$ is injective on $\mathcal{X}_{\text{1-var}}$, hence distinguishes paths that differ by a time-change. Removing such a big class of invariances (the set of time-changes is infinite-dimensional) is a powerful tool and has received much attention in engineering [56] where this is known as "time-warping"; in the above discussed discrete case of strings, this is known as global alignment [43]. The same theoretical guarantees discussed in Sect. 4 translate to the corresponding statements about functions or probability measures on such equivalence classes.

Weak and Adapted Topologies. The signature MMD is a metric on the space of probability measures on either $\mathcal{X}_{\text{seq}}$ or $\mathcal{X}_{\text{paths}}$. Arguably the most important topology on the space of probability measures is the weak topology and we discussed above when the signature resp. signature MMD metrizes this metric. However, the weak topology completely ignores the filtration of a stochastic process. The need for finer topologies for laws of stochastic processes that take the filtration into account has been understood a long time ago [38] and have recently seen a revived interest [53]. These finer topologies run under the name of higher rank topologies (the higher the rank, the more of the filtration structure is taken into account). In [4], higher rank signatures are introduced which take filtrations into account and allow us to metrize these higher rank topologies. The same methodology that constructs the signature kernel given a static kernel can be carried out by using the higher-rank signatures and [39, 58] use the PDE solver to estimate the associated MMD in various machine learning tasks. See chapter "Adapted Topologies and Higher-Rank Signatures" (Adapted Topologies) for further details.

Rough Paths. Throughout we made the assumption that our paths are of bounded variation, that is we defined $\mathsf{k}_{\mathsf{Sig}}$ on $\mathcal{X}_{1\text{-var}}$. However, many stochastic processes have sample paths that are of unbounded variation, and thus are not elements of $\mathcal{X}_{1\text{-var}}$. This is fairly straightforward to address with tools from stochastic analysis and rough path theory. There are two ways to think about this:

(i) In practice we are given sequences and not paths, and these can always be embedded as bounded variation path, $\mathcal{X}_{\text{seq}} \hookrightarrow \mathcal{X}_{\text{paths}}$.

(ii) Even though we can embed any sequence as a bounded variation path as in (i), this can become flawed in the high-frequency scaling limit since the iterated integrals might not converge. A more radical approach is to define the signature kernel for rough paths. This requires us to prescribe the first iterated integrals since these are not well-defined as Riemann–Stieltjes integrals. In [42, Appendix B] this was done for geometric rough paths, which are those for which piecewise linear approximations converge[13] and hence allow for canonical choices of the higher iterated integrals. This covers semimartingales and large classes of other popular stochastic processes. The same algorithm applies if other choices of the higher order integrals

[13] In the notation of [42, Appendix B], $\mathsf{k}^{+}_{(d,m)}$ with $d = \lfloor p \rfloor$ where p denotes the p-variation of the rough path. In our current notation d, m are parts of the parameter $\theta_{algebra} \in \Theta_{algebra}$ of $\mathsf{k}_{\mathsf{Sig}}$.

are prescribed[14] but these are typically not available in the usual sequence datasets.

Beyond (Geometric) Rough Paths. Throughout we focused on the classical signature but for many applications, non-geometric or branched rough paths provide a different way to describe a path as a sequence of tensors. Much of the discussion we had so far can be translated to branched rough paths with minor modification, that is one can again define a "signature kernel" between branched rough paths.[15] One can even go beyond that and argue that the essential structure is that of embedding a sequence into a non-commutative algebra in which we "stitch" the sequence elements together by the non-commutative multiplication in the algebra. That this multiplication is non-commutative is essential since this algebraically captures the sequence order. From this point of view, the signature and kernel that we discussed so far realizes this abstract method by instantiating it with the shuffle Hopf algebra and using as embedding the Lie group exponential; branched rough paths replace the shuffle algebra by the Connes–Kreimer algebra. Following this approach, we can simply parametrize the algebraic structure and learn the best choice from the data. The underlying theory is beyond the current scope but we refer to [68] for a demonstration of how parametrizing the underlying algebraic structure and going beyond geometric rough paths can improve benchmark performance, often significantly, and how to deal with the computational challenges. Similarly, see the discussion about the non-geometric signature kernel[16] in [42, Remark 4.5]. Besides improving performance, such variations lead to interesting and challenging algebraic question which no longer can be treated within a Hopf algebraic framework. We refer to [20] for a detailed study of the underlying algebraic structures.

Evolving (Non-Euclidean) Data. The signature kernel transforms any static kernel on the domain $\mathcal{X}$ into a kernel on the corresponding space of sequences $\mathcal{X}_{\text{seq}}$, allowing us to design kernels for objects evolving in general (non-Euclidean) spaces. One example arises from the field of topological data analysis, in which a method called persistent homology allows us to quantify the multi-scale topological structure‘ of data such as point clouds in highly nonlinear objects called persistence diagrams [19]. In [14], static persistence diagrams are transformed into paths, and the signature kernel allows us to capture the topological properties of a static point cloud. Given an evolving point cloud, for instance representing a swarm of agents, we obtain a sequence of persistence

[14] For example, to describe a " rough" path in d dimensions, one could consider sequences $\mathbf{x} = (\mathbf{x}_0, \ldots, \mathbf{x}_L)$ in the state space $\mathcal{X} = G_{m,d}$, the group-like elements; informally $\mathbf{x}_i$ are the first m "iterated integrals" of the underlying path on the interval $[t_i, t_{i+1}]$. In the notation of Sect. 7.3, producing such group-like elements from raw sequential data could be regarded as a preprocessing step.

[15] This is in the ArXiv version v1 of [16, Appendix D]. Appendix D is not part of the published version [15] of [16] for brevity.

[16] Denoted as $\mathrm{k}^{+}_{(d,m)}$ when $d \neq m$ in the notation of [42].

diagrams which summarize the dynamic topological structure of the data, and signature kernels again provide an effective method to perform learning tasks such as regression [30]. Another example is based on the observation that the definition of the signature only requires an inner product on the derivatives of a sequence. For paths evolving on Lie groups, where derivatives lie in the associated Lie algebra, one can formulate the signature kernel in exactly the same way [44]. Finally, another case where a general approach to sequences can be studied with the signature kernel is to analyze classic deep learning architectures via the attractive theoretical properties of the signature kernel; for example, [26] studies RNNs in terms of the signature kernel, and [52] study scaling of Resnets from this perspective.

Likelihood-free Inference. Given a parametrized statistical model, the likelihood function captures its fit to data. Often the underlying models are complex which leads to intractable likelihood functions and makes it infeasible to find the optimal parameter. So called likelihood-free inference addresses these challenges and [23] combine the signature kernel and its MMD with density ratio estimation to develop a likelihood-free inference approach; see also [24] for its use in approximate Bayesian computation.

Random Approximations. Kernel methods allow the implicit use of a high- or infinite-dimensional feature map $\varphi : \mathcal{X} \to H$ on a dataset $\{\mathbf{x}_1, \ldots, \mathbf{x}_N\} \subset \mathcal{X}$ of N points, by replacing the computation and storage of the features $\varphi(\mathbf{x}_1), \ldots, \varphi(\mathbf{x}_N) \in H$ by the evaluation and storage of the Gram matrix $\left(\mathrm{k}(\mathbf{x}_i, \mathbf{x}_j)\right)_{i,j=1,\ldots,N}$. However, even if the kernel is cheap to evaluate, the storage cost now scales quadratically in N which can be prohibitive for many datasets. Moreover, the signature kernel scales quadratically in the sequence length which results in quadratic complexity in both, the number of data points and the sequence length. A classic way kernel learning deals with this is to use randomness to construct lower-dimensional approximations such as the Nyström method which constructs a low-rank approximation to the full Gram matrix by random subsampling; see [42] for an algorithm that does this for the signature kernel. Another way to achieve this is by using *random Fourier features*. The starting point is that different feature maps can give rise to the same kernel,

$$\mathrm{k}(\mathbf{x}, \mathbf{y}) = \langle \varphi(\mathbf{x}), \varphi(\mathbf{y}) \rangle = \langle \tilde{\varphi}(\mathbf{x}), \tilde{\varphi}(\mathbf{y}) \rangle$$

where $\varphi : \mathcal{X} \to H$ and $\tilde{\varphi} : \mathcal{X} \to \tilde{H}$ may not even take values in the same Hilbert space. Random Fourier features construct for a given kernel k associated with a feature map φ a *low-dimensional and random* feature map $\tilde{\varphi} : \mathcal{X} \to \mathbb{R}^d$ such that

$$\mathrm{k}(\mathbf{x}, \mathbf{y}) \approx \langle \tilde{\varphi}(\mathbf{x}), \tilde{\varphi}(\mathbf{y}) \rangle$$

with high probability. Random Fourier features have become a mainstay in kernel learning but do rely on Bochner's theorem which does not immediately extend to kernels on non-linear domains such as the signature kernel (recall that there is no natural addition on $\mathcal{X}_{\text{seq}}$). Nevertheless, [70] mimicks a similar argument to construct *"Random Fourier Signature Features"*. This allows us to combine the strengths of the signature kernel with *linear complexity* in both the number of sequences and sequence length by using an approximation that holds with high probability. Experiments show that this is in a general a favourable tradeoff and allows us to tackle challenging time-series datasets. We refer the reader to the User's Guide to `KSig`, [66], for a detailed discussion.

Learning to Forget. The signature (kernel) provides a global summary of a sequence, giving equal weight to events that happened recently and in distant history. Sometimes this can be a strength, but sometimes one needs to emphasize recent events; especially, in long time series. So far, this has mostly been addressed by ad hoc approaches, but a principled, data-driven Bayesian approach of "learning to forget" was developed in [67].

Acknowledgments HO and DL were supported by the Hong Kong Innovation and Technology Commission (InnoHK Project CIMDA). HO was also supported by the EPSRC grant Datasig [EP/S026347/1].

References

1. R. Adler, J. Taylor, *Random Fields and Geometry*. Springer Monographs in Mathematics (Springer, New York, 2009)
2. H. Andrès, A. Boumezoued, B. Jourdain, Signature-based validation of real-world economic scenarios. *arXiv preprint arXiv:2208.07251* (2022)
3. C. Bahlmann, B. Haasdonk, H. Burkhardt, Online handwriting recognition with support vector machines-a kernel approach, in *Proceedings Eighth International Workshop on Frontiers in Handwriting Recognition*. (IEEE, New York, 2002), pp. 49–54
4. P. Bonnier, C. Liu, H. Oberhauser, Adapted Topologies and Higher Rank Signatures. Ann. Appl. Probab. **33**(3), 2136–2175 (2023)
5. P. Bonnier, H. Oberhauser, Signature cumulants, ordered partitions, and independence of stochastic processes. Bernoulli **26**(4), 2727–2757 (2020)
6. R.W. Brockett, Volterra series and geometric control theory. Automatica **12**(2), 167–176 (1976)
7. H. Buehler, B. Horvath, T. Lyons, I.P. Arribas, B. Wood, A data-driven market simulator for small data environments. *arXiv preprint arXiv:2006.14498* (2020)
8. T. Cass, T. Lyons, X. Xu, Weighted signature kernels. Ann. Appl. Probab. **34**(1A), 585–626 (2024)
9. T. Cass, W.F. Turner, Topologies on unparameterised path space. J. Funct. Anal. **286**(4), 110261 (2024)
10. K.-T. Chen, Iterated integrals and exponential homomorphisms†. Proc. Lond. Math. Soc. **s3-4**(1), 502–512 (1954)
11. K.-T. Chen, Integration of paths, geometric invariants and a generalized Baker- Hausdorff formula. Ann. Math. **65**(1), 163–178 (1957)
12. K.-T. Chen, Integration of paths—a faithful representation of paths by noncommutative formal power series. Trans. Am. Math. Soc. **89**(2), 395–407 (1958)
13. I. Chevyrev, T. Lyons, Characteristic functions of measures on geometric rough paths. Ann. Probab. **44**(6), 4049–4082 (2016)

14. I. Chevyrev, V. Nanda, H. Oberhauser, Persistence paths and signature features in topological data analysis. IEEE Trans. Pattern Anal. Mach. Intell. **42**, 192–202 (2018)
15. I. Chevyrev, H. Oberhauser, Signature moments to characterize laws of stochastic processes. J. Mach. Learn. Res. **23**(176), 1–42 (2022)
16. I. Chevyrev, H. Oberhauser, Signature Moments to Characterize Laws of Stochastic Processes. *arXiv preprint 1810.10971.v1* (2022)
17. F. Cucker, S. Smale, On the mathematical foundations of learning. Bull. Am. Math. Soc. **39**(1), 1–49 (2002)
18. M. Cuturi, J.-P. Vert, Ø. Birkenes, T. Matsui, A kernel for time series based on global alignments, in *2007 IEEE International Conference on Acoustics, Speech and Signal Processing—ICASSP '07*, vol. 2 (2006), pp. II–413–II–416
19. T.K. Dey, Y. Wang, *Computational Topology for Data Analysis* (Cambridge University, Cambridge, 2022)
20. J. Diehl, K. Ebrahimi-Fard, N. Tapia, Generalized iterated-sums signatures. J. Algebra **632**, 801–824 (2023)
21. P. Drineas, M.W. Mahoney, N. Cristianini, On the Nyström method for approximating a gram matrix for improved kernel-based learning. J. Mach. Learn. Res. **6**(12), 2153–2175 (2005)
22. R.M. Dudley, Sample functions of the gaussian process, in *Selected works of RM Dudley* (Springer, Berlin, 2010), pp. 187–224
23. J. Dyer, P. Cannon, S.M. Schmon, Amortised likelihood-free inference for expensive time-series simulators with signatured ratio estimation, in *International Conference on Artificial Intelligence and Statistics* (2022)
24. J. Dyer, J. Fitzgerald, B. Rieck, S.M. Schmon, Approximate bayesian computation for panel data with signature maximum mean discrepancies, in *NeurIPS 2022 Temporal Graph Learning Workshop* (2022)
25. A. Fermanian, Functional linear regression with truncated signatures. J. Multivar. Anal. **192**, 105031 (2020)
26. A. Fermanian, P. Marion, J.-P. Vert, G. Biau, Framing RNN as a kernel method: a neural ode approach. Adv. Neural Inf. Proces. Syst. **34**, 3121–3134 (2021)
27. M. Fliess, Fonctionnelles causales non linéaires et indéterminées non commutatives. Bulletin de la Société Mathématique de France **109**, 3–40 (1981)
28. D. Garreau, W. Jitkrittum, M. Kanagawa, Large sample analysis of the median heuristic. *arXiv:1707.07269 [math, stat]* (2018)
29. R. Giles, A generalization of the strict topology. Trans. Am. Math. Soc. **161**, 467–474 (1971)
30. C. Giusti, D. Lee, Signatures, lipschitz-free spaces, and paths of persistence diagrams. SIAM J. Appl. Algebra Geom. **7**(4), 828–866 (2023)
31. A. Gretton, K.M. Borgwardt, M.J. Rasch, B. Schölkopf, A. Smola, A kernel two-sample test. J. Mach. Learn. Res. **13**, 723–773 (2012)
32. A. Gretton, K. Fukumizu, C. Teo, L. Song, B. Schölkopf, A. Smola, A kernel statistical test of independence, in *Advances in Neural Information Processing Systems*, vol. 20 (2007)
33. A. Gretton, D. Sejdinovic, H. Strathmann, S. Balakrishnan, M. Pontil, K. Fukumizu, B.K. Sriperumbudur, Optimal kernel choice for large-scale two-sample tests, in *Advances in Neural Information Processing Systems*, ed. by F. Pereira, C. Burges, L. Bottou, K. Weinberger, vol. 25 (Curran Associates, Inc., New York, 2012)
34. B. Hambly, T. Lyons, Uniqueness for the signature of a path of bounded variation and the reduced path group. Ann. Math. **171**(1), 109–167 (2010)
35. F.R. Hampel, *Robust Statistics: The Approach Based on Influence Functions* (Wiley, New York, 1986)
36. D. Haussler et al., *Convolution Kernels on Discrete Structures*. Technical report (Citeseer, New York, 1999)
37. C.C. Heyde, On a property of the lognormal distribution. J. R. Stat. Soc. Ser. B Methodol. **25**(2), 392–393 (1963)
38. D.N. Hoover, H.J. Keisler, Adapted probability distributions. Trans. Am. Math. Soc. **286**, 159–201 (1984)

39. B. Horvath, M. Lemercier, C. Liu, T. Lyons, C. Salvi, Optimal Stopping via Distribution Regression: A Higher Rank Signature Approach. *arXiv:2304.01479 [math]* (2023)
40. P.J. Huber, E.M. Ronchetti, *Robust Statistics* (Wiley, New York, 2011)
41. P. Kidger, J. Foster, X. Li, H. Oberhauser, T.J. Lyons, Neural SDES as infinite-dimensional GANS, in *International Conference on Machine Learning* (PMLR, New York, 2021), pp. 5453–5463
42. F.J. Kiraly, H. Oberhauser, Kernels for sequentially ordered data. J. Mach. Learn. Res. **20**(31), 1–45 (2019)
43. J.B. Kruskal, An overview of sequence comparison: Time warps, string edits, and macromolecules. SIAM Rev. **25**(2), 201–237 (1983)
44. D. Lee, R. Ghrist, Path Signatures on Lie Groups. *arXiv:2007.06633 [cs, math, stat]* (2020)
45. M. Lemercier, C. Salvi, T. Cass, E.V. Bonilla, T. Damoulas, T.J. Lyons, Siggpde: scaling sparse gaussian processes on sequential data, in *International Conference on Machine Learning* (PMLR, New York, 2021), pp. 6233–6242
46. C.S. Leslie, R. Kuang, Fast string kernels using inexact matching for protein sequences. J. Mach. Learn. Res. **5**, 1435–1455 (2004)
47. C. Litterer, H. Oberhauser, On a Chen–Fliess approximation for diffusion functionals. Monatsh. Math. **175**(4), 577–593 (2014)
48. H. Lodhi, C. Saunders, J. Shawe-Taylor, N. Cristianini, C.J.C.H. Watkins, Text classification using string kernels. J. Mach. Learn. Res. **2**, 419–444 (2002)
49. T. Lyons, Z. Qian, *System Control and Rough Paths* (Clarendon, Oxford, 2007)
50. C. Moler, C. Van Loan, Nineteen dubious ways to compute the exponential of a matrix, twenty-five years later. SIAM Rev. **45**(1), 3–49 (2003)
51. K. Muandet, K. Fukumizu, B. Sriperumbudur, B. Schölkopf, et al., Kernel mean embedding of distributions: A review and beyond. Found. Trends® Mach. Learn. **10**(1-2), 1–141 (2017)
52. N. Muca Cirone, M. Lemercier, C. Salvi, Neural signature kernels as infinite-width-depth-limits of controlled ResNets, in *Proceedings of the 40th International Conference on Machine Learning*, ed. by A. Krause, E. Brunskill, K. Cho, B. Engelhardt, S. Sabato, and J. Scarlett. *Proceedings of Machine Learning Research*, vol. 202 (PMLR, New York, 2023), pp. 25358–25425
53. G. Pammer, A note on the adapted weak topology in discrete time. *arXiv preprint arXiv:2205.00989* (2022)
54. C.E. Rasmussen, C.K.I. Williams, *Gaussian Processes for Machine Learning*. Adaptive Computation and Machine Learning Series (MIT Press, Cambridge, 2005)
55. C. Reutenauer, *Free Lie Algebras*. London Mathematical Society Monographs. (Oxford University, Oxford, 1993)
56. H. Sakoe, S. Chiba, Dynamic programming algorithm optimization for spoken word recognition. IEEE Trans. Acoust. Speech Signal Process. **26**(1), 43–49 (1978)
57. C. Salvi, T. Cass, J. Foster, T. Lyons, W. Yang, The signature Kernel is the solution of a Goursat PDE. SIAM J. Math. Data Sci. **3**(3), 873–899 (2021)
58. C. Salvi, M. Lemercier, C. Liu, B. Horvath, T. Damoulas, T. Lyons, Higher order kernel mean embeddings to capture filtrations of stochastic processes, in *Advances in Neural Information Processing Systems*, ed. by M. Ranzato, A. Beygelzimer, Y. Dauphin, P. Liang, J.W. Vaughan, vol. 34 (Curran Associates, Inc., New York, 2021), pp. 16635–16647
59. A. Schell, A Robustness Analysis of Blind Source Separation. *arXiv:2303.10104 [math, stat]* (2023)
60. A. Schell, H. Oberhauser, Nonlinear independent component analysis for discrete-time and continuous-time signals. Ann. Stat. **51**(2), 487–518 (2023)
61. M. Schetzen, *The Volterra and Wiener Theories of Nonlinear Systems* (Krieger Publication, New York, 2006)
62. B. Scholkopf, A.J. Smola, *Learning with Kernels: Support Vector Machines, Regularization, Optimization, and Beyond* (MIT Press, New York, 2018).
63. H. Shimodaira, K.-i. Noma, M. Nakai, S. Sagayama, Dynamic time-alignment kernel in support vector machine. Adv. Neural Inf. Proces. Syst. **14** (2001)

64. C.-J. Simon-Gabriel, A. Barp, B. Schölkopf, L. Mackey, Metrizing weak convergence with maximum mean discrepancies. *arXiv preprint arXiv:2006.09268* (2020)
65. B.K. Sriperumbudur, K. Fukumizu, A. Gretton, G.R.G. Lanckriet, B. Schölkopf, Kernel choice and classifiability for RKHS embeddings of probability distributions, in *NIPS* (2009)
66. C. Tóth, D.J.D. Cruz, H. Oberhauser, A User's Guide to KSIG: GPU-Accelerated Computation of the Signature Kernel. *arxiv:2501.07145* (2025)
67. C. Tóth, M. Adachi, M.A. Osborne, H. Oberhauser, Learning to Forget: Bayesian Time Series Forecasting using Recurrent Sparse Spectrum Signature Gaussian Processes, in *Proceedings of The 28th International Conference on Artificial Intelligence and Statistics*, ed. by Y. Li, S. Mandt, S. Agrawal, E. Khan, vol. 258 (Proceedings of Machine Learning Research, PMLR) pp. 4654–4662 (2025). https://raw.githubusercontent.com/mlresearch/v258/main/assets/toth25b/toth25b.pdf. https://proceedings.mlr.press/v258/toth25b.html
68. C. Tóth, P. Bonnier, H. Oberhauser, Seq2Tens: An efficient representation of sequences by low-rank tensor projections, in *International Conference on Learning Representations* (2021)
69. C. Tóth, H. Oberhauser, Bayesian learning from sequential data using Gaussian Processes with Signature Covariances, in *Proceedings of International Conference on Machine Learning (ICML)*, vol. 1 (2020)
70. C. Tóth, H. Oberhauser, Z. Szabó, Random Fourier Signature Features. SIAM J. Math. Data Sci. **7**(1), 329–354 (2025). https://doi.org/10.1137/23M1620478

Part II
Applications of Signature Methods in Finance

Market Generators: A Paradigm Shift in Financial Modeling

Blanka Horvath, Jonathan Plenk, and Milena Vuletić

Abstract Market Generators are a rapidly evolving class of neural-network-based models to simulate financial market behavior, offering a powerful alternative to classical stochastic models. These deep learning models are trained to encode the underlying distribution of financial data and generate new synthetic market scenarios from the learned distribution. Though the expressions "Market Generator" and its related form "Market Simulator" have only entered the vocabulary of financial modeling around 2019, by today, the modeling techniques related to them have already grown into an area in their own right. This growing interest is matched by a dramatic rise in research, publications and an accelerating rate of innovation in the ambient technological arena of generative modeling. Recently, this trend has culminated in the emergence of large generative networks, particularly GPT-type language models, which are proving to be driving one of the biggest disruptions in the history of technology.

1 Introduction

This chapter is dedicated to providing a brief synopsis of the evolution of Market Generator models, embedding their emergence against the backdrop of the development of deep learning over the past decade. The chapter outlines the distinct challenges involved in generating various types of financial data, and the aspects which set this task apart from other forms of synthetic data generation.

It addresses how these models compare to existing techniques in finance including classical stochastic modeling and numerical techniques, and highlights the relevance of path-wise methods (such as rough path signatures) when taking the generative route. It also aims to unravel relationships (common characteristics

B. Horvath (✉) · J. Plenk · M. Vuletić
University of Oxford, Mathematical Institute and Oxford Man Institute of Quantitative Finance, Oxford, UK
e-mail: blanka.horvath@maths.ox.ac.uk; jonathan.plenk@maths.ox.ac.uk; milena.vuletic@maths.ox.ac.uk

C. Bayer et al. (eds.), *Signature Methods in Finance*, Springer Finance Lecture Notes, https://doi.org/10.1007/978-3-031-97239-3_4

and differences) in generative models proposed as Market Generators to date, to the most recent technological advances in Generative AI including the recently emerged foundation models. The main value of this lies in contrasting finance-specific objectives with general AI trends, illustrating barriers and bottlenecks for applying the off the shelf AI solutions directly in financial contexts.

A critical challenge in generative modeling lies in evaluating the quality of generated data, as no universally accepted framework exists for assessing and validating outputs produced by trained models, especially in financial settings which present distinctive challenges. Metrics to evaluate the success of generative models in financial tasks are often the centerpiece of the success of the Market Generator as well as associated downstream tasks.

In this piece we show how signature methods can provide strikingly effective tools to work with time series, data-streams and path-valued distributions in an efficient and versatile way, that is future-proof for the directions where generative AI is headed. While the field is expected to advance rapidly at the time of writing, these distinctions between broader GenAI trends and finance-specific requirements, rooted in insights from decades of quantitative analysis and risk management, are likely to endure.

1.1 Generative Models, Generative AI, Market Generators, and Related (Recent) Developments

The concept of Generative AI (GenAI) has rapidly evolved, especially since the successes of large language models (LLMs) like GPT 3.5 and 4. Although these models have different architectures, they share common features. This section provides a brief overview of the main concepts and terminologies related to generative modeling and distinguishes some concepts and methods which are already adopted in finance, and the ones currently emerging in the field.

> Generative AI

Generative AI refers to a type of artificial intelligence that can create (generate) new data, such as text, images, videos, music, or code, resembling the patterns and structure of the data it was trained on, typically *in response to prompts*.

The key elements are:

1. Generation of *novel* outputs which exhibit similar characteristics to the training data.
2. Reliance on ***generative models*** to learn underlying patterns of the training data.

(continued)

3. Conditioning on *prompts* or inputs (text or sketches) to guide the generation process.

Examples of generative AI applications include (but are not limited to):

- Image generation: Creating realistic images of people, objects, or scenes.
- Video synthesis: Creating videos of people or objects performing actions.
- Text generation: Producing articles, stories, poems, or code snippets.
- Music composition: Generating original pieces of music in different styles.
- Drug discovery: Generating new molecules with desired properties.

Recent breakthroughs in the field have the potential to drastically change the way we approach content creation.

> **Foundation Models**
> The related term **foundation model** was coined by researchers[1] in August 2021 to mean *any model that is trained on broad data that can be adapted (e.g., fine-tuned) to a wide range of "downstream tasks"*. The term is often used to describe generative AI models capable of performing a wide variety of general tasks (such as generating text and images) as well as specific *downstream tasks* within these areas.

Generative AI relies on Generative Models for the generation of new data.

> **Generative Models**
> **Generative Models** are deep learning models that can be trained to capture the underlying patterns and structure of input data in order to generate new, similar data. There are many different approaches (EBMs, VAEs, GANs, normalizing flows, diffusion models, see below) for learning how to generate synthetic data statistically resembling key characteristics of the training data. These models are capable of creating *new data*, such as images, text, music, or other formats of content.

Deepfakes[2] (a portmanteau of "deep learning" and "fake") are synthetic media productions which have been digitally created or manipulated to convincingly

[1] At the Stanford Institute for Human-Centered Artificial Intelligence's (HAI) Center for Research on Foundation Models (CRFM).

[2] The term, coined in 2017 by an anonymous Reddit user.

replace one person's likeness with that of another. The term has been expanded to cover any videos, pictures, or audio made with AI to appear deceptively real, by fabricating observably realistic content of individuals who (1) do not exist or (2) did not participate in the captured content.

Generative Modeling Avenues. The Main Approaches to Generative Modeling

There are several different types of generative models, each with its own approach to generating data:[3]

1. **Energy-Based Models (EBMs)** follow the idea of parameterizing the unnormalized negative log-likelihood ("Energy") with a neural network, $p_\theta(x) = \frac{1}{Z_\theta} e^{-E_\theta(x)}$. The model can be trained by maximizing the likelihood of the given samples. The difficulty lies in approximating the normalization factor ("partition function") $Z_\theta = \int e^{-E_\theta(x)} dx$. Alternatively, the network can be trained by "score-matching", which amounts to matching the score $\nabla_x \log p_\theta(x)$ to the score of the empirical data distribution. One advantage of EBMs is the access to the likelihood function, however the drawback is that sampling from the model (e.g. via Langevin sampling or other likelihood-based sampling methods) is computationally expensive.
2. **Generative Adversarial Networks (GANs)** consist of two neural networks, a generator and a discriminator, that compete against each other. The generator tries to create artificial data that looks real, while the discriminator tries to distinguish between real and generated data. This competition drives both networks to improve until the generator can produce convincingly realistic data.
3. **Neural Stochastic Differential Equations (Neural SDEs)** are continuous-time generative models for sequential data. The mean and volatility parameter of the SDE which determines the probability distribution of the generated time series are parametrized by a neural network.
4. **Variational Autoencoders (VAEs)** are a type of generative model that use an encoder to compress input data into a lower-dimensional latent space and a decoder to reconstruct the original data from the latent space. The latent space captures the underlying structure of the data, allowing the decoder to generate new data by sampling from that space.

(continued)

[3] For further details see for example: Bishop & Bishop, Section 16.4.

5. **Deep Autoregressive Models**[4] These models generate discrete data sequentially, one element at a time, based on the previous elements. The discrete probability distribution is typically defined by applying the softmax function to the output layer of the neural network. They are commonly used for tasks such as text generation (e.g. in LLMs), where the next word in a sentence is predicted based on the previous words. Another use case is generating pictures, where the probability distribution of a pixel (whose value is discrete) depends on the values of close-by pixels.
6. **Flow-Based Models (Normalizing Flows)** learn a series of invertible transformations that can transform simple probability distributions into complex ones. They are often used for tasks where understanding the transformation of data is crucial. There are many choices for parameterizing the invertible transformations with neural networks. A big advantage is that one has access to both the likelihood and an efficient way to generate samples.
7. **Diffusion models** are a class of generative models that (like the continuous-time version of a Flow-based model) learn to iteratively refine noise into samples from the target probability distribution. This so-called reverse diffusion process (reversing the forward diffusion process which turns the data distribution into noise) is a neural SDE, whose parameters are learned by score-matching for different noise levels. When combined with LLMs, these have achieved state-of-the-art results for picture generation based on text input.

> Market Generator, Market Simulator

The term **Market Generator**[5] refers to numerical techniques that rely on ***generative models*** for the purpose of ***synthetic market data*** generation. In essence, Market Generators are designed to capture and ***replicate the probability distribution*** of the input data via a generative model and to produce new data samples, whose distribution aligns with the captured distribution.

The term **Market Simulator** was initially used interchangeably with the term Market Generator. More recently, it is increasingly used to refer to models that generate market data by simulating ***interactions*** between different market participants.

[4] Note, that these should not be confused with Linear Autoregressive models (e.g. AR(1)) which are used to model time series with a continuous state space.

[5] It was coined in 2019 in the paper by Kondratyev and Schwarz.

1.2 Historical Background: Development of Neural Networks and Generative Models

Historical Development: The Rise of Deep Generative Models

- **The Resurgence of Deep Learning:** Beginning in the late 2000s, the resurgence of deep learning drove progress and research in image classification, speech recognition, natural language processing and other tasks. Neural networks in this era were typically trained as discriminative models, due to the challenges of generative modeling.
- **VAEs and GANs:** Introduced in 2013, VAEs were among the first deep-learning models to successfully generate realistic images and speech. According to Akash Srivastava, an expert on generative AI at the MIT-IBM Watson AI Lab, "VAEs opened the floodgates to deep generative modeling by making models easier to scale."[6] This milestone laid the foundation for today's generative AI technologies. In 2014, GANs further advanced generative capabilities by introducing adversarial training, enabling the creation of highly realistic and complex image data. Together, VAEs and GANs marked a turning point in deep generative modeling, transitioning from simple discriminative outputs (e.g., class labels) to the generation of new images.
- **Transformers and LLMs:** In 2017, Google introduced the Transformer architecture in the landmark paper "Attention Is All You Need". It combines the encoder-decoder architecture with a text-processing mechanism called attention, revolutionizing language model training. OpenAI's GPT (Generative Pre-trained Transformer) series built on this, with GPT-1 (2018) demonstrating pre-training, GPT-2 (2019) showcasing generalization, and GPT-3 (2020) solidifying foundation models. ChatGPT, launched in 2022, brought generative AI to the mainstream, making advanced language models widely accessible. For further developments and current challenges see the Appendix.
- **Diffusion Models and Generative AI for Art:** The release of DALL-E in 2021 showcased the ability to generate art from natural language prompts using a VAE-based approach, but it had limitations in image quality and fidelity. In contrast, diffusion models, introduced around the same time, achieved outputs of much higher quality. These advancements powered DALL-E 2, which significantly surpassed its predecessor in generating detailed, realistic, and coherent images. Diffusion models quickly became the backbone for other AI art tools like Stable Diffusion and MidJourney, further revolutionizing AI-generated art and creative industries.

[6] https://research.ibm.com/blog/what-is-generative-AI

1.3 Where Quantitative Finance Goals Deviate from Traditional Applications of Generative Models

Generative models excel in tasks such as photo and language generation, but their application in financial data generation, particularly for quantitative finance, presents unique challenges. It may seem natural (and tempting) to directly apply off-the-shelf generative modeling techniques from other domains to these areas of finance, but there are some differences to be mindful of when doing so.

For example, there is a difference between permissible *invariance*s in visual tasks like image or video generation and financial data generation. These invariances associated with the data are often reflected in the (optimal) architecture of the associated neural network performing the task. In visual domains, transformations such as scaling, rotating, mirroring, or translating an image generally preserve its essential identity—for example, an image of a dog remains identifiable. However, analogous operations in financial time series can drastically alter their structure. Consequently, off-the-shelf solutions developed for domains where these invariances are central, can be erroneous when transferred to financial settings without further attention.

Another difference is the relevance of the underlying distribution in the training data. This is reflected in the fact that **Market Generators** explicitly place an **emphasis on the distribution of the input and output data**, while in other generative applications this is often not needed: in many generative applications, the goal is to produce *one* realistic output at a time, e.g. one realistic image. The quality of each of the outputs can be evaluated individually, e.g. by visual inspection, or by the plausibility of the produced sequence or coherence of the text.

In financial modeling, however, the probabilistic composition of the training set is central, in fact it is typically the main target of modeling. Based on observed data, models are used to infer the underlying distribution of the training data and therefore, outputs of Market Generators are typically probabilistic. Multiple outputs will form an (empirical) distribution that can be evaluated based on its similarity/score or/distance to the (empirical or believed) underlying distribution of the training data. This wouldn't be feasible with a single observation value: for instance, if the market generator is tasked with the objective to approximate a Gaussian distribution, it would be hard to tell—based on a single sample—whether the objective has been met successfully.

In many machine learning domains working with time series forecasting (TSF), there is a reflex to gravitate towards point estimates, but in financial settings there are a number of compelling reasons to target probabilistic outputs instead, see Sect. 2.1 below. Another point is that, while in physics and experimental sciences it is often possible to repeat the experiment multiple times and extract the underlying dynamics from multiple observations, in financial settings it is rarely possible to repeat the same exact experiment with real data, and this often results in worse signal-to-noise ratios. Since market data is prone to frequent regime shifts,

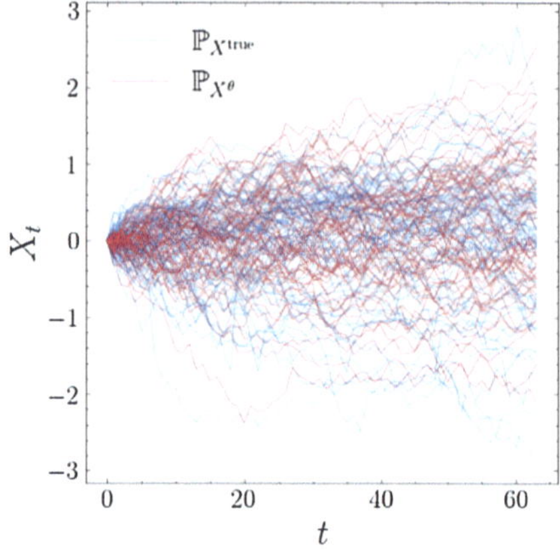

Fig. 1 A display of target paths $\mathbb{P}_{X^{\text{true}}}$ versus synthetic paths generated by a market generator $\mathbb{P}_{X^\theta}$ with model parameters θ shows the challenge of evaluating the model visually through samples

performance on (ever-changing) real data is a volatile metric of the quality of modeling success.[7]

The aforementioned differences in the nature of the data and intended application domain are also reflected in differences in the appropriate choice of loss functions and *quality evaluation metrics* for the generative task: the evaluation of the quality of outputs is rarely as simple as visual inspection,[8] see for example Fig. 1 in Sect. 3.1 below. Similarly, in regard to off-the-shelf loss functions, [8] lays out multiple examples where generative models trained under 'traditional' loss functions are not guaranteed to perform well in typical financial tasks such as portfolio construction. The construction of suitable loss functions and evaluation metrics are often the centerpiece of successful modeling with generative models in financial settings. Signature methods (as outlined in Sect. 3.1) provide a universal framework that allows us to bypass the limitations of working with time-series in a return-by-return fashion and to model data-streams as path-valued distributions instead, thereby unlocking versatile tools for modeling, high-fidelity features, and more efficient training.

2 Synthetic Data Generation in Finance: Classical Models and Generative Models

Though the emergence of generative models is a development of this past decade and the research on *Market Generators* (as defined above) has only gained traction in recent years, there is a far longer history of numerical modeling and synthetic

[7] A way out is to repeat experiments on synthetic data in controlled environments (for example on stochastic models).

[8] Even though we occasionally may be able to identify wildly unrealistic outputs by visual inspection.

data generation in mathematical finance. The aim of simulating realistic (probable) future scenarios of market activity has a history that spans over almost five decades.[9] Models that proxy for the dynamic evolution of asset prices have evolved to reflect more and more realistic properties (stylized facts) that are universally present in any data stream aiming to represent financial markets. These models can broadly be grouped into two classes:

1. **Continuous-time stochastic models**: This started with the Black-Scholes model, modeling the asset price process via a Geometric Brownian motion. These models are used for the pricing and hedging of derivatives. With options becoming liquid assets, and exotic derivatives being created, much more sophisticated models were developed to capture more aspects of the market (such as including stochastic volatility). The (few) parameters of these models are calibrated to the current prices of the market ("$\mathbb{Q}$-world").
2. **Discrete-time models for forecasting**: These models are used to forecast the mean and/or volatility of the asset returns. Standard models include Linear Autoregressive models (AR-models), and models capturing heteroscedasticity (e.g. GARCH-models). The parameters of these morels are calibrated to historical time series of prices ("$\mathbb{P}$-world").

2.1 Market Generators in Finance: Can Generative Models Add Value?

Financial time series simulation is a central topic since it bypasses rigidities imposed by parametric models and extends the limited real data for training and evaluation of DNN models. The fidelity and flexibility of the output is also crucial for enhancing data privacy and it has proven to add value for validating models, backtesting, and addressing issues related to data scarcity, e.g. in extreme scenarios.
Generators for Time Series Forecasting: Traditional time series models like ARIMA-GARCH have limitations when it comes to capturing non-linear relationships and adapting to changing market conditions. A probabilistic Market Generator offers a more flexible and powerful alternative, for time series generation in the $\mathbb{P}$-measure as demonstrated in [56, 61]. For example the probabilistic Market Generator in [56] allows for the simultaneous prediction of both the conditional mean and variance of time series data, which enables the generation of realistic synthetic financial market time series. For this purpose the objective in [56] is to create synthetic financial data that preserves key statistical properties of real market data while balancing performance and interpretability, making it suitable for regulated industries such as finance.

[9] Recall that 2023 marked the 50-year anniversary of the Black-Scholes Model.

> **Example: Time Series Modeling with Market Generators FIN-GAN**
FIN-GAN [61] explores the use of GANs for probabilistic forecasting of financial time series. Traditionally, this task has been achieved using methods that rely solely on point estimates. Classical machine learning and time series approaches do not allow for probabilistic forecasts or have strong assumptions on the form of the future distribution of the target variable.
FIN-GAN is a customised conditional GAN, which utilises ForGAN [37] architecture and is trained via a PnL-based loss function. The input condition a_t in the FIN-GAN case are the previous L values of the returns time series X_t, and the output is a return scenario for the next time step $G(a_t, Z)$, given Gaussian noise Z. As forecasting is mainly performed for trading purposes, FIN-GAN uses the distribution of $G(a_t, Z)$ to adjust the trade size based on the uncertainty in the direction of the move.

FIN-GAN is benchmarked against a classical deep learning method for time series forecasting, LSTM [29], and with a standard time series model ARIMA [59].

Empirical studies on equity returns demonstrate that FIN-GAN attains superior performance to the benchmarks.

Further Market Generators for time series forecasting include [24, 37, 56].

Forecasting the probability distribution of future scenarios, rather than just point estimates, is beneficial in low signal-to-noise environments even in scenarios where one was traditionally interested only in point estimates (the most-likely next-step). Schwarz takes this view a step further and argues in [56], that probabilistic outputs are more desirable. An example is synthetic data for market making, where probabilistic forecasts can enhance (both human and algorithmic) trading strategies, allowing for more profitable and risk-aware trading.

> **Example: A Case for Probabilistic outputs**
A market maker wants to place bids and ask quotes in a way that minimizes the probability of a loss while recycling acquired inventory. This requires both a liquidity model (predicting volume traded) and a model which forecasts the probability distribution of future returns. Trading strategies incorporating such probabilistic forecasts are superior to those relying on point estimates alone, since they allow traders to quantify confidence in their trades, determine optimal trade sizes, and improve risk management.

Generators for Stochastic Modeling
Similarly to the above, Market Generators grant added flexibility to fit more complex distributions, compared to traditional models for modeling with stochastic models. If one uses a (classical) parametric model to simulate (future) market scenarios, one first needs to settle on a model described by a set of (possibly quite restrictive) assumptions on the underlying distribution which is governed by a **(pre-)fixed set of parameters**. Then, when fitting a parametric model to the data, one selects those parameters that bring the model predictions closest to the observations in the target data. In data-driven, non-parametric modeling (such as generative models) the number and scale of model parameters is not pre-set but, the complexity and nature of the data at hand influences them for optimal fit.

An advantage of **(non-parametric) generative modeling** is that it can produce high-fidelity data, even in highly complex scenarios, without the need to make prior (restrictive) assumptions about the structure of the underlying distribution, contrary to parametric models.

Training There are multiple ways of training Market Generators; one appealing option is to do this directly on the level of samples, by comparing the samples produced by the model and samples in the training data, computing a similarity score resulting from this comparison, and backpropagating that through the neural network. Indeed, in this way the training/fitting of the Market Generator can be done[10] *without* explicit knowledge of the underlying distribution, which permits more modeling freedom if desired, using sample-based similarity metrics (such as MMD) as objective functions and backpropagation.

A further valuable advantage of Generative Models is the versatility of possibilities for **conditioning**: One of the particularly appealing features of Generative models is that they (almost always) come with a *conditonable* counterpart: VAE $\Rightarrow$ CVAE (Conditional Variational Auto Encoder), GAN $\Rightarrow$ CGAN (Conditional Generative Adversarial Network) and so on. Conditioning enables us to generate new samples from a distribution, *conditional* on a few requirements or on context.[11] While conditioning in parametric models was limited to the few parameters of the model, this feature of generative AI opens up a whole new range of possibilities for conditioning on more states on the world, which may be relevant to the future probability distribution of the considered asset.

Some Use Cases Where Such Flexibility Can Add Value One of the **first**[12] and primary directives in synthetic financial data generation at the onset of this line of research was the need for protecting the privacy of customers and entities

[10] In most cases of generative approaches (including GANs, VAEs, or Diffusion Models).

[11] This has previously appeared in the context of other AI applications like Image Generation too, for example when creating an image of Mona Lisa, but conditioned on the requirement that it should appear in Pop Art style.

[12] An early 2020 Study by JP Morgan summarized some of the initial compelling reasons for the use of generative models in finance, see https://www.jpmorgan.com/technology/artificial-intelligence/initiatives/synthetic data/synthetic equity market-data.

involved with banks, while still making it possible to share such data outside of the organisations that generate it: especially for research, risk assessment and auditing purposes of new technologies. Financial data contains some of the most sensitive and personally identifiable attributes of customers. Due to such privacy concerns and regulatory restrictions, financial institutions are often limited in sharing real data both internally and externally. Synthetic data, which is generated to mimic real financial data while protecting privacy, is proposed as a solution.

Secondly, synthetic data can be used to train data-hungry Deep Learning models on a richer set of scenarios, than real data can provide. This can be a trading strategy with the goal of maximizing profits (finding alpha), in portfolio management [8], or a hedging objective ("deep hedging" [4]). In the latter, the idea is to use a Market Generator, and train a strategy on a rich but highly realistic set of synthetic price path scenarios, with the objective of minimizing variance or a risk measure of choice. This leads to improved hedging strategies in real world scenarios. In this context, [8] points out the relevance of the initial sample size of the training data and some pitfalls of generating more data than originally available.[13]

Finally, generative models can be seen as extensions of classical parametric models. They can learn and replicate complex and nonlinear dependencies better than classical parametric models. Synthetic data can be used to create highly synthetic scenarios which are useful for risk management, scenario analysis, stress testing, and back-testing of trading strategies. This is especially important when the original data is scarce. Originally, parametric models where used for this purpose.[14] Generative models make it possible to make use of past datasets and condition them to more contemporary market conditions, or to include possible scenarios that are realistic, but may not have happened yet. They can faithfully replicate any parametric (target) model but they can also be used for high-dimensional (or otherwise highly complex) scenarios where parametric models are not available.

There are a number of examples where the data is highly complex in finance, where a more flexible model is beneficial. Most notably, these complexities come with unique challenges. Tackling these efficiently requires specialized architectures, and bespoke loss functions (objective functions, discriminators) during training. Indeed, performance evaluation metrics become a centerpiece of financial modeling with Market Generators, as we will see in the two examples below and in later sections. The complexities are dependent on the *asset class considered*, as each brings special challenges with it. For example, modeling dynamics of the implied volatility surface involves considering intricate no-arbitrage conditions.

[13] This is discussed in chapter MMD in greater detail.

[14] Such as the Pareto distribution, or assuming normal multivariate returns for risk management.

> **Example: High Dimensional Target Data of with Complex Dynamics Volatility Surface Modeling (Vol-GAN) [60]: Learning to Simulate SPX Implied Volatility Scenarios**
VOLGAN [60] of Cont and Vuletić is a generative model for arbitrage-free implied volatility surfaces. The implied volatility surface, which summarizes the cross-section of option prices across strikes and maturities, gives a snapshot of the state of the options market. Any model of implied volatility should appropriately capture the co-movements of implied volatilities across moneyness and time-to-maturity, reproduce the empirically observed dynamics of implied volatilities [15] and the underlying, be able to capture the smile, skew, and term structure, and satisfy arbitrage constraints [22, 25]. Given the high dimensionality of the volatility surface and the complexity of its dynamics, it is challenging to capture all these properties in a parametric model.
The model: VOLGAN is a custom conditional GAN with a smoothness penalty [33, 54] incorporated into the loss function, and arbitrage penalty scenario re-weighting [16]. VOLGAN receives as input

- the implied volatility surface at the previous date,
- the two previous underlying returns,
- the realized volatility from the previous period,

and Gaussian noise. It outputs (joint) scenarios for

- the return of the underlying asset and
- the implied volatility surface

for the next period, along with a set of weights (probabilities) associated with these scenarios. The outputs of the generator described above are a-priori not guaranteed to satisfy the static arbitrage constraints: call prices should be decreasing and convex in strike and increasing in time to maturity. To correct for this, methodology from [16] is applied to re-weight the one-day-ahead scenarios generated by the GAN.

This example shows that a carefully crafted loss function is central for the success of the Market Generator, and it is also a central ingredient for a number of downstream tasks related to evaluating similarity of distributions in the target application space. Further Market Generators for volatility surfaces include, but are not limited to [9, 11, 13, 49, 63, 64].

The following paper addresses returns distributions, and again its success in efficiently capturing tail events is due to a well-chosen loss function. It replaces the classical loss functions used in GANs with a financially relevant one based on the joint elicitability of a risk measure couple, thereby forcing the model to better learn the tails of return distributions.

> **Example: Modeling Extreme and Rare Events: TAIL-GAN [17]**
TAIL-GAN [17] (Cont, Cucuringu, Xu and Zhang) is a custom (unconditional) GAN designed for tail risk scenario simulations, with the aim of being used for estimating Value-at-Risk (VaR) and Expected Shortfall (ES) for both static and dynamic portfolios.[15] Gneiting [26] and Weber [62] show ES is not elicitable,[16] whereas VaR at level $\alpha \in (0, 1)$ is elicitable for random variables with a unique α-quantile, however, the pair $(\text{VaR}_\alpha(\mu), \text{ES}_\alpha(\mu))$ is jointly elicitable. TAIL-GAN uses the result from [23] which gives a family of score functions which are strictly consistent for $(\text{VaR}_\alpha(\mu), \text{ES}_\alpha(\mu))$. Therefore, TAILGAN constructs a min-max game for the generator and the discriminator such that the equilibrium point is given precisely by Value-at-Risk and Expected Shortfall.

2.2 *(Why) Is There a Paradigm Shift Towards Generative Models? Where Is Generative AI Headed?*

While a complete *paradigm shift might not have happened yet*, the increasing adoption, technological advancements, and evolving regulatory landscape suggest that generative models are poised to play a more significant role in financial modeling in the future. The extent to which they transform the industry will depend on how well these challenges are addressed and how effectively generative models can demonstrate their value in real-world financial applications. Additionally, the recent emergence of foundation models and their convenient use in all applications, including finance (see Appendix for some examples), points in the direction that this shift in the near and foreseeable future will be continuing towards generative AI and sequence to sequence models.

[15] Recall that is X denotes the PnL of a portfolio at a certain horizon, under probability density μ, Value-at-Risk and Expected Shortfall at level $\alpha \in (0, 1)$ are defined by

$$\text{VaR}_\alpha(\mu) := \inf\{x \in \mathbb{R} : \mu(X \leq x) \geq \alpha\}, \text{ and } \text{ES}_\alpha(\mu) := \frac{1}{\alpha}\int_0^\alpha \text{VaR}_\beta(\mu)d\beta. \tag{1}$$

[16] Recall that a statistical functional $T : \mathcal{F} \mapsto \mathbb{R}$ defined on a set of distributions $\mathcal{F}$ on $\mathbb{R}^d$ is *elicitable* if there is a score function $S(x, y)$ such that

$$T(\mu) = \arg\min_x \int S(x, y)\mu(dy), \tag{2}$$

for any $\mu \in \mathcal{F}$. The elicitability of a functional T implies that it might be obtained through machine learning by using the loss $S(x, y)$.

? Where is Generative AI Headed?
Will generative models be *bigger* or *smaller* than today?
Trend towards bigger models

- Until recently, a **dominant trend** in generative AI has been **growing scale**, with larger models trained on **ever-growing datasets**. Scaling laws allow researchers to estimate how powerful a new, larger model (whether larger in size or trained on more data) will be, before investing in the tremendous computing resources it takes to train them.
- There is a continued interest in the *emergent capabilities arising when a model reaches a certain size*. Examples include *glimmers of logical reasoning*. It's not just the model's architecture that causes these skills to emerge but its scale.

Trend towards smaller models:
More recent trends point toward ***smaller models*** trained on ***domain-specific data***. We expect models to shrink in size, speeding up tuning and inference.

- **Goal: Reducing Carbon Footprint of Generative Models**
  Researchers are increasingly focused on developing energy-efficient methods to train, tune, and run AI models to lower costs and reduce AI's enormous carbon footprint. Smaller models and specialization comes with advantages; a smaller model is vastly cheaper and less carbon-intensive to run.
  Example: One encouraging finding is that effective AI models can be a lot smaller than they are today. DeepMind recently showed in [30] that a smaller model trained on more data could outperform a model four times larger trained on less data.[17]

- **Better performance with specialization:** *Smaller models* trained on more *domain-specific data* can often outperform larger, general-purpose models.
  "When you want specific advice, it may be better to ask a domain expert for help rather than trying to find the single smartest person you know" (David Cox, IBM).
  Example: PubMedGPT 2.75B is a relatively small model, trained on biomedical abstracts. Researchers at Stanford found that it could answer medical questions significantly better than a generalist model the same size. Their work suggests that smaller, domain-specialized models may be the right choice when domain-specific performance is important.

[17] The key to this was to scale parameters and data proportionally: Double the size of your model, then double the size of your training data for optimal performance.

We believe the future holds both generalist models with more capabilities, but also more specialization with smaller, more efficient models, via dynamically navigating the weight-composition of specialist models according to their relevance in the given setting. Two promising approaches are using mixture of experts as in DeepSeekMoE [21] or using filtering as in Filtered-not-Mixed [55].

2.3 A Paradigm Shift Towards Path-Wise Modeling, and Sequence-to-Sequence Models

The concept of generative AI has evolved, especially after the successes of LLMs like GPT 3.5 and 4. Though based on different Generative modeling cores and different architectures, these models share a common feature: they are probabilistic, autoregressive sequence models, where an efficient methodology to capture path-wise effects and relationships in the data-stream is poised to become increasingly important. Here, we give a quick outlook into two generative modeling examples where *signature methods* have demonstrated significant advantages, making the generation task more efficient. See more in Sect. 3.1.

> **Deep Signature Transforms [34]**
Kidger et al. [34] describes the signature as an infinite graded sequence of statistics, which can uniquely characterize a stream of data up to a negligible equivalence class. As introduced in Chaps. "A Primer on the Signature Method in Machine Learning" and "An Introduction to Tensors for Path Signatures", the signature transform can be seen as a feature transformation, on top of which a model may be built. A striking property of the signature transform is the efficiency with which it can capture the relevant information in a data stream. The paper **Deep signature transforms** proposes an approach which combines the advantages of the signature transform with modern deep learning frameworks for efficient generation of data streams, including financial time series.[18] This idea has been further refined in the *Rough Transformers* paper [46], where the authors address the struggles of traditional recurrent models with real-world time series data, which often exhibit long-range dependencies and irregular sampling intervals. The Rough Transformer makes use of signature attention to capture multi-scale dependencies efficiently while reducing computational costs.

[18] See https://github.com/patrick-kidger/Deep-Signature-Transforms for the code.

Another route to explaining this viewpoint's efficiency in capturing time series is provided by neural controlled differential equations (CDEs), which can be seen as the continuous-time analogue of recurrent neural networks and offer a memory-efficient continuous time way to model functions of potentially irregular time series. The paper [35] demonstrates how to represent the input signal of streamed data over small time intervals through its log signature, which are statistics describing how the signal drives a CDE. This is the approach for solving rough differential equations (RDEs). This view offers significant training speed-ups, improvements in model performance, and reduced memory requirements compared to existing approaches.

In a related theme, [41] aproximates SDEs using RNNs with log signature features.

Trends from the very recent past indicate that Deep **Structured state-space models (SSMs)**[19] deliver exceptional performance across application domains while offering reduced training and inference costs compared to attention-based transformers.

> **The Efficiency of Deep Selective State-Space Models Has Theoretical Foundations, Closely Connected with Rough Path Theory [12]**
(SSMs) are increasingly recognized as effective approaches for modeling sequential data. Recent advances show that incorporating multiplicative interactions between inputs and hidden states into the linear recurrence mechanisms powering SSMs—exemplified by architectures like GateLoop, Mamba, and GLA—enables these models to outperform attention-based foundation models, both in accuracy and computational efficiency, even at billion-parameter scales. The work [12], explains the empirical success of selective SSMs such as Mamba but also establishes a rigorous basis for understanding the expressive potential of future SSM variants, providing theoretical foundation for these observations rooted in Rough Path Theory.

3 The Roles and Advantages of Signature Methods for Generative Modeling of Financial Data Streams

Signature-Based Modelling and Generation of Paths Signature-based techniques recently emerged as leading machine learning technology for learning time series data. They offer a mathematically precise framework for capturing interactions within complex, evolving data streams. This precision lends itself to effective numerical methods for analyzing streamed data, particularly in contexts

[19] Such as S4, building on the foundational work of Gu et al. [28].

where the data is irregular [32, 43], non-stationary [5, 48], and of moderate dimensionality and sample size, see [42] for a comprehensive survey.

Below we highlight two compelling examples of models using the signature in the path generation process and remark that a full list would go hopelessly beyond the limitations of this chapter. In the next sections, we will look at how the signature can be used to discriminate between the generated paths.

Example: Sig-SDEs for Quantitative Finance [1]
Arribas et al. [1] proposes a data-driven model selection by integrating a classical quantitative setup with a generative modeling approach, leveraging the properties of the signature-transform. The framework provides a new perspective on SDEs and exotic financial products that depend, in a non-linear way, on the whole trajectory of asset prices. It further allows to simulate future possible market scenarios and consistently calibrate under the pricing measure $\mathbb{Q}$ and real-world measure $\mathbb{P}$.

Further papers in this direction include [19, 20, 32, 53], and many more.

Example: Signature-Based Models: Theory and Calibration [18]
Cuchiero et al. [18] presents asset price models whose dynamics are described by linear functions of the (time-extended) signature of a primary underlying process.[20] In this framework, available classical models can be approximated arbitrarily well and the model's parameters can be learned from all sources of available data by simple methods. Results presented in [18] include conditions guaranteeing absence of arbitrage in the associated model, as well as tractable option pricing formulas for *'sig-payoffs'*.

For more on this topic, see [18] and Chap. "Signature-Based Models in Finance" of this volume.

3.1 *Challenges and Pitfalls in Performance Evaluation of Market Generators*

A critical challenge in generative modeling lies in evaluating the quality of generated data, as no universally accepted framework exists for assessing and validating

[20] Any *continuous* market-inferred process from Brownian motion to multidimensional semimartingales.

outputs produced by trained models, cf. [8, 24]. Since **(a)** mathematical modeling of financial markets and associated simulation of market data has existed before the onset of market generation and since **(b)** ML researchers have used evaluation metrics for generative models outside of the domain of market generation, there are several potentially useful evaluation metrics available. But our task has special characteristics and (as we have established already) deviations from these domains, which may deem some of established evaluation metrics suboptimal, insufficient,[21] or otherwise unsuitable for Market Generators. Correspondingly, in this section we highlight some of the historically available metrics and discuss how the aforementioned differences in goals and settings manifest *with regard to performance evaluation metrics*. We discuss challenges in modeling and performance evaluation (a) **due to differences compared to other AI applications** or (b) **due to their differences compared to classical, (continuous-time) stochastic models** discussed in previous sections, and some challenges that have to do (c) more inherently with the purpose[22] of Market Generators to generate more volume of high-fidelity training date at instances where access to data is limited.
While it is straightforward to assess how realistic the output is in a model for photo generation, model evaluation is more complex in financial settings, since targets are distributional and path-valued. The evaluation of the quality of outputs will depend on the application of the model and will likely be investigated through tests tailored to the context.

We now break down challenges in each step of the generation[23] pipeline:

i. After observing a (limited) number of instances of the underlying probability distribution, the goal is to infer this distribution from the available observations.
ii. During the training phase, the underlying distribution is reconstructed via approximation by a generative model from which it can generate (any number of) new samples.
iii. Once the new samples have been generated it remains to verify that output is of good quality, e.g. that genuinely new data has been created and it is hard (impossible) to tell synthetic data apart from real data.

Challenges:

i. To infer the underlying distribution in neural network training, for step (i.) we implicitly *assume* that there is *one common* underlying probability distribution that our target data is sampled from ($\Rightarrow$ **(a)**). In a time series setting this translates to some form of temporal stability in the data. This amounts to the

[21] E.g. Due to the relative flexibility of the generative model in comparison with traditional models.

[22] Use-Case 2 in Sect. 2.1.

[23] For the related task of time series forecasting with point estimates, though the output is typically single-element-valued, the computation of this output nevertheless requires information about the underlying distribution of the data generating process, which needs to be inferred from observations. If this distribution is difficult to infer, the same challenges apply as explained in the MG context.

assumption of some form of stationarity,[24] but in reality these assumptions are not necessarily fulfilled, or only for very limited periods.[25] Financial time series are known to be highly **non-stationary**. As a result, inferring the underlying distribution (step i.) may suffer from the challenges of **low signal-to-noise ratio**, limited availability of data[26] or having to infer the underlying distribution from a **very limited number of samples** (before significant distributional changes happen).

ii. In order to train an appropriate model that can approximate an underlying distribution with high fidelity, one needs a **flexible (data-driven)** model (typically the richer the characteristics of the data, the more parameters needed) which is ***parsimonious*** enough that it can be trained reliably well on the limited data. Hence steps (i.) and (ii.) are in **conflict** with each other. While the limited data available may have been sufficient to calibrate the few parameters of classical models, this may be a challenge in the generative setting ($\Rightarrow$ **(b)**) if the number of parameters becomes large. A careful choice of features is beneficial and can keep the number of parameters at bay. On a different note, it is desirable to be able to train models **without explicitly specifying the underlying distribution.**[27]

iii. For performance evaluation, established targets from classical models may not carry over to all generative modeling settings or may become only necessary but not sufficient characteristics in the more flexible setting of Market Generators:

   - **Stylized facts** [14] used to be the bar that any new model needed to reach. These universal properties remain relevant and any Market Generator violating these can be deemed flawed. However, passing this requirement, without further ones, does not necessarily guarantee a good Market Generator in this setting of increased flexibility.
   - **QQ plot and common summary statistics**: though a sufficiently flexible generative model with sufficient training data can be expected to successfully approximate any target (marginal) distribution, a good performance at a finite number of (discretely observed) marginal distributions may have been very informative in classical settings, but for multiple types of generative models it would still leave the model unspecified between observation steps. Similar ideas hold for other typical summary statistics.

Usefulness of the synthetic data for the target application: Generated data should be as useful as the real data when used for the downstream task.

---

[24] Weak or strong.

[25] For the related task of time series forecasting (with point estimates) though the output is single-element-valued, the computation of this output nevertheless requires the underlying distribution of the data generating process to be inferred from observations.

[26] Financial time series data often suffer from limited length and quality.

[27] See the section below, for possibilities of doing so without explicit knowledge of the target distribution.

- For the use case of **data privacy** protection: We refer for the corresponding tools from differential privacy and recent adaptations to [2].
- If the use case is to enrich the training dataset for a neural network (e.g. performing trading [8] or hedging [6]) some consider the **'train on synthetic test on real' score**:[28] For example, in the case of deep hedging this would translate (as the name suggests) to re-training the hedging network on synthetic data and evaluating the profit and loss distribution of the hedging engine on real data, as a measure of success for the Market Generator's output. This means of evaluation is potentially expensive as depending on the application retraining and re-running of the model is computationally expensive (and potentially risky). Therefore, efficiently computable proxy metrics for the training data whose similarity can be translated to similar hedging strategies (hence similar PnL performance on real data) are desirable.
- For finite-dimensional distributions,[29] the most commonly used[30] distributional metrics are Wasserstein distances: e.g. *'Earthmover distance'*; Information theoretic divergences: e.g. Kullback-Leibler Divergence, or Maximum Mean Discrepancy (MMD)-type metrics, see [27]. If the target distribution is not explicitly known one can

  - **either** approximate distributions from observed samples and then work with the **similarity metrics for inferred distributions**,
  - **or** directly work on the level of samples with **two sample tests** to verify if the generated and the training samples originate from the same distribution (H_0) or not (H_A) without having to explicitly specify what that common distribution is at any point (Kolmogorov-Smirnov two sample test, or MMD-two sample test). The power of these tests depends on the sample size and the probability of a type 2 error is large if the sample size is small. See [8] and Chap. "Signature Maximum Mean Discrepancy Two-Sample Statistical Tests" of this volume for more on this.
  - **Scoring rules** assign a numerical score based on the predictive distribution and on the event or value that materializes [26]. Similarly to the idea of log-likelihood type evaluation methods, which evaluate the likelihood of real world observations using a density estimated from the generated data, scoring rules [26] can be used in settings where single observations in the real data are scored against an estimated density.

[28] We point to [3] and the webpage https://deepintostreams.github.io/Time-Series-Generative-Models.github.io/ curated by Hao Ni, Hang Lou, Baoren Xiao, and Jiajie Tao that records the relevant literature on time series generation.

[29] Signature (kernel) methods provide a means to establish such metrics beyond the finite-dimensional case to path-valued settings, as the following sections show.

[30] Among many available ones, see [3].

3.2 The Signature-Kernel-Backed Discriminator and Performance Evaluator for Market Generators

Signature (kernel) methods are appealing for data streams as they provide an elegant framework to consider distributions on path space. Denote by $\mathcal{X}$ the space of continuous paths of bounded variation from $[0, T]$ to $\mathbb{R}^{d_x}$. For any random variable X with values on $\mathcal{X}$, let $\mathbb{P}_X := \mathbb{P} \circ X^{-1}$ denote its law. Consider a target distribution $\mathbb{P}_{X^{\mathrm{true}}}(\cdot)$ on path space. The goal is to train a generative model such that the generated law $\mathbb{P}_{X^\theta}$ is as close as possible $\mathbb{P}_{X^{\mathrm{true}}}$, for some appropriate notion of *"closeness"*.

In the following examples, this notion of closeness will be derived from relevant properties of the *expected signature* of a stochastic process. The expected signature (see Section 6 in [42]) of a process X_t is the natural generalisation of the moments of a distribution (generalised to the process X_t) and the *exepcted log signature* $\log S(X)$ provides the natural generalisation of the cumulants of a distribution (to X_t). Hence, under appropriate assumptions[31] these quantities can be used to identify (characterize) the law of the stochastic process in question (See [42], p. 40).

First Generation of Signature-Based Market Generator Models, Based on Truncated Signature Kernels The two following examples[32] use the *truncated Signature Maximum Mean Discrepancy (Sig-MMD)* metric to evaluate the similarity of generated (synthetic) paths versus target (real) paths, which is closely related to the Wasserstein-1 metric on the relevant path space as indicated in [48].

> Example: Generating the Signatures of Log-Returns with VAEs (Using Truncated sig-MMD)

In [5], signatures were used in combination with VAEs to generate financial time series when the available training data is scarce. The methodology contains 5 main steps:

Step 1 **Data extraction from the time series:** Given a sample path from a data stream, subdivide the full time series into equal length intervals of (i) 1 day, (ii) 5 days (a business week), (iii) 20 days (a month)

Step 2 **Preprocessing the data:** To obtain training data from the resulting path segments, calculate log-returns $r(t, \Delta) := \log(S_{t+\Delta}/S_t)$ for $\Delta \in \{1, 5, 20\}$. Then, convert the obtained data samples into truncated log signatures, applying the lead-lag transformation. This will enable a path-wise generation process.

(continued)

[31] Assumptions and the extent to which the expected signature of a stochastic process X_t determines the law of X_t may be found in [10].

[32] Among possibly several more, similar ones.

Step 3 **Creating and training the VAE network:** After splitting the historical data into training/testing/validation sets, train a CVAE (conditioning on relevant market conditions) on the log signatures.

Step 4 **Postprocessing of the outputs of the VAE:** At this stage, one can either convert the generated log signatures back into paths, or use them directly. In order to invert signatures into paths, [5] propose a possible method to do so. In [44] a method to directly price and hedge derivatives using the signature of a path was proposed.

Step 5 **Performance evaluation:** Using signature-based similarity metrics, one can evaluate the performance of the generative process.

A crucial insight is at Step 5, using the signature viewpoint, where the performance evaluation is done via the properties of the expected signature.

> **Example: Signature Wasserstein-1 GAN, (Using Truncated sig-MMD)**

In [47], the authors propose to use the truncated (Conditional) Sig-Wasserstein Distance (i.e. the Signature MMD) as objective for a Market Generator. They propose different types of time series generators, a neural SDE, and a so-called LogSig-RNN. The LogSig-RNN can be seen as a higher-order Taylor approximation of SDEs by combining Recurrent Neural Networks (RNNs) with the log signature: Consider a $\mathbb{R}^e$-valued time-homogeneous SDE

$$dX_t = \mu(X_t)dt + \sigma(X_t) \circ dW_t = f(X_t) \circ d\mathbf{W}_t, \tag{3}$$

where $f : x \times (s, w) \mapsto \mu(x)s + \sigma(x)w$ is a vector field. Locally approximating X via the step-M Taylor expansion yields for $s < t$:

$$X_t - X_s \approx \sum_{m=1}^{M} f^{\circ m}(X_s) \int_{s<s_1<\cdots<s_m<t} d\mathbf{W}_{s_1} \otimes \cdots \otimes d\mathbf{W}_{s_m}, \tag{4}$$

where $f^{\circ m} : \mathbb{R}^e \to L\left((\mathbb{R}^{d+1})^{\otimes m}, \mathbb{R}^e\right)$ is defined inductively via $f^{\circ 1} = f$, $f^{\circ m+1} = D(f^{\circ m})f$.[33] Pasting these local step-M Taylor approximations

[33] $D(g)$ denotes the Differential of a function g.

together, yields a numerical approximation scheme: Consider time partitions for the Taylor expansion and X, $\Pi_h = (u_j)_{j=0}^{N_1}$ and $\Pi_X = (t_k)_{k=0}^{N_2}$, with $\Pi_h \subseteq \Pi_X$. Define $\hat{X}$ on Π_X inductively. Let $\hat{X}_0 = x_0$. For t_k, find j with $t_k \in (u_{j-1}, u_j]$ and define

$$\hat{X}_{t_k} := \hat{X}_{u_{j-1}} + \sum_{m=1}^{M} f^{\circ m}(\hat{X}_{u_{j-1}}) \int_{u_{j-1}<s_1<\cdots<s_m<t_k} d\mathbf{W}_{s_1} \otimes \cdots \otimes d\mathbf{W}_{s_m} \tag{5}$$

$$=: F\left(\hat{X}_{u_{j-1}}, S^M(\mathbf{W}_{u_{j-1},t_k})\right) =: \tilde{F}\left(\hat{X}_{u_{j-1}}, \mathrm{LogSig}^M(\mathbf{W}_{u_{j-1},t_k})\right), \tag{6}$$

by using the step-M Taylor expansion of X_{t_k} around u_{j-1}. Approximating $\tilde{F}$ leads to a generalized Logsig-RNN model [41]. This shows the use of signatures in the Market Generator, and in the discriminator used to train the parameters of the generator.

In [48], a conditional version of this is proposed.

The ***Signature*** **Maximum Mean Discrepancy** metric is formed by combining the Maximum Mean Discrepancy (MMD) metric [27][34] with the signature kernel [10].[35] To approximate the signature kernel,[36] the previous two examples use the *truncated signature*. While in [48] the Sig-MMD is back-propagated in the objective, in [5] a *Sig-MMD* informed *two sample test* is performed[37] to determine whether the simulated data and the training data have been sampled from the same (H_0) underlying distribution (stochastic process) or from different ones (H_A).

Truncated signature kernels do not take into consideration the entire (infinite) signature vector, they are truncated at a given (finite) level, and benefit from the property that the error between the full and truncated versions decays factorially in the level of truncation (see Lemma 2.3.1 in [7]).

[34] Introduced originally for finite-dimensional distributions.

[35] Which permits to extend the former to path-valued applications.

[36] The signature kernel

$$k_{\text{sig}} : \mathcal{X} \times \mathcal{X} \to \mathbb{R}, \quad k_{\text{sig}}(x, y) := \langle S(x), S(x) \rangle, \tag{7}$$

is a symmetric positive definite function. It can be shown, that the signature kernel is characteristic for every compact $\mathcal{K} \subset \mathcal{X}$, i.e. $\mathcal{P}(\mathcal{K}) \ni \mathbb{P} \mapsto \int k_{\text{sig}}(x, \cdot)\mathbb{P}(dx) \in \mathcal{H}$ is injective. In particular, this implies that the signature kernel is cc-universal, i.e. for every compact $\mathcal{K} \subset \mathcal{X}$, the linear span of the set of path functionals $\{k_{\text{sig}}(x, \cdot) : x \in \mathcal{K}\}$ is dense in $C(\mathcal{K})$ in the topology of uniform convergence.

[37] See the Chap. "Signature Maximum Mean Discrepancy Two-Sample Statistical Tests" of this volume.

The disadvantage of these first-generation type of (truncated signature kernel) models is that computing signature kernels by taking inner products of truncated signatures relies on some means of computing iterated integrals (to compute elements of the signature vector). This can introduce computational *bottlenecks*, especially in financial settings, where the target distributions are *high-dimensional*. Another disadvantage of these first-generation models there is no universal consensus about the ideal truncation level, which can be highly dependent on the situation at hand. **The second generation of Market Generators with signature kernel** bypass the need to compute truncated signatures or any entries of the signature vector explicitly: In [51] the authors provided a kernel trick proving that the signature kernel satisfies

$$k_{\text{sig}}(x, y) = f(T, T) \quad \text{where} \quad f(s, t) = 1 + \int_0^s \int_0^t f(u, v)\langle dx_u, dy_u\rangle, \tag{8}$$

which reduces to a linear hyperbolic PDE when x and y are almost everywhere differentiable.[38] Not only does this remove the necessity to justify the level of truncation which had been a frequent question in first generation models, but it also alleviates the computational bottleneck associated with the computation of high number of iterated integrals.[39] The second generation models scale better to high dimensional scenarios, see [31]. There is a more sophisticated version of the second generation models where the kernel trick is used in conjunction with **higher rank signatures** see [38, 52] and [57] as well as the Chap. "Adapted Topologies and Higher-Rank Signatures" of this volume.

The Third Generation of Market Generators with Signature Kernel In situations where we expect to see a very low number of observations (single observations) which are to be evaluated against a backdrop of distribution, scoring rules have an advantage over MMD-type distances, see [27]. The signature kernel score $\phi_{\text{sig}} : \mathcal{P}(\mathcal{X}) \times \mathcal{X} \to \mathbb{R}$ is defined through

$$\phi_{\text{sig}}(\mathbb{P}, y) := \mathbb{E}_{x,x'\sim\mathbb{P}}[k_{\text{sig}}(x, x')] - 2\mathbb{E}_{x\sim\mathbb{P}}[k_{\text{sig}}(x, y)]. \tag{9}$$

It can be shown, that for any compact $\mathcal{K} \subset \mathcal{X}$, ϕ_{sig} is a strictly proper kernel score relative to $\mathcal{P}(\mathcal{K})$, i.e. $\mathbb{E}_{y\sim\mathbb{Q}}[\phi_{\text{sig}}(\mathbb{Q}, y)] \leq \mathbb{E}_{y\sim\mathbb{Q}}[\phi_{\text{sig}}(\mathbb{P}, y)]$ for all $\mathbb{P}, \mathbb{Q} \in \mathcal{P}(\mathcal{K})$, with equality if and only if $\mathbb{P} = \mathbb{Q}$. Let $\mathbb{P} \in \mathcal{P}(\mathcal{X})$ and $y \in \mathcal{X}$. Given m sample paths $\{x^i\}_{i=1}^m \sim \mathbb{P}$, then

[38] Here, we denote by $\mathcal{H}$ the unique reproducing kernel Hilbert space (RKHS) of k_{sig} and endow $\mathcal{X}$ with a topology with respect to which the signature is continuous and denote by $\mathcal{P}(\mathcal{X})$ the set of Borel probability measures on $\mathcal{X}$, see [32] for full details.

[39] It is not possible to give a universally valid number to answer at what level of truncation bottlenecks would start causing bottlenecks, since the computational complexity depends on an interplay of multiple factors: The length of the path segments, the special dimension of the setting, the number of the paths considered and the level of truncation. See Chap. "Signature Maximum Mean Discrepancy Two-Sample Statistical Tests" for the interplay of the first 3 factors.

$$\hat{\phi}_{\text{sig}}(\mathbb{P}, y) := \frac{1}{m(m-1)} \sum_{j \neq i} k_{\text{sig}}(x^i, x^j) - \frac{2}{m} \sum_i k_{\text{sig}}(x^i, y) \tag{10}$$

is a consistent and unbiased estimator of ϕ_{sig}. In such cases a *divergence operator* can be associated with the scoring rule. Note, that in this case the signature kernel score is closely related to the Sig-MMD, the latter can be rewritten as:

$$\mathcal{D}_{k_{\text{sig}}}(\mathbb{P}, \mathbb{Q})^2 := \mathbb{E}_{y \sim \mathbb{Q}}[\phi_{\text{sig}}(\mathbb{P}, y)] + \mathbb{E}_{y, y' \sim \mathbb{Q}}[k_{\text{sig}}(y, y')]. \tag{11}$$

The associated kernel scoring rule benefits from better computational efficiency, while maintaining a close connection to the sig-kernel-MMD metric.

Another successfully used notion of distance between a point and a distribution in this context is Mahalanobis distance [42, 45].

> Example: Non-adversarial Training of Neural SDEs with Signature Kernel Scores

The innovation in [32], is Signature Kernel Scores are used, in order to train Neural SDEs.[40] Usually, neural SDEs are trained adversarially as infinite-dimensional GANs, see [36]. The above metrics however provides all the tools needed to train the Neural SDE model using signature kernel scores. The training objective is given by

$$\min_{\theta} \mathcal{L}(\theta), \quad \text{where} \quad \mathcal{L}(\theta) := \mathbb{E}_{y \sim \mathbb{P}_{X^{\text{true}}}}[\phi_{\text{sig}}(\mathbb{P}_{X^\theta}, y)]. \tag{13}$$

Note, that this is equivalent to minimizing $\mathcal{D}_{k_{\text{sig}}}(\mathbb{P}_{X^\theta}, \mathbb{P}_{X^{\text{true}}})$. $\mathcal{L}(\theta)$ is approximated by solving a system of linear PDEs depending on sample paths from the neural SDE. Training the model involves backpropagating through the SDE solver and the PDE solver. In [32] conditional Neural SDEs were applied to simulate FX pairs. It is shown, that the conditional generator exhibits the capability to produce conditional distributions that frequently encompass the observed path. Furthermore, these generated distributions capture certain distinctive characteristics of financial markets, such as martingality, mean reversion, or leverage effects when applicable.

[40] Recall that a Neural SDE is a model of the form

$$Y_0 = \xi_\theta(a), \quad dY_t = \mu_\theta(t, Y_t)dt + \sigma_\theta(t, Y_t) \circ dW_t, \quad X_t^\theta = A_\theta Y_t + b_\theta. \tag{12}$$

Under appropriate regularity assumptions, there is a strong solution $Y : [0, T] \to \mathbb{R}^{d_y}$ of this Stratanovich SDE.

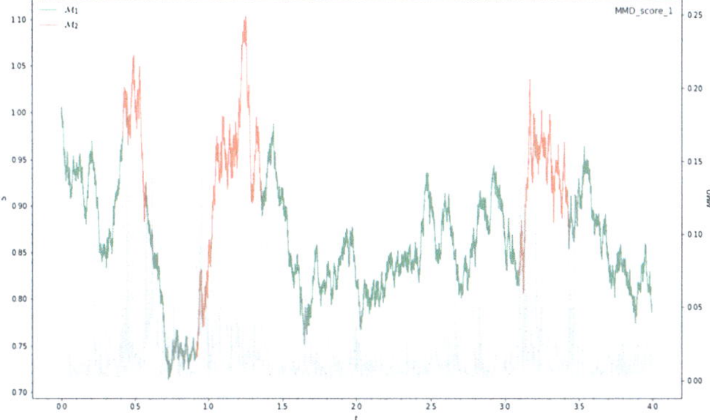

Fig. 2 Regime detection via the sig-kernel-MMD discriminator

3.3 A Good Discriminator Opens the Door to Further Associated Goals

It is not only Market Generation that suffers from the modeling challenges mentioned in previous sections. These challenges often propagate to other, related tasks as well. We have established in earlier sections that a centerpiece of a good Market Generator in financial contexts is the right performance evaluation metric, or discriminator. While it is rarely the case that this performance evaluation metric can be used off-the-shelf for other tasks, it can be the ideal starting point for other tasks (here referred to as 'downstream tasks') with the right amount of fine-tuning and care.

Distribution Regression Distribution regression refers to the supervised learning problem where labels are only available for groups of inputs instead of individual inputs. In [39] Lemercier et al. achieve a significant leap forward in this domain by establishing the distribution regression framework for data streams.

Anomaly, Outlier, and Change Detection [31] Anomaly detection focuses on determining whether a specific observation is significantly different from a reference set of observations considered normal. Change detection is a related idea without a reference to a corpus of normality.

> Example: Second-Generation sig-Kernel MMD Two-Sample Test to Indicate Distributional Changes in the Data [31]

Figure 2 shows a second-generation sig-kernel-MMD discriminator, as a two sample test across time, which is fine-tuned and enhanced for the purpose

(continued)

of identification of distributional (regime-) changes in sequential data, see [31, 32]. The grey line shows the value of the test statistic, where a higher value implicates a larger Sig-MMD distance in the samples, which indicates that the two samples are less likely to have been sampled from the same distribution (H_0). The time series is marked red if this test statistic surpasses a given threshold, indicating a regime change.

> **Example: Downstream Application of the Discriminator of VOLGAN [60]**
> The trained discriminator of the VOLGAN might be used for the downstream task of detecting extreme market events.
> Figure 3 contains discriminator scores on the training and testing data: Most notably, the data from start of the Covid-19 pandemic is highlighted as unusual.

Appendix: Recent Developments and Challenges in Generative AI

Generative AI and foundation models hold enormous potential to create new capabilities and value for enterprise. However, it also can introduce new risks, be they legal, financial, reputational or societal. In this section we provide some further insights into these aspects.

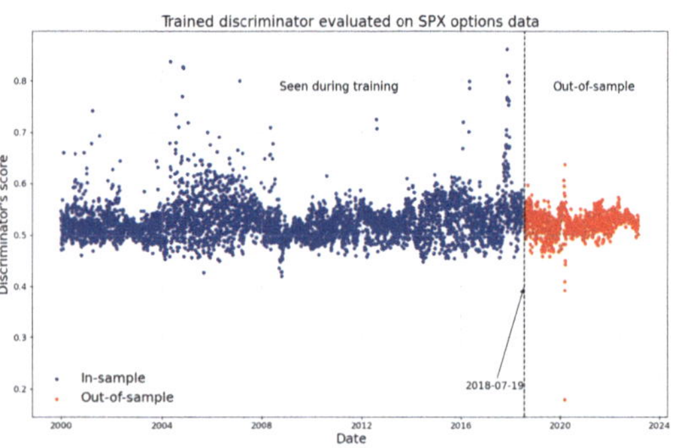

Fig. 3 Discriminator scores for VOLGAN

On the Use of Foundation Models in Finance (by Raeid Saqur)

> **Recent Developments in the Use of Foundation Models in Finance**
>
> The application of foundation models, specifically Pre-trained Language Models (PLMs) and Large Language Models (LLMs), in finance has evolved considerably over recent years. The journey began with the adaptation of foundational natural language processing models like BERT to the financial domain. **FinBERT** was one of the first attempts to bring PLMs into finance by pre-training BERT with large-scale financial texts. **BloombergGPT** marked a significant milestone in this evolution by introducing the first financial LLM with 50 billion parameters, pre-trained using both general and financial datasets [65]. Although BloombergGPT remains closed-source, its development demonstrated the effectiveness of scaling LLMs by using domain-specific training data. Meta AI's LLaMA series, specifically **LLaMA-13B**, emerged as the first open-source LLM to attract considerable attention in financial research, opening up opportunities for extensive experimentation and domain-specific fine-tuning [58].
>
> Foundation models in finance were first used for analyzing financial news or sentiment. They have since begun to assume roles, eg. in forecasting market trends. **TradingGPT**, a novel multi-agent system, leverages LLMs with a layered memory architecture [40] enabling trading agents to make more nuanced financial decisions by considering historical context, market dynamics, and individual risk profiles. More recently, from 2023 onward, the synergistic merging of deep reinforcement learning (RL) methods with large language model (LLM) alignment has gained momentum, exemplified by **InstructGPT** [50], also known as **ChatGPT**. In the realm of explainable financial forecasting, the work by Xinli-Yu et al. [66] investigates the capabilities of LLMs in explainable financial time series forecasting. Utilizing NASDAQ-100 stock data, company metadata, and economic/financial news, this study employs GPT-4 and Open LLaMA models for forecasting, demonstrating that LLMs can outperform traditional models like ARMA-GARCH in both accuracy and explanatory power.

There is no question that the future of foundation models in finance holds plenty of promise as researchers and practitioners refine these models, focusing on the nuanced needs of financial markets while leveraging advances in reinforcement learning, retrieval augmentation, and domain adaptation. The implementation of foundation models in finance continues to expand, with an increasing emphasis on integrating structured financial data and incorporating real-time multi-modal data for comprehensive analysis.

General Challenges with Generative AI

Environmental Impact and Global AI Divide

- Infrastructure requirements: Building a foundation model from scratch is expensive, requires enormous resources and has a large environmental footprint.
- Access to large and expensive quantities of computing power has become a prerequisite for developing advanced general-purpose AI. This has led to a growing dominance of large technology companies in general-purpose AI development.
- The AI R&D divide often overlaps with existing global socioeconomic disparities, potentially exacerbating them.

Exacerbation of Societal Biases Through Malrepresentation in Training Datasets

Models can pick up undesirable characteristics of the data they were trained on. To avoid this, developers should carefully filter training data and encode specific norms into their models. For Market Generators a malrepresentation in the training data of the true target distribution can lead to flawed forecasts, modeling errors and arbitrage. In more general-purpose AI systems this leads to ingrained inequalities in race, gender etc.

Malicious Use of Generative Models Through Fake Content aka "deepfakes"

On an individual scale this can involve **'phishing'** attacks or **impersonation** without consent. On a societal scale, realistic fake content can be used in "political disinformation campaigns" or combined with 'microtargeting' for conditioning content generation to match personal preferences.

Risks from Malfunctions Due to Inadequate Use

- Product functionality issues occur when there is confusion or misinformation about what a general-purpose AI model or **system is capable of**. These addressing assumptions is also central for risk management in financial modeling.
- As **conditions change**, systems (algorithms) may not be robust enough to generalize to the new environment and the outputs may no longer be valid.

References

1. I.P. Arribas, C. Salvi, L. Szpruch, Sig-SDEs model for quantitative finance, in *Proceedings of the First ACM International Conference on AI in Finance* (2020), pp. 1–8
2. T. Balch, V.K. Potluru, D. Paramanand, M. Veloso, Six levels of privacy: a framework for financial synthetic data. Preprint. arXiv:2403.14724 (2024)
3. A. Borji, Pros and cons of GAN evaluation measures. Comput. Vis. Image Underst. **179**, 41–65 (2019)
4. H. Buehler, L. Gonon, J. Teichmann, B. Wood, Deep hedging. Quant. Financ. **19**(8), 1271–1291 (2019)
5. H. Buehler, B. Horvath, T. Lyons, I.P. Arribas, B. Wood, A data-driven market simulator for small data environments. Preprint. arXiv:2006.14498 (2020)

6. H. Buehler, P. Murray, M.S. Pakkanen, B. Wood, Deep hedging: learning to remove the drift under trading frictions with minimal equivalent near-martingale measures. Preprint. arXiv:2111.07844 (2021)
7. T. Cass, S. Crisopher, Lecture notes on rough paths and applications to machine learning. Preprint. arxiv.2404.06583 (2024)
8. A.R. Cetingoz, C.-A. Lehalle, Synthetic data for portfolios: a throw of the dice will never abolish chance. Preprint. arXiv:2501.03993 (2025)
9. J. Chen, J.C. Hull, Z. Poulos, H. Rasul, A. Veneris, Y. Wu, A variational autoencoder approach to conditional generation of possible future volatility surfaces. Available at SSRN (2023). https://ssrn.com/abstract=4628457 or http://dx.doi.org/10.2139/ssrn.4628457
10. I. Chevyrev, H. Oberhauser, Signature moments to characterize laws of stochastic processes. J. Mach. Learn. Res. **23**(176), 1–42 (2022)
11. V. Choudhary, S. Jaimungal, M. Bergeron, FuNVol: multi-asset implied volatility market simulator using functional principal components and neural SDEs. Quant. Financ. **24**(8), 1077–1103 (2024). https://doi.org/10.1080/14697688.2024.2396977
12. N.M. Cirone, A. Orvieto, B. Walker, C. Salvi, T. Lyons, Theoretical foundations of deep selective state-space models. Preprint. arXiv:2402.19047 (2024)
13. S.N. Cohen, C. Reisinger, S. Wang, Arbitrage-free neural-SDE market models. Appl. Math. Financ. **30**(1), 1–46 (2023)
14. R. Cont, Empirical properties of asset returns: stylized facts and statistical issues. Quant. Financ. **1**(2), 223 (2001)
15. R. Cont, J. da Fonseca, Dynamics of implied volatility surfaces. Quant. Financ. **2**(1), 45–60 (2002)
16. R. Cont, M. Vuletic, Simulation of arbitrage-free implied volatility surfaces. Appl. Math. Financ. **30**, 94–121 (2023)
17. R. Cont, M. Cucuringu, R. Xu, C. Zhang, Tail-GAN: nonparametric scenario generation for tail risk estimation. Preprint. arXiv:2203.01664 (2022)
18. C. Cuchiero, G. Gazzani, S. Svaluto-Ferro, Signature-based models: theory and calibration. SIAM J. Financ. Math. **14**(3), 910–957 (2023)
19. C. Cuchiero, S. Svaluto-Ferro, J. Teichmann, Signature SDEs from an affine and polynomial perspective. Preprint. arXiv:2302.01362 (2023)
20. C. Cuchiero, E. Flonner, K. Kurt, Robust financial calibration: a bayesian approach for neural sdes. Preprint. arXiv:2409.06551 (2024)
21. D. Dai, C. Deng, C. Zhao, R. Xu, H. Gao, D. Chen, J. Li, W. Zeng, X. Yu, Y. Wu, et al. DeepSeekMoE: towards ultimate expert specialization in mixture-of-experts language models. Preprint. arXiv:2401.06066 (2024)
22. M.H. Davis, D.G. Hobson, The range of traded option prices. Math. Financ. **17**(1), 1–14 (2007)
23. T. Fissler, J.F. Ziegel, et al. Higher order elicitability and osband's principle. Ann. Stat. **44**(4), 1680–1707 (2016)
24. W. Fu, A. Hirsa, J. Osterrieder, Simulating financial time series using attention. Preprint. arXiv:2207.00493 (2022)
25. S. Gerhold, I.C. Gülüm, Consistency of option prices under bid–ask spreads. Math. Financ. **30**(2), 377–402 (2020)
26. T. Gneiting, Making and evaluating point forecasts. J. Am. Stat. Assoc. **106**(494), 746–762 (2011)
27. A. Gretton, K.M. Borgwardt, M.J. Rasch, B. Schölkopf, A. Smola, A kernel two-sample test. J. Mach. Learn. Res. **13**(1), 723–773 (2012)
28. A. Gu, K. Goel, C. Ré, Efficiently modeling long sequences with structured state spaces. Preprint. arXiv:2111.00396 (2021)
29. S. Hochreiter, J. Schmidhuber, Long short-term memory. Neural Comput. **9**, 1735–1780 (1997)
30. J. Hoffmann, S. Borgeaud, A. Mensch, E. Buchatskaya, T. Cai, E. Rutherford, D.d.L. Casas, L.A. Hendricks, J. Welbl, A. Clark, et al., Training compute-optimal large language models. Preprint. arXiv:2203.15556 (2022)

31. Z. Issa, B. Horvath, Non-parametric online market regime detection and regime clustering for multidimensional and path-dependent data structures. Preprint. arXiv:2306.15835 (2023)
32. Z. Issa, B. Horvath, M. Lemercier, C. Salvi, Non-adversarial training of Neural SDEs with signature kernel scores. Published: 21 Sept 2023, Last Modified: 22 Jan 2024 NeurIPS 2023 poster
33. N. Jackson, E. Suli, S. Howison, Computation of deterministic volatility surfaces. J. Computational Financ. **2**(2), 5–32 (1999)
34. P. Kidger, P. Bonnier, I. Perez Arribas, C. Salvi, T. Lyons, in *Proceedings of the 33rd International Conference on Neural Information Processing Systems*, Article No.: 279, 3105–3115 (2019)
35. P. Kidger, J. Morrill, J. Foster, T. Lyons, Neural controlled differential equations for irregular time series. Adv. Neural Inf. Proces. Syst. **33**, 6696–6707 (2020)
36. P. Kidger, J. Foster, X. Li, T.J. Lyons, Neural SDEs as infinite-dimensional GANs, in *International Conference on Machine Learning*. PMLR (2021), pp. 5453–5463
37. A. Koochali, P. Schichtel, A. Dengel, S. Ahmed, Probabilistic forecasting of sensory data with generative adversarial networks–ForGAN. IEEE Access **7**, 63868–63880 (2019)
38. M. Lemercier, T. Lyons, A high order solver for signature kernels. Preprint. arXiv:2404.02926 (2024)
39. M. Lemercier, C. Salvi, T. Damoulas, E. Bonilla, T. Lyons, Distribution regression for sequential data, in *International Conference on Artificial Intelligence and Statistics*. PMLR (2021), pp. 3754–3762
40. Y. Li, Y. Yu, H. Li, Z. Chen, K. Khashanah, TradingGPT: multi-agent system with layered memory and distinct characters for enhanced financial trading performance. Papers 2309.03736, arXiv.org (2023)
41. S. Liao, T. Lyons, W. Yang, H. Ni, Learning stochastic differential equations using RNN with log signature features. Preprint. arXiv:1908.08286 (2019)
42. T. Lyons, A.D. McLeod, Signature methods in machine learning. Preprint. arXiv:2206.14674 (2024)
43. T. Lyons, H. Ni, H. Oberhauser, A feature set for streams and an application to high-frequency financial tick data, in *Proceedings of the 2014 International Conference on Big Data Science and Computing* (2014), pp. 1–8
44. T. Lyons, S. Nejad, I. Perez Arribas, Non-parametric pricing and hedging of exotic derivatives. Appl. Math. Financ. **27**(6), 457–494 (2020)
45. P.C. Mahalanobis, On the generalised distance in statistics. Proc. Natl. Inst. Sci. India **2**(1), 49–55 (1936)
46. F. Moreno-Pino, A. Arroyo, H. Waldon, X. Dong, Á. Cartea, Rough transformers for continuous and efficient time-series modelling. Preprint. arXiv:2403.10288 (2024)
47. H. Ni, L. Szpruch, M. Sabate-Vidales, B. Xiao, M. Wiese, S. Liao, Sig-Wasserstein GANs for time series generation, in *Proceedings of the Second ACM International Conference on AI in Finance* (2021), pp. 1–8
48. S. Liao, H. Ni, M. Sabate-Vidales, L. Szpruch, M. Wiese, B. Xiao, Sig-Wasserstein GANs for conditional time series generation. Math. Financ. **34**(2), 622–670 (2024). (Special Issue on Machine Learning in Finance)
49. B. Ning, S. Jaimungal, X. Zhang, M. Bergeron, Arbitrage-free implied volatility surface generation with variational autoencoders. SIAM J. Financ. Math. **14**(4), 1004–1027 (2023)
50. L. Ouyang, J. Wu, X. Jiang, D. Almeida, C.L. Wainwright, P. Mishkin, C. Zhang, S. Agarwal, K. Slama, A. Ray, et al., Training language models to follow instructions with human feedback (2022). Preprint. arXiv:2203.02155 (2022)
51. C. Salvi, T. Cass, J. Foster, T. Lyons, W. Yang, The signature kernel is the solution of a Goursat PDE. SIAM J. Math. Data Sci. **3**(3), 873–899 (2021)
52. C. Salvi, M. Lemercier, C. Liu, B. Horvath, T. Damoulas, T. Lyons, Higher order kernel mean embeddings to capture filtrations of stochastic processes. Adv. Neural Inf. Proces. Syst. **34**, 16635–16647 (2021)

53. C. Salvi, M. Lemercier, A. Gerasimovics, Neural stochastic PDEs: resolution-invariant learning of continuous spatiotemporal dynamics. Adv. Neural Inf. Proces. Syst. **35**, 1333–1344 (2022)
54. B.H. Sana, R. Cont, Recovering volatility from option prices by evolutionary optimization. J. Comput. Financ. **8**(4), 43–76 (2005)
55. R. Saqur, A. Kratsios, F. Krach, Y. Limmer, J.-J. Tian, J. Willes, B. Horvath, F. Rudzicz, Filtered not mixed: stochastic filtering-based online gating for mixture of large language models. Preprint. arXiv:2406.02969 (2024)
56. C. Schwarz, Interpretable genai: Synthetic financial time series generation with probabilistic LSTM. Available at SSRN 4877007 (2024)
57. J. Tao, H. Ni, C. Liu, High rank path development: an approach of learning the filtration of stochastic processes. Preprint. arXiv:2405.14913 (2024)
58. H. Touvron, T. Lavril, G. Izacard, X. Martinet, M.-A. Lachaux, T. Lacroix, B. Rozière, N. Goyal, E. Hambro, F. Azhar, et al., LLaMA: open and efficient foundation language models. Preprint. arXiv:2302.13971 (2023)
59. R.S. Tsay, *Analysis of Financial Time Series* (John Wiley & Sons, Hoboken, 2005)
60. M. Vuletić, R. Cont, VolGAN: a generative model for arbitrage-free implied volatility surfaces. App. Math. Financ. **31**(4), 203–238 (2024). https://doi.org/10.1080/1350486X.2025.2471317
61. M. Vuletić, F. Prenzel, M. Cucuringu, Fin-GAN: forecasting and classifying financial time series via generative adversarial networks. Quant. Financ. **24**(2), 175–199 (2024). https://doi.org/10.1080/14697688.2023.2299466
62. S. Weber, Distribution-invariant risk measures, information, and dynamic consistency. Math. Financ. Int. J. Math. Stat. Financ. Econ. **16**(2), 419–441 (2006)
63. M. Wiese, L. Bai, B. Wood, H. Buehler, Deep hedging: learning to simulate equity option markets. Preprint. arXiv:1911.01700 (2019)
64. M. Wiese, B. Wood, A. Pachoud, R. Korn, H. Buehler, P. Murray, L. Bai, Multi-asset spot and option market simulation. Preprint. arXiv:2112.06823 (2021)
65. S. Wu, O. Irsoy, S. Lu, V. Dabravolski, M. Dredze, S. Gehrmann, P. Kambadur, D. Rosenberg, G. Mann, BloombergGPT: a large language model for finance. Preprint. arXiv:2303.17564 (2023)
66. X. Yu, Z. Chen, Y. Ling, S. Dong, Z. Liu, Y. Lu, Temporal data meets LLM – explainable financial time series forecasting. Papers 2306.11025, arXiv.org (2023)

Signature Maximum Mean Discrepancy Two-Sample Statistical Tests

Andrew Alden, Blanka Horvath, and Zacharia Issa

Abstract Maximum Mean Discrepancy (MMD) is a widely used concept in machine learning research which has gained popularity in recent years as a highly effective tool for comparing (finite-dimensional) distributions. Since it is designed as a kernel-based method, the MMD can be extended to path space valued distributions using the signature kernel. The resulting signature MMD (sig-MMD) can be used to define a metric between distributions on path space. Similarly to the original use case of the MMD as a test statistic within a two-sample testing framework, the sig-MMD can be applied to determine if two sets of paths are drawn from the same stochastic process. This chapter is dedicated to understanding the possibilities and challenges associated with applying the sig-MMD as a statistical tool in practice. We introduce and explain the sig-MMD, and provide easily accessible and verifiable examples for its practical use. We present examples that can lead to Type 2 errors in the hypothesis test, falsely indicating that samples have been drawn from the same underlying process (which generally occurs in a limited data setting). We then present techniques to mitigate the occurrence of this type of error.

A. Alden (✉)
Department of Informatics, King's College London, London, UK
e-mail: andrew.alden@kcl.ac.uk

B. Horvath
Mathematical Institute, University of Oxford, Oxford, UK

Oxford-Man Institute of Quantitative Finance, University of Oxford, Oxford, UK
e-mail: blanka.horvath@maths.ox.ac.uk

Z. Issa
Department of Mathematics, King's College London, London, UK
e-mail: zacharia.issa@kcl.ac.uk

C. Bayer et al. (eds.), *Signature Methods in Finance*, Springer Finance Lecture Notes, https://doi.org/10.1007/978-3-031-97239-3_5

1 Introduction

The seminal paper *A Kernel Two-Sample Test* [21] lays the foundation for establishing a versatile and robust framework for analysing and comparing distributions. This framework has been assessed and benchmarked against established classical two-sample tests in the finite-dimensional setting (such as Kolmogorov-Smirnov test) where it demonstrated superior or comparable performance [12, 19, 21]. It has also been shown to be well-suited for machine learning (ML) applications [2, 10, 28, 33, 49]. This framework is therefore particularly well-known for its applications to analysing finite-dimensional distributions in a ML context.

The test statistic which underpins this framework is based on computing the largest average difference between distributions, over functions in the unit ball of a Reproducing Kernel Hilbert Space (RKHS). This concept has been named the kernel Maximum Mean Discrepancy (MMD) [21]. On an appropriate (rich enough) space, the kernel MMD determines a metric between probability distributions. Using a signature kernel (k_{Sig}) (cf [8, 12, 31, 39]), the aforementioned MMD can be established for measures on path space defined over the set $\mathcal{X}_{1-\mathrm{var}}$. Similar to the originally proposed framework of the MMD (cf [21]), its path-valued counterpart (sig-MMD) can also be applied as a test statistic in a two-sample hypothesis test to determine if two collections of time series are drawn from different stochastic processes. If $\mathbf{X}$ and $\mathbf{Y}$ are two stochastic processes with laws $\mathbb{P}_{\mathbf{X}} = \mathbb{P} \circ \mathbf{X}^{-1}$ and $\mathbb{P}_{\mathbf{Y}} = \mathbb{P} \circ \mathbf{Y}^{-1}$ respectively, the two-sample test is described by the following null hypothesis (H_0) and alternative hypothesis (H_1):

$$H_0 \colon \mathbb{P}_{\mathbf{X}} = \mathbb{P}_{\mathbf{Y}} \text{ against } H_1 \colon \mathbb{P}_{\mathbf{X}} \neq \mathbb{P}_{\mathbf{Y}}.$$

The sig-MMD has facilitated the use of distribution regression on path space [2, 27, 40] and the training of generative models for time series[1] [7, 28].

Since the path signature is infinite dimensional, the extension of these concepts to path space using the (full) signature within numerical algorithms is non-trivial. Recent literature focuses on the numerical computation of the signature kernel. These include (i) truncating the signature [31], (ii) using a kernel trick by means of an associated partial differential equation (PDE) to work with the full signature [39, 40], and (iii) using random Fourier features [38, 47]. This chapter aims to demonstrate how to use the sig-MMD in practice with specific focus on effectively navigating the challenges that may emerge in this context. We give practical insights for setting up the associated ML algorithms that can save the interested user considerable time when implementing these tools. We also suggest hyperparameter settings and discuss common pitfalls. We provide simple and easy-to-verify examples together with explanatory discussions which help build intuition surrounding these tools.

[1] For further details see Chap. 4 (Market Generation).

As mentioned in the chapter on signature kernels (Chap. 3 (The Signature Kernel)), in practice, the numerical approximation of the signature kernel is based on various hyperparameters. We study the statistical power of the two-sample hypothesis test as a function of the following subset of hyperparameters:

- **Static Kernel Parameters** Θ_k: Instead of working directly with the original paths, the paths can be embedded in an infinite dimensional feature space using another kernel [31, 39].
- **Truncation Parameter** $\mathbb{N}_\infty$: Since the signature of a path is an infinite dimensional object, it is not possible to work with the full signature. To work directly with the signature, one must truncate the signature by using the first $M \in \mathbb{N}$ terms of the signature. However, when using kernel methods, a kernel trick can be used to work with the full signature. In this case, computing the signature kernel amounts to solving a PDE [40]. Therefore, when using the signature, one works with either an imperfect representation of the signature through its truncated version or the full signature by way of the kernel trick.
- **Preprocessing Parameters** $\Theta_{\text{preprocess}}$: Preprocessing the paths by using lead-lag transformations [11, 17, 26, 46], adding a time-component [11, 46], and by standardising dimensions[2] [28] can be performed.
- **Normalisation Parameters** $\Theta_{\text{normalisation}}$: Normalising the signature kernel results in a more robust statistic [12]. Moreover, by rescaling inner products, a generalised form of the signature kernel is obtained through which further sig-MMDs can be constructed [8, 9].
- **Time Discretisation Parameters** Θ_{discrete}: In practice, computations are performed on discrete data. The discretisation of the time interval can sometimes be controlled. Finer grids provide better approximations of the stochastic process [13] whilst also increasing the computational burden. In addition, if working with the full signature through the PDE approach, the PDE is solved on a discrete grid which itself is controlled through a hyperparameter.

The following sections are aimed at quantifying and interpreting the behaviour of the sig-MMD test statistic in the context of the two-sample hypothesis test. This behaviour is mostly characterised by the above parameters and therefore a discussion of the role of these parameters in improving performance on the hypothesis test is provided. The variance of the sig-MMD plays a significant role in characterising this behaviour. This variance is closely linked with the number of sample paths used. Although this can sometimes be directly addressed through sampling more samples from the processes and/or collecting additional data if the paths correspond to a physical system, one cannot disregard the computational bottlenecks associated with the sig-MMD. Therefore, increasing the sample size is generally not a feasible solution. We provide insights into stabilising the errors associated with the sig-MMD in the two-sample hypothesis testing framework in a high variance setting.

[2] If the path is multi-dimensional, this involves standardising the individual dimensions. This could be done by subtracting the mean to have each dimension centred around 0.

Specifically, we focus on the following:

1. **Hypothesis Testing Errors Associated with the Sig-MMD**: In most cases, the standard two-sample test using the sig-MMD on the space of stochastic processes has a high probability of a Type 2 error (acceptance of H_0 when H_1 is true) [3]. We explore techniques for reducing this. We show that scaling of the input paths plays a critical role in stabilising the statistical test. The effect of scaling on the signature kernel was first studied in [8, 9]. They showed that this has the equivalent effect of weighting inner products[3] [8, Lemma 2.9], [9, Lemma 2.10]. By weighting inner products, we can zoom in and focus on specific moments of the processes and tailor the two-sample hypothesis test to the specific processes. We leverage the relationship between path scaling and the decaying absolute value of the graded signature terms to construct a two-sample test statistic targeted at improving the test power. In practice, the true distributions of the stochastic processes are not observed and instead, empirical distributions are used to approximate the true distributions. Since sample sizes affect the quality of the approximation, we study the likelihood of Type 2 errors occurring across various sample sizes. We also study the effect of path scaling on the likelihood of a Type 1 error occurring.
2. **Information Associated with the Signature Level Terms**: We study the effects the different levels of the signature have on the statistical power of the two-sample test. In particular, we focus on the trade-off between the level of truncation ($\mathbb{N}_\infty$) and the information embedded within the higher-order terms. Using the kernel trick allows us to study the effects of including the full signature when computing the sig-MMD. We then need to account for the factorial decay of the signature terms to ensure that higher-level terms are contributing to the final value of the sig-MMD.

2 General Signature MMD

The path signature is well suited for working with data streams as a feature map for learning tasks. As a transform, the signature is invariant to reparametrisation, filtering out symmetries [39], and a continuous real-valued function on path space can be approximated by a linear function of the signature arbitrarily well [17, 31]. This result is analogous to the classical result that a continuous real-valued function defined on a closed interval can be approximated arbitrarily well by a polynomial function. In our case, the signature corresponds to a polynomial on path space. In terms of learning tasks, instead of fitting a non-linear function on path space, a linear function (i.e. linear regression) with the (truncated) signature terms as regressors can be used.

[3] The choice of path scaling is contained within the set $\Theta_{\text{normalisation}}$.

When a path $\mathbf{x}$ has finite p-variation ($p \geq 1$), the signature terms ($\Phi_{\text{Sig},m}(\mathbf{x})$) decay factorially according to the uniform estimate [39], [46, Theorem 3.2], [45, Lemma 5.1]

$$\left|\left|\Phi_{\text{Sig},m}(\mathbf{x})\right|\right| \leq \frac{||\mathbf{x}||_p^m}{m!} \tag{1}$$

where $||\cdot||_p$ denotes the p-variation metric. Computationally, working with the signature is challenging as it is infinite dimensional, and so, a large proportion of signature-based ML approaches use the truncated signature. In terms of absolute value, only a small amount of information is lost by truncating due to the factorial decay [33, 36, 46]. However, we show that higher-level terms play a critical role in capturing distribution differences through the sig-MMD. In a regression setting, specific signature terms can be re-weighted to account for the factorial decay. In the case of the sig-MMD, this is a non-trivial task. This plays a key role in this chapter.

2.1 Sig-MMD

Let H be a RKHS associated with kernel k. Also, let $\mathcal{F}$ be a unit ball in H defined over a compact metric space $\mathcal{X}$. Given two distributions $\mu, \nu \in \mathcal{P}(\mathcal{X})$, the MMD between μ and ν is defined as [21]

$$d_k(\mu, \nu) := \sup_{f \in \mathcal{F}} \left| \int_{\mathcal{X}} f(\mathbf{x})\, \mu(d\mathbf{x}) - \int_{\mathcal{X}} f(\mathbf{x})\, \nu(d\mathbf{x}) \right|.$$

If k is a continuous and characteristic kernel, then d_k is a metric [21, Theorem 5]. If H is the RKHS associated with the signature kernel k_{Sig},[4] the corresponding MMD is the sig-MMD. Since the signature kernel is characteristic [12, 40] and universal [12] when restricted to a compact subset $\mathcal{K} \subset \mathcal{X}_{1-\text{var}}$, the sig-MMD is a metric on the space of distributions of stochastic processes defined over paths in $\mathcal{K}$. Throughout this chapter, $\mathcal{K} \subset \mathcal{X}_{1-\text{var}}$ denotes a compact subset of $\mathcal{X}_{1-\text{var}}$.

Throughout this chapter, we consider processes in continuous time. However, in practice, we only have access to discrete datapoints. A path is constructed from the data using linear interpolation. The linear interpolated paths are of bounded variation and belong to the set $\mathcal{X}_{1-\text{var}}$. For this reason, throughout this chapter we assume that all paths are elements of $\mathcal{X}_{1-\text{var}}$ even though their continuous counterpart may not be.

Given two stochastic processes $\mathbf{X}$ and $\mathbf{Y}$, to compute the sig-MMD between $\mathbb{P}_{\mathbf{X}}$ and $\mathbb{P}_{\mathbf{Y}}$, it is first re-formulated in terms of Kernel Mean Embeddings (KMEs). The KME of a stochastic process maps the distribution of the stochastic process to an

[4] Given $\mathbf{x}, \mathbf{y} \in \mathcal{X}_{1-\text{var}}$ the signature kernel is defined as $k_{\text{Sig}}(\mathbf{x}, \mathbf{y}) := \langle \Phi_{\text{Sig}}(\mathbf{x}), \Phi_{\text{Sig}}(\mathbf{y}) \rangle$.

element in a RKHS. Let H_{Sig} be the RKHS with kernel k_{Sig} restricted to paths in the compact subset $\mathcal{K}$. Since the signature kernel is characteristic when restricted to the compact subset $\mathcal{K}$, each distribution defined over paths in $\mathcal{K}$ is embedded as a unique element within the RKHS H_{Sig}. Formally, the KME[5] is given by the mapping[6]

$$\mu_{\mathbf{X}} : \mathcal{P}(\mathcal{K}) \mapsto H_{\text{Sig}}$$

defined by

$$\mu_{\mathbf{X}} : \mathbb{P}_{\mathbf{X}} \mapsto \mathbb{E}_{\mathbf{X} \sim \mathbb{P}_{\mathbf{X}}} \left[k_{\text{Sig}} (\mathbf{X}, \cdot) \right].$$

Then, the sig-MMD is given by [21, Lemma 4]

$$\begin{aligned} d_{k_{\text{Sig}}} (\mathbb{P}_{\mathbf{X}}, \mathbb{P}_{\mathbf{Y}})^2 &= \| \mu_{\mathbf{X}} - \mu_{\mathbf{Y}} \|^2_{H_{\text{Sig}}} \\ &= \left\| \mathbb{E}_{\mathbf{X} \sim \mathbb{P}_{\mathbf{X}}} \left[k_{\text{Sig}} (\mathbf{X}, \cdot) \right] - \mathbb{E}_{\mathbf{Y} \sim \mathbb{P}_{\mathbf{Y}}} \left[k_{\text{Sig}} (\cdot, \mathbf{Y}) \right] \right\|^2_{H_{\text{Sig}}} \end{aligned} \tag{2}$$

with $\|\cdot\|_{H_{\text{Sig}}}$ being the norm induced by the inner product of the RKHS H_{Sig}.

The KME formulation provided in Eq. (2) shows that the sig-MMD is the norm of the difference between the embeddings of the laws of the stochastic processes. Given independent collections $\mathbf{X}, \mathbf{X}' \sim \mathbb{P}_{\mathbf{X}}$ and $\mathbf{Y}, \mathbf{Y}' \sim \mathbb{P}_{\mathbf{Y}}$, by expanding the norm into inner products and using the reproducing property of the kernel operator, an alternative formulation of the sig-MMD is given by [21, Lemma 6]

$$\begin{aligned} d_{k_{\text{Sig}}} (\mathbb{P}_{\mathbf{X}}, \mathbb{P}_{\mathbf{Y}})^2 &= \mathbb{E}_{\mathbf{X}, \mathbf{X}' \sim \mathbb{P}_{\mathbf{X}}} \left[k_{\text{Sig}} \left(\mathbf{X}, \mathbf{X}' \right) \right] + \mathbb{E}_{\mathbf{Y}, \mathbf{Y}' \sim \mathbb{P}_{\mathbf{Y}}} \left[k_{\text{Sig}} \left(\mathbf{Y}, \mathbf{Y}' \right) \right] \\ &\quad - 2\mathbb{E}_{\mathbf{X} \sim \mathbb{P}_{\mathbf{X}}, \mathbf{Y} \sim \mathbb{P}_{\mathbf{Y}}} \left[k_{\text{Sig}} (\mathbf{X}, \mathbf{Y}) \right]. \end{aligned} \tag{3}$$

2.2 Empirical Estimator

To compute the sig-MMD between two distributions, the expectations in Eq. 3 are approximated using empirical samples. Suppose N independent samples

$$\mathbf{X}^{1:N} = \left(\mathbf{X}^1, \cdots, \mathbf{X}^N \right) = \left(\mathbf{X}^i \right)_{i=1}^N \text{ with } \mathbf{X}^i \sim \mathbb{P}_{\mathbf{X}}$$

and M independent samples

[5] Further details on KMEs can be found in [21, 35, 40, 43].

[6] Since for practical purposes we work with elements from the set $\mathcal{X}_{\text{Seq}}$, the mapping μ extends to this set by means of the natural inclusion $\mathcal{X}_{\text{Seq}} \hookrightarrow \mathcal{K} \subset \mathcal{X}_{1-\text{var}}$ induced through linear interpolation.

$$\mathbf{Y}^{1:M} = \left(\mathbf{Y}^1, \cdots, \mathbf{Y}^M\right) = \left(\mathbf{Y}^j\right)_{j=1}^{M} \text{ with } \mathbf{Y}^j \sim \mathbb{P}_\mathbf{Y}$$

are available. N, M are the sample sizes (batch sizes) and throughout this chapter, sample size and batch size are used interchangeably.

Given empirical samples, the MMD can be computed using either a biased or an unbiased estimator [21]. In the numerical examples presented in Sect. 4, we compare the performance of the two estimators when performing the two-sample hypothesis test. We show that in certain cases, the biased estimator has superior performance over the unbiased estimator. A biased estimator is given by

$$\begin{aligned}\widehat{d^b_{k_{\text{Sig}}}}\left(\mathbb{P}_\mathbf{X}, \mathbb{P}_\mathbf{Y}\right)^2 = &\frac{1}{N^2}\sum_{i,j=1}^{N} k_{\text{Sig}}\left(\mathbf{X}^i, \mathbf{X}^j\right) + \frac{1}{M^2}\sum_{i,j=1}^{M} k_{\text{Sig}}\left(\mathbf{Y}^i, \mathbf{Y}^j\right)\\ &- \frac{2}{NM}\sum_{i,j=1}^{N,M} k_{\text{Sig}}\left(\mathbf{X}^i, \mathbf{Y}^j\right).\end{aligned} \tag{4}$$

An unbiased estimator is given by

$$\begin{aligned}\widehat{d_{k_{\text{Sig}}}}\left(\mathbb{P}_\mathbf{X}, \mathbb{P}_\mathbf{Y}\right)^2 = &\frac{1}{N(N-1)}\sum_{\substack{i,j=1\\ i\neq j}}^{N} k_{\text{Sig}}\left(\mathbf{X}^i, \mathbf{X}^j\right) + \frac{1}{M(M-1)}\sum_{\substack{i,j=1\\ i\neq j}}^{M} k_{\text{Sig}}\left(\mathbf{Y}^i, \mathbf{Y}^j\right)\\ &- \frac{2}{NM}\sum_{i,j=1}^{N,M} k_{\text{Sig}}\left(\mathbf{X}^i, \mathbf{Y}^j\right).\end{aligned} \tag{5}$$

The bias present in the biased estimator is caused by the cross-terms $k_{\text{Sig}}\left(\mathbf{X}^i, \mathbf{X}^i\right)$ and $k_{\text{Sig}}\left(\mathbf{Y}^i, \mathbf{Y}^i\right)$ in the computation of the sample averages.

2.3 General Signature Kernels

The authors of [8, 9] extend the signature kernel to more generic kernels. A weight function is applied to the level terms of the signature to weight their relative importance in the kernel computation. Let $\phi\colon \mathbb{N}\cup\{0\} \mapsto \mathbb{R}^+$ be a weight function and $\mathbf{s} = (\mathbf{s}_0, \mathbf{s}_1, \cdots), \mathbf{t} = (\mathbf{t}_0, \mathbf{t}_1, \cdots) \in \prod_{m\geq 0}\left(\mathbb{R}^d\right)^{\otimes m}$. We then define the bilinear form

$$\langle \mathbf{s}, \mathbf{t}\rangle_\phi := \sum_{m=0}^{\infty} \phi(m)\,\langle \mathbf{s}_m, \mathbf{t}_m\rangle_m$$

where

$$\langle \mathbf{s}_m, \mathbf{t}_m \rangle_m := \sum_{i_1,\cdots,i_m=1}^{d} \mathbf{s}_m^{i_1,\cdots,i_m} \mathbf{t}_m^{i_1,\cdots,i_m}.$$

Given two paths $\mathbf{x}, \mathbf{y} \in \mathcal{X}_{1-\text{Var}}$ and a weight function $\phi \colon \mathbb{N} \cup \{0\} \mapsto \mathbb{R}^+$, the ϕ-signature kernel is defined as

$$k_{\text{Sig}}^{\phi}(\mathbf{x}, \mathbf{y}) := \langle \Phi_{\text{Sig}}(\mathbf{x}), \Phi_{\text{Sig}}(\mathbf{y}) \rangle_{\phi}.$$

The following lemma provides a necessary condition on the weight function ϕ to guarantee that the ϕ-signature kernel is well-defined [8, Lemma 2.4], [9, Lemma 2.5].

Lemma 2.1 *The ϕ-signature kernel is well-defined provided the function $\phi \colon \mathbb{N} \cup \{0\} \mapsto \mathbb{R}^+$ is such that the series*

$$\sum_{m=0}^{\infty} C^m \phi(m) (m!)^{-2}$$

is summable for every $C > 0$.

Using the ϕ-signature kernel, a general version of the sig-MMD can be defined. The ϕ-MMD between the distributions of stochastic processes $\mathbf{X}, \mathbf{Y}$ is given by

$$\begin{aligned} d_{k_{\text{Sig}}^{\phi}}(\mathbb{P}_{\mathbf{X}}, \mathbb{P}_{\mathbf{Y}})^2 &= \mathbb{E}_{\mathbf{X},\mathbf{X}' \sim \mathbb{P}_{\mathbf{X}}}\left[k_{\text{Sig}}^{\phi}(\mathbf{X}, \mathbf{X}')\right] + \mathbb{E}_{\mathbf{Y},\mathbf{Y}' \sim \mathbb{P}_{\mathbf{Y}}}\left[k_{\text{Sig}}^{\phi}(\mathbf{Y}, \mathbf{Y}')\right] \\ &\quad - 2\mathbb{E}_{\mathbf{X} \sim \mathbb{P}_{\mathbf{X}}, \mathbf{Y} \sim \mathbb{P}_{\mathbf{Y}}}\left[k_{\text{Sig}}^{\phi}(\mathbf{X}, \mathbf{Y})\right]. \end{aligned} \tag{6}$$

Since the mapping $\mu_{\phi} \colon \mathcal{P}(\mathcal{K}) \mapsto H_{\phi}$[7] is injective [9], the ϕ-MMD is a metric. If ϕ is the mapping $\phi \colon \mathbb{N} \cup \{0\} \mapsto \{1\}$, the ϕ-signature kernel is k_{Sig}.

2.4 Example of General Signature Kernels

The ϕ-signature kernel provides a general framework for expanding the set of available signature-based kernels. We now provide a specific example of the ϕ-signature kernel and its broader implications on the ϕ-MMD. This ϕ-MMD is used for the purpose of performing the two-sample hypothesis test in Sect. 4.

[7] H_{ϕ} is the RKHS with kernel k_{Sig}^{ϕ}.

2.4.1 Scalar Multiplication

Let $\mathbf{s} = (\mathbf{s}_0, \mathbf{s}_1, \cdots,) \in \prod_{m\geq 0} \left(\mathbb{R}^d\right)^{\otimes m}$ and consider the mapping

$$\gamma_\theta \colon \prod_{m\geq 0} \left(\mathbb{R}^d\right)^{\otimes m} \to \prod_{m\geq 0} \left(\mathbb{R}^d\right)^{\otimes m} \text{ defined by } \gamma_\theta(\mathbf{s}) = \sum_{m\geq 0} \theta^m \mathbf{s}_m$$

for some scalar $\theta \in \mathbb{R}^+$. Incorporating this within the bilinear form $\langle \cdot, \cdot \rangle_\phi$, we have that [8, Lemma 2.9]

$$\langle \gamma_\theta \mathbf{s}, \mathbf{t} \rangle_{\phi(m)\mapsto 1} = \langle \mathbf{s}, \gamma_\theta \mathbf{t} \rangle_{\phi(m)\mapsto 1} = \langle \mathbf{s}, \mathbf{t} \rangle_{\phi(m)\mapsto \theta^m}.$$

Applying the homomorphism γ_θ to elements of the extended tensor algebra is equivalent to scaling the individual bilinear forms $\langle \cdot, \cdot \rangle_m$ within the overall computation of the bilinear form $\langle \cdot, \cdot \rangle_\phi$. In our specific context, **s** and **t** are signatures corresponding to two paths **x** and **y** respectively. In this case, the following holds [8, Corollary 2.10]

$$k_{\text{Sig}}(\theta \mathbf{x}, \mathbf{y}) = k_{\text{Sig}}(\mathbf{x}, \theta \mathbf{y}) = k_{\text{Sig}}\left(\sqrt{\theta}\mathbf{x}, \sqrt{\theta}\mathbf{y}\right) = k^\phi_{\text{Sig}}(\mathbf{x}, \mathbf{y})$$

where ϕ is the weight function $\phi_\theta \colon \mathbb{N} \cup \{0\} \to \mathbb{R}$ defined by $\phi(m) = \theta^m$. Multiplying paths by a constant scalar $\sqrt{\theta}$ equates to using a ϕ-signature kernel with weight function ϕ_θ. Therefore, if both paths are multiplied by the scalar $\sqrt{\theta}$, the inner product $\langle \cdot, \cdot \rangle_m$ corresponding to level m is scaled by a factor of θ^m. Through path scaling, either one of the following three scenarios holds:

1. If $\theta < 1$, a higher weighting is attributed to lower-level terms;
2. If $\theta = 1$, the original signature kernel is used; or
3. If $\theta > 1$, a higher weighting is attributed to higher-level terms.

In terms of ϕ-MMD, by re-weighting the level terms, we can target the ϕ-MMD to focus on particular distributional properties. For example, by setting a high value for θ we would be prioritising higher-order moments over lower-order moments in the computation of the ϕ-MMD. The opposite effect occurs if we use values for θ which are less than 1. A key difference between the ϕ-MMD and the ϕ-signature kernel is that in the case of the ϕ-signature kernel, rescaling one of the paths by θ is equivalent to scaling both paths by $\sqrt{\theta}$. This is not the case for the ϕ-MMD. The ϕ-MMD is a function of kernels of the forms

- $k^\phi_{\text{Sig}}\left(\mathbf{X}, \mathbf{X}'\right)$ • $k^\phi_{\text{Sig}}\left(\mathbf{Y}, \mathbf{Y}'\right)$ • $k^\phi_{\text{Sig}}(\mathbf{X}, \mathbf{Y})$

with inputs being $\mathbb{P}_\mathbf{X}, \mathbb{P}_\mathbf{Y}$. If we scale $\mathbb{P}_\mathbf{X}$ by θ, we would be re-weighting the terms $k^\phi_{\text{Sig}}\left(\mathbf{X}, \mathbf{X}'\right)$, $k^\phi_{\text{Sig}}(\mathbf{X}, \mathbf{Y})$, and $k^\phi_{\text{Sig}}\left(\mathbf{Y}, \mathbf{Y}'\right)$ by θ^2, θ, and 1 respectively. Hence, inconsistent scalings are being applied and, more importantly, different ϕ-signature kernels are being used to compute the ϕ-MMD. We need to scale both inputs and scaling by $\sqrt{\theta}$ is equivalent to re-weighting the signature terms of level m by θ^m.

2.5 Detailed Construction of the ϕ-MMD

We reformulate the computation of the ϕ-MMD as a sum over *level terms*—the signature terms corresponding to a specific level of the signature. We use this reformulation to quantify the contribution of the level terms to the ϕ-MMD.

Definition 2.2 An alphabet $\mathcal{A}_d$ is a set comprising of the d letters $1, \cdots, d$. A word of length k is a sequence of k letters from the alphabet (repetitions are allowed). The set of all words of length k over alphabet $\mathcal{A}_d$ is denoted by $\mathcal{W}_k(\mathcal{A}_d)$.

For any two d-dimensional stochastic processes $\mathbf{X}, \mathbf{Y}$ and any $m \in \mathbb{N}$, define the function $\Lambda_m : \mathcal{P}(\mathcal{K}) \times \mathcal{P}(\mathcal{K}) \mapsto \mathbb{R}$ by

$$\Lambda_m(\mathbb{P}_\mathbf{X}, \mathbb{P}_\mathbf{Y}) := \sum_{i_1,\cdots,i_m=1}^{d} \mathbb{E}_{\mathbf{X}\sim\mathbb{P}_\mathbf{X}}\left[\Phi_{\mathrm{Sig},m}(\mathbf{X})^{i_1,\cdots,i_m}\right]\mathbb{E}_{\mathbf{Y}\sim\mathbb{P}_\mathbf{Y}}\left[\Phi_{\mathrm{Sig},m}(\mathbf{Y})^{i_1,\cdots,i_m}\right].$$

Λ_m is the level-m contribution to the expected signature kernel. This can be used to define the level-m contribution to the ϕ-MMD. The level-m contribution is the mapping $\Gamma_m^\phi : \mathcal{P}(\mathcal{K}) \times \mathcal{P}(\mathcal{K}) \mapsto \mathbb{R}$ defined by

$$\Gamma_m^\phi(\mathbb{P}_\mathbf{X}, \mathbb{P}_\mathbf{Y}) := \phi(m)\left[\Lambda_m(\mathbb{P}_\mathbf{X}, \mathbb{P}_\mathbf{Y}) - 2\Lambda_m(\mathbb{P}_\mathbf{X}, \mathbb{P}_\mathbf{Y}) + \Lambda_m(\mathbb{P}_\mathbf{X}, \mathbb{P}_\mathbf{Y})\right].$$

Using the level-m contribution Γ_m^ϕ, the ϕ-MMD can be reformulated as

$$d_{k_{\mathrm{Sig}}^\phi}(\mathbb{P}_\mathbf{X}, \mathbb{P}_\mathbf{Y})^2 = \sum_{m\geq 0} \Gamma_m^\phi(\mathbb{P}_\mathbf{X}, \mathbb{P}_\mathbf{Y}).$$

By restricting the sum to the first k levels, we obtain the truncated ϕ-MMD at level k.

Suppose we have N independent and identically distributed (i.i.d.) samples $\mathbf{X}^{1:N}$ and M i.i.d. samples $\mathbf{Y}^{1:M}$. Let $\delta_{(\cdot)}$ denote the Dirac measure and for two paths $\mathbf{X}, \mathbf{Y} \in \mathcal{K}$ denote $\Lambda_m(\delta_\mathbf{X}, \delta_\mathbf{Y})$ by $\Lambda_m(\mathbf{X}, \mathbf{Y})$. The expectations Λ_m can be estimated using an unbiased or biased sample average. The unbiased estimators are given by

$$\begin{aligned}\widehat{\Lambda_m}\left(\delta_{\mathbf{X}^{1:N}}, \delta_{\mathbf{Y}^{1:M}}\right) &= \frac{1}{NM}\sum_{i,j=1}^{N,M} \widehat{\Lambda_m}\left(\mathbf{X}^i, \mathbf{Y}^j\right) \\ &= \frac{1}{NM}\sum_{i,j=1}^{N,M}\sum_{r_1,\cdots,r_m=1}^{d} \Phi_{\mathrm{Sig},m}\left(\mathbf{X}^i\right)^{r_1,\cdots,r_m}\Phi_{\mathrm{Sig},m}\left(\mathbf{Y}^j\right)^{r_1,\cdots,r_m},\end{aligned}$$

and

$$\begin{aligned}
&\widehat{\Lambda_m}\left(\delta_{\mathbf{X}^{1:N}}, \delta_{\mathbf{X}^{1:N}}\right) \\
&= \frac{1}{N(N-1)} \sum_{\substack{i,j=1 \\ i \neq j}}^{N} \widehat{\Lambda_m}\left(\mathbf{X}^i, \mathbf{X}^j\right) \\
&= \frac{1}{N(N-1)} \sum_{\substack{i,j=1 \\ i \neq j}}^{N} \sum_{r_1,\cdots,r_m=1}^{d} \Phi_{\mathrm{Sig},m}\left(\mathbf{X}^i\right)^{r_1,\cdots,r_m} \Phi_{\mathrm{Sig},m}\left(\mathbf{X}^j\right)^{r_1,\cdots,r_m}.
\end{aligned}$$

The biased estimator is denoted by $\widehat{\Lambda_m^b}$ and is defined by taking the sum over all terms i, j including the cross terms. Using the empirical estimator for the Λ_m terms, an approximator for the level-m contribution to the ϕ-MMD is given by

$$\begin{aligned}
\widehat{\Gamma_m^\phi}\left(\delta_{\mathbf{X}^{1:N}}, \delta_{\mathbf{Y}^{1:M}}\right) = \phi(m)\Big[&\widehat{\Lambda_m}\left(\delta_{\mathbf{X}^{1:N}}, \delta_{\mathbf{X}^{1:N}}\right) - 2\widehat{\Lambda_m}\left(\delta_{\mathbf{X}^{1:N}}, \delta_{\mathbf{Y}^{1:M}}\right) \\
&+ \widehat{\Lambda_m}\left(\delta_{\mathbf{Y}^{1:M}}, \delta_{\mathbf{Y}^{1:M}}\right)\Big].
\end{aligned}$$

Finally, an unbiased estimator for the ϕ-MMD is

$$\widehat{d_{k_{\mathrm{Sig}}^\phi}}\left(\delta_{\mathbf{X}^{1:N}}, \delta_{\mathbf{Y}^{1:M}}\right)^2 = \sum_{m \geq 0} \widehat{\Gamma_m^\phi}\left(\delta_{\mathbf{X}^{1:N}}, \delta_{\mathbf{Y}^{1:M}}\right)$$

with the biased estimator being

$$\widehat{d^b_{k_{\mathrm{Sig}}^\phi}}\left(\delta_{\mathbf{X}^{1:N}}, \delta_{\mathbf{Y}^{1:M}}\right)^2 = \sum_{m \geq 0} \widehat{\Gamma_m^{\phi,b}}\left(\delta_{\mathbf{X}^{1:N}}, \delta_{\mathbf{Y}^{1:M}}\right).$$

Whenever the weight function ϕ is the constant function $\phi(m) = 1$ for all m, the ϕ superscript is omitted from the notation.

By reformulating the sig-MMD in terms of level contributions, we can better understand the role the individual levels play in the computation of the sig-MMD. Since the level-K term of the expected signature is associated with the Kth moment of the distribution of the stochastic process, differences between the Kth moment of the distributions start featuring in the computation of the sig-MMD as from the level-K term Γ_K^ϕ. Suppose the two distributions have equal moments up to the $(K-1)$th moment and differ in their Kth moment. Since the signature terms decay factorially, the terms $\widehat{\Gamma_m}$ are of order $(m!)^{-2}$ for all $m \leq K$. Hence, terms corresponding to equal moments contribute more to the final value of the sig-MMD and therefore, using the sig-MMD as a statistic on which to perform the two-sample hypothesis test could result in a high probability of a Type 2 error occurring. To counteract the effect of the factorial decay, the ϕ-MMD can be used to re-weight the level contributions. Generally, by using constant mappings $\phi\colon m \mapsto \theta^m$ for some

$\theta \in \mathbb{R}^+$, higher-order properties can be captured since higher weightings are given to higher-level signature terms (Sect. 2.4.1). Numerical examples illustrating this are provided in Sect. 4.

Constant mappings can be regarded as a change of units. For example, if the sample paths correspond to percentage returns and are provided within a normalised range, changing the units to percentages corresponds to the ϕ-MMD with $\theta = 100^2$. As shown in Sect. 4, scalings have an effect on test performance. Consequently, the test performed with normalised returns could result in a different conclusion than if returns provided as a percentage were used. In contrast to other test statistics (such as the Kolmogorov-Smirnov test statistic [24]), test performance with respect to the sig-MMD is not scale invariant. On the contrary, the scaling factor should be optimised since it plays an important role in test performance.

3 Two-Sample Hypothesis Testing

The standard two-sample hypothesis test between stochastic processes [12] is used to test whether two collections of time series stem from the same distribution. The null hypothesis under the two-sample test is

$$H_0 \colon \mathbb{P}_{\mathbf{X}} = \mathbb{P}_{\mathbf{Y}}$$

with the alternative hypothesis being

$$H_1 : \mathbb{P}_{\mathbf{X}} \neq \mathbb{P}_{\mathbf{Y}}.$$

For the remainder of this section, we fix the distributions $\mathbb{P}_{\mathbf{X}}$ and $\mathbb{P}_{\mathbf{Y}}$. We describe the procedure and the key components required to perform the two-sample test.

Since the MMD is a metric on the space of Borel probability measures on stochastic processes [12, 21], it is used as the test statistic for the hypothesis test. Let $\widehat{c}_{1-\alpha}$ be the $(1-\alpha)$-quantile of the empirical MMD under the null hypothesis. The null hypothesis is rejected with significance α if $\widehat{d^2_{k_{\text{Sig}}}}(\mathbb{P}_{\mathbf{X}}, \mathbb{P}_{\mathbf{Y}}) < \widehat{c}_{1-\alpha}$.

In two-sample testing, a Type 1 error occurs when the statistical test determines that the two samples have different underlying distributions when the null hypothesis holds. A Type 2 error occurs when the test concludes that the two samples arise from the same stochastic process (distribution) even though this is not the case.

To quantify the likelihood of a Type 2 error occurring, we need to approximate the distribution of the sig-MMD under both the null and alternative hypotheses. Suppose the empirical MMD has distribution $\widehat{F_{H_0}}$ under the null hypothesis and $\widehat{F_{H_1}}$ under the alternative hypothesis. For a given level α, the probability of a Type 2 error is

$$\mathbb{P}\left[\text{Type 2 error}\right] = \widehat{F_{H_1}}\left(\widehat{c}_{1-\alpha}\right). \tag{7}$$

To perform the two-sample hypothesis test and compute the probabilities of a Type 1 error and Type 2 error, the following are needed:

1. An estimate for the $(1-\alpha)$-quantile of the null distribution; and
2. The distribution of the MMD under the alternative hypothesis.

We describe two techniques for approximating the distributions; bootstrapping and asymptotic analysis. The former is based on re-sampling from the stochastic processes until an 'accurate'[8] approximation of the MMD distribution can be established. The latter approach relies heavily on U-statistics [1, 4, 16, 25, 37, 42].

3.1 Bootstrapping

Let $\mathbf{X}^{1:N'}$ and $\mathbf{Y}^{1:M'}$ be two independent collections of paths sampled from $\mathbb{P}_X$ and $\mathbb{P}_Y$ respectively with values in $\mathbb{R}^d$. To sample the sig-MMD under the null distribution, we sample batches of size N_1, N_2 from the N' available. This procedure is repeated B times to construct the collection

$$\left\{\left(\mathbf{X}_i^{1:N_1}, \mathbf{X}_i^{1:N_2}\right)\right\}_{i=1}^{B}.$$

The empirical sig-MMD between these samples is calculated to obtain B distances under the null hypothesis. These B distances are used to construct the empirical null distribution. Repeating a similar procedure, the distribution of the test statistic under the alternative hypothesis is approximated by sampling B collections $\left\{\left(\mathbf{X}^{1:N_1}, \mathbf{Y}^{1:M_1}\right)\right\}_{i=1}^{B}$ with $(M_1 < M')$ and then computing the test statistic over every collection. The value of B is closely related to the quality of the approximating distribution.

Permutation testing [12] is an alternative approach to sampling the test statistic under the null hypothesis. To perform such test, the sample paths $\mathbf{X}^{1:N'}$ and $\mathbf{Y}^{1:M'}$ are pooled together. Sampling the empirical sig-MMD under the null is performed by computing the empirical sig-MMD between two separate collections of size N', M' respectively. These collections are sampled uniformly from all possible permutations of the pooled dataset into two groups of size N' and M'.

[8] Generally, this is a function of the number of samples. However, if large sample sizes are available, it may not be computationally feasible to use all samples.

3.2 U-Statistics

Let $\mathcal{M}$ be a compact metric space and consider a distribution $\mu \in \mathcal{P}(\mathcal{M})$. Suppose we have samples $x_1, \cdots, x_N \sim \mu$. Let $h\colon \mathcal{M}^r \to \mathbb{R}$ be a kernel function with $r \leq N$. The U-statistic [1, 4, 16, 25, 37, 42] with kernel h is defined as

$$U_N := \frac{(N-r)!}{N!} \sum_{(i_1,\cdots,i_r)\in \mathbf{P}_{r,N}} h\left(x_{i_1}, \cdots, x_{i_r}\right)$$

where $\mathbf{P}_{r,N}$ denotes the set of all $N!/\left(N-r\right)!$ permutations of size r chosen from $\{1, \cdots, N\}$. If h is symmetric, then

$$U_N = \frac{1}{\binom{N}{r}} \sum_{(i_1,\cdots,i_r)\in \mathbf{C}_{r,N}} h\left(x_{i_1}, \cdots, x_{i_r}\right)$$

where $\mathbf{C}_{r,N}$ denotes the set of all combinations of r integers from $\{1, \cdots, N\}$ such that $i_1 \leq \cdots \leq i_r$. For example, if $r = 1$, the sample mean $N^{-1}\sum_i x_i$ is a U-statistic. By setting h to be a linear combination of ϕ-signature kernels, asymptotic results regarding limiting distributions of U-statistics can be applied to the ϕ-MMD.

3.3 Asymptotic Analysis: Null Distribution

To perform the two-sample hypothesis test and quantify the probabilities of Type 1 and Type 2 errors occurring, an estimate for the threshold value $\widehat{c}_{1-\alpha}$ is needed. Since $\widehat{c}_{1-\alpha}$ is the α-quantile of the null distribution $\widehat{F_{H_0}}$, by approximating the null distribution, the $(1-\alpha)$-quantile can then be computed.

Suppose we have N samples $\mathbf{X}^{1:N}, \mathbf{Y}^{1:N}$. Define $\mathbf{Z}^i := \left(\mathbf{X}^i, \mathbf{Y}^i\right)$. Another unbiased estimator for $d_{k_{\text{Sig}}}\left(\mathbb{P}_\mathbf{X}, \mathbb{P}_\mathbf{Y}\right)^2$ is given by the one sample U-statistic [19, Lemma 7]

$$\widehat{d_{k_{\text{Sig}},2}}\left(\mathbb{P}_\mathbf{X}, \mathbb{P}_\mathbf{Y}\right)^2 = \frac{1}{N\left(N-1\right)} \sum_{\substack{i,j=1 \\ i\neq j}}^{N} h\left(\mathbf{Z}^i, \mathbf{Z}^j\right)$$

where $h\left(\cdot,\cdot\right)$ is defined by

$$h\left(\mathbf{Z}^i, \mathbf{Z}^j\right) := k_{\text{Sig}}\left(\mathbf{X}^i, \mathbf{X}^j\right) + k_{\text{Sig}}\left(\mathbf{Y}^i, \mathbf{Y}^j\right) - k_{\text{Sig}}\left(\mathbf{X}^i, \mathbf{Y}^j\right) - k_{\text{Sig}}\left(\mathbf{X}^j, \mathbf{Y}^i\right).$$

The difference between $\widehat{d_{k_{\text{Sig}},2}}$ and $\widehat{d_{k_{\text{Sig}}}}$ (Eq. (5)) is that the terms $k\left(\mathbf{X}^i, \mathbf{Y}^i\right)$ are not included in the computation of $\widehat{d_{k_{\text{Sig}},2}}$ whilst they are accounted for when computing

$\widehat{d_{k_{\text{Sig}}}}$. As described by the authors of [19], assuming $\mathbb{E}\left[h^2\right] < \infty$, when applied to larger sample sizes, the null distribution can be approximated by an infinite weighted sum of shifted independent χ_1^2 random variables as follows

$$N\widehat{d_{k_{\text{Sig}},2}}\left(\mathbb{P}_{\mathbf{X}}, \mathbb{P}_{\mathbf{Y}}\right)^2 \sim \sum_{i=1}^{\infty} \lambda_i \left[z_i^2 - 2\right]$$

where z_i are i.i.d. mean-zero normally distributed random variables with variance 2. The weight λ_l is the solution to the eigenvalue problem

$$\int_{\mathcal{K}} \tilde{k}\left(\tilde{\mathbf{X}}, \mathbf{X}'\right) \psi_l\left(\tilde{\mathbf{X}}\right) d\mathbb{P}_{\mathbf{X}}\left(\tilde{\mathbf{X}}\right) = \lambda_l \psi_l\left(\mathbf{X}'\right)$$

where

$$\begin{aligned}\tilde{k}\left(\tilde{\mathbf{X}}, \mathbf{X}'\right) := k_{\text{Sig}}\left(\tilde{\mathbf{X}}, \mathbf{X}'\right) - \mathbb{E}_{\mathbf{X}_1}\left[k_{\text{Sig}}\left(\tilde{\mathbf{X}}, \mathbf{X}_1\right)\right] \\ - \mathbb{E}_{\mathbf{X}_2}\left[k_{\text{Sig}}\left(\mathbf{X}_2, \mathbf{X}'\right)\right] + \mathbb{E}_{\mathbf{X}_1, \mathbf{X}_2}\left[k_{\text{Sig}}\left(\mathbf{X}_1, \mathbf{X}_2\right)\right].\end{aligned}$$

This approach does not necessitate the same number of samples from the two distributions and it can be computationally intensive as it is based on matrix computations [3]. Alternatively, the null distribution of the biased MMD can be approximated using two different techniques, both based on low-order moments of the empirical MMD. One of these approaches uses Pearson curves [19–21]. The other approach is computationally more efficient and is based on approximating the null distribution of the biased MMD using a gamma distribution [20]. The authors of [29] show that

$$N\widehat{d_{k_{\text{Sig}}}^b}\left(\mathbb{P}_{\mathbf{X}}, \mathbb{P}_{\mathbf{Y}}\right)^2 \sim \frac{x^{\tau-1} e^{-x/\psi}}{\psi^{\tau} \Gamma(\tau)}$$

where $\Gamma(\cdot)$ is the gamma function and

$$\tau := \frac{\mathbb{E}\left[\widehat{d_{k_{\text{Sig}}}^b}\left(\mathbb{P}_{\mathbf{X}}, \mathbb{P}_{\mathbf{Y}}\right)^2\right]^2}{\operatorname{Var}\left(\widehat{d_{k_{\text{Sig}}}^b}\left(\mathbb{P}_{\mathbf{X}}, \mathbb{P}_{\mathbf{Y}}\right)^2\right)} \quad \text{and} \quad \psi := \frac{N \operatorname{Var}\left(\widehat{d_{k_{\text{Sig}}}^b}\left(\mathbb{P}_{\mathbf{X}}, \mathbb{P}_{\mathbf{Y}}\right)^2\right)}{\mathbb{E}\left[\widehat{d_{k_{\text{Sig}}}^b}\left(\mathbb{P}_{\mathbf{X}}, \mathbb{P}_{\mathbf{Y}}\right)^2\right]}.$$

When closed-form solutions are available for truncated estimates of the two moments, these closed-form solutions are used. When these are not available, empirical estimates are constructed using bootstrapping.

3.4 Asymptotic Analysis: Alternative Distribution

If $\mathbb{E}\left[h^2\right] < \infty$, the distribution of the squared MMD under the alternative hypothesis converges in distribution to a Gaussian according to [19, Theorem 8]

$$\sqrt{N}\left(\widehat{d_{k_{\text{Sig}},2}}\left(\mathbb{P}_{\mathbf{X}}, \mathbb{P}_{\mathbf{Y}}\right)^2 - d_{k_{\text{Sig}}}\left(\mathbb{P}_{\mathbf{X}}, \mathbb{P}_{\mathbf{Y}}\right)^2\right) \to \mathcal{N}\left(0, \sigma^2\right)$$

where

$$\sigma^2 := 4\left(\mathbb{E}_{\mathbf{Z}}\left[\mathbb{E}_{\mathbf{Z}'}\left[h\left(\mathbf{Z}, \mathbf{Z}'\right)\right]^2\right] - \mathbb{E}_{\mathbf{Z},\mathbf{Z}'}\left[h\left(\mathbf{Z}, \mathbf{Z}'\right)\right]^2\right)$$

and $\mathcal{N}\left(0, \sigma^2\right)$ denotes a Gaussian random variable with mean 0 and variance $\sigma^2 > 0$.

4 Numerical Examples

In this section we carry out numerical simulations to show the effect different signature kernels have on reducing errors when performing two-sample hypothesis testing. In certain scenarios, the original signature kernel and standard two-sample hypothesis test perform well. In our simulations, we use parametric models to construct examples in which the probability of Type 2 errors occurring are high. This was done to illustrate the importance of re-weighting higher signature levels. The probabilities of a Type 2 error occurring were computed by approximating the distributions of the test statistic under the null and alternative hypotheses using bootstrapping. To compute the probabilities of a Type 1 error occurring, the same procedure as for Type 2 errors was applied, however in this case, both collections of paths were simulated from a stochastic model with the same parameters. When the parameters are chosen such that they are very close, very few ϕ-signature kernels would reduce the probabilities of errors and more adhoc weight functions may be required. In some situations more sample paths will be needed. In practice, linearly interpolated sample paths are used instead of Brownian motion paths. Although Brownian motion paths do not belong to the set $\mathcal{K}$, the linearly interpolated paths belong to the set $\mathcal{K}$. Also, the sample paths converge to Brownian motion sample paths (Donsker's invariance principle [13]) as the mesh size tends to 0. Therefore, the linearly interpolated paths adhere to the framework described in Sect. 2.

Throughout the experiments, the level was set to $\alpha = 0.05$. Simulations were mainly run on a GPU[9] for computational efficiency. The code used to generate these results can be found at the following repository: https://github.com/Andrew-Alden/SignatureMMDTesting.

[9] GPUs were provided by the King's College London CREATE environment [30].

4.1 Scaled Brownian Motion

Consider the time interval $[0, T]$ and two scaled Brownian motions $\mathbf{S}$ and $\mathbf{G}$ with dynamics described by the stochastic differential equations (SDEs)

$$d\mathbf{S}_t = \sigma dW_t^1 \text{ and } d\mathbf{G}_t = \beta dW_t^2.$$

The parameters $\sigma, \beta > 0$ control the volatility of the processes and $(W_t^1), (W_t^2)$ are independent Brownian motions.[10] Define the time-augmented[11] processes $\mathbf{X}, \mathbf{Y}$ as $\mathbf{X}_t := (t, \mathbf{S}_t)$ and $\mathbf{Y}_t := (t, \mathbf{G}_t)$. The two-sample hypothesis test is performed on the stochastic processes $\mathbf{X}$ and $\mathbf{Y}$.

The processes $\mathbf{S}$ and $\mathbf{G}$ are simulated using the Euler discretisation scheme

$$\mathbf{S}_{t_i} = \mathbf{S}_{t_{i-1}} + \sigma\sqrt{h}Z, \quad Z \sim \mathcal{N}(0, 1), \quad t_i - t_{i-1} = h.$$

It is assumed[12] that $\mathbf{S}_0 = \mathbf{G}_0 = 0$ and we set $T = 1$. Since these processes are driftless,

$$\mathbb{E}[\mathbf{S}_t] = \mathbb{E}[\mathbf{G}_t] = 0$$

for all $t \in [0, T]$. If $\sigma \neq \beta$, the processes $\mathbf{S}$ and $\mathbf{G}$ differ from their second moment.[13] Therefore, the sig-MMD should distinguish between these processes. The empirical probabilities of a Type 2 error and a Type 1 error occurring were computed using the empirical distributions of the sig-MMD. To compute probabilities of Type 1 errors occurring, both collections of sample paths were sampled from $\mathbf{X}$ (i.e. H_0).

In the simulations, σ was set to 0.2, β was set to 0.3, and the batch size was set to 128. The probability of a Type 2 error occurring was 72.6% with the biased estimator and 85.8% with the unbiased estimator. These probabilities correspond to the large overlap in the histograms in Fig. 1. To better understand the underlying cause of the high probability 85.8%, the level contributions (Γ_m^ϕ) were plotted. As can be noted in Fig. 2, the area of overlap decreases as the signature level increases. As can be seen in the plot of the level-1 contribution (Γ_1^ϕ), the histograms completely overlap. This occurs because the processes are driftless. However, from the second level onwards, the histograms start to gradually separate (region of overlap decreases). Due to the factorial decay of the absolute value of the signature terms, even though the histograms start to separate, the level with the largest

[10] For further details on Brownian motions and their expected signature see [18, 34].

[11] Adding the time component removes the possibility of tree-like path occurrences.

[12] This does not affect the results. If $\mathbf{S}_0, \mathbf{G}_0 \neq 0$ consider the processes $\mathbf{S}' := \mathbf{S} - \mathbf{S}_0$ and $\mathbf{G}' := \mathbf{G} - \mathbf{G}_0$.

[13] $\mathbb{E}[\mathbf{S}_t^2] \neq \mathbb{E}[\mathbf{G}_t^2]$ for all $t \in [0, T]$.

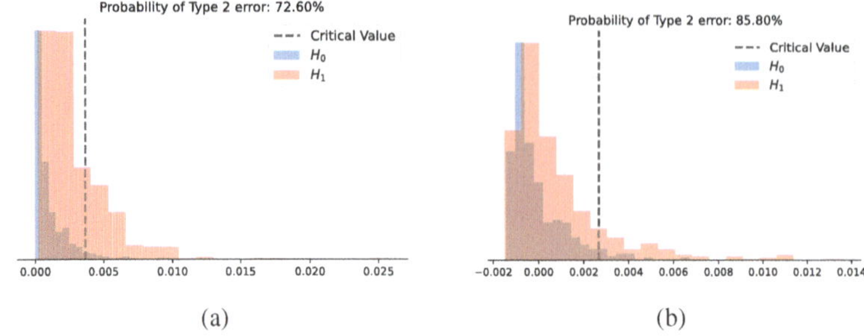

Fig. 1 Null and alternative distributions of the (squared) sig-MMD between two scaled Brownian motions. Batch size of 128 was used and 500 independent simulations were run. (**a**) Biased. (**b**) Unbiased

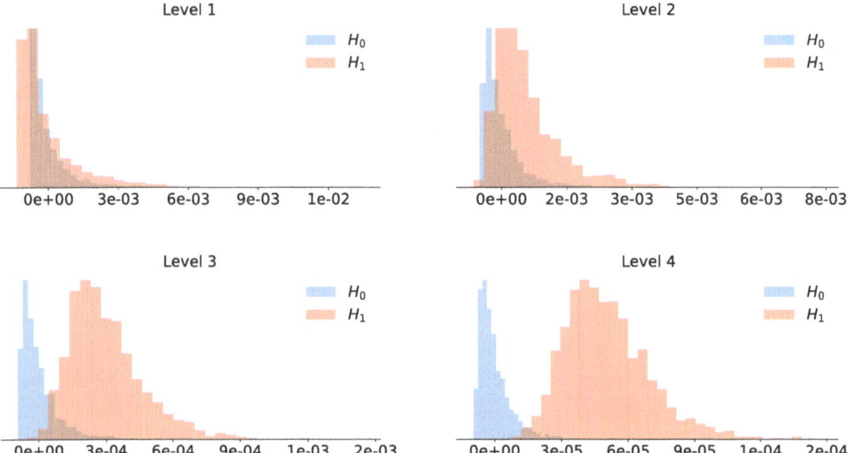

Fig. 2 Level contributions to the sig-MMD between two scaled Brownian motions. Batch size of 128 was used and 2,048 independent simulations were run. No scaling was applied and the unbiased estimator was used

contribution to the MMD (first level) corresponds to the plot in which the histograms completely overlap. As a result, the probability of a Type 2 error occurring is high.

By using general signature kernels, the test statistic is tailored to target specific moment(s) when performing the two-sample hypothesis test. In this particular example, scaling the paths allows us to focus on the second moment of the processes. In our simulations, we tested various scaling factors. Scaling was applied to the processes **S**, **G** and not to the time-augmented processes. Therefore, the time-index still ran from $[0, T]$. This was done because the time-index was used to add a time structure to the process. We could have set T in the time-augmented process to be the inverse of the scaling factor. In this case, multiplying the entire process including the time-index by the scaling factor would result in time starting at 0 and ending at $T = 1$. This is equivalent to our approach of not scaling the time index.

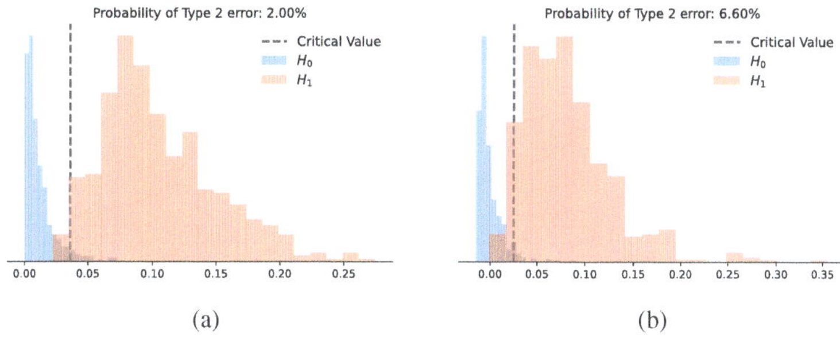

Fig. 3 Null and alternative distributions of the (squared) ϕ-MMD between two scaled Brownian motions. Batch size of 128 was used and 500 independent simulations were run. A scaling of 3 was applied. (**a**) Biased. (**b**) Unbiased

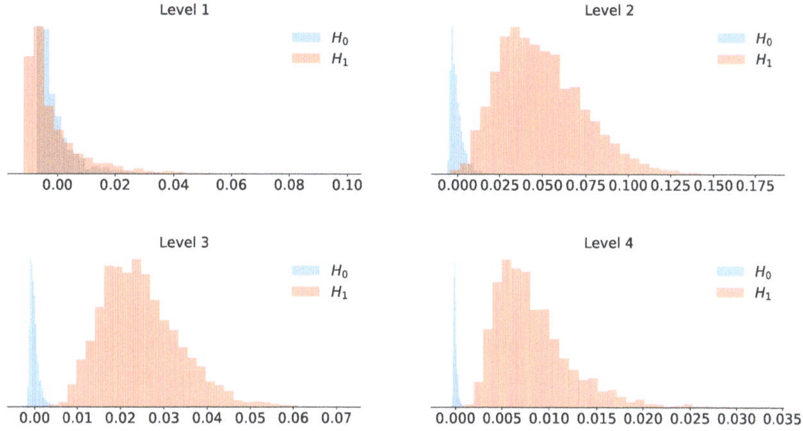

Fig. 4 Level contributions to the ϕ-MMD between two scaled Brownian motions. Batch size of 128 was used and 2,048 independent simulations were run. A scaling of 3 was applied and the unbiased estimator was used

When re-running all simulations with scaled paths using a scaling factor of 3, the probability of a Type 2 error occurring drops to 2.0% (Fig. 3a) when using the biased estimator and 6.6% (Fig. 3b) when using the unbiased estimator. Scaling the paths by a factor of 3 is equivalent to scaling the parameters σ, β by a factor of 3. The scaling amplifies the differences between distributions. The level contributions to the ϕ-MMD are plotted in Fig. 4. The separation of histograms is evident as the level increases. A crucial difference between the histograms in Figs. 2 and 4 is that, when scaling is applied (Fig. 4), the absolute value of the contribution is larger for the scaled version than for the standard sig-MMD. Hence, higher-level terms which better capture distributional differences contribute more to the final test statistic than when using the standard sig-MMD.

In the above example, the batch size was kept fixed at 128. It is important to study the effect of scaling paths on the probability of a Type 2 error occurring as a function

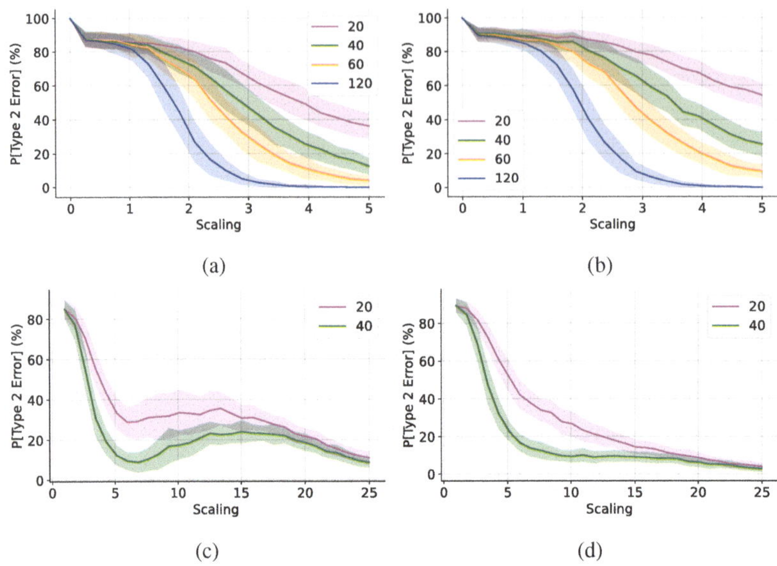

Fig. 5 Probability of a Type 2 error occurring as a function of scaling factor and batch size between two scaled Brownian motions. Solid line corresponds to the mean and the shaded region corresponds to one standard deviation away from the mean. (**a**) Biased. (**b**) Unbiased. (**c**) Biased—extreme scalings. (**d**) Unbiased—extreme scalings

of the batch size. In Fig. 5, the probability of a Type 2 error occurring is plotted as a function of the batch size and the scaling factor. Figure 5a corresponds to the biased estimator for the ϕ-MMD and Fig. 5b corresponds to the unbiased estimator. For low batch sizes, scaling reduces the probability of a Type 2 error occurring. However, it does not reduce it to levels below 30% unless extreme scalings are used (Fig. 5c, d). Also, although the impact of scaling is consistent across the biased and the unbiased estimator, the effect takes place earlier in the biased version for scaling values less than 5. Therefore, for a given batch size, the probability of a Type 2 error occurring is most likely lower when using the biased estimator as opposed to the unbiased estimator if no extreme scaling is applied.

These results demonstrate the importance of calibrating the signature kernel to focus on specific moments of the distributions. Apart from focusing on Type 2 errors, we ensured that the probability of a Type 1 error occurring did not increase as the scaling is altered. We did this by plotting the distributions of the probability of a Type 1 error occurring across various scaling factors (Fig. 6). These plots show that this probability is independent of batch size and it averages at around 5% in every scenario, corresponding to the level set for the test. These plots confirm that scaling does not negatively effect the probability of a Type 1 error occurring. This is consistent across both the biased and the unbiased estimator.

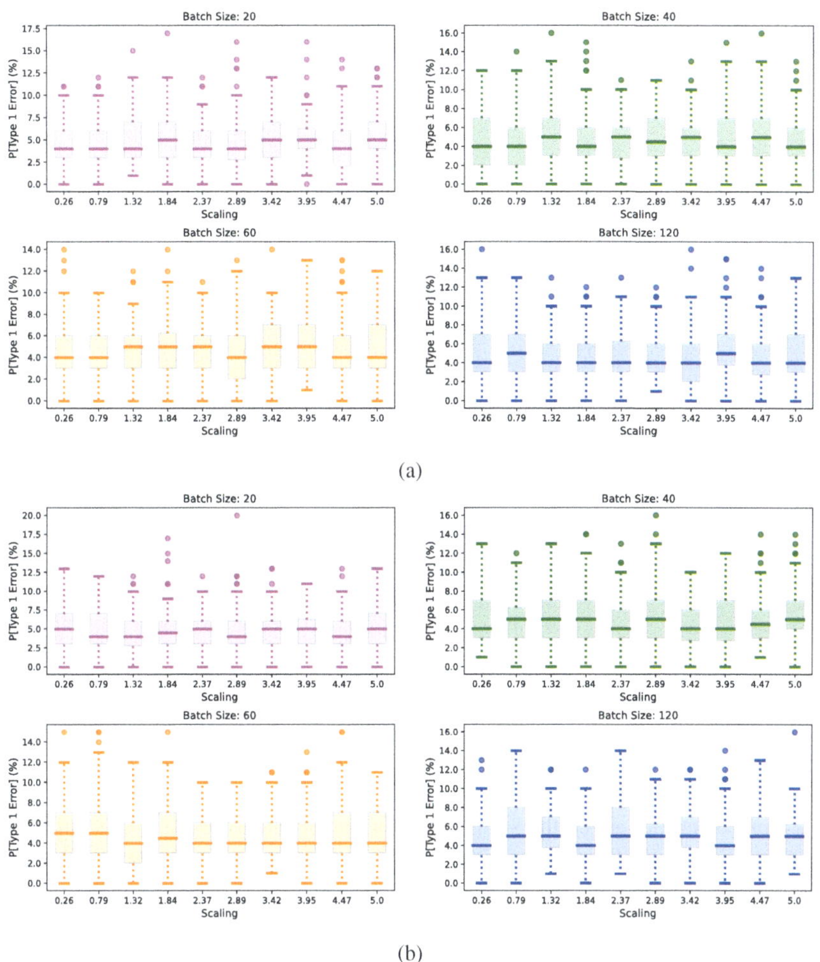

Fig. 6 Probability of a Type 1 error occurring as a function of sample size and scaling factor between two scaled Brownian motions. (**a**) Biased. (**b**) Unbiased

We now study the effect of path scalings on higher dimensional paths. Consider the processes $\mathbf{X} = \left(\mathbf{X}_t = \left(t, S_t^1, \cdots, S_t^{d-1}\right)\right)_t$ and $\mathbf{Y} = \left(\mathbf{Y}_t = \left(t, G_t^1, \cdots, G_t^{d-1}\right)\right)_t$. Each process pair $\left(\mathbf{S}^j, \mathbf{G}^j\right)$ is a scaled Brownian motion with parameters $\left(\sigma_j, \beta_j\right)$. We set $d = 5$ and used the following parameter values

- $(\sigma_1, \beta_1) = (0.2, 0.3)$;
- $(\sigma_2, \beta_2) = (0.5, 0.6)$;
- $(\sigma_3, \beta_3) = (0.7, 0.8)$; and
- $(\sigma_4, \beta_4) = (0.1, 0.2)$.

Table 1 Likelihood of a Type 2 error occurring as dimension and batch size increase

Dimension	Sample size	Bias scaling (%)	Bias no scaling (%)	Unbiased scaling (%)	Unbiased no scaling (%)
3	16	77.8	90.5	87.8	92.2
	32	72.1	89.9	84.6	92.6
	64	58.2	87.4	75.3	90.5
	128	33.2	84.0	50.3	88.1
	256	6.80	70.2	15.2	76.1
4	16	77.4	86.8	88.2	92.3
	32	68.4	85.4	83.9	91.3
	64	53.6	83.4	75.6	89.3
	128	31.0	73.8	57.2	83.5
	256	8.13	52.4	24.5	68.8
5	16	61.4	86.6	88.3	92.4
	32	49.4	85.3	83.2	91.3
	64	28.9	82.0	72.8	89.1
	128	6.9	73.5	42.6	84.5
	256	0.1	49.3	4.4	67.1

Table 1 contains the probability of a Type 2 error occurring for various batch sizes and dimensions. A scaling factor of 2 was applied to compute the probabilities. Scaling the paths does reduce the probability of a Type 2 error. In addition, scaled paths always had a lower probability than their original (unscaled) counterpart. More importantly, as the batch size increases, the probability does decrease when no scaling is applied since the sig-MMD is a metric and converges as sample size increases. The main effect of scaling the paths in this case is capturing distributional differences with smaller batch sizes. This is crucial when working in low-data regimes and/or working with limited computational resources.

4.2 *Autoregressive Time Series Models*

This example focuses on autoregressive conditional heteroskedasticity (ARCH) processes. Let $\mathbf{r} = (\mathbf{r}_t)$ be a stochastic process with conditional volatility modelled as a general autoregressive conditional heteroskedastic (GARCH) process [6, 15], a class of ARCH models [14]. The process $\mathbf{r}$ is described by the equations

$$\begin{aligned}
\mathbf{r}_t &= \mu + \epsilon_t \\
\sigma_t^2 &= \omega + \alpha_1 \epsilon_{t-1}^2 + \beta_1 \sigma_{t-1}^2 \\
\epsilon_t &= \sigma_t z_t \\
z_t &\sim \mathcal{N}(0, 1) .
\end{aligned}$$

Table 2 GARCH model parameters

Process	$\mu\ (\times 10^{-3})$	$\omega\ (\times 10^{-3})$	$\alpha_1\ (\times 10^{-2})$	$\beta_1\ (\times 10^{-2})$
$\mathbf{r}_1$	1.0	3.8	4	4.2
$\mathbf{r}_2$	5.0	5.3	8.0	1.0

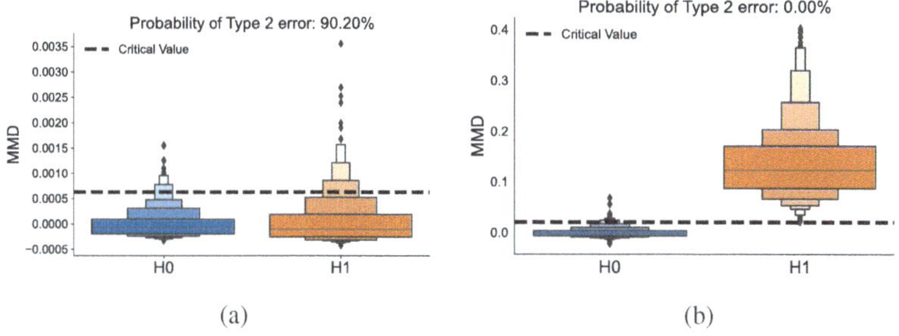

(a) (b)

Fig. 7 Null and alternative distributions of the ϕ-MMD between two GARCH models. Batch size of 128 was used and 500 independent simulations were run. (**a**) Without scaling. (**b**) Scaling of 5.5 was applied

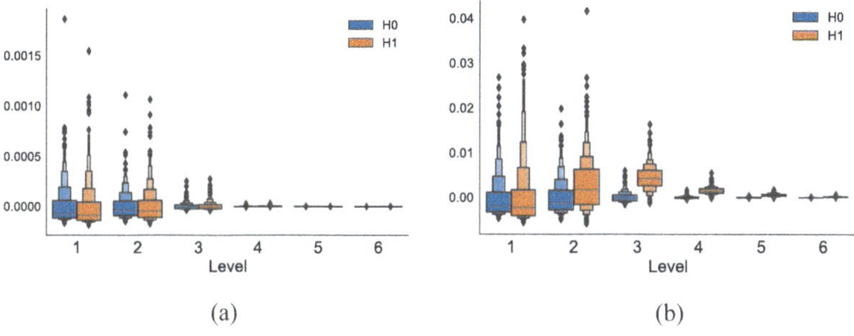

(a) (b)

Fig. 8 Null and alternative distributions of the level contributions between two GARCH models. Batch size of 128 was used and 500 simulations were run. (**a**) Without scaling. (**b**) Scaling of 5.5 was applied

Two processes $\mathbf{r}_1, \mathbf{r}_2$ were simulated using the parameters in Table 2. As was done in the previous example (Sect. 4.1), the probability of a Type 2 error occurring in the two-sample hypothesis test was calculated through re-sampling techniques. Using the unbiased estimator, the computation resulted in a probability of 90.2% (Fig. 7a). Once again, path scaling was sufficient to reduce this probability. When a scaling of 5.5 was applied, the probability of a Type 2 error occurring dropped to 0.0% (Fig. 7b).

Once again, the empirical distributions of the level contributions to the sig-MMD in the case of two GARCH models contained significant overlap (Fig. 8a). When a scaling of 5.5 was applied, the distribution of level terms contained sufficient separation (Fig. 8b). Another important difference between the distributions of the level terms with and without scaling is that when scaling is used, the level

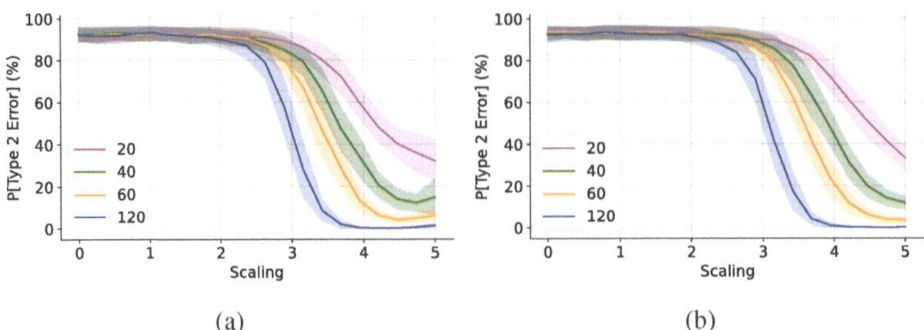

Fig. 9 Probability of a Type 2 error occurring between two GARCH models as a function of sample size. (**a**) Biased. (**b**) Unbiased

contributions in the case of the alternative hypothesis have larger absolute value. As previously discussed, this contributed to reducing the probability of a Type 2 error occurring.

All GARCH simulations were performed using a batch size of 128. We once again plotted the probability of a Type 2 error occurring as the batch size increases and scaling factor increases. This is provided in Fig. 9. Scaling does reduce the probability of a Type 2 error occurring. Comparing Fig. 9 with Fig. 5, we find that scaling has the common effect of reducing the probability of a Type 2 error occurring. However, a difference between the two is the scaling factor needed to reduce this probability. In the case of scaled Brownian motions, the probability started reducing when scalings slightly larger than 1 were considered. In the case of GARCH models, a high probability was maintained up until a scaling of around 3.5. The need for more aggressive scaling when working with GARCH models was confirmed in the distribution of the level terms depicted in Fig. 10. Figure 10a shows the first level contribution as a function of the scaling. Scaling does not alter the average contribution under either hypothesis. It mainly affects the variance of the level contribution. When considering the fourth and sixth level terms (Fig. 10b, c respectively), besides effecting the variance the scaling also alters the mean contribution. However, this effect occurs at high scaling values. For example, in Fig. 10b there is a noticeable separation in distributions when considering scaling factors greater than 4.0.

We once again study the effect of scaling on the probability of a Type 1 error occurring. The distribution of the probability of a Type 1 error occurring using the ϕ-MMD as a function of various scaling factors and batch sizes is plotted in Fig. 11. On average, the probability remained stable at around the 5% value.

In this example, differences between processes were difficult to capture. This was confirmed by the level contributions plotted in Figs. 8 and 10 and the plot showing the effect of scaling on the probability of a Type 2 error occurring (Fig. 9). Although the processes have different first two moments, higher-level terms of the signature were needed to capture differences.

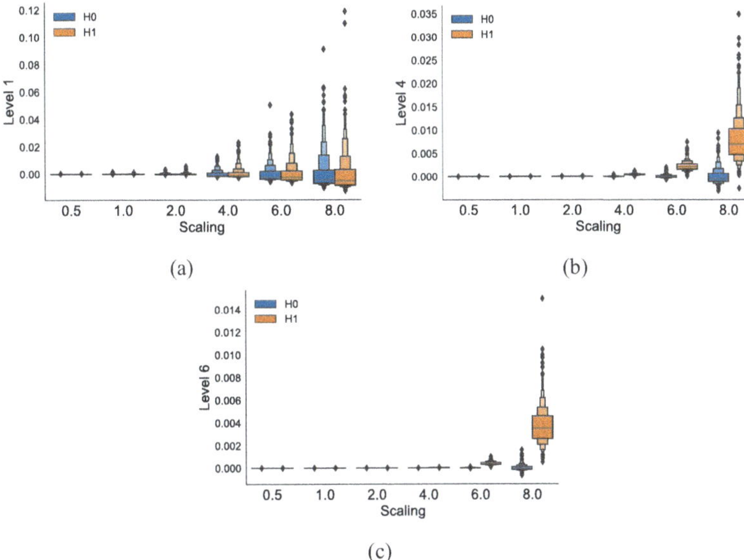

Fig. 10 Distribution of the level terms between two GARCH models as a function of scaling factor. Unbiased estimator was used. (**a**) Level 1. (**b**) Level 4. (**c**) Level 6

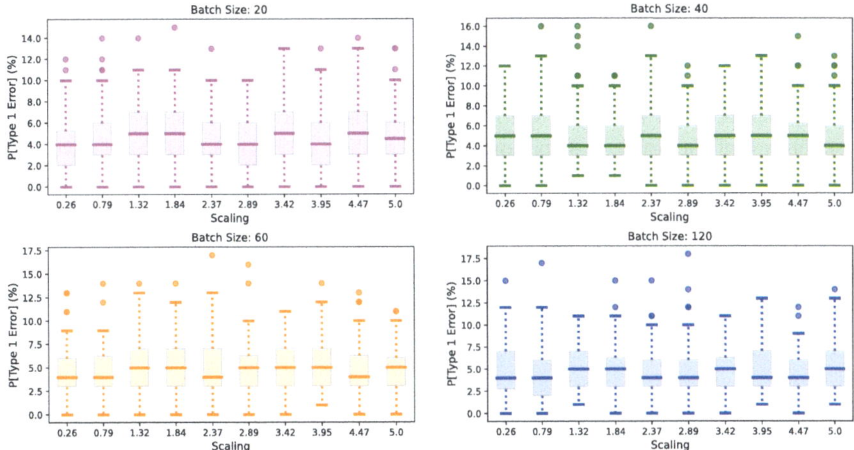

Fig. 11 Probability of a Type 1 error occurring between two GARCH models as a function of sample size and scaling factor. Unbiased estimator was used

4.3 Mixture Models

To understand the effect of scaling on the Type 2 error of the statistical test between two stochastic processes which differ in their third moment, we use a mixture of a geometric Brownian motion (GBM) [5] and an Ornstein-Uhlenbeck (OU) process

[48]. The GBM is described by the SDE

$$d\mathbf{S}_t = \mu \mathbf{S}_t dt + \sigma \mathbf{S}_t dW_t$$

and the OU process is described by the SDE

$$d\mathbf{G}_t = -\theta \mathbf{G}_t dt + \tilde{\sigma} dB_t$$

where $\mu, \theta, \sigma, \tilde{\sigma} > 0$ and $\{W_t\}, \{B_t\}$ are Brownian motions. Consider the process

$$d\mathbf{X}_t = d\mathbf{S}_t + d\mathbf{G}_t$$

and assume $\{W_t\}, \{B_t\}$ are uncorrelated Brownian motions. By choosing specific parameter values, the marginal distributions of X_T, Y_T with $T = 1$ satisfy

$$\mathbb{E}[X_T] \approx \mathbb{E}[Y_T], \quad \mathbb{E}\left[X_T^2\right] \approx \mathbb{E}\left[Y_T^2\right], \quad \text{and} \quad \mathbb{E}\left[X_T^3\right] \neq \mathbb{E}\left[Y_T^3\right].$$

We simulate processes according to the parameters specified in Table 3. We use paths of length 30 and a batch size of 128. The probability of a Type 2 error occurring was 94.6%. This corresponds to the overlap in distributions in Fig. 12a. Scaling the paths did not reduce this probability (Fig. 12b). To understand the reason for this, we plot the level contributions. In Fig. 13a, at each level, there is overlap between distributions. Increasing the batch size to 512 and 2,000, the distributions diverge at levels 3, 4, and 5. Therefore, the sig-MMD can distinguish between these processes, but it cannot distinguish when using a few paths. Re-plotting the distribution of the level contributions with a scaling of 2 applied (Fig. 14), we still

Table 3 Mixture model parameters used to perform simulations

Process	μ	θ	σ	$\tilde{\sigma}$	G_0	S_0
X	0.3	0.3	0.5	0.5	0.75	1.0
Y	0.3	0.3	0.3	0.84	0.75	1.0

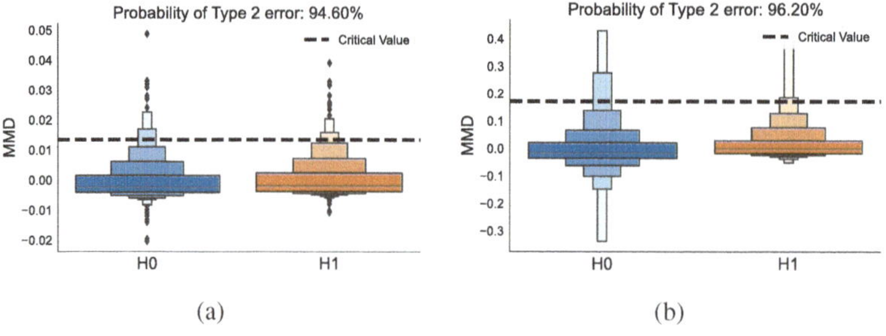

(a) (b)

Fig. 12 Null and alternative distributions of the ϕ-MMD between two mixture models. Batch size of 128 and the unbiased estimator were used. (**a**) No scaling. (**b**) Scaling of 2.0 was applied

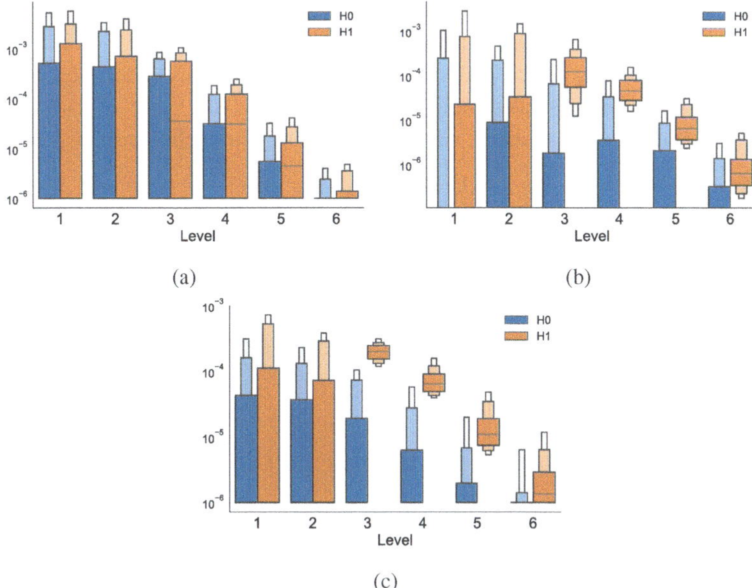

Fig. 13 Null and alternative distributions of the level contributions between two mixture models as a function of batch size. 100 independent simulations were run. The level contributions are plotted on a logarithmic scale. (**a**) Batch size of 128. (**b**) Batch size of 512. (**c**) Batch size of 2,000

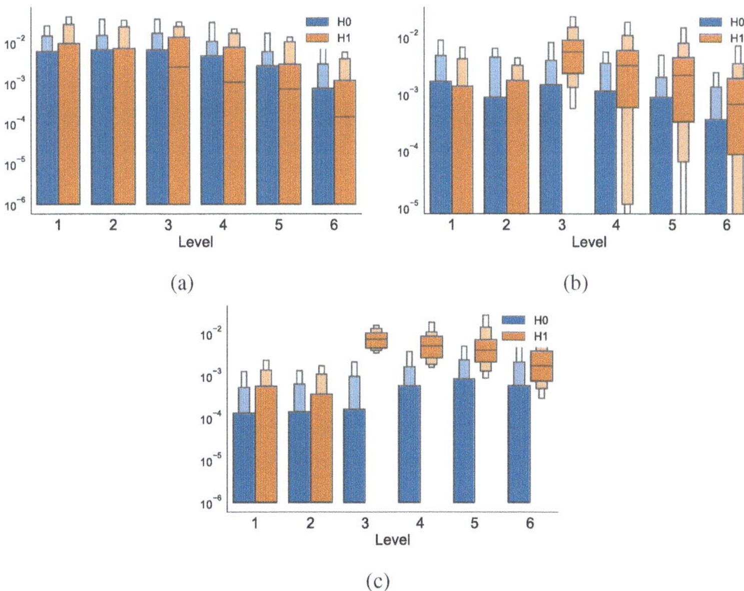

Fig. 14 Null and alternative distributions of the level contributions between two mixture models as a function of batch size. A scaling of 2 was applied. 100 independent simulations were run. The values are plotted on a logarithmic scale. (**a**) Batch size of 128. (**b**) Batch size of 512. (**c**) Batch size of 2,000

note significant overlap between the null and alternative distributions at a batch size of 128. By increasing the batch size we have less overlap at levels 3, 4, 5, and 6. As a result of scaling, the absolute value of the level contributions under the alternative hypothesis do not decrease as rapidly compared to Fig. 13.

To overcome these challenges, we use three additional techniques; namely

Path Standardisation We standardise the paths using the technique adopted in [28]. If $\mathbf{X}_t$ is a path, we consider the standardised path

$$\widehat{\mathbf{X}}_t := (\mathbf{X}_t - \mu_T) / \sigma_T$$

where μ_T, σ_T are the mean and standard deviation of $\mathbf{X}$ at the terminal time.

Lead-Lag Transformation Let $0 = t_0 < t_{1/2} < \cdots < t_{l-1/2} < t_l = T$ be a partition of $[0, T]$. The lead-lag transformation [3, 23, 46] of a 1-dimensional path $\mathbf{X}_t$ is the 2-dimensional path $\left(\mathbf{X}_t^{\text{Lead}}, \mathbf{X}_t^{\text{Lag}}\right)$ defined by linear interpolation on the points

$$\mathbf{X}_{t_{i/2}}^{\text{Lead}} := \begin{cases} \mathbf{X}_{t_j} & \text{if } i = 2j \\ \mathbf{X}_{t_{j+1}} & \text{if } i = 2j+1 \end{cases}, \quad \mathbf{X}_{t_{i/2}}^{\text{Lag}} := \begin{cases} \mathbf{X}_{t_j} & \text{if } i = 2j \\ \mathbf{X}_{t_j} & \text{if } i = 2j+1 \end{cases}.$$

Feature Space Transformation To gain more expressive power, the paths are lifted to the infinite dimensional Hilbert space $H'\left(\sigma_{\text{RBF}}^2\right)$ associated with the radial basis function (RBF) kernel with smoothing parameter $\sigma_{\text{RBF}} > 0$. For $\mathbf{x}, \mathbf{y} \in \mathbb{R}^d$, the RBF kernel is defined by

$$k(\mathbf{x}, \mathbf{y}; \sigma_{\text{RBF}}) := \exp\left(-\frac{||\mathbf{x} - \mathbf{y}||_2^2}{\sigma_{\text{RBF}}^2}\right)$$

where $||\cdot||_2$ denotes the Euclidean norm. Using the Riesz representation theorem, there exists a mapping $\zeta_{\sigma_{\text{RBF}}} : \mathbb{R}^d \to H'$ such that

$$k(\mathbf{x}, \mathbf{y}; \sigma_{\text{RBF}}) = \langle \zeta_{\sigma_{\text{RBF}}}(\mathbf{x}), \zeta_{\sigma_{\text{RBF}}}(\mathbf{y}) \rangle_{H'}.$$

The path $\mathbf{X}_t$ is lifted to the infinite dimensional path $k_{\mathbf{X}} : t \mapsto \zeta_{\sigma_{\text{RBF}}}(\mathbf{X}_t)$. Since the lifted path is infinite dimensional, it is not possible to work directly with this path. However, using the kernel trick, we work directly with the RBF kernel instead. Computing the signature kernel between two lifted paths can be done using either dynamic programming as described in [31], or directly through the PDE approach [39]. An alternative to directly computing the RBF kernel consists of approximating the RBF kernel using random Fourier features [38, 47]. An important consideration when working with lifted paths is that scaling the original paths does not necessarily translate to weighting the level contributions to the signature kernel since the signature kernel is now computed between the lifted paths. Therefore, we need to ensure that we are scaling the path $\zeta_\sigma(\mathbf{X}_t)$ and not the path $\mathbf{X}_t$.

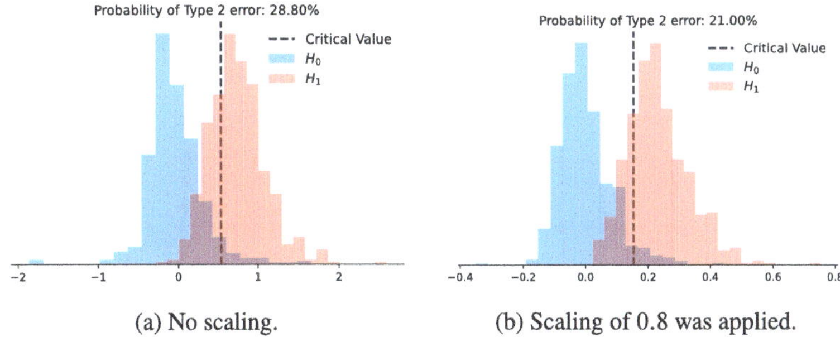

(a) No scaling. (b) Scaling of 0.8 was applied.

Fig. 15 Null and alternative distributions of the ϕ-MMD between two mixture models. Path standardisation and lead-lag transformations were performed. The RBF smoothing parameter was set to $\sigma_{\text{RBF}} = \sqrt{0.5}$. Batch size of 128 was used. (**a**) No scaling. (**b**) Scaling of 0.8 was applied

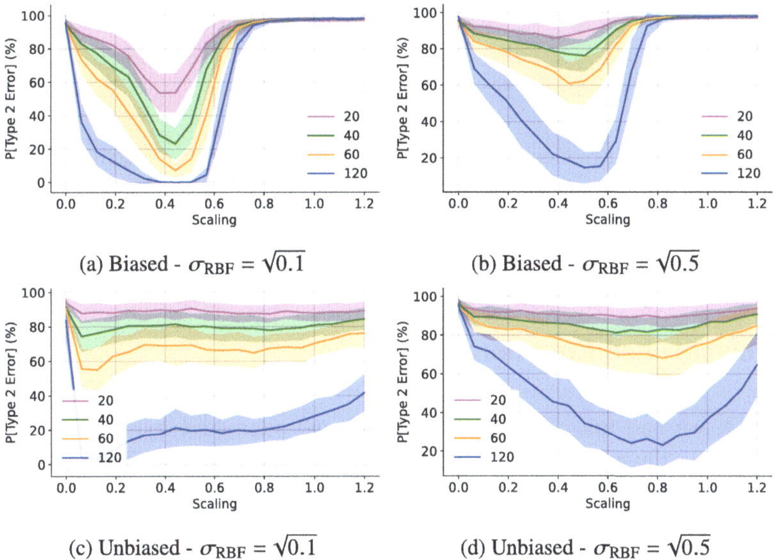

(a) Biased - $\sigma_{\text{RBF}} = \sqrt{0.1}$ (b) Biased - $\sigma_{\text{RBF}} = \sqrt{0.5}$

(c) Unbiased - $\sigma_{\text{RBF}} = \sqrt{0.1}$ (d) Unbiased - $\sigma_{\text{RBF}} = \sqrt{0.5}$

Fig. 16 Probability of a Type 2 error occurring between two mixture models as a function of sample size. (**a**) Biased—$\sigma_{\text{RBF}} = \sqrt{0.1}$. (**b**) Biased—$\sigma_{\text{RBF}} = \sqrt{0.5}$. (**c**) Unbiased—$\sigma_{\text{RBF}} = \sqrt{0.1}$. (**d**) Unbiased—$\sigma_{\text{RBF}} = \sqrt{0.5}$

Keeping the batch size fixed at 128, the paths were first standardised using the above procedure. Performing the hypothesis test with lead-lag paths lifted to the space $H'(0.5)$, the probability of a Type 2 error occurring reduced from 94.6% to 28.8% (Fig. 15a). By applying scaling, this was further reduced to 21.0%.

To understand the relationship between path scalings and Type 2 error applied to lifted paths, we plot the probability of a Type 2 error as a function of the scaling factor and the batch size (Fig. 16). Whenever the probability was significantly reduced, the curve of probabilities had a parabolic shape. A difference between the plots in Fig. 16 and the plots in Figs. 5 and 9 is that, in Figs. 5 and 9 there

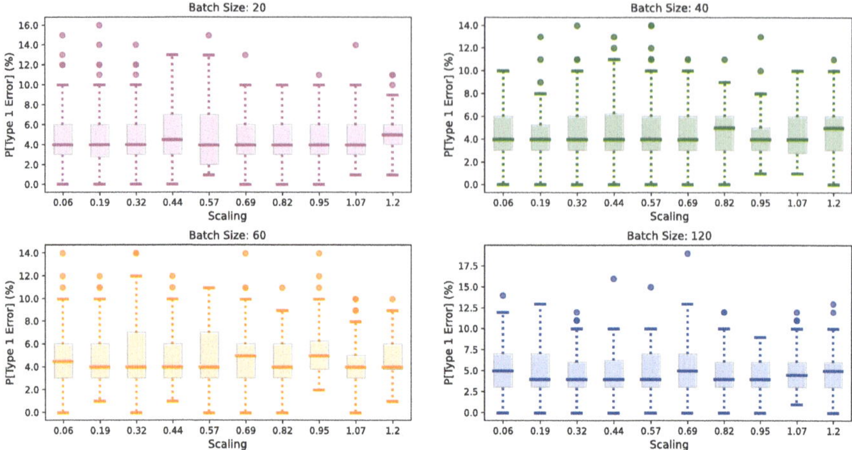

Fig. 17 Probability of a Type 1 error occurring between two mixture models as a function of sample size and scaling factor. Biased estimator with $\sigma_{\text{RBF}} = \sqrt{0.5}$

was a scaling which reduced the probability of Type 2 error occurring. This was not necessarily the case when performing the two-sample hypothesis test between mixture models (see Fig. 16d). Moreover, we once again show that the probability of a Type 1 error occurring is unaffected by the scaling factor used (Fig. 17).

5 Uncontrolled Environment

In Sect. 4, we had full control over the null and alternative hypotheses, i.e. we knew whether the null or alternative hypothesis held prior to performing the test. In practice, we are not in a controlled environment and need to determine which hypothesis holds. As demonstrated through the simulations, finding a robust test setup suitable in all cases is not possible. We presented various techniques for tailoring the two-sample test depending on the collections of sample paths. In practice, finding an appropriate setup is similar to hyperparameter optimisation in a ML context. As is the case in most ML tasks, task performance is highly dependent on specific configurations (learning rates, network architecture, etc.). We propose optimising the scaling factor in a similar way, mimicking the optimisation procedure for certain hyperparameters in a ML context. Since we are focused on two-sample testing, the hyperparameters are optimised for a low probability of a Type 2 error occurring. We demonstrated various test configurations and techniques[14] which prove to be effective in the context of two-sample testing using path signatures. Since we have shown that the probability of a Type 1 error occurring is largely unaffected across all techniques presented, by iterating over the various techniques,

[14] There could be other techniques. We presented the ones we find most useful in practice.

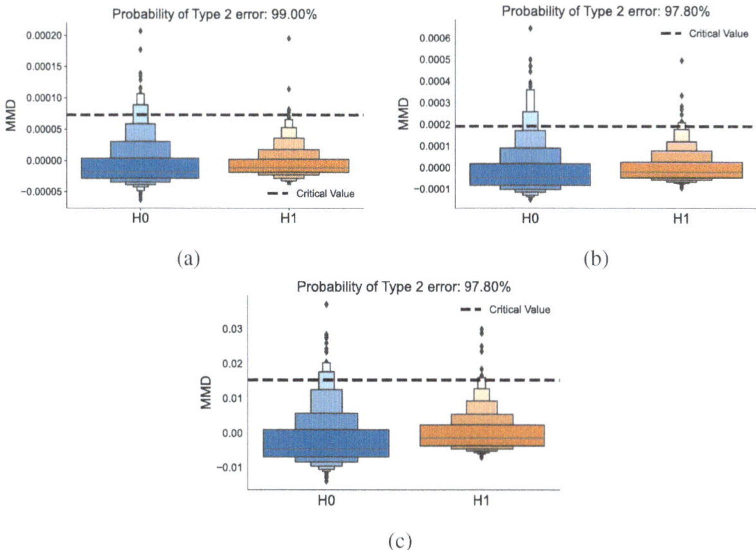

Fig. 18 Null and alternative distributions of the ϕ-MMD. 500 independent simulations were run. Batch size of 128 and unbiased estimator were used. (**a**) No scaling. (**b**) $\sigma_{\text{RBF}} = 1$. (**c**) $\sigma_{\text{RBF}} = 1$, scaling $= 5$

we are optimising the test for Type 2 errors whilst maintaining a low rate of Type 1 errors. When performing hyperparameter optimisation in the context of two-sample hypothesis testing, one would need to split the data into two parts. The first part is used to fine-tune the hyperparameters with respect to some criterion and the second part of the data is used to perform the hypothesis test [21, 22, 32, 41, 44].

We now demonstrate how to apply these techniques in an uncontrolled environment. We would like to determine whether the time series of returns corresponding to baskets of assets have the same distribution. Each basket consists of assets from the same sector. The sectors we consider are the technology sector and consumer based sector. We use 5 assets from each sector and for each asset we compute the percentage return. The return time series corresponding to each asset is split into multiple time series, each consisting of 15 consecutive observations (approximately 3 weeks of returns).[15] Aggregating all length 15 time series of all assets within the same basket (sector) constitutes the available samples from the stochastic process corresponding to that basket of assets. We use data from 10 September 2014 to 8 September 2024. This is an uncontrolled setting since we do not know in advance whether the returns have the same distribution. The data was split into two parts according to the ratio 80:20. 80% of the data was used to calibrate the hyperparameters of the test. We first start the calibration procedure by using the raw returns without any scaling applied. The probability of a Type 2 error occurring in this setup was calculated at 99.0% (Fig. 18a). Performing a feature transformation

[15] The sub-sampled time series have no overlapping observations.

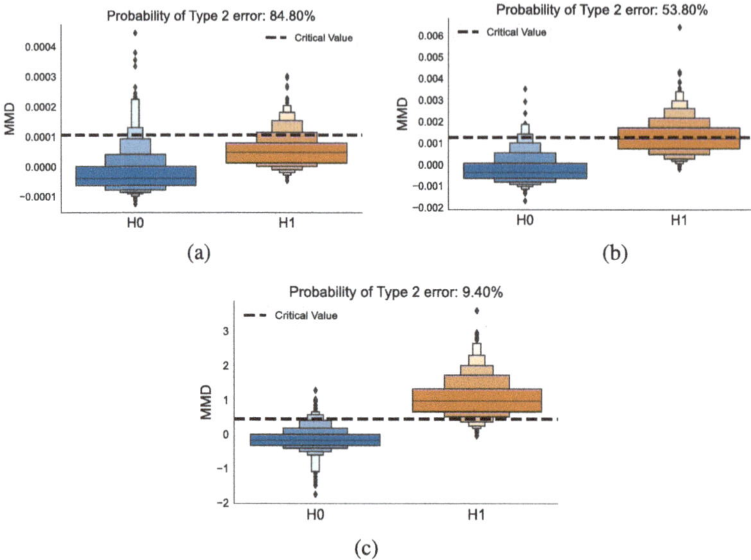

Fig. 19 Null and alternative distributions of the ϕ-MMD after applying lead-lag transformation. 500 independent simulations were run. Batch size of 128 and unbiased estimator were used. (**a**) No scaling applied. (**b**) $\sigma_{\text{RBF}} = \sqrt{0.5}$. (**c**) $\sigma_{\text{RBF}} = \sqrt{0.5}$, scaling = 5

using a RBF kernel with smoothing parameter set to 1, we then performed the two-sample test again. This feature transformation had no effect on the probability of a Type 2 error occurring (Fig. 18b). This was also the case when a scaling of 5 was applied (Fig. 18c).

Repeating the same tests after transforming the time series using the lead-lag transform (Fig. 19), we get a significant reduction in probability of Type 2 errors occurring when scaling was applied after a feature space transformation. Performing a permutation test on the remaining 20% of the data after applying a lead-lag transformation, using an RBF kernel with smoothing parameter of $\sqrt{0.5}$, and applying a scaling of 5, the hypothesis test concluded that the distributions are not equal.

6 Conclusion

This chapter focuses on two-sample hypothesis testing between stochastic processes. Whilst the sig-MMD has proven to be an effective tool for two-sample hypothesis testing, careful consideration is needed to fine-tune the test hyperparameters to the samples available. We explored techniques to make the test more robust; namely

1. Path scaling;
2. Lead-lag transformation;
3. Standardisation; and
4. Feature space transformation using the RBF kernel.

The relationship between the batch size and probability of a Type 2 error occurring was also explored. In addition, we also showed that path scalings do not adversely affect the probability of a Type 1 error occurring.

A contribution made through this work is quantifying the amount of information carried by the level terms of the signature. We showed that the level terms of the ϕ-MMD incorporate specific distributional properties which, when analysed, can be used to target particular moments using tailored ϕ-signature kernels. This guides the design of the two-sample test using the ϕ-MMD. We show that higher-level signature terms contain information regarding distributional moments which is crucial for reducing Type 2 errors in the two-sample testing framework.

In summary, the power of the test is a function of the; truncation level, batch size, feature space (RBF kernel), and ϕ-signature kernel used. When these are configured appropriately, the two-sample hypothesis test using the ϕ-MMD as a test statistic is a powerful statistical tool.

Acknowledgments This work was supported by the UK Engineering and Physical Sciences Research Council (EPSRC) under Grant [EP/T517963/1]. The code is publicly available and can be found at the repository: https://github.com/Andrew-Alden/SignatureMMDTesting.

References

1. J. Ai, O. Kuželka, Y. Wang, Hoeffding–Serfling inequality for U-statistics without replacement. J. Theor. Probab. **36**, 390–408 (2023)
2. A. Alden, C. Ventre, B. Horvath, G. Lee, Model-agnostic pricing of exotic derivatives using signatures, in *Proceedings of the Third ACM International Conference on AI in Finance*, ICAIF '22, New York (Association for Computing Machinery, New York, 2022), pp. 96–104
3. H. Andrès, A. Boumezoued, B. Jourdain, Signature-based validation of real-world economic scenarios. ASTIN Bull. **54**(2), 410–440 (2024)
4. E. Beutner, H. Zähle, Deriving the asymptotic distribution of U- and V-statistics of dependent data using weighted empirical processes. Bernoulli **18**(3), 803–822 (2012)
5. F. Black, M. Scholes, The pricing of options and corporate liabilities. J. Polit. Econ. **81**(3), 637–654 (1973)
6. T. Bollerslev, Generalized autoregressive conditional heteroskedasticity. J. Econ. **31**(3), 307–327 (1986)
7. H. Bühler, B. Horvath, T. Lyons, I.P. Arribas, B. Wood, A data-driven market simulator for small data environments. Preprint. arXiv:2006.14498 (2020). https://arxiv.org/pdf/2006.14498
8. T. Cass, T. Lyons, X. Xu, General signature kernels. Preprint. arXiv:2107.00447 (2021). https://arxiv.org/pdf/2107.00447
9. T. Cass, T. Lyons, X. Xu, Weighted signature kernels. Ann. Appl. Probab. **34**(1A), 585–626 (2024)
10. B.-E. Chérief-Abdellatif, P. Alquier, MMD-bayes: robust Bayesian estimation via maximum mean discrepancy, in *Proceedings of The 2nd Symposium on Advances in Approximate Bayesian Inference*, vol. 118. *Proceedings of Machine Learning Research (PMLR)*. PMLR (2020), pp. 1–21

11. I. Chevyrev, A. Kormilitzin, A primer on the signature method in machine learning. Preprint. arXiv:1603.03788 (2016). https://arxiv.org/pdf/1603.03788
12. I. Chevyrev, H. Oberhauser, Signature moments to characterize laws of stochastic processes. J. Mach. Learn. Res. **23**(176), 1–42 (2022)
13. M.D. Donsker, *An Invariance Principle for Certain Probability Limit Theorems* (American Mathematical Society, Memoirs, Providence, 1951)
14. R.F. Engle, Autoregressive conditional heteroscedasticity with estimates of the variance of United Kingdom inflation. Econometrica **50**(4), 987–1007 (1982)
15. R.F. Engle, A.J. Patton, What good is a volatility model? Quant. Financ. **1**(2), 237–245 (2001)
16. T.S. Ferguson, U-statistics notes for statistics 200C. UCLA (2005). https://www.math.ucla.edu/~tom/Stat200C/Ustat.pdf
17. A. Fermanian, Embedding and learning with signatures. Comput. Stat. Data Anal. **157** (2021)
18. X. Geng, H. Ni, C. Wang, Expected signature on a Riemannian manifold and its geometric implications. Preprint. arXiv:2407.13086 (2024). https://arxiv.org/pdf/2407.13086
19. A. Gretton, K. Borgwardt, M. Rasch, B. Schölkopf, A. Smola, A Kernel method for the two-sample-problem, in *Advances in Neural Information Processing Systems*, vol. 19 (MIT Press, Cambridge, 2006)
20. A. Gretton, K. Fukumizu, Z. Harchaoui, B.K. Sriperumbudur, A fast, consistent Kernel two-sample test, in *Advances in Neural Information Processing Systems*, vol. 22 (Curran Associates, Inc., Red Hook, 2009)
21. A. Gretton, K. Borgwardt, M. Rasch, B. Schölkopf, A. Smola, A Kernel two-sample test. J. Mach. Learn. Res. **13**(25), 723–773 (2012)
22. A. Gretton, D. Sejdinovic, H. Strathmann, S. Balakrishnan, M. Pontil, K. Fukumizu, B.K. Sriperumbudur, Optimal kernel choice for large-scale two-sample tests, in *Advances in Neural Information Processing Systems*, vol. 25 (Curran Associates, Inc., Red Hook, 2012)
23. L.G. Gyurkó, T. Lyons, M. Kontkowski, J. Field, Extracting information from the signature of a financial data stream. Preprint. arXiv:1307.7244 (2014). https://arxiv.org/pdf/1307.7244
24. D. Harrison, D. Sutton, P. Carvalho, M. Hobson, Validation of Bayesian posterior distributions using a multidimensional Kolmogorov–Smirnov test. Mon. Not. R. Astron. Soc. **451**(3), 2610–2624 (2015)
25. W. Hoeffding, A class of statistics with asymptotically normal distribution. Ann. Math. Stat. **19**(3), 293–325 (1948)
26. B. Hoff, The Brownian Frame Process as a Rough Path. PhD thesis, University of Oxford (2005)
27. B. Horvath, M. Lemercier, C. Liu, T. Lyons, C. Salvi, Optimal stopping via distribution regression: a higher rank signature approach. Preprint. arXiv:2304.01479 (2023). https://arxiv.org/pdf/2304.01479
28. Z. Issa, B. Horvath, M. Lemercier, C. Salvi, Non-adversarial training of Neural SDEs with signature kernel scores, in *Advances in Neural Information Processing Systems*, vol. 36 (Curran Associates, Inc., Red Hook, 2023)
29. N.L. Johnson, S. Kotz, N. Balakrishnan, *Continuous Univariate Distributions*, vol. 1, 2nd edn. (John Wiley and Sons, Hoboken, 1994)
30. King's Computational Research, Engineering and Technology Environment (CREATE) (2022). King's College London
31. F.J. Kiraly, H. Oberhauser, Kernels for sequentially ordered data. J. Mach. Learn. Res. **20**(31), 1–45 (2019)
32. J.M. Kübler, W. Jitkrittum, B. Schölkopf, K. Muandet, A witness two-sample test, in *Proceedings of The 25th International Conference on Artificial Intelligence and Statistics*, vol. 151. *Proceedings of Machine Learning Research (PMLR)*. PMLR (2022), pp. 1403–1419
33. M. Lemercier, C. Salvi, T. Damoulas, E. Bonilla, T. Lyons, Distribution regression for sequential data, in *Proceedings of the 24th International Conference on Artificial Intelligence and Statistics*, vol. 130. *Proceedings of Machine Learning Research (PMLR)*. PMLR (2021), pp. 3754–3762
34. T. Lyons, H. Ni, Expected signature of Brownian motion up to the first exit time from a domain. Ann. Probab. **43**(5), 2729–2762 (2015)

35. K. Muandet, K. Fukumizu, B. Sriperumbudur, B. Schölkopf, *Kernel Mean Embedding of Distributions: A Review and Beyond*, Now Foundations and Trends (2017)
36. H. Ni, L. Szpruch, M. Sabate-Vidales, B. Xiao, M. Wiese, S. Liao, Sig-wasserstein gans for time series generation, in *Proceedings of the Second ACM International Conference on AI in Finance*, ICAIF '21, New York (Association for Computing Machinery, New York, 2022)
37. D. Pati, U-statistics. https://ani.stat.fsu.edu/%7Edebdeep/ustat.pdf
38. A. Rahimi, B. Recht, Random features for large-scale Kernel machines, in *Advances in Neural Information Processing Systems*, vol. 20 (Curran Associates, Inc., Red Hook, 2007)
39. C. Salvi, T. Cass, J. Foster, T. Lyons, W. Yang, The signature Kernel is the solution of a goursat PDE. SIAM J. Math. Data Sci. **3**(3), 873–899 (2021)
40. C. Salvi, M. Lemercier, C. Liu, B. Horvath, T. Damoulas, T. Lyons, Higher order kernel mean embeddings to capture filtrations of stochastic processes, in *35th Conference on Neural Information Processing Systems*, vol. 34. NeurIPS 2021 (2021)
41. A. Schrab, I. Kim, M. Albert, B. Laurent, B. Guedj, A. Gretton, MMD aggregated two-sample test. J. Mach. Learn. Res. **24**(194), 1–81 (2023)
42. J. Shao, U-and V-statistics. UW-Madison (Statistics) (2018). https://pages.stat.wisc.edu/~shao/stat709/stat709-18.pdf
43. C.-J. Simon-Gabriel, B. Schölkopf, Kernel distribution embeddings: universal kernels, characteristic kernels and kernel metrics on distributions. J. Mach. Learn. Res. **19**(44), 1–29 (2018)
44. D.J. Sutherland, H. Tung, H. Strathmann, S. De, A. Ramdas, A.J. Smola, A. Gretton, Generative models and model criticism via optimized maximum mean discrepancy, in *5th International Conference on Learning Representation*. ICLR 2017 (2017)
45. L. Terry, Rough paths, signatures and the modelling of functions on streams, in *Proceedings of the International Congress of Mathematicians* (2014)
46. A. McLeod, L. Terry, Signature methods in machine learning. EMS Surv. Math. Sci. (2025). https://doi.org/10.4171/EMSS/95
47. C. Tóth, H. Oberhauser, Z. Szabó, Random fourier signature features. SIAM J. Math. Data Sci. **7**(1), 329–354 (2025). https://doi.org/10.1137/23M1620478
48. G.E. Uhlenbeck, L.S. Ornstein, On the theory of Brownian motion. Phys. Rev. **36**(5), 823–841 (1930)
49. P. Zhang, X. Chen, L. Zhao, W. Xiong, T. Qin, T.-Y. Liu, Distributional reinforcement learning for multi-dimensional reward functions, in *Advances in Neural Information Processing Systems*, vol. 34, pp. 1519–1529 (Curran Associates, Inc., Red Hook, 2021)

Signature and the Functional Taylor Expansion

Bruno Dupire and Valentin Tissot-Daguette

Abstract This chapter links the path signature to the functional Itô calculus (Dupire, Functional Itô calculus. SSRN (2009). Republished in Quantitative Finance 19(5), 721–729, 2019). The junction is the functional Taylor expansion (FTE), discussed in Sect. 5, which is a powerful tool to approximate functionals after an observed path.

In particular, the FTE decomposes the functional from future scenarios. In risk analysis, the FTE leads to new Greeks, one of particular importance being the Lie bracket between the space and time functional derivatives, which we call *Libra*. Once paired with the Lévy area, the Libra can be used to speed up the computation of risk measures such as Value at Risk and Expected Shortfall as explained in Sect. 6.1. In Sect. 6.3, we explain how the FTE can quantify the hedging error of replicating portfolios in a robust manner. In the context of financial derivatives, we demonstrate in Sect. 6.4 how the FTE generates approximations of exotic payoffs. The relevance of the FTE for cubature methods is finally outlined in Sect. 6.5.

1 Introduction

The Taylor expansion is a central tool in calculus. It requires two essential ingredients:

(i) The derivatives of the target function at a given point.
(ii) A collection of basis or primitive functions.

In $\mathbb{R}$, the Taylor expansion reads

$$f(x+y) = f(x) + f'(x)y + \frac{1}{2}f''(x)y^2 + \dots, \tag{1}$$

B. Dupire · V. Tissot-Daguette (✉)
Quantitative Research, Bloomberg, New York, NY, USA
e-mail: bdupire@bloomberg.net; vtissotdague@bloomberg.net

C. Bayer et al. (eds.), *Signature Methods in Finance*, Springer Finance Lecture Notes, https://doi.org/10.1007/978-3-031-97239-3_6

and (i), (ii), correspond respectively to the derivatives $\{f'(x), f''(x), \ldots\}$ and monomials $\{1, y, y^2, \ldots\}$. We may therefore ask ourselves: what are the analogues of (i), (ii) in the path space?

With regards to (i), the *functional Itô calculus* [8] provides convenient notions of derivatives in the path space, recalled in Sect. 3. Moreover, as mentioned several times in this book, the signature can be regarded as a generalization of the monomials in path space. In fact, the signature pairs well with the functional derivatives and is therefore a good candidate for (ii). This is supported by our main result, the *functional Taylor expansion* (FTE), confirming that signature functionals are the building blocks of path dependence. We shall see in Theorem 5.1 that the FTE admits the general expression

$$f(X * Y) = \sum_I \Delta_I f(X) S^I(Y), \tag{2}$$

where I is a *word* in the alphabet $\{0, 1\}$, where 0 refers to the time and 1 to the space variable. Also, $\Delta_I f$, S^I are respectively the functional derivative and signature associated to I. Note that the FTE echoes the classical Taylor expansion in (1).

Taylor expansions are classically used to either provide a local approximation of functions using a small truncation order or express real analytic functions as a power series similar to (1) and (2). Herein, we will focus on the former, with applications given in Sect. 6, and refer the interested reader to [9, Section 3.1.5] for the latter.

Notations We fix throughout a time window $T > 0$ and consider the bundle

$$\Lambda := \bigcup_{t \in [0,T]} \Lambda_t,$$

where for each $t \in [0, T]$, Λ_t is the space of càdlàg paths (right continuous with left limit) on $[0, t]$ and taking values in $\mathbb{R}$. Also, write $X_{\mathcal{T}} = (X_t)_{t \in \mathcal{T}}$, $\mathcal{T} \subseteq [0, T]$, so that $X_{[0,s]} \in \Lambda_s$ is the restriction of X to $[0, s]$, $s \leq t$.[1]

A *functional* f is a map from the path space Λ to $\mathbb{R}$. We say that f is Λ-*continuous* if it is continuous with respect to the following metric,

$$d_\Lambda(X, Y) = \sup_{0 \leq u \leq t} |X_{u \wedge s} - Y_u| + t - s, \quad X \in \Lambda_s,\ Y \in \Lambda_t, \quad s \leq t. \tag{3}$$

Let $* : \Lambda^2 \to \Lambda$ be the path concatenation operation that preserves continuity as seen in chapter "A Primer on the Signature Method in Machine Learning". If $\mathbb{Q}$ is a probability measure on Λ and $f : \Lambda \to \mathbb{R}$, the *conditioned expectation* under $\mathbb{Q}$ of

[1] Some of the usual notations in the functional Itô calculus have been changed for consistency with other chapters of the book.

$f(Z)$, $Z \in \Lambda_t$, given $Z_{[0,s]} = X \in \Lambda_s$ is defined as

$$\mathbb{E}^{\mathbb{Q}}[f(Z) \mid X] = \int_{\Lambda_{t-s}} f(X * Y)\mathbb{Q}(dY). \tag{4}$$

Unless stated otherwise, the letter X is used for observed paths, Y represents future scenarios, and Z are random paths.

An implementation of the financial examples can be found in the GitHub repository https://github.com/valentintissot/Signature-and-FTE.

2 Functionals in Finance and Embeddings

Functionals express path dependence which is ubiquitous in finance, through the definition of exotic options (even vanillas; see Sect. 6.2) or the dynamics of the underlying asset price. For completeness, we recall from the previous section that:

Definition 2.1 A *functional* is a map $f : \Lambda \to \mathbb{R}$.

To better distinguish functionals on fixed and varying horizons (i.e., on Λ), we also introduce the following notion.

Definition 2.2 Given $T > 0$, a *T-functional* is a map $g : \Lambda_T \to \mathbb{R}$.[2]

In finance, functionals and T-functionals play different roles. First, T-functionals typically represent the payoff of a path-dependent claim with maturity T. Depending on the term sheet of g, it is usually not possible to specify the reward value having only observed the price path up to some intermediate time. For instance, the payoff of a forward start call option with maturity T and strike fixing date $\tau \in (0, T)$ reads

$$g(X) = (X_T - X_\tau)^+, \quad X \in \Lambda_T.$$

Then g is a T-functional whilst the price of the option before maturity (that depends on the model) is a functional. On the one hand, one can directly define a T-functional from a functional $f : \Lambda \to \mathbb{R}$ by defining the restriction $g = f|_{\Lambda_T}$ of f to Λ_T. For the reverse direction, we introduce the notion of *embeddings*.

Definition 2.3 We say that $f : \Lambda \to \mathbb{R}$ *embeds* $g : \Lambda_T \to \mathbb{R}$ into the space of functionals if f coincides with g on Λ_T, that is $f(X) = g(X)$ for all $X \in \Lambda_T$.

Note that there are infinitely many functionals that embed a given T-functional. A natural candidate is the *intrinsic functional*, defined as

[2] We note that "functionals" classically refer to T-functionals in calculus and functional analysis.

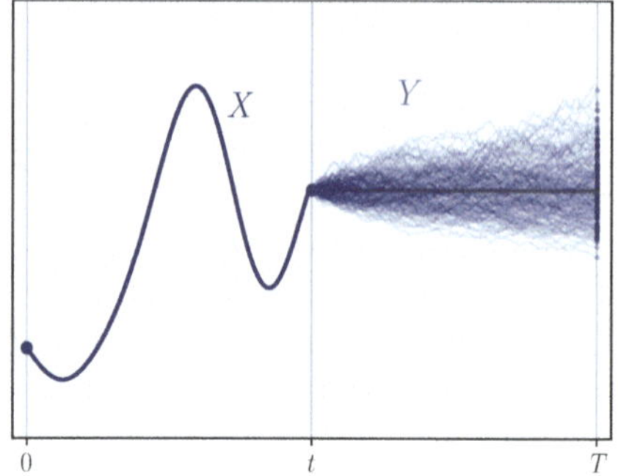

Fig. 1 Illustration of the Black-Scholes embedding (6) with $\sigma = 25\%$. The intrinsic functional ($\sigma = 0\%$) corresponds to a flat extension of X (solid line at level X_t)

$$f_0(X) = g\big(X^{*(T-t)}\big), \quad X \in \Lambda_t, \tag{5}$$

with the *flat extension* $X_s^{*\delta t} = X_{s\wedge t}$, $s \leq t + \delta t$, which is central in Definition 3.1 when introducing the functional time derivative. In other words, the enlarged path $X^{*(T-t)} \in \Lambda_T$ is obtained by flat extending X until T; see Fig. 1. One family of particular interest in finance consists of the *Black-Scholes embeddings* $g \mapsto f_\sigma$, defined as

$$f_\sigma(X) = \mathbb{E}^{\mathbb{Q}_\sigma}[g(X * Y)] = \int_{\Lambda_{T-t}} g(X * Y)\mathbb{Q}_\sigma(dY), \quad \sigma \geq 0, \; X \in \Lambda_t, \tag{6}$$

where the concatenated process $Y \in \Lambda_{T-t}$ is, under $\mathbb{Q}_\sigma$, a geometric Brownian motion with zero drift and volatility parameter σ. Concretely, $Y_t = Y_0 e^{\sigma W_t - \frac{1}{2}\sigma^2 t}$, where W is an $\mathbb{Q}_\sigma$-Brownian motion. Note that the initial value Y_0 is irrelevant since Y is concatenated to X in a continuous way. If $X \in \Lambda_T$, then $Y = \emptyset$ in (6) which implies that $f_\sigma(X) = g(X)$. In other words, f_σ coincides with g on Λ_T in line with Definition 2.3. In financial terms and absent of interest rates, $f_\sigma(X)$ gives the *price* of the claim g in the Black-Scholes model $\mathbb{Q}_\sigma$ having observed X.

Another possibility is to use *Bachelier embeddings* where $\mathbb{Q}_\sigma$ in (6) is replaced by the Wiener measure scaled by the (Bachelier) volatility σ. Put differently, $Y = Y_0 + \sigma W$, where W is an $\mathbb{Q}_\sigma$-Brownian motion.

The usefulness of the Black-Scholes and Bachelier embeddings will be revealed in Sect. 6 when discussing financial applications.

Exercise 2.4 Let $g(X) = \max_{t \leq T} X_t - X_T$ which delivers the *drawdown* of $X \in \Lambda_T$.

(a) Compute the Bachelier embedding $f_\sigma(X)$ explicitly for arbitrary $\sigma > 0$.
(b) What happens to $f_\sigma(X)$ when $\sigma \downarrow 0$? In other words, what is the intrinsic functional of g?

3 Functional Itô Calculus

As recalled in the introduction, any Taylor expansion requires a notion of derivative in the corresponding ambient space. When dealing with paths, natural candidates are the derivatives arising from the *functional Itô calculus* [8], which we now outline.

Definition 3.1 The *functional time derivative* of $f : \Lambda \to \mathbb{R}$ is given for $X \in \Lambda_t$ by

$$\Delta_t f(X) = \lim_{\delta t \downarrow 0} \frac{f(X^{*\delta t}) - f(X)}{\delta t}, \qquad X_s^{*\delta t} = X_{s \wedge t}, \qquad s \leq t + \delta t, \quad (7)$$

when the limit exists. Moreover, the *functional space derivative* of $f : \Lambda \to \mathbb{R}$ is

$$\Delta_x f(X) = \lim_{h \to 0} \frac{f(X^h) - f(X)}{h}, \qquad X_s^h = \begin{cases} X_s, & s < t, \\ X_s + h, & s = t, \end{cases} \qquad s \leq t. \quad (8)$$

The time and space functional derivatives are respectively illustrated in Figs. 2 and 3.

Unlike classical calculus, Δ_x and Δ_t need not commute in general, i.e. $\Delta_{tx} f \neq \Delta_{xt} f$ where $\Delta_{tx} f = \Delta_t(\Delta_x f)$ and similarly for $\Delta_{xt} f$. In particular, the *Lie Bracket*

$$Lf(X) := (\Delta_{xt} - \Delta_{tx}) f(X), \quad (9)$$

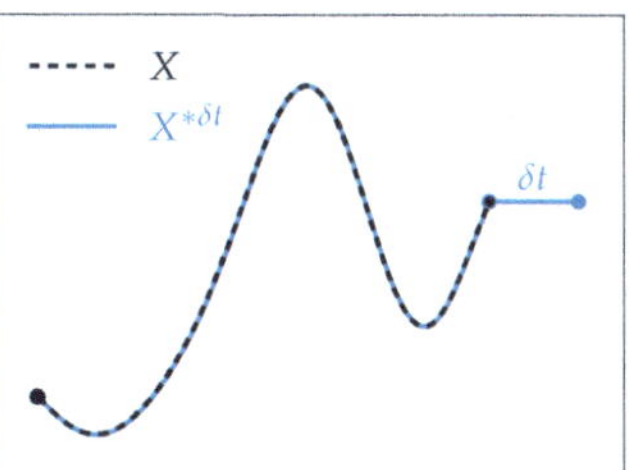

Fig. 2 The functional time derivative $\Delta_t f(X)$, defined in (7), is obtained by introducing a flat extension of the path at the spot level X_t

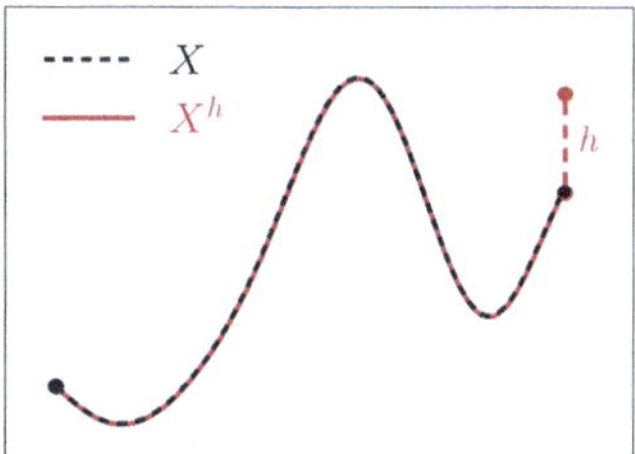

Fig. 3 The functional space derivative $\Delta_x f(X)$ seen in (8) is obtained by shifting the spot value of X while leaving the time unchanged

may well be nontrivial; see Exercise 3.2 for a counterexample. Under the following technical condition,

$$\lim_{h\to 0}\sup_{0\le\delta t\le T-t}\left|\frac{f((X^{*\delta t})^h)-f(X^{*\delta t})}{h}-\Delta_x f(X^{*\delta t})\right|=0,\quad \forall X\in\Lambda_t,\tag{10}$$

it is shown in [16, Lemma 3.2] that Lf can be expressed as the double limit

$$Lf(X)=\lim_{h,\delta t\to 0}\frac{f((X^h)^{*\delta t})-f((X^{*\delta t})^h)}{h\delta t}.\tag{11}$$

We gather from (11) that the Lie bracket thus captures the instantaneous path-dependence, i.e., how the order in which a path is shifted in space and extended in time affects a functional.

Exercise 3.2 Consider the functional $f(X)=\int_0^t X_s ds$, $X\in\Lambda_t$.

(a) What are $\Delta_t f$ and $\Delta_x f$?
(b) Compute the Lie bracket $L=\Delta_{xt}-\Delta_{tx}$ of f and conclude that Δ_t, Δ_x do not commute in general.
(c) Construct a genuinely path-dependent functional whose Lie bracket is zero.

3.1 Functional Stratonovich Formula

As shown in [8], the functional derivatives can be used to generalize Itô's formula to the path space. In line with the path signature and the functional Taylor expansion in Sect. 5, we here present the analogue of the functional Itô formula in terms of Stratonovich integrals. We also adopt a pathwise framework as in [9, 14].

Let $\Pi=(\Pi^N)$ be a sequence of partitions of $[0,T]$ with vanishing mesh size, e.g., the regular partition $\Pi^N=\{Tn/N : n=0,\dots,N\}$. We say that $X\in\Lambda_t$ is a Π-*integrator* if it is continuous and its quadratic variation process along Π, namely

$$\langle X\rangle_s^\Pi := \lim_{N\uparrow\infty}\sum_{t_n\in\Pi_s^N}(X_{t_n}-X_{t_{n-1}})^2,\quad \Pi_s^N:=\{t_n\wedge s\mid t_n\in\Pi^N\},\tag{12}$$

is well-defined, finite, and continuous. We write Λ_t^Π for the set of Π-integrators in Λ_t and $\Lambda^\Pi := \bigcup_{t\in[0,T]}\Lambda_t^\Pi$. Important examples of Π-integrators are sample paths of continuous semimartingales. Indeed, it is shown in [14] that for every sequence of partitions Π, almost all trajectories of continuous semimartingales belong to Λ^Π. We also note that Λ^Π contains all continuous functions of bounded variation, in particular smooth paths. Next, we define

$$\bar\Lambda^\Pi=\bigcup_{t\in[0,T]}\bar\Lambda_t^\Pi,\quad \bar\Lambda_t^\Pi=\Lambda_t^\Pi+\mathcal{J}_t,$$

where $\mathcal{J}_t$ contains all piecewise constant paths in Λ_t taking finitely many values. The enlarged space $\bar{\Lambda}^\Pi$ is needed to compute the functional space derivative (8) which entails a discontinuity at the final time. We now state a generalization of the Stratonovich formula to path functionals. For a proof, see [9, Theorem 3.2].

Theorem 3.3 (Functional Stratonovich Formula) *Let $X \in \Lambda_t^\Pi$ for some $t \in [0, T]$. If $\Delta_t f$, $\Delta_x f$, $\Delta_{xx} f$ exist and are Λ-continuous, then*

$$f(X) = f(X_0) + \int_0^t \Delta_t f(X_{[0,s]})ds + \int_0^t \Delta_x f(X_{[0,s]}) \circ dX_s, \tag{13}$$

where the last integral is defined as a limit of Riemann sums along Π, namely

$$\int_0^t \Delta_x f(X_{[0,s]}) \circ dX_s = \lim_{N\uparrow\infty} \sum_{t_n \in \Pi_t^N} \frac{\Delta_x f(X_{[0,t_{n-1}]}) + \Delta_x f(X_{[0,t_n]})}{2} (X_{t_n} - X_{t_{n-1}}).$$

The next example shows that (13) is consistent with the standard Stratonovich formula.

Example 3.4 Let $X \in \Lambda_t^\Pi$ and the path-independent functional $f(X) = \phi(t, X_t)$, $\phi \in C^{1,2}([0, T] \times \mathbb{R})$. It is easily seen that $\Delta_t f = \partial_t \phi$, $\Delta_x f = \partial_x \phi$, $\Delta_{xx} f = \partial_{xx} \phi$. We can thus invoke Theorem 3.3 to recover the classical Stratonovich formula, namely,

$$\phi(t, X_t) = \phi(0, X_0) + \int_0^t \partial_t \phi(s, X_s)ds + \int_0^t \partial_x \phi(s, X_s) \circ dX_s.$$

The above theorem allows, in passing, to construct pathwise Stratonovich integrals $\int \varphi(X) \circ dX$ where φ is continuously differentiable (i.e., φ is Λ-continuous and $\Delta_t \varphi$, $\Delta_x \varphi$ exist and are Λ-continuous). Indeed, assume that $X_0 = 0$ for simplicity and define $\Phi(X) = \int_0^{X_t} \varphi(X^{-h})dh$, $X \in \bar{\Lambda}_t^\Pi$. Then it follows from Definition 3.1 that $\Delta_x \Phi = \varphi$. We can thus apply Theorem 3.3 to Φ and obtain

$$\Phi(X) = \int_0^t \Delta_t \Phi(X_{[0,s]})ds + \int_0^t \varphi(X_{[0,s]}) \circ dX_s. \tag{14}$$

Noting also that $\Delta_t \Phi(X) = \int_0^{X_t} \Delta_t \varphi(X_t^{-h})dh$, we can rearrange (14) to arrive at

$$\int_0^t \varphi(X_{[0,s]}) \circ dX_s = \int_0^{X_t} \varphi(X^{-h})dh - \int_0^t \int_0^{X_s} \Delta_t \varphi(X_{[0,s]}^{-h})dh\, ds. \tag{15}$$

In particular, when $\Delta_t \varphi = 0$, we simply have $\int_0^t \varphi(X_s) \circ dX_s = \int_{X_0}^{X_t} \varphi(X_t^{(\varepsilon)})\, d\varepsilon$. The Stratonovich integral in (15) is therefore well-defined pathwise. Incidentally, this allows us to properly define signature functionals introduced in the next section.

It is counterintuitive that the existence of the Stratonovich integral in (15) requires some conditions on the functional derivatives of the integrand φ (namely that $\Delta_t\varphi, \Delta_x\varphi$ exist and are Λ-continuous). Assuming that φ is merely Λ-continuous turns out not to be sufficient in general, as shown in the next example.

Example 3.5 Consider the functional $\varphi(X) = \int_0^t \kappa(t-s) \circ dX_s$ with the fractional kernel $\kappa(\delta) = \delta^{H-1/2}$, $H > 0$. It is easily seen that φ is Λ-continuous, while $\Delta_x\varphi(X) = \lim_{s\uparrow t} \kappa_H(t-s) = \infty$ when $H < 1/2$. On the other hand, it is shown in [3] that $\int \varphi(X) \circ dX$ diverges when X is a typical Brownian path (which belongs to Λ^{Π} almost surely). This confirms that the Λ-continuity of φ is typically not enough to define the pathwise Stratonovich integral in (15).

4 Signature Functionals

In this chapter, the signature is regarded as a collection of functionals as we now explain. Introduce the extended path (t, X) and write $X_t^0 = t$, $X_t^1 = X_t$ for convenience. Recall also the set $\mathcal{W}$ seen in chapter "A Primer on the Signature Method in Machine Learning" containing all the words $I = (i_1, \ldots, i_k)$, $k \geq 0$, herein on the alphabet $\{0, 1\}$. Then the *signature* is given by the collection $S = \{S^I : \bar{\Lambda}^{\Pi} \to \mathbb{R} : I \in \mathcal{W}\}$, where $S^{\emptyset} \equiv 1$ and for $|I| := \text{length}(I) \geq 1$,

$$S^I(X) := S(X)_{0,t}^I = \int_0^t \int_0^{t_k} \cdots \int_0^{t_2} \circ dX_{t_1}^{i_1} \cdots \circ dX_{t_k}^{i_k}, \quad X \in \bar{\Lambda}_t^{\Pi}. \tag{16}$$

The symbol $\circ$ indicates that the integrals in (16) are in the Stratonovich sense. The signature $S^I(X)$ is well-defined pathwise for all $X \in \bar{\Lambda}^{\Pi}$, $I \in \mathcal{W}$ as a consequence of Theorem 3.3; see [9, Proposition 3.5].

Remark 4.1 We stress that it is necessary to enlarge X, here one-dimensional, with the time itself to enrich the signature. In the alphabet $\{1\}$, S boils down to the scaled monomials $1, X_t - X_0, \frac{(X_t-X_0)^2}{2!}, \frac{(X_t-X_0)^3}{3!}, \ldots$ so it would not carry any information about the path apart from the increment $X_t - X_0$. This is in contrast with the multi-dimensional case as in chapter "A Primer on the Signature Method in Machine Learning" where a path is described as a curve parametrized by t, e.g., $t \mapsto (X_t^1, X_t^2) \in \mathbb{R}^2$ as depicted in the right panel of Fig. 4, and the signature is typically rich enough in this formulation.

Moreover, adding time makes the signature map $X \mapsto S(X)$ injective. We may therefore wonder how to reconstruct the unique path (when it exists) associated to a given sequence of signature elements. We refer the interested reader to Appendix for further details.

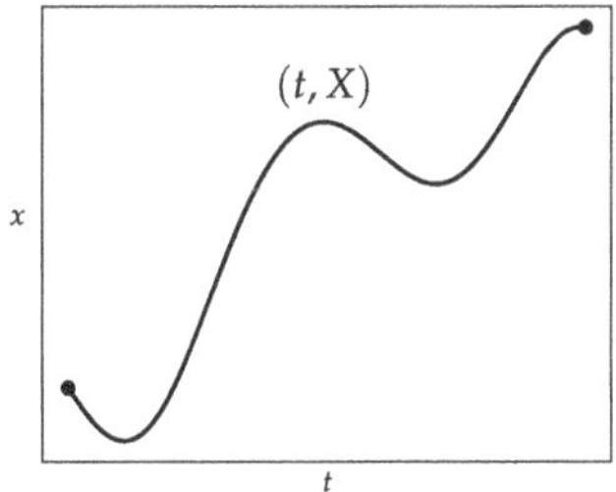

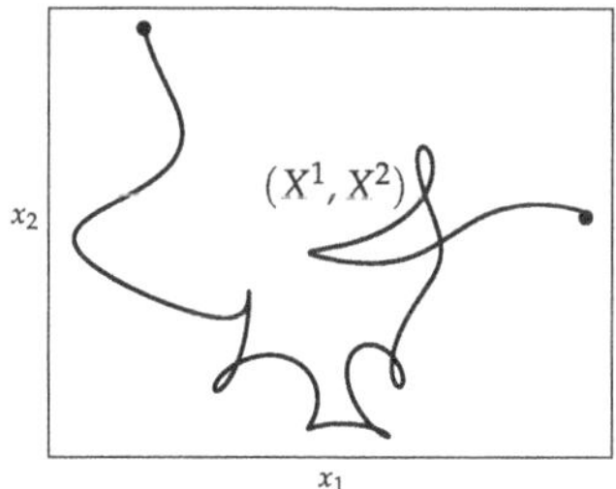

Fig. 4 Contrast between (t, X) paths (left) and $(X^1, X^2) = (X_t^1, X_t^2)_{t\in[0,T]}$ (right)

It turns out that the signature is compatible with the functional derivatives in the following sense: for fixed $I = (i_1, \ldots, i_k)$, then

$$\Delta_t S^I = \begin{cases} S^{(i_1,\ldots,i_{k-1})}, & i_k = 0, \\ 0, & i_k = 1, \end{cases} \qquad \Delta_x S^I = \begin{cases} S^{(i_1,\ldots,i_{k-1})}, & i_k = 1, \\ 0, & i_k = 0. \end{cases} \tag{17}$$

From the above recursions, we deduce a more general property in part (a) of the next Proposition. We also collect other properties of the signature, proved in [9].

Proposition 4.2 *The following properties holds:*

(a) *Let $I = (i_1, \ldots, i_k) \in \mathcal{W}$. Then $\Delta_J S^I \neq 0$ if and only if $J = (i_\ell, \ldots, i_k)$, for some $\ell \leq k$. If so, then $\Delta_J S^I = S^{(i_1,\ldots,i_{\ell-1})}$.*

(b) *The signature functionals are linearly independent, i.e., for all finite subset $\mathcal{W}' \subseteq \mathcal{W}$, and coefficients $(c_I)_{I\in W'} \subseteq \mathbb{R}$, then*

$$\sum_{I\in\mathcal{W}'} \lambda_I S^I(X) = 0 \quad \forall\, X \in \bar{\Lambda}^\sqcap \quad \Longrightarrow \quad \lambda_I = 0 \quad \forall\, I \in \mathcal{W}'.$$

(c) *The collection of T-functionals $\mathcal{A}^k = \{S^I : \bar{\Lambda}_T^\sqcap \to \mathbb{R} : |I| \leq k\}$ is linearly dependent.*

(d) *For all $k \in \mathbb{N}$, the family of T-functionals $\mathcal{B}^k = \{S^I : \bar{\Lambda}_T^\sqcap \to \mathbb{R} : I = (i_1, \ldots, i_k, 1), k < k\}$ is linearly independent. Moreover,* $\text{span}(\mathcal{B}^k) = \text{span}(\mathcal{A}^k)$.

A proof of part (d) in the above Proposition can be found in [9, Appendix A.3]. The family $\mathcal{B}^k$ yields a basis for T-functionals spanned by the signature functionals up to order k. Moreover, it is incremental in the sense that $\mathcal{B}^k$ contains $\mathcal{B}^{k-1}$ (and in turn $\mathcal{B}^k$ for all $k \leq k-2$). The nested structure is of practical convenience as increasing the truncation level k does not alter the elements already present in the basis.

Exercise 4.3 Prove parts (a), (b), (c) of Proposition 4.2. *Hint:* Use (a) to show (b).

5 Functional Taylor Expansion

We now iterate the functional Stratonovich formula established in Theorem 3.3 to derive the functional Taylor expansion (FTE). As we shall see, the signature functionals appear naturally. We first state the FTE in general form (Theorem 5.1) and later discuss some important consequences. Write $\Delta_0 = \Delta_t$, $\Delta_1 = \Delta_x$ (recall that $0 \sim t$, $1 \sim x$) and define the higher order functional derivatives as $\Delta_I = \Delta_{I_1} \cdots \Delta_{I_k}$, e.g., $\Delta_{110} f = \Delta_x(\Delta_x(\Delta_t f))$. Note also that $\Delta_\emptyset f = f$. Next, let $|I|_0$ (respectively $|I|_1$) denote the number of 0's (resp. 1's) in $I \in \mathcal{W}$ and introduce for $k, \ell \in \mathbb{N}$ and $\Omega \subseteq \Lambda$,

$$\mathbb{C}^{k,\ell}(\Omega) = \{f : \Omega \to \mathbb{R} \, : \, \Delta_I f \ \text{ is } \Lambda\text{-continuous } \forall I \in \mathcal{W}_{k,\ell}\},$$

$$\mathcal{W}_{k,\ell} = \{I \in \mathcal{W} \, : \, |I|_0 \leq k, \ |I|_1 \leq \ell, \ |I| \leq k \vee \ell\}.$$

If $k = \ell$, we simply write $\mathbb{C}^k(\Omega) = \mathbb{C}^{k,k}(\Omega)$. We are now set to state our main theorem.

Theorem 5.1 (Functional Taylor Expansion (FTE)) *Let $f \in \mathbb{C}^{k,k+1}(\Lambda)$ and fix $X \in \Lambda_s$, $Y \in \Lambda_t^{\Pi}$ with $s, t \in [0, T]$, $s + t \leq T$. Then,*

$$f(X * Y) = \sum_{|I|<k} \Delta_I f(X) S^I(Y) + R_k(X, Y), \tag{18}$$

$$R_k(X, Y) = \sum_{|I|=k} \int_0^t \int_0^{t_k} \cdots \int_0^{t_2} \Delta_I f(X * Y_{[0,t_1]}) \circ dY_{t_1}^{i_1} \cdots \circ dY_{t_k}^{i_k}. \tag{19}$$

Proof See [9]. □

The concatenated path $X * Y$ appearing in the above theorem is illustrated in Fig. 5. As can be seen, the length of future path Y may vary.

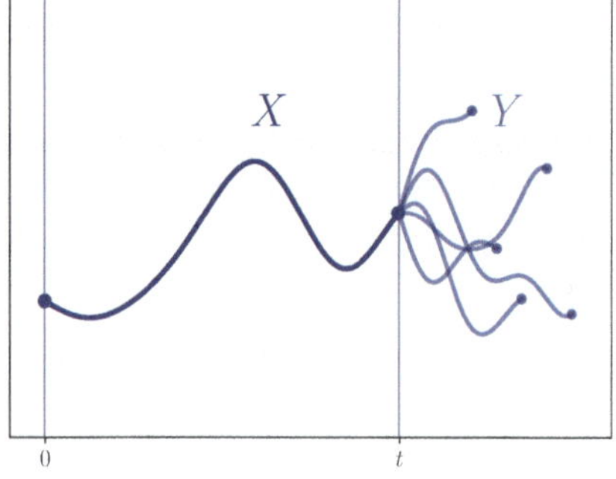

Fig. 5 Illustration of the concatenated path $X * Y$ in the functional Taylor expansion (Theorem 5.1). Five realizations of Y with various length

A related result can be found in Litterer and Oberhauser [18, Theorem 14], yet there are two important differences. First, the expansion in [18] is established for Itô diffusions, while our expansion is made in the pathwise setting introduced by Föllmer [14]. Second, the partition of the words is different: to properly reflect the scale of diffusions, [18] use the weighted length $\|I\| = 2|I|_0 + |I|_1$ instead of $|I|$ to truncate the expansion. The literature on functional expansions will be further discussed in Sect. 8.

Interestingly, the functional Taylor expansion can be used to recover Chen's identity seen in chapter "A Primer on the Signature Method in Machine Learning".

Corollary 5.2 (Chen's Identity) *Let $I = (i_1, \dots, i_k)$. Then for all $X \in \Lambda_s^\Pi$, $Y \in \Lambda_t^\Pi$,*

$$S^I(X * Y) = (S(X) \otimes S(Y))^I = \sum_{\ell=0}^{k} S^{(i_1,\dots,i_\ell)}(X)\ S^{(i_{\ell+1},\dots,i_k)}(Y), \tag{20}$$

where we write $(i_1, \dots, i_0) = (i_{k+1}, \dots, i_k) = \varnothing$.

Proof Let us apply the functional Taylor expansion of order $k+1$ to the signature functional $f = S^I$ defined on $\bar{\Lambda}^\Pi$. This gives the expression

$$S^I(X * Y) = \sum_{|J| \le k} \triangle_J S^I(X) S^J(Y) + R_{k+1}(X, Y),$$

provided that S^I satisfies the regularity condition in Theorem 5.1 which we now verify. Using Proposition 4.2 (a), then $\triangle_J S^I(X)$ equals $S^{(i_1,\dots,i_\ell)}$ if $J = (i_{\ell+1}, \dots, i_k)$, $\ell \le k$ (where we recall that $(i_1, \dots, i_0) = (i_{k+1}, \dots, i_k) = \varnothing$), and is zero otherwise. In particular, $\triangle_J S^I = 0$ for all word J of length greater than k, so the remainder $R_{k+1}(X, Y)$ vanishes. We gather that S^I is smooth and conclude from Theorem 5.1 that

$$S^I(X * Y) = \sum_{\ell=0}^{k} S^{(i_1,\dots,i_\ell)}(X)\ S^{(i_{\ell+1},\dots,i_k)}(Y),$$

which is Chen's identity. □

Oftentimes, we simply expand a functional around the initial value of the path. Given $X \in \bar{\Lambda}_t^\Pi$, this amounts to choosing $X = X_0$, $Y = X$ in Theorem 5.1, leading to the so-called *Functional Maclaurin expansion.*

Corollary 5.3 (Functional Maclaurin Expansion) *Let f be a functional in $\mathbb{C}^{k,k+1}(\bar{\Lambda}^\Pi)$. Then for all $t \in [0, T]$, the following expansion holds,*

$$f(X) = \sum_{|I| < k} \triangle_I f(X_0) S^I(X) + R_k(X), \quad X \in \bar{\Lambda}_t^\Pi, \tag{21}$$

$$R_k(X) = \sum_{|I|=k} \int_0^t \int_0^{t_k} \cdots \int_0^{t_2} \Delta_I f(X_{[0,t_1]}) \circ dX_{t_1}^{i_1} \cdot \cdots \cdot \circ dX_{t_k}^{i_k}. \tag{22}$$

Formula (21) suggests that the signature can be used to approximate sufficiently regular functionals. Indeed, applying a first order functional Maclaurin expansion (Corollary 5.3) to the remainder R_k, namely $R_k(X) \approx \sum_{|I|=k} \Delta_I f(X_0) S^I(X)$, yields the approximation,

$$f(X) \approx f^k(X) := \sum_{|I| \le k} \lambda^I S^I(X), \quad \lambda^I = \Delta_I f(X_0), \quad X \in \bar{\Lambda}_t^{\Pi}. \tag{23}$$

In other words, f^k approaches f via a linear combination of signature functionals. Naturally, the approximation in (23) is valid only if the truncation error $f(X) - f^k(X)$ is small. For explicit remainder estimates, see [9, Section 3.1.5] and [18, 19] in the context of diffusions.

In finance, f^k is referred to as *signature payoff* [1, 20] or simply *sig-payoff* [6, 7]. By applying the Stone–Weierstrass theorem, it can be shown that signature payoffs form a family of continuous functionals which is dense in compact subsets of Λ [1, 6, 7, 20]. While this guarantees the existence of a signature payoff arbitrarily close to a target functional, the former is not given explicitly. In contrast, the functional Taylor expansion provides a recipe to approximate functionals as hinted by (23) and further explained in Sect. 6.4.

Another avenue is to use a regression approach to compute the coefficients λ^I, as carried out in [1, 20] and illustrated in Sect. 6.2. In this case, the regression is performed using paths from the reference measure, or from a pool of models (e.g. Black-Scholes with varying volatility) to improve the robustness of the coefficients. On the other hand, the FTE approach is model-free as the Taylor expansion is done pathwise. In Sect. 6.4, we shall see that an embedding must be nevertheless introduced in (23) when f is a T-functional, yet the auxiliary model that defines the embedding can be much simpler than the reference model. Moreover, Taylor expansions are incremental in the sense that increasing the order does not change the previously computed coefficients. This is not the case for regressions, where adding features affects all coefficients.

Exercise 5.4 Let $f(X) = \int_0^t X_u^2 du$, $X \in \Lambda_t$.

(a) Write down explicitly the FTE of f with $k = 1$ and arbitrary paths X, Y.
(b) Compute the functional Taylor approximation f^k (given in (23)) with $k = 3$. What do you observe?

Note: If X represents the volatility process of some asset, say S, then $f(X)$ is the total realized variance of S.

6 Applications

6.1 Risk Analysis: A New Greek

Let us see how the general formulation of the FTE (Theorem 5.1) can be used for local approximation of functionals in portfolio risk management. Concretely, fix $X \in \Lambda_t$ and scenario $Y \in \Lambda_{\delta t}$, and increment $\delta y = Y_{\delta t} - Y_0$. Notice that $\delta y = S^{(1)}(Y)$. Then, a second order functional Taylor expansion of f after X gives (writing $\Delta_I f = \Delta_I f(X)$),

$$f(X * Y) \approx f + \Delta_t f \delta t + \Delta_x f \delta y + \Delta_{xx} f \, \frac{\delta y^2}{2} + \Delta_{tt} f \, \frac{\delta t^2}{2} \tag{24}$$

$$+ \Delta_{tx} f \; S^{(0,1)}(Y) + \Delta_{xt} f \; S^{(1,0)}(Y). \tag{25}$$

The terms on the right side of (24) are classical, while the cross terms in (25) are new. Indeed, the latter cannot be grouped together since the functional derivatives Δ_t, Δ_x do not commute in general; see again Exercise 3.2. Nevertheless, we can introduce the symmetric-antisymmetric decomposition,

$$\Delta_{xt} = \partial_{tx} + \frac{1}{2}L, \qquad \Delta_{tx} = \partial_{tx} - \frac{1}{2}L,$$

with the mixed derivative $\partial_{tx} := \frac{\Delta_{xt}+\Delta_{tx}}{2}$, and Lie bracket $L = \Delta_{xt} - \Delta_{tx}$ seen in (9). As $S^{(0,1)}(Y) + S^{(1,0)}(Y) = \delta t \delta y$, we can thus rewrite (25) as

$$\Delta_{tx} f \; S^{(0,1)}(Y) + \Delta_{xt} f \; S^{(1,0)}(Y) = \partial_{tx} f \; \delta t \delta y + Lf \mathcal{A}(Y), \tag{26}$$

where $\mathcal{A}(Y)$ is the *Lévy area* of the enlarged path (t, Y) as seen in chapter "A Primer on the Signature Method in Machine Learning", namely

$$\mathcal{A}(Y) = \frac{1}{2}(S^{(1,0)}(Y) - S^{(0,1)}(Y)) = \int_0^{\delta t} (Y_u - Y_0)du - \frac{\delta t \delta y}{2}. \tag{27}$$

We recall that it represents the signed area enclosed between the graph of $Y - Y_0$ and the segment $\{(u, Y_0 + u\delta y) : u \leq \delta t\}$. In the present context, $\mathcal{A}(Y)$ gauges the *convexity* of the path Y. Indeed, $\mathcal{A}(Y) < 0$ if $u \mapsto Y_u$ is convex and $\mathcal{A}(Y) > 0$ if $u \mapsto Y_u$ is concave; see Fig. 6. Expression (26) may be visually explained as follows:

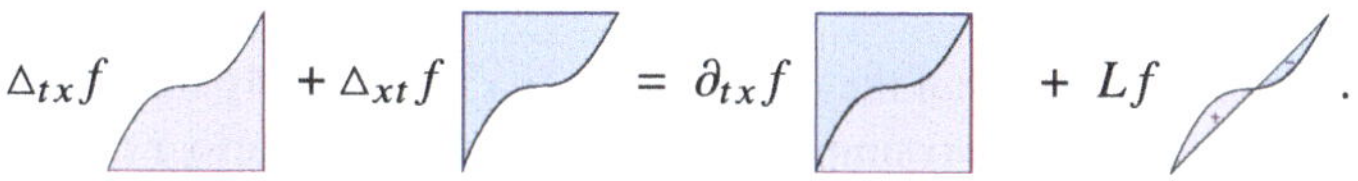

$\Delta_{tx} f$ + $\Delta_{xt} f$ = $\partial_{tx} f$ + Lf .

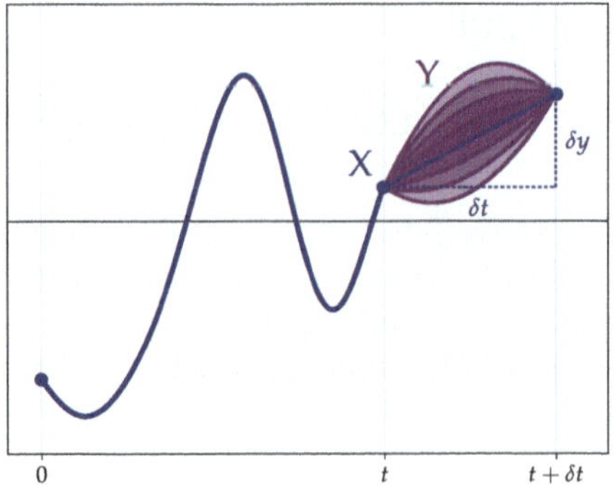

Fig. 6 Observed path $X \in \Lambda_t$ and scenario $Y \in \Lambda_{\delta t}$, $\delta y = Y_{\delta t} - Y_0$ (several realizations). The Lévy Area $\mathcal{A}(Y)$ corresponds to the signed area between the graph of $Y - Y_0$ and the diagonal $\{(u, Y_0 + u\delta y) : u \leq \delta t\}$

Moving to applications in risk management, let $f(X)$ represent the time t value of a portfolio contingent upon $X \in \Lambda_t$, we can break down the profit and loss (P&L) between t and $t + \delta t$ in terms of the path-dependent sensitivities $\Theta = \Delta_t f(X)$, $\Delta = \Delta_x f(X)$, $\Gamma = \Delta_{xx} f(X)$ and a new Greek stemming from the Lie Bracket,

$$\mathcal{L} = Lf(X) = \Delta_{xt} f(X) - \Delta_{tx} f(X), \tag{28}$$

which we call *Libra*. Indeed, we obtain from (24) and (26) the P&L decomposition,

$$\text{P\&L} = f(X * Y) - f(X) \approx \Theta\delta t + \Delta\delta y + \Gamma\frac{\delta y^2}{2} + \mathcal{L}\mathcal{A}(Y). \tag{29}$$

While Θ, Δ, Γ are paired with the increments $\delta t, \delta y$, the last term in (29) is genuinely path-dependent as it entails the Lévy area $\mathcal{A}(Y)$. If the future path is a bridge, that is $\delta y = 0$, then (29) simply reads P&L $= \Theta\delta t + \mathcal{L}\mathcal{A}(Y)$. Hence the convexity of the future path Y also contributes to the profit-and-loss on top of the usual time decay.

For a given observed path X and price functional f, the Greeks can be precomputed offline and used thereafter to approximate the P&L efficiently. Schematically,

$$(f, X) \overset{\text{offline}}{\longrightarrow} (\Theta, \Delta, \Gamma, \mathcal{L}) \overset{\text{online}}{\longrightarrow} \text{P\&L} \approx \Theta\delta t + \Delta\delta y + \Gamma\frac{\delta y^2}{2} + \mathcal{L}\mathcal{A}(Y). \tag{30}$$

The functional Greeks in the offline phase can be estimated using finite difference (forward in t, centered in x). In particular, the Libra $\mathcal{L}$ is approximated by the ratio

$$\frac{f((X^h)^{*\delta t'}) - f((X^{*\delta t'})^h) - f((X^{-h})^{*\delta t'}) + f((X^{*\delta t'})^{-h})}{2h\delta t'}, \tag{31}$$

for small $\delta t', h > 0$. Note that the online phase in (30) is nearly instantaneous as it only requires to compute two simple quantities for each scenario Y: the increment δy and the Lévy area $\mathcal{A}(Y)$. The approximated P&L can then be translated into risk

Table 1 Option portfolio value and Greeks in Example 6.1

$f(X)$	Θ	Δ	Γ	$\mathcal{L}$
0.273	− 3.864	0.089	0.023	0.229

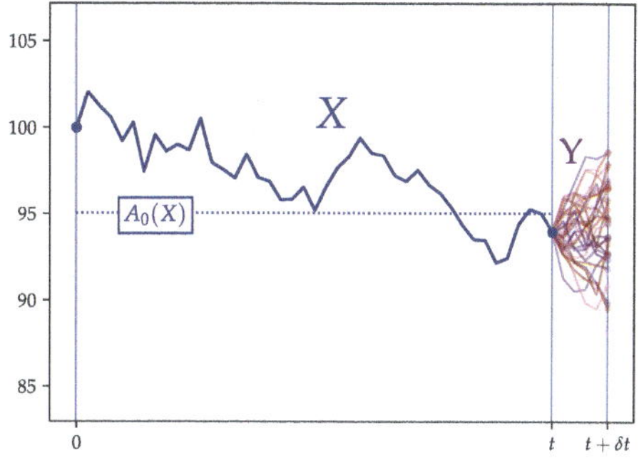

Fig. 7 Observed path X, intrinsic time average $A_0(X) = A(X) + X_t(1 - t/T)$, and scenarios Y where $T = 6$ months, $t = 2$ months, $\delta t = 1$ week

measures, such as Value at Risk (VaR) and Expected Shortfall (ES) as illustrated in the next example.

Example 6.1 Consider a (simplified) portfolio consisting of a long position in one at-the-money Asian call options as in Sect. 6.4.2, that is

$$g(X) = (A(X) - X_0)^+, \quad A(X) = \frac{1}{T}\int_0^T X_u du, \quad X \in \Lambda_T.$$

Introduce the value functional $f(X) = \mathbb{E}^{\mathbb{Q}}[g(Z)|X]$, $X \in \Lambda_t$, where $\mathbb{Q}$ is the measure associated to the Black-Scholes model with $\sigma = 20\%$ and zero interest rates. Suppose also that $X_0 = 100$, $T = 0.5$, $t = T/3$ (4 months to maturity) and $\delta t = \frac{5}{252}$ (one-week horizon). We then approximate the one-week P&L of the portfolio as in (29) assuming daily observations of the underlying.[3] More precisely, we simulate 1000 scenarios $Y = (Y_{t_0}, \ldots, Y_{t_5})$ and compute the Lévy area $\mathcal{A}(Y)$ by discretizing the integral in (27).

The portfolio value and Greeks are reported in Table 1 for a fixed path $X \in \Lambda_t$ depicted in Fig. 7. Since the average value of the transformed path $(X^h)^{*\delta t}$ is larger than the one of $(X^{*\delta t})^h$ when h is positive and g is nondecreasing in the average functional A, it is then expect from (31) that the Libra $\mathcal{L}$ is positive. This is confirmed in Table 1, and we conclude without much surprise that the convexity of Y, captured by $\mathcal{A}(Y)$, has a negative impact on the value of the option portfolio.

Table 2 shows the Value at Risk (VaR) as well as Expected Shortfall (ES) at the 95% and 99% level, respectively. The small approximation error is offset by a significant speedup. Indeed, the run time for the exact calculation of the P&L (to

[3] See https://github.com/valentintissot/Signature-and-FTE, file `NewGreek.ipynb` for implementation details.

Table 2 Comparison of value at risk and expect shortfall in Example 6.1 using Monte Carlo simulation and the FTE

	Value at risk		Expect shortfall	
Level	95%	99%	95%	99%
Exact	– 0.237	– 0.255	– 0.249	– 0.261
FTE	– 0.248	– 0.252	– 0.251	– 0.254

retrieve VaR and ES) is 52.50 seconds, while the FTE approximation in (30) (offline and online phase) takes 0.49 seconds.

6.2 Vanilla Options Are Path-Dependent

In this section, we make a somewhat counterintuitive statement: the Libra $\mathcal{L}$ of vanilla options may be nonzero. Put differently, vanilla options are path-dependent!

We start with a simple example. Let $T > 0$ and suppose that we are given a call option with strike $K > 0$ and maturity $\tau \in (0, T)$. If $\mathbb{Q}$ is a risk-neutral measure, then the value of the option is

$$f(X) = \begin{cases} \mathbb{E}^{\mathbb{Q}}[(Z_\tau - K)^+ \mid X] & t \leq \tau, \\ (X_\tau - K)^+, & t \geq \tau, \end{cases} \qquad X \in \Lambda_t, \tag{32}$$

where in the case $t \geq \tau$, we assume that the proceeds from the payoff is reinvested until time t with zero interest rates. Suppose that the price process Z is Markovian under $\mathbb{Q}$, so there exists a function $c : [0, T] \times \mathbb{R} \to \mathbb{R}$ such that $f(X) = c(t, X_{t\wedge\tau})$, with $c(t, x) = (x - K)^+$ if $t \geq \tau$. We also assume that the Libra can be expressed as a double limit as in (11), namely $Lf(X) = \lim_{h,\delta t \to 0} \frac{1}{h\delta t}[f((X^h)^{*\delta t}) - f((X^{*\delta t})^h)]$. When $t \leq \tau$, the Markov representation of f implies that for all $\delta t < \tau - t$,

$$f((X^h)^{*\delta t}) = c(t + \delta t, X_t + h) = f((X^{*\delta t})^h).$$

If $t > \tau$, then $f((X^h)^{*\delta t}) = f((X^{*\delta t})^h) = (X_\tau - K)^+$ as neither a spatial bump nor a flat extension after the expiration τ changes the payoff of the option. Hence $Lf(X) = 0$ for all $X \in \Lambda_t$ with $t \neq \tau$. Finally when $t = \tau$ and $X_\tau > K$, observe that for $h \neq 0$ small enough,

$$f((X^h)^{*\delta t}) - f((X^{*\delta t})^h) = (X_\tau + h - K) - (X_\tau - K) = h \neq 0. \tag{33}$$

We conclude that $\mathcal{L} = Lf(X) \neq 0$ (in fact, $\mathcal{L} = \infty$) when $X \in \Lambda_\tau$. Indeed, the value of the payoff is unchanged if we first time extend the path since the maturity has then passed. On the other hand, starting with a spatial bump does affect the payoff when the spot price X_τ is in-the-money; See also Fig. 8.

As announced at the beginning of this section, the Libra of vanilla options is indeed *nonzero* in general. We also stress that even when $\tau = T$, a vanilla call

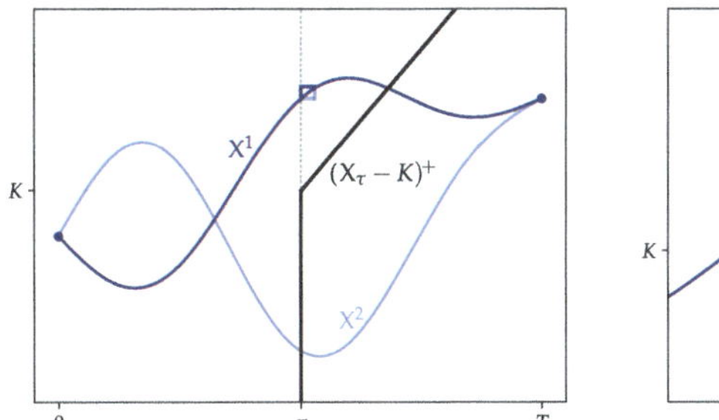

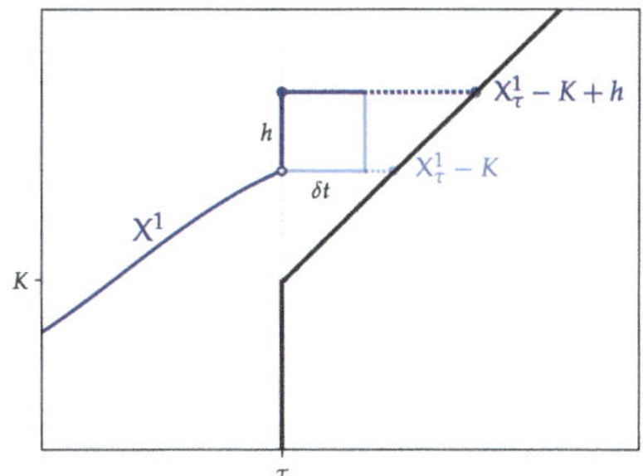

Fig. 8 Vanilla call option expiring at $\tau < T$. The Libra associated to the in-the-money path (X^1), as shown on the right panel (zoom of left panel)

option can have nonzero Libra if the underlying process is non-Markovian under $\mathbb{Q}$. This situation notably arises in path-dependent or rough volatility models.

We end this section with a more general example.

Example 6.2 Consider a portfolio of vanilla call options of the form

$$g(X) = \int_0^T \int_{\mathbb{R}_+} \pi_{K,\tau}(X_\tau - K)^+ dK d\tau, \quad X \in \Lambda_T, \tag{34}$$

for some weights $(\pi_{K,\tau})$. Given a risk-neutral measure $\mathbb{Q}$, then the value of the portfolio having observed $X \in \Lambda_t$ is

$$f(X) = \int_0^t \int_{\mathbb{R}_+} \pi_{K,\tau}(X_\tau - K)^+ dK d\tau + \int_t^T \int_{\mathbb{R}_+} \pi_{K,\tau} \mathbb{E}^{\mathbb{Q}}[(Z_\tau - K)^+ \mid X] dK d\tau.$$

Again, the price process Z is Markovian under $\mathbb{Q}$. Hence for all $(K, \tau) \in \mathbb{R}_+ \times [0, T]$, there exists $c_{K,\tau} : [0, T] \times \mathbb{R} \to \mathbb{R}$ such that $\mathbb{E}^{\mathbb{Q}}[(Z_\tau - K)^+ \mid X] = c_{K,\tau}(t, X_t)$ with $c_{K,\tau}(t, x) = (x - K)^+$ when $t \geq \tau$. Assuming that the Libra can be expressed as a double limit as in (11), we then observe that

$$\begin{aligned} f((X^h)^{*\delta t}) = &\int_0^t \int_{\mathbb{R}_+} \pi_{K,\tau}(X_\tau - K)^+ dK d\tau \\ &+ \int_t^{t+\delta t} \int_{\mathbb{R}_+} \pi_{K,\tau}(X_t + h - K)^+ dK d\tau \\ &+ \int_{t+\delta t}^T \int_{\mathbb{R}_+} \pi_{K,\tau} c_{K,\tau}(t + \delta t, X_t + h) dK d\tau, \end{aligned}$$

and similarly,

$$f((X^{*\delta t})^h) = \int_0^t \int_{\mathbb{R}_+} \pi_{K,\tau}(X_\tau - K)^+ dK d\tau + \int_t^{t+\delta t} \int_{\mathbb{R}_+} \pi_{K,\tau}(X_t - K)^+ dK d\tau$$

$$+ \int_{t+\delta t}^T \int_{\mathbb{R}_+} \pi_{K,\tau} c_{K,\tau}(t + \delta t, X_t + h) dK d\tau.$$

The Libra is therefore given by

$$\mathcal{L} = \lim_{h,\delta t \to 0} \frac{f((X^h)^{*\delta t}) - f((X^{*\delta t})^h)}{h \delta t}$$

$$= \lim_{h,\delta t \to 0} \frac{1}{\delta t} \int_t^{t+\delta t} \int_{\mathbb{R}_+} \pi_{K,\tau} \frac{(X_t + h - K)^+ - (X_t - K)^+}{h} dK d\tau$$

$$= \int_0^{X_t} \pi_{K,t} dK.$$

As can be seen, $\mathcal{L}$ is nonzero and corresponds to the cumulative portfolio weight of in-the-money options (i.e., the strike is below the spot X_t). Furthermore, rewriting the portfolio as

$$g(X) = \int_0^T \phi(\tau, X_\tau) d\tau,$$

where $\phi(\tau, x) = \int_{\mathbb{R}_+} \pi_{K,\tau}(x - K)^+ dK$ is the aggregate vanilla payoff at time τ, then the Libra becomes $\mathcal{L} = \partial_x \phi(t, X_t)$ and represents the sensitivity of the t-section $\phi(t, \cdot)$ with respect to the spot level x.

6.3 *Static Hedging*

The FTE possesses also direct applications to static hedging. Consider a contingent claim $g : \Lambda_T \to \mathbb{R}$ with price functional $f \in \mathbb{C}^{k,k+1}(\Lambda)$, that is $f(X) = \mathbb{E}^{\mathbb{Q}}[g(Z)|X]$ for a risk-neutral measure $\mathbb{Q}$. Suppose we are given tradable instruments $\varphi_1, \ldots, \varphi_M$ and we associate to any weight vector $\pi = (\pi_1, \ldots, \pi_M) \in \mathbb{R}^M$ the portfolio

$$\varphi(X) = \sum_{m=1}^M \pi_m \varphi_m(X).$$

We then choose $\pi_1, \ldots, \pi_M$ so as to match the functional derivatives of f up to some order $k \geq 0$, namely

$$\Delta_I f(X_0) = \Delta_I \varphi(X_0) = \sum_{m=1}^{M} \pi_m \Delta_I \varphi_m(X_0), \quad \forall\, |I| \leq k. \tag{35}$$

The weights (π_m) are thus obtained by solving a linear system, where we assume that a solution exists. Using the FTE, we are able to quantify the hedging error $g-\varphi$. Indeed, applying Corollary 5.3 to the difference $f - \varphi$ gives

$$f(X) - \varphi(X) = \sum_{|I|<k} \Delta_I (f - \varphi)(X_0) S^I(X) + R_k(X) = R_k(X), \quad \forall\, X \in \bar{\Lambda}^{\Pi},$$

with remainder given in (22). If f, φ belong to the space spanned by the signature functionals up to order k, we would obtain a perfect hedge, i.e. $R_k \equiv 0$. Since $f = g$ at maturity, we obtain in particular that

$$g(X) = \varphi(X) \text{ for } \textit{all } X \in \bar{\Lambda}_T^{\Pi}.$$

When $R_k \neq 0$, the hedging error $|g - \varphi|$ can be estimated as we now explain. In line with financial markets, we assume that X is the piecewise interpolation of tick data and is in particular Lipschitz continuous. If furthermore, there exists a constant $C < \infty$ such that $\sup_{|I|=k} |\Delta_I (f - \varphi)(X)| \leq C$ for all Lipschitz path X in Λ_t, then Proposition 3.16 in [9] gives $R_k(X) = \frac{O(t^k)}{k!}$. Choosing $t = T$, the hedging error for the claim g is therefore bounded by

$$|g(X) - \varphi(X)| \leq |R_k(X)| = \frac{O(T^k)}{k!}, \quad \forall\, X \in \{Y \in \Lambda_T : \text{Lipschitz}\}.$$

We emphasize that the above upper bound holds pathwise, matching the needs of exotic option traders to be protected against future scenarios.

6.4 *Pricing of Exotic Options*

Let $\mathbb{Q}$ be a risk-neutral measure and $g : \Lambda_T \to \mathbb{R}$ a path-dependent payoff such that $\mathbb{E}^{\mathbb{Q}}[|g(Z)|]$ is finite. Regarding the signature functionals as hedging instruments, we can derive from the FTE a nearly replicating portfolio using (S^I). The initial price is then approximated by taking expectation with respect to $\mathbb{Q}$. However, the FTE can only be applied to functionals, not T-functionals such as g. A remedy is to employ embeddings (Sect. 2) as we now explain. Fix a functional $f : \Lambda \to \mathbb{R}$ such that $f = g$ on Λ_T, e.g. using Black-Scholes or Bachelier embeddings. Provided that

$f \in \mathbb{C}^{k,k+1}(\Lambda)$ for some $k \geq 0$, then Corollary 5.3 gives as in (23),

$$g^k(X) = \sum_{|I| \leq k} \lambda^I S^I(X), \quad \lambda^I = \Delta_I f(X_0). \tag{36}$$

By the law of one price, we must have that

$$p^k = \mathbb{E}^{\mathbb{Q}}[g^k(Z)] = \sum_{|I| \leq k} \lambda^I \mathbb{E}^{\mathbb{Q}}[S^I(Z)]. \tag{37}$$

Importantly, the "model" $\mathbb{Q}$ *only* materializes through the expected signature, which can be precomputed once and for all. This becomes greatly advantageous over a straightforward Monte Carlo approach when the simulation of paths under $\mathbb{Q}$ is costly. We next illustrate the procedure with examples.

6.4.1 Vanilla Call Option

Let $X_0 = 100$ and consider a three-month at-the-money vanilla call option, that is $g(X) = (X_T - X_0)^+$, $T = \frac{1}{4}$, in the Black-Scholes model with zero interest rates and volatility $\sigma_{\text{BS}} = 20\%$. As the Black-Scholes price $p^{\text{BS}} := \mathbb{E}^{\mathbb{Q}}[(Z_T - X_0)^+]$ is readily available, this example permits to verify the pricing procedure in (36) and (37). The reader is nevertheless encouraged to explore more complex risk-neutral dynamics of the underlying. Given $X \in \Lambda_t$, introduce the Bachelier embedding $f_\sigma(X) = \mathbb{E}^{\mathbb{Q}_\sigma}[(Z_T - X_0)^+ \mid X] = p^{\text{B}}(t, X_t, \sigma)$, where $p^{\text{B}}(t, x, \sigma)$ is the price given $X_t = x$ of a call with maturity T in the Bachelier model with volatility[4] σ and zero interest rates. For completeness, recall that

$$p^{\text{B}}(t, x, \sigma) = \bar{\sigma}\phi\left(\frac{x - X_0}{\bar{\sigma}}\right) + (x - X_0)\Phi\left(\frac{x - X_0}{\bar{\sigma}}\right), \quad \bar{\sigma} = \sigma\sqrt{T - t}, \tag{38}$$

where ϕ, Φ is respectively the PDF and CDF of $\mathcal{N}(0, 1)$. Writing $p_\sigma(t, x) = p^{\text{B}}(t, x, \sigma)$, we obtain $\Delta_t f_\sigma = \partial_t p_\sigma$, $\Delta_x f_\sigma = \partial_x p_\sigma$ and the same holds for higher derivatives.

Owing to the functional Feynman-Kac formula [8], the price functional f_σ in (6) solves the path-dependent partial differential equation (PPDE)[5]

$$\Delta_t f_\sigma + \frac{1}{2}\sigma^2 \Delta_{xx} f_\sigma = 0, \text{ on } \Lambda_- := \bigcup_{t<T} \Lambda_t, \quad f_\sigma = g, \text{ on } \Lambda_T. \tag{39}$$

[4] We recall that the Bachelier volatility is expressed in *dollar amount*. That is, as $X_0 = 100$, a 20% Black-Scholes volatility would roughly correspond to a Bachelier volatility equal to $\sigma = 20$.

[5] To be more precise (39) is satisfied in the *topological support* of the measure $\mathbb{Q}_\sigma$, which consists of all continuous paths starting at zero in light of the Stroock-Varadhan support theorem [21].

In terms of p_σ, the above PPDE therefore translates to

$$\left(\partial_t + \frac{1}{2}\sigma^2\partial_{xx}\right)p_\sigma(t,x) = 0, \quad t < T, \qquad p_\sigma(T,x) = (x - X_0)^+. \tag{40}$$

Writing, with a slight abuse of notations, $f_\sigma = f_\sigma(X_0)$, $p_\sigma = p_\sigma(0, X_0)$, and $S^I = S^I(X)$ for fixed $X \in \Lambda_T$, then a Taylor expansion of order $k = 2$ reads

$$\begin{aligned}(X_T - X_0)^+ &\approx f_\sigma S^{\emptyset} + \Delta_t f_\sigma S^{(0)} + \Delta_x f_\sigma S^{(1)} \\ &\quad + \Delta_{tt} f_\sigma S^{(0,0)} + \Delta_{xt} f_\sigma S^{(1,0)} + \Delta_{tx} f_\sigma S^{(0,1)} + \Delta_{xx} f_\sigma S^{(1,1)} \\ &= p_\sigma + \partial_t p_\sigma\ T + \partial_x p_\sigma\ (X_T - X_0) \qquad (41) \\ &\quad + \frac{1}{2}\partial_{tt} p_\sigma\ T^2 + \partial_{tx} p_\sigma\ T(X_T - X_0) + \frac{1}{2}\partial_{xx} p_\sigma\ (X_T - X_0)^2, \qquad (42)\end{aligned}$$

where we used that $\Delta_{tx} f_\sigma = \Delta_{xt} f_\sigma = \partial_{tx} p_\sigma$ and $(S^{(1,0)} + S^{(0,1)})(X) = T(X_T - X_0)$, in the last equality. Omitting the higher order Greeks $\partial_{tt} p_\sigma$, $\partial_{tx} p_\sigma$, this gives

$$(X_T - X_0)^+ = p_\sigma + \partial_t p_\sigma T + \partial_x p_\sigma (X_T - X_0) + \frac{1}{2}\partial_{xx} p_\sigma (X_T - X_0)^2 + o(T). \tag{43}$$

The call payoff is therefore replicated using a parabola in X_T with coefficients depending on the Bachelier volatility σ as seen in the left panel of Fig. 9.[6] As Taylor approximations are inherently local, the above replication deteriorates as X_T moves away from X_0 or for longer maturity T. Note that the Greeks $\partial_t p_\sigma$, $\partial_x p_\sigma$, ... are available in closed form in the Bachelier model, hence there is no numerical error induced by the computation of the FTE coefficients themselves in this simple case. For more general payoff, one can use finite difference.

In Fig. 9, we also report the approximation obtained from a regression approach similar to [1, 6, 20]. That is, we simulate trajectories $(X^j)_{j=1}^J$ with $J = 1000$ from the reference model $\mathbb{Q}$ and perform the regression

$$\begin{aligned}(X_T^j - X_0)^+ &= \beta^{\emptyset} + \beta^{(0)}T + \beta^{(1)}(X_T^j - X_0) \\ &\quad + \beta^{(1,1)}\frac{(X_T^j - X_0)^2}{2} + \varepsilon_j, \quad j = 1, \dots, J.\end{aligned}$$

As can be observed in Fig. 9, the fitted parabola is close to the Taylor expansion with Bachelier volatility $\sigma = 20$.

We next take expectation with respect to the martingale measure $\mathbb{Q}$ and compare the estimated prices p_σ^k given in (37) with $k = 0, 1, 2$, respectively. In view of (41),

[6] An implementation is given in the Jupyter notebook `PricingExotics.ipynb` in https://github.com/valentintissot/Signature-and-FTE.

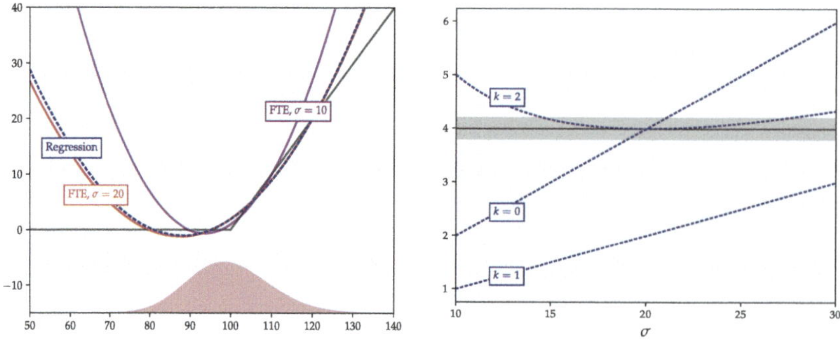

Fig. 9 *Left:* Approximation of $g(X) = (X_T - X_0)^+$ using the FTE with Bachelier embedding for varying volatility σ compared to the regression approach. The risk-neutral density of X_T is also displayed (bottom). *Right:* Price (dashed line) as function of σ and truncation order k. The solid line is the exact Black-Scholes price, surrounded by a 5% error band

we have $p_\sigma^0 = p_\sigma$ which is precisely the call price in the Bachelier model. It is therefore intuitive to align the volatility of the Bachelier embedding with the Black-Scholes volatility, even though the higher order terms would eventually compensate. Next, taking $k = 1$ gives

$$p_\sigma^1 = p_\sigma + \partial_t p_\sigma T + \partial_x p_\sigma \mathbb{E}^{\mathbb{Q}}[Z_T - X_0] = p_\sigma + \partial_t p_\sigma T. \tag{44}$$

where $\mathbb{E}^{\mathbb{Q}}[Z_T - X_0] = 0$ follows from the martingality of Z under the risk-neutral measure $\mathbb{Q}$. Moving finally to the second order, we obtain from (42) that

$$\begin{aligned} p_\sigma^2 &= p_\sigma^1 + \frac{1}{2}\partial_{tt} p_\sigma T^2 + \frac{1}{2}\partial_{xx} p_\sigma \mathbb{E}^{\mathbb{Q}}[(Z_T - X_0)^2] \\ &= p_\sigma + \partial_t p_\sigma T + \frac{1}{2}\partial_{xx} p_\sigma \mathbb{V}^{\mathbb{Q}}[Z_T] + o(T), \end{aligned} \tag{45}$$

with the risk-neutral variance $\mathbb{V}^{\mathbb{Q}}[Z_T] = \mathbb{E}^{\mathbb{Q}}[(Z_T - X_0)^2]$. We note that the incremental nature of Taylor expansions is computationally convenient as terms of lower orders can be directly added. The effect of σ on the price is illustrated on the right panel of Fig. 9. We observe that it is indeed beneficial to choose σ close to the scaled Black-Scholes volatility $X_0\sigma_{\text{BS}} = 20$.

6.4.2 Asian Call Option

For fixed maturity $T > 0$, consider an at-the-money Asian call option with payoff

$$g(X) = (A(X) - X_0)^+, \quad A(X) = \frac{1}{T}\int_0^T X_t dt, \quad X \in \Lambda_T. \tag{46}$$

Similar to the "better Asian PDE" example in [8], define the intrinsic time average

$$A_0(X) := \mathbb{E}^{\mathbb{Q}_\sigma}[A(Z) \mid X] = \frac{1}{T}\int_0^t X_u du + X_t(1 - t/T), \quad X \in \Lambda_t, \tag{47}$$

where the value of σ is here irrelevant. If $Z = X * Y$ with $X \in \Lambda_t$ and $Y \in \Lambda_{T-t}$ such that $Y_0 = 0$, then

$$A(Z) = \frac{1}{T}\int_0^t X_u du + \frac{1}{T}\int_t^T (X_t + Y_u)du = A_0(X) + \frac{1}{T}\int_t^T Y_u du.$$

Hence, the Bachelier embedded functional associated to g reads ($\rightarrow$ Exercise 6.3),

$$f_\sigma(X) = p^{\text{B}}(t, A_0(X), \Sigma(t, \sigma)), \quad \Sigma(t, \sigma) = \frac{\sigma}{\sqrt{3}}(1 - t/T). \tag{48}$$

with the Bachelier price p^{B} given in (38). As in [8], the functional A_0 is introduced to avoid a numerically undesirable drift term in the associated pricing PDE, otherwise present if $f_\sigma(X)$ is expressed in terms of $\int_0^t X_u du$. Next, we compute the FTE coefficients $\lambda_\sigma^I = \Delta_I f_\sigma(X_0)$. It is easily seen ($\rightarrow$ exercise) that $\Delta_t A_0(X) = 0$, while $\Delta_x A_0(X) = 1 - t/T$. Writing $p^{\text{B}} = p^{\text{B}}(0, X_0, \Sigma(0, \sigma))$ (note that $A_0(X_0) = X_0$), the first order FTE coefficients read

$$\lambda_\sigma^{(0)} = \partial_t p^{\text{B}} - \frac{\sigma}{\sqrt{3}T}\partial_\sigma p^{\text{B}}, \quad \lambda_\sigma^{(1)} = (1 - t/T)|_{t=0}\, \partial_x p^{\text{B}} = \partial_x p^{\text{B}}.$$

It can be shown after rearranging ($\rightarrow$ exercise) that

$$g(X) = p_\sigma + \partial_t p_\sigma\, T + \partial_x p^{\text{B}}(A(X) - X_0) + \frac{1}{2}\partial_{xx} p^{\text{B}}\, (X_T - X_0)^2 + o(T).$$

Since $A(X)$ is the average value of the asset over the horizon $[0, T]$, then $A(X) - X_0$ captures the *trend* of X compared to the terminal value X_0. As in Sect. 6.2, the partial derivatives $\partial_t p^{\text{B}}, \partial_x p^{\text{B}}, \ldots$ can be computed explicitly from (48). Again, $\mathbb{Q}$ is the risk-neutral distribution in the Black-Scholes model with volatility $\sigma_{\text{BS}} = 20\%$ and $T = \frac{1}{4}$ (three-month maturity). Choosing $\sigma = X_0\sigma_{\text{BS}}$ gives a value of 2.321 using the FTE, which is close to the exact Black-Scholes price (equal to 2.303).

Exercise 6.3

(a) Derive formula (48). *Hint*: Recall that $\int_0^t W_u du \sim \mathcal{N}(0, t^3/3)$ if W is Brownian motion.
(b) Verify that $da_t = \sigma(1 - t/T)dW_t$, where $a_t = A_0(Z_{[0,t]})$ and $dZ_t = \sigma dW_t$.
(c) Derive using (b) the PDE solved by p_σ and compare it with the path-dependent PDE (39).

Exercise 6.4 Consider a floating strike Asian put option where $g(X) = (X_T - A(X))^+$ and $A : \Lambda_T \to \mathbb{R}$ given in (46).

1. Show that $f_\sigma(X) = p_\sigma(t, X_t, A_0(X))$ for some function p_σ to be determined.
   Hint: Compute the joint distribution of $(W_s, \int_0^s W_u du)$.
2. Compute the functional derivatives $\Delta_I f_\sigma$, where $I \in \{0, 1, 00, 01, 10, 11\}$ and write down the Taylor-Maclaurin approximation f_σ^k for $k = 2$.
3. Write a code that returns the replication price $p_\sigma^2 = \mathbb{E}^{\mathbb{Q}}[f_\sigma^2(Y)]$ as function of σ, where $\mathbb{Q}$ is an arbitrary model.

6.4.3 Discussion: Barrier and Lookback Options

Barrier and Lookback claims are contingent upon the range of the underlying. For instance, a down-and-out digital option pays a notional amount (say, $1) if the stock price has stayed above a given barrier level throughout the life of the contract and zero otherwise. Another example is a floating strike lookback put option that delivers at expiration the realized drawdown $g(X) = \max_{t \le T} X_t - X_T$ seen in Exercise 2.4.

Similar to the previous example, we may apply the FTE to an embedding f_σ of g and obtain an approximation of the initial price. However, f_σ is typically not regular enough when g depends on the maximum (or minimum) of the path. Indeed, taking the drawdown example and $\mathbb{Q}_\sigma$ to be the measure associated to the Black-Scholes model with volatility σ, it can be shown that $f_\sigma(X) = p_\sigma(t, X_t, \max_{0 \le s \le t} X_s)$, $X \in \Lambda_t$, where $p_\sigma : [0, T] \times \mathbb{R}^2 \to \mathbb{R}$ is a smooth function. However, the functional $f_\sigma(X)$ is *not* smooth when the spot value equals the maximum. One remedy is to regularize the maximum as in [8, Example 2] and perform an expansion of the regularized functional. We note that the application of the procedure (36) and (37) to range-based options is still subject to ongoing research.

6.5 A Glimpse into Cubature Methods

Cubature is the analogue of Gaussian quadrature for multiple integrals. In short, a cubature rule is described by a collection of nodes $x_1, \dots, x_M \in \mathbb{R}^d$ and weights $\lambda_1, \dots, \lambda_M$ such that

$$\int_{\mathbb{R}^d} p_I(x) dx = \sum_{m=1}^{M} \lambda_m p_I(x_m), \tag{49}$$

for a collection of polynomials (p_I). Put differently, the polynomials have to be exactly integrated under the discrete measure $\sum_{m=1}^{M} \lambda_m \delta_{x_m}$. Similarly, *cubature on Wiener space* [17, 19] requires that the signature elements, up to a prescribed order, are "priced" perfectly under a discrete measure. The use of cubature methods in option pricing is well-established for non-exotic payoffs [5, 10], but little is known in the path-dependent case (see also the discussion in [2]). In this section, we briefly address the latter by connecting cubature on the path space Λ with the FTE.

Let $\mathbb{Q}$ denote the Wiener measure such that the canonical process Z on Λ_T is Brownian motion. Akin to (49), we say that $(X, \lambda) \in (\Lambda_T \times \mathbb{R})^M$ is a *cubature rule* of order k if

$$\mathbb{E}^{\mathbb{Q}}[S^I(Z)] = \sum_{m=1}^{M} \lambda_m S^I(X^m), \quad \forall |I| \le k. \tag{50}$$

Setting $\hat{\mathbb{Q}} = \sum_{m=1}^{M} \lambda_m \delta_{X^m}$ where δ_X is the Dirac measure at some path X, we can then estimate the expectation of some $\mathbb{Q}$-integrable T-functional $G : \Lambda_T \to \mathbb{R}$ as

$$\mathbb{E}^{\mathbb{Q}}[G(Z)] \approx \mathbb{E}^{\hat{\mathbb{Q}}}[G(Z)] = \sum_{m=1}^{M} \lambda_m G(X^m). \tag{51}$$

In option pricing, the functional G would correspond to the composition $G = g \circ \Phi$, where $g : \Lambda_T \to \mathbb{R}$ is the payoff of an exotic option and $\Phi : \Lambda_T \to \Lambda_T$ is the so-called *Itô map* linking Brownian motion to the underlying price process. For instance, in the Black-Scholes model with zero interest rate, volatility σ and initial asset value equal to one, then $\Phi(X) = e^{\sigma X_T - \frac{1}{2}\sigma^2 T}$. The price of the claim would then be approximated by the linear combination in (51).

In view of (50), it is expected that the FTE plays a central role in cubature methods. Indeed, if $F \in \mathbb{C}^{k,k+1}(\Lambda)$ is an embedding of G, then $F^k(X) = \sum_{|I| \le k} \Delta_I F(X_0) S^I(X)$ seen in (23) satisfies $\mathbb{E}^{\mathbb{Q}}[f^k(Z)] = \mathbb{E}^{\hat{\mathbb{Q}}}[f^k(Z)]$ by linearity and (50). This leads to the following upper bound,

$$|\mathbb{E}^{\mathbb{Q}}[f(Z)] - \mathbb{E}^{\hat{\mathbb{Q}}}[f(Z)]| \le |\mathbb{E}^{\mathbb{Q}}[R^k(Z)] - \mathbb{E}^{\hat{\mathbb{Q}}}[R^k(Z)]|, \quad R^k = f - f^k.$$

The above pricing error is therefore controlled as soon as the remainder R^k is small in magnitude. See [9, Section 3.1.5] for precise remainder estimates and [5, 19] for a comprehensive treatment of cubature methods.

7 Extension to the Multi-Dimensional Setting

The functional Taylor has been introduced for paths with one space dimension. However, the multi-dimensional setting is by no means irrelevant. In short, the path space becomes

$$\Lambda^d := \bigcup_{t \in [0,T]} \Lambda_t^d, \quad \Lambda_t^d = \{X_u = (X_u^1, \dots, X_u^d),\ u \le t : X^i \in \Lambda_t,\ i = 1, \dots, d\},$$

and functionals are maps from Λ^d to $\mathbb{R}$. The signature is defined as in (16), with words in the alphabet $\{0, 1, ..., d\}$. This general setting allows to tackle multi-asset exotic options, where the payoff is a T-functional $g : \Lambda_T^d \to \mathbb{R}$. Examples include Asian basket options, where g depends on a weighted average of the form $\frac{1}{T}\int_0^T \sum_{i=1}^d w_i X_t^i dt$.

The functional space derivatives $\Delta_{x_1}, ..., \Delta_{x_d}$ again do not commute with Δ_t, while they actually commute with each other. Consequently, some terms in the functional Taylor expansion can be grouped together (e.g. $\Delta_{x_1x_2} f S^{(1,2)}$ and $\Delta_{x_2x_1} f\ S^{(2,1)}$), leading to simplified approximations of multi-asset exotic payoffs. Towards a fully general expansion, the functional space derivatives need to be modified so as to satisfy $\Delta_{x_ix_j} f \neq \Delta_{x_jx_i} f$ when $i \neq j$. This goes, nevertheless, beyond the scope of this chapter.

8 Bibliographical Notes

More details on the functional Itô calculus can be found in the original paper [8] and [16] focusing on the calculation of Greeks. In particular, the authors express the functional derivative of options in terms of Malliavin weights, giving an alternative approach to finite difference seen in Sect. 6.1. For Stone-Weierstrass type results in the path space, see [18] for continuous paths of bounded variation and in [6, 7] for paths emanating from general semimartingales. Applications to finance, including regression approaches, can be found in [1, 6, 20].

Related generalizations of Taylor's theorem were introduced by Fliess [11–13] for T-functionals and continuously differentiable paths and more recently by Litterer and Oberhauser [18], called *Chen-Fliess approximation*. The latter formulation also uses the functional derivatives in [8] and assumes that the paths under consideration solve a stochastic differential equation with smooth coefficients. The pathwise Itô calculus was introduced by Föllmer in his seminal work [14]. The pathwise Stratonovich formula in Theorem 3.3 is similar to the approach in [4] where a pathwise functional Itô formula is proved.

Acknowledgments Research partially supported by the Bloomberg Quantitative Finance Ph.D. Fellowship (second author).

We would like to thank Harald Oberhauser, Josef Teichmann, Christa Cuchiero, Peter Friz, as well as our colleagues Bryan Liang and Guixin Liu at Bloomberg for fruitful discussions.

Appendix: Path Reconstruction

An important property of the signature is that it uniquely characterizes a path, up to the so-called *tree-like equivalence* [15]. Here, the enlargement (t, X) and monotonicity of $t \mapsto t$ rules out the existence of tree-like paths, which are specific loops [15]. Hence, there is exactly one path associated to a given collection of signatures $\{S^I(X) : I \in \mathcal{W}\}$ and we may wonder how to rebuild it. In fact, a single "diagonal" in the signature tree is enough, as we now outline. See [23, Section 2.3] for additional details.

Set $T = 1$, $X_0 = 0$ for simplicity and focus on paths on the full horizon $[0, 1]$. Let $\mathcal{H} = L^2([0, 1])$ be the space of square integrable functions on $[0, 1]$ with inner product $(h_1, h_2) = \int_0^1 h_1(t)h_2(t)dt$. If $X, Y \in \Lambda_1 \cap \mathcal{H}$, we also write $(X, Y) = \int_0^1 X_t Y_t dt$. We start with a simple but crucial observation.

Lemma 8.1 *Consider the Legendre words $I_k := (1, 0, \ldots, 0) \in \{0, 1\}^{k+2}$, $k \geq 0$. Then for every $X \in \bar{\Lambda}_1^{\sqcap} \cap \mathcal{H}$,*

$$S^{I_k}(X) = \int_0^1 X_t \frac{(1-t)^k}{k!} dt = \frac{1}{k!}(X, r_k), \quad r_k(t) = (1-t)^k. \tag{52}$$

Proof This is a straightforward adaptation of [23, Lemma 2.7]. □

Since the time-reversed monomials r_k in (52) form a total basis of $\mathcal{H}$, we conclude from the above lemma states that the sequence $(S^{I_k}(X))_{k \geq 0}$ characterizes the path X in its entirety. However, nothing is said (yet) about how the path is reconstructed. One possibility, which we follow, is to carry out a Gram-Schmidt procedure that linearly transforms the time-reversed monomials into an orthonormal basis (ONB) and use Hilbert projections.

Let (q_k) be the (normalized) *Legendre polynomials* on $[0, 1]$ given by (see [22]),

$$q_k = \sum_{\ell \leq k} a_{k,\ell}\, m_\ell\,, \quad m_\ell(t) = t^\ell, \quad a_{k,\ell} = \sqrt{2k+1}(-1)^{k+l}\binom{k}{\ell}\binom{k+\ell}{\ell}. \tag{53}$$

Then (q_k) is an ONB of $\mathcal{H}$, hence for all $X \in \Lambda_1 \cap \mathcal{H}$, the projected path

$$X_t^K := \sum_{k \leq K} (X, q_k) q_k(t), \quad t \in [0, T], \tag{54}$$

converges to X in $\mathcal{H}$, that is

$$\|X - X^K\|_{\mathcal{H}}^2 := \int_0^1 (X_t - X_t^K)^2 dt = \sum_{k > K} (X, q_k)^2 \to 0 \text{ as } K \to \infty.$$

Note that (54) gives an explicit recipe to reconstruct the original path. What remains is to express the L^2 products (X, q_k) in terms of $(S^{I_k}(X))$. This is immediate once observing that

$$(X, m_k) = \int_0^1 X_t (1-(1-t))^k dt = \sum_{\ell=0}^{k} \frac{(X, r_\ell)}{\ell!} \frac{k!(-1)^\ell}{(k-\ell)!} = k! \sum_{l=0}^{k} S^{I_\ell}(X) \frac{(-1)^\ell}{(k-\ell)!}.$$

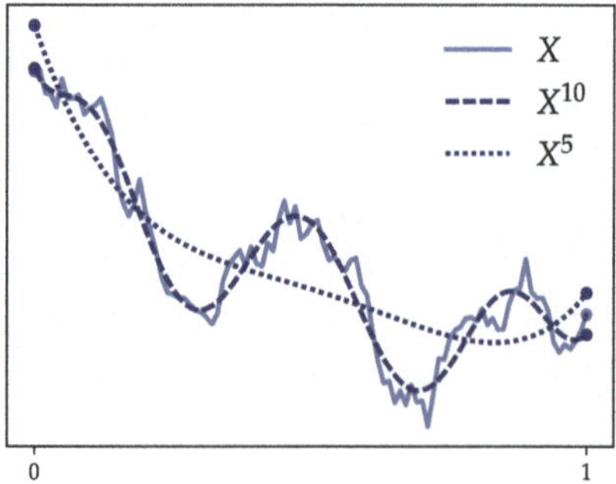

Fig. 10 Path reconstruction. Brownian path X and projection X^K, $K \in \{5, 10\}$

Thereafter, the L^2 products (X, q_k) can be finally retrieved from $(X, m_\ell)_{\ell \leq k}$ by linearity. The algorithm is summarized below[7] and illustrated in Fig. 10 when $K \in \{5, 10\}$. This procedure can be easily generalized to the multidimensional case. Indeed, each component $i = 1, \ldots, d$ can be reconstructed separately from the words $(i, 0, \ldots, 0) \in \{0, 1\}^{k+2}$, $k \geq 0$.

Algorithm: Path Reconstruction from the Signature of Legendre Words
Given: $S^{I_0}(X), \ldots, S^{I_K}(X)$

I. For $k = 0, \ldots, K$, **compute:**

1. L^2 products with monomials: $(X, m_k) = k! \sum_{\ell=0}^{k} S^{I_\ell}(X) \frac{(-1)^\ell}{(k-\ell)!}$.
2. L^2 products with Legendre polynomials (53): $(X, q_k) = \sum_{\ell=0}^{k} a_{k,\ell}(X, m_k)$.

II. **Return** $X^K := \sum_{k \leq K} (X, q_k) q_k$.

References

1. I.P. Arribas, C. Salvi, L. Szpruch, Sig-SDEs model for quantitative finance, in *Proceedings of the First ACM International Conference on AI in Finance*, ICAIF '20 (Association for Computing Machinery, 2020)
2. C. Bayer, P.K. Friz, Cubature on wiener space: Pathwise convergence. Appl. Math. Optim. **67**, 261–278 (2012)
3. C. Bayer, P.K. Friz, P. Gassiat, J. Martin, B. Stemper, A regularity structure for rough volatility. Math. Finance **30**(3), 782–832 (2020)
4. R. Cont, Functional Itô calculus and functional Kolmogorov equations, in *Stochastic Integration by Parts and Functional Itô Calculus*, ed. by V. Bally, L. Caramellino, R. Cont. Advanced Courses in Mathematics - CRM Barcelona (Birkauser, Cham, 2016)

[7] An implementation can be found in https://github.com/valentintissot/Signature-and-FTE, file `PathReconstruction.ipynb`.

5. D. Crisan, K. Manolarakis, C. Nee, Cubature methods and applications, in *Paris-Princeton Lectures on Mathematical Finance 2013* (2013), pp. 203–316
6. C. Cuchiero, G. Gazzani, S. Svaluto-Ferro, Signature-based models: Theory and calibration. SIAM J. Financ. Math. **14**(3), 910–957 (2023)
7. C. Cuchiero, F. Primavera, S. Svaluto-Ferro, Universal approximation theorems for continuous functions of càdlàg paths and Lévy-type signature models. arXiv:2208.02293 (2023)
8. B. Dupire, Functional Itô calculus. SSRN (2009). Republished in Quantitative Finance **19**(5), 721–729 (2019)
9. B. Dupire, V. Tissot-Daguette, Functional expansions. arXiv:2212.13628 (2022)
10. D. Filipović, M. Larsson, S. Pulido, Markov cubature rules for polynomial processes. Stoch. Process. Appl. **130**(4), 1947–1971 (2020)
11. M. Fliess, Fonctionnelles causales non linéaires et indéterminées non commutatives. Bull. Soc. Math. France **109**, 3–40 (1981)
12. M. Fliess, On the concept of derivatives and Taylor expansions for nonlinear input-output systems, in *The 22nd IEEE Conference on Decision and Control* (1983), pp. 643–646
13. M. Fliess, Vers une notion de dérivation fonctionnelle causale. Ann. Inst. Henri Poincaré C Anal. Non Linéaire **3**, 67–76 (1986)
14. H. Föllmer, Calcul d'Itô sans probabilités. Séminaire Probab. Strasbourg **15**, 143–150 (1981)
15. B. Hambly, T. Lyons, Uniqueness for the signature of a path of bounded variation and the reduced path group. Ann. Math. **171**(1), 109–167 (2010)
16. S. Jazaerli, Y.F. Saporito, Functional Itô calculus, path-dependence and the computation of greeks. Stoch. Process. Appl. **127**(12), 3997–4028 (2017)
17. S. Kusuoka, Approximation of expectation of diffusion process and mathematical finance, in *Taniguchi Conference on Mathematics Nara '98* (Mathematical Society of Japan, 2001), pp. 146–165
18. C. Litterer, H. Oberhauser, On a Chen-Fliess approximation for diffusion functionals. Monatsh. Math. **175**(4), 577–593 (2014)
19. T. Lyons, N. Victoir, Cubature on Wiener space. Proc. Math. Phys. Eng. Sci. **460**(2041), 169–198 (2004)
20. T.J. Lyons, S. Nejad, I.P. Arribas, Numerical method for model-free pricing of exotic derivatives using rough path signatures. Appl. Math. Finance **26**(6), 583–597 (2019)
21. D.W. Stroock, S.R.S. Varadhan, *On the Support of Diffusion Processes with Applications to the Strong Maximum Principle* (University of California Press, Berkeley, 1972), pp. 333–360
22. G. Szegö, *Orthogonal Polynomials*, vol. XXII, 4th edn. (American Mathematical Society, 1975)
23. V. Tissot-Daguette, Projection of functionals and fast pricing of exotic options. SIAM J. Financ. Math. **13**(2), SC74–SC86 (2022)

Signature-Based Models in Finance

Christa Cuchiero, Guido Gazzani, Janka Möller, and Sara Svaluto-Ferro

Abstract We consider two classes of asset price models where either the price or the volatility dynamics are described by a linear function of the (time extended) signature of a primary process, in general a multidimensional continuous semimartingale. These model classes are universal in the sense that classical models can be approximated arbitrarily well or are simply nested in our setup. Under the additional assumption that the primary process is polynomial, we obtain tractable option pricing formulas for so-called sig-payoffs in the first class and closed form expressions for the VIX squared and the log-price in the second one. In both cases the signature samples can be easily precomputed, hence the calibration task can be split into an offline sampling and a standard optimization. We present several applications, in particular the successfully solved joint SPX/VIX calibration problem.

1 Introduction

The present chapter is based on the two articles [13] and [11].

In recent years, the traditional approach of calibrating a few well-interpretable parameters has given way to learning the model's characteristics as a whole, by leveraging all available sources of data. Consequently, overparametrized models have become increasingly important. This shift has opened the door to more robust

---

The present work was initiated when the second author was affiliated to École des Ponts ParisTech, CERMICS Lab, Marne la Vallée Cedex 2, France.

C. Cuchiero · J. Möller
Department of Statistics and Operations Research, Vienna University, Wien, Austria
e-mail: christa.cuchiero@univie.ac.at; janka.moeller@univie.ac.at

G. Gazzani (✉) · S. Svaluto-Ferro
Department of Economics, University of Verona, Verona, Italy
e-mail: guido.gazzani@univr.it; sara.svalutoferro@univr.it

C. Bayer et al. (eds.), *Signature Methods in Finance*, Springer Finance Lecture Notes, https://doi.org/10.1007/978-3-031-97239-3_7

and data-driven model selection mechanisms. At the same time the choice of admissible models must, however, be limited to those adhering to first principles such as "no-arbitrage". Finding classes of dynamic processes that are data-driven *and* satisfy such well-established theoretical principles can be achieved by relying on different universal approximation theorems.

One class of such financial models are so-called *signature SDEs*, as considered in [12, 37]. These are Itô-diffusions, where the characteristics are linear or real-analytic functions of the *signature* of some (properly extended) driving Brownian motion or the process itself. The models that we consider in this chapter can be embedded in the framework of signature SDEs and serve as particularly tractable examples thereof.

In the first part (Sect. 3) of the present chapter our goal is to provide a data-driven, universal, tractable and easy to calibrate asset price model. For the sake of exposition we shall assume throughout that the asset S is one-dimensional. We model S as a linear function of the signature of—what we call—*primary process*. This primary process can either be a classical driving signal, e.g., a Brownian motion or a polynomial process [9], but also a general d-dimensional continuous semimartingale $X = (X_t^1, \dots, X_t^d)_{t\geq 0}$, usually augmented with time t. The corresponding time-extended process is denoted by $(\widehat{X}_t)_{t\geq 0} = (t, X_t^1, \dots, X_t^d)_{t\geq 0}$ and its signature by $\widehat{\mathbb{X}}$. The asset S is then modeled/approximated via a process $S_n(\ell)$ defined as

$$S_n(\ell)_t := \ell(\widehat{\mathbb{X}}_t), \tag{1}$$

where ℓ is a linear map of the signature of $\widehat{X}$ up to some level of truncation $n \in \mathbb{N}$ which needs to be inferred from data (see Definition 3.2 for further details). Note that the parameters of X are prespecified beforehand and can thus be seen—in analogy to machine learning terminology—as *hyperparameters*. This is crucial as it allows to split the calibration task into precomputable samples of the signature and a standard optimization to find the parameters of the linear map. This is one of the attractive features, further ones are summarized subsequently.

No Arbitrage We will present in Sect. 3.2, that (1) can also be expressed in terms of stochastic integrals, whence is straightforward to deduce no-arbitrage conditions.

Universality We refer the reader to [11, Example 3.15], where we show that classical stochastic volatility models with sufficiently regular coefficients can be arbitrarily well approximated by models of form (1).

Option Pricing Formulas for Sig-Payoffs By approximating (path dependent) payoffs via so-called sig-payoffs of the form $L(\widehat{\mathbb{S}}_T(\ell))$ (see also [34]), where $\widehat{\mathbb{S}}$ denotes the signature of $(\widehat{S}_t)_t := (t, S_t)_t$ and L a linear map, we show that this kind of approximate option pricing reduces to the computation of the expected signature of $\widehat{X}$. These formulas lead to high tractability whenever $\mathbb{E}[\widehat{\mathbb{X}}_T]$ is easy to compute. This is the case for all polynomial process as analyzed in [12] and in Sect. 2 for the truncated signature.

Calibration to Options As mentioned above, when calibrating to the market's volatility surface (see Sect. 3.5), we precompute Monte Carlo samples of $\widehat{\mathbb{X}}$ and are then only left with finding the parameters ℓ, which is subject to a standard optimization. We perform for both simulated and real market data (S&P 500 index) a full calibration to the volatility surface, and show that this is not only highly accurate but also very fast (see [11] for details, in particular in the case of time dependent parameters).

The second part (Sect. 4) treats a slightly different class of signature-based models and constitutes a novel contribution for joint VIX and S&P 500 modeling as well as calibration. In the context of volatility modeling the joint calibration to SPX[1] and VIX options is still considered a rather hard problem which has become increasingly important over the past years. However it is worth to mention that significant progress has been made recently (see [13] and [17]). We address the reader to [13, 17, 28] for an extensive literature review.

Inspired by the previous model, we now study a stochastic volatility model for the discounted price process $S = (S_t)_{t\geq 0}$, namely

$$dS_t(\ell) = S_t(\ell)\sigma_t^S(\ell)dB_t,$$

with some initial condition $S_0 \in \mathbb{R}_+$, standard Brownian motion B, and volatility process σ^S satisfying

$$\sigma_t^S(\ell) := \ell(\widehat{\mathbb{X}}_t), \tag{2}$$

for a linear map ℓ of the signature $\widehat{\mathbb{X}}_t$ of a time extended d-dimensional continuous semimartingale X, which takes here the role of the primary process. We thus assume that the signature of $\widehat{X}$ serves as a linear regression basis for the volatility process, while the parameters of the linear map ℓ have to be learned from (option price) data.

Also in this case the modeling framework can be seen as *universal* in the class of continuous non-rough stochastic volatility, which is a consequence of the universal approximation properties of the signature. Besides that it truly nests several classical models (see Remark 4.3). and incorporates both, purely Markovian (in (S, X)) and path-dependent ones. Moreover, it provides for the first time a signature-based approach for pricing VIX options and highly accurate joint calibration results, as illustrated in Sect. 4.2. For the latter we exploit the following mathematical and numerical properties.

- Setting $Z := (X, B)$, then not only $\sigma^S(\ell)$ but also the log-price $\log(S(\ell))$ is a linear function of the signature of $\widehat{Z}$. Therefore no (Euler) simulation scheme

[1] Note that the SPX is a theoretical index tracking the value of the S&P 500 without holding an actual portfolio of the constituent stocks. In the following we shall use the terms SPX and S&P 500 interchangeably.

is needed to sample the price process, leading to immediate computational advantages.

- If we additionally assume X to be a polynomial process (see Definition 2.1 and [9, 19]), we obtain an analytic expression for the VIX, which only involves the computation of a matrix exponential. This follows from the property that the truncated signature of a polynomial process is again a polynomial process (see Sect. 2).
- We can apply a Monte Carlo approach for option pricing *and* calibration where we are able to generate the signature samples of $\widehat{Z}$ offline and independently of the model parameters to be optimized. As in the previous model the calibration task can thus be split into an offline sampling procedure and a standard optimization, since the latter does not require re-sampling for updated model parameters ℓ.

The codes for the first part are available at GuidoGazzani-ai/sigsde_calibration, while an implementation of the model for the joint calibration can be found in GuidoGazzani-ai/_jointcalib_sigsde or janka-moeller/joint_calib_SPX_VIX. Moreover another implementation of the expected signature of polynomial process is available at github.com/sarasvaluto/AffPolySig. The first repository refers to [11], the other three have been introduced in [13]. We additionally share a Colab notebook where the reader can benefit from an implementation of the building blocks of signature-based models as here introduced.

1.1 Notation

Let us first fix the notation used throughout the chapter. Recall the notions introduced in chapter "A Primer on the Signature Method in Machine Learning", in particular the *algebra of formal power series* which we will also refer to as *extended tensor algebra* and denote by $T((\mathbb{R}^d))$. For a multi-index $I := (i_1, \dots, i_n) \in \{1, \dots, d\}^n$ we set $|I| := n$. We also consider the empty index $I := \emptyset$ and set $|I| := 0$. For each index I we then define

$$I' := \begin{cases} (i_1, \dots, i_{|I|-1}) & \text{if } |I| \geq 2, \\ \emptyset & \text{if } |I| = 1, \\ 0 & \text{if } |I| = 0. \end{cases} \tag{3}$$

Similarly we set $I'' := (I')'$ if $|I| \neq 0$ and $I'' = 0$ if $|I| = 0$. Let us denote by $e_\emptyset$ the basis element corresponding to $(\mathbb{R}^d)^{\otimes 0}$, then each element $\mathbf{a} \in T((\mathbb{R}^d))$ can be expressed as

$$\mathbf{a} = \sum_{|I| \geq 0} \mathbf{a}_I e_I,$$

for a collection of $\mathbf{a}_I \in \mathbb{R}$. Moreover a vectorization of the elements of the truncated tensor algebra $T^{(n)}(\mathbb{R}^d)$ whose elements are of the form

$$\mathbf{u} = \sum_{0 \leq |I| \leq n} \mathbf{u}_I e_I$$

will prove useful. To this end, we introduce the isomorphism $\mathbf{vec} : T^{(n)}(\mathbb{R}^d) \to \mathbb{R}^{d_n}$ and an arbitrary but fixed injective labeling function $\mathscr{L} : \{I : |I| \leq n\} \longrightarrow \{1, \ldots, d_n\}$, such that

$$\mathbf{vec}(\mathbf{u}) := \sum_{|I| \leq n} e_{\mathscr{L}(I)} \mathbf{u}_I, \tag{4}$$

with $d_n := \frac{d^{n+1}-1}{d-1}$.

We stick to the convention of mathematical finance, that the process $S = (S_t)_{t \geq 0}$ denotes the price process. Therefore, we denote the signature of a process X by $\mathbb{X}$. Moreover, for a multi-index I and $0 \leq s \leq t$, we denote the corresponding increment of a component of the signature by its projection on e_I that is $\langle e_I, \mathbb{X}_{s,t} \rangle$. To make the connection to the notation introduced in chapter "A Primer on the Signature Method in Machine Learning": for a multi-index I and times $0 \leq s \leq t$ we denote the corresponding element of the signature of a process X by

$$\langle e_I, \mathbb{X}_{s,t} \rangle = S(X)^I_{s,t},$$

and write $\mathbb{X}_t := \mathbb{X}_{0,t}$. We emphasize again that from now on we will only use the letter S to denote the price process and never the signature of a process. Recall that for any $t \geq 0$ the signature of an $\mathbb{R}^d$-dimensional process $X = (X)_{t \geq 0}$ it holds that $\mathbb{X}_t \in T((\mathbb{R}^d))$ at any time $t \geq 0$. We denote its canonical projection to $T^{(n)}(\mathbb{R}^d)$ by $\mathbb{X}^n$ and refer to it as the *truncated signature*.

In this chapter, we assume the stochastic processes to be continuous semimartingales. Recall that the elements of the signature of a continuous semimartingale $X = (X)_{t \geq 0}$ can be defined recursively as

$$\langle e_\emptyset, \mathbb{X}_{s,t} \rangle := 1, \qquad \langle e_I, \mathbb{X}_{s,t} \rangle := \int_s^t \langle e_{I'}, \mathbb{X}_{s,r} \rangle \circ \mathrm{d}X^{i_n}_r,$$

for each $I = (i_1, \ldots, i_n)$, $I' = (i_1, \ldots, i_{n-1})$ and $0 \leq s \leq t$, where $\circ$ denotes the Stratonovich integral.

Definition 1.1 (Shuffles) Set $I = (i_1, \ldots, i_n)$ and $J = (j_1, \ldots, j_m)$ for some $n \in \mathbb{N}$ and $m > 0$. We define the shuffle $\sqcup\!\sqcup$ and half-shuffle $\widetilde{\sqcup\!\sqcup}$ as $e_I \sqcup\!\sqcup e_\emptyset := e_I$, $e_I \widetilde{\sqcup\!\sqcup} e_\emptyset := 0$, respectively and

$$e_I \sqcup\!\sqcup e_J := (e_{I'} \sqcup\!\sqcup e_J) \otimes e_{i_n} + (e_I \sqcup\!\sqcup e_{J'}) \otimes e_{j_m}, \quad e_I \widetilde{\sqcup\!\sqcup} e_J := (e_I \sqcup\!\sqcup e_{J'}) \otimes e_{j_m}.$$

Remark 1.2 Using shuffle and half-shuffle one can recover linear representations of more complex operations on the signature. In particular, for any pair of multi-indices I, J and each $t \geq 0$ it holds that (see e.g. [11])

$$\langle e_I, \mathbb{X}_t \rangle \langle e_J, \mathbb{X}_t \rangle = \langle e_I \sqcup\!\sqcup e_J, \mathbb{X}_t \rangle \quad \text{and} \quad \int_0^t \langle e_I, \mathbb{X}_s \rangle \circ \mathrm{d} \langle e_J, \mathbb{X}_s \rangle = \langle e_I \widetilde{\sqcup\!\sqcup} e_J, \mathbb{X}_t \rangle.$$

Choosing for example $I = (1)$ and $J = (2)$ we get that

$$e_{(1)} \sqcup\!\sqcup e_{(2)} = (e_\emptyset \sqcup\!\sqcup e_{(2)}) \otimes e_{(1)} + (e_{(1)} \sqcup\!\sqcup e_\emptyset) \otimes e_{(2)} = e_{(2,1)} + e_{(1,2)}$$

and $e_{(1)} \widetilde{\sqcup\!\sqcup} e_{(2)} = (e_{(1)} \sqcup\!\sqcup e_\emptyset) \otimes e_{(2)} = e_{(1,2)}$. Setting $X_0 = 0$ this corresponds to

$$\langle e_{(1)}, \mathbb{X}_t \rangle \langle e_{(2)}, \mathbb{X}_t \rangle = X_t^1 X_t^2 = \int_0^t X_s^2 \circ dX_s^1 + \int_0^t X_s^1 \circ dX_s^2 = \langle e_{(2,1)} + e_{(1,2)}, \mathbb{X}_t \rangle$$

and $\int_0^t \langle e_{(1)}, \mathbb{X}_s \rangle \circ d \langle e_{(2)}, \mathbb{X}_t \rangle = \int_0^t X_s^1 \circ dX_s^2 = \langle e_{(1,2)}, \mathbb{X}_t \rangle$, respectively.

2 Expected Signature of Polynomial Processes

Consider the $\mathbb{R}^d$-valued process $Y = (Y_t)_{t \geq 0}$ given by

$$\mathrm{d}Y_t = b(Y_t)\mathrm{d}t + \sqrt{a(Y_t)}\mathrm{d}W_t, \qquad Y_0 = y_0, \tag{5}$$

for some d-dimensional Brownian motion W and some maps $a : \mathbb{R}^d \to \mathbb{S}_+^d$ (where $\mathbb{S}_+^d$ denotes the set of symmetric positive definite $d \times d$ matrices) and $b : \mathbb{R}^d \to \mathbb{R}^d$. Set $\sigma(Y_t) := \sqrt{a(Y_t)}$. Assume then that $Y = (Y_t)_{t \geq 0}$ is a polynomial process as defined in the next definition and denote by $\mathbb{Y}$ the corresponding signature.

Definition 2.1 (Polynomial Process) A process $Y = (Y_t)_{t \geq 0}$ satisfying (5) is called polynomial process if a_{ij} is a polynomial of degree at most 2 and b_j is a polynomial of degree at most 1 for each $i, j \in \{1, \ldots, d\}$.

Various representations of the conditional expected signature of Y and analogous quantities, in particular for Brownian motion, can be found in [3, 5, 18, 32, 33]. We also refer to [15, 20] for the case of Lévy processes. Our approach aligns with [12] and is grounded in the classical theory of polynomial processes (see [9] and [19]). Although the framework of general signature SDEs considered in [12] requires results for infinite-dimensional stochastic processes (see, for example, [8, 14]), the current assumption of Y being a polynomial process allows us to remain within the finite-dimensional setting.

Lemma 2.2 *Let $(Y_t)_{t \geq 0}$ be a polynomial process satisfying* (5). *The corresponding drift and diffusion coefficients b and a can then be written as*

$$b_j(y) = b_j^c + \sum_{k=1}^d b_j^k y_k \qquad \textit{and} \qquad a_{ij}(y) = a_{ij}^c + \sum_{k=1}^d a_{ij}^k y_k + \sum_{k,h=1}^d a_{ij}^{kh} y_k y_h,$$

for some $b_j^c, b_j^k, a_{ij}^c, a_{ij}^k, a_{ij}^{kh} = a_{ij}^{hk} \in \mathbb{R}$, *with* $i, j = 1, \ldots, d$. *Moreover,*

$$b_j(Y_t) = \langle \mathbf{b}_j, \mathbb{Y}_t^1 \rangle \qquad \textit{and} \qquad a_{ij}(Y_t) = \langle \mathbf{a}_{ij}, \mathbb{Y}_t^2 \rangle$$

for

$$\mathbf{b}_j = \Big(b_j^c + \sum_{k=1}^d b_j^k Y_0^k\Big) e_\emptyset + \sum_{k=1}^d b_j^k e_k \qquad \textit{and}$$

$$\mathbf{a}_{ij} = \Big(a_{ij}^c + \sum_{k=1}^d a_{ij}^k Y_0^k + \sum_{k,h=1}^d a_{ij}^{kh} Y_0^k Y_0^h\Big) e_\emptyset + \sum_{k=1}^d \Big(a_{ij}^k + 2\sum_{h=1}^d a_{ij}^{kh} Y_0^h\Big) e_k$$
$$+ \sum_{k,h=1}^d a_{ij}^{kh} e_k \sqcup\!\sqcup\, e_h.$$

Note that the upper index on Y_0^k *and* Y_0^h *refers to the components of* Y *and not to powers.*

Proof The first representation follows by the definition of polynomial processes, according to which b and a are polynomials of degree at most 1 and 2, respectively. For the second representation it then suffices to note that $\langle e_\emptyset, \mathbb{Y}_t^1 \rangle = \langle e_\emptyset, \mathbb{Y}_t^2 \rangle = 1$, $\langle e_k, \mathbb{Y}_t^1 \rangle = \langle e_k, \mathbb{Y}_t^2 \rangle = (Y_t^k - Y_0^k)$, and $\langle e_k \sqcup\!\sqcup\, e_h, \mathbb{Y}_t^2 \rangle = (Y_t^k - Y_0^k)(Y_t^h - Y_0^h)$. □

Lemma 2.3 *Let* $(Y_t)_{t \geq 0}$ *be a polynomial process satisfying* (5). *Then the truncated signature* $(\mathbb{Y}_t^n)_{t \geq 0}$ *is a polynomial process and for each* $|I| \leq n$ *it holds that*

$$\langle e_I, \mathbb{Y}_t^n \rangle = \int_0^t \langle L e_I, \mathbb{Y}_s^n \rangle \mathrm{d}s + \int_0^t \langle e_{I'}, \mathbb{Y}_s^n \rangle \sigma_{i_{|I|}}(Y_s) \mathrm{d}W_s,$$

where $L : T((\mathbb{R}^d)) \to T((\mathbb{R}^d))$ *satisfies* $L(T^{(n)}(\mathbb{R}^d)) \subseteq T^{(n)}(\mathbb{R}^d)$ *and is given by*

$$L e_I = e_{I'} \sqcup\!\sqcup\, \mathbf{b}_{i_{|I|}} + \frac{1}{2} e_{I''} \sqcup\!\sqcup\, \mathbf{a}_{i_{|I|-1} i_{|I|}}, \tag{6}$$

for **b** *and* **a** *as in Lemma 2.2.*

Proof Let $\sigma_j(Y_t)$ denote the j-th row of $\sigma(Y_t)$. Using the definition of the signature, the Stratonovich integral and the shuffle property for each $|I| \geq 0$ we can compute

$$
\begin{aligned}
\langle e_I, \mathbb{Y}_t\rangle &= \int_0^t \langle e_{I'}, \mathbb{Y}_s\rangle \circ \mathrm{d}\langle e_{i_{|I|}}, Y_s\rangle \\
&= \int_0^t \langle e_{I'}, \mathbb{Y}_s\rangle \mathrm{d}\langle e_{i_{|I|}}, Y_s\rangle + \frac{1}{2}\int_0^t \langle e_{I''}, \mathbb{Y}_s\rangle \mathrm{d}[\langle e_{i_{|I|-1}}, Y_s\rangle, \langle e_{i_{|I|}}, Y_s\rangle]_s \\
&= \int_0^t \langle e_{I'}, \mathbb{Y}_s\rangle\langle \mathbf{b}_{i_{|I|}}, \mathbb{Y}_s\rangle \mathrm{d}s + \int_0^t \langle e_{I'}, \mathbb{Y}_s\rangle \sigma_{i_{|I|}}(Y_s)\mathrm{d}W_s \\
&\quad + \frac{1}{2}\int_0^t \langle e_{I''}, \mathbb{Y}_s\rangle\langle \mathbf{a}_{i_{|I|-1}, i_{|I|}}, \mathbb{Y}_s\rangle \mathrm{d}s \\
&= \int_0^t \langle e_{I'} \sqcup\!\sqcup\, \mathbf{b}_{i_{|I|}}, \mathbb{Y}_s\rangle \mathrm{d}s + \int_0^t \langle e_{I'}, \mathbb{Y}_s\rangle \sigma_{i_{|I|}}(Y_s)\mathrm{d}W_s \\
&\quad + \frac{1}{2}\int_0^t \langle e_{I''} \sqcup\!\sqcup\, \mathbf{a}_{i_{|I|-1} i_{|I|}}, \mathbb{Y}_s\rangle \mathrm{d}s \\
&= \int_0^t \langle L e_I, \mathbb{Y}_s\rangle \mathrm{d}s + \int_0^t \langle e_{I'}, \mathbb{Y}_s\rangle \sigma_{i_{|I|}}(Y_s)\mathrm{d}W_s.
\end{aligned}
$$

Fix $|I| \leq n$. Since $|I \sqcup\!\sqcup\, J| = |I| + |J|$ we get that $L(T^{(n)}(\mathbb{R}^d)) \subseteq T^{(n)}(\mathbb{R}^d)$ and thus that the corresponding drift components are linear maps in $\mathbb{Y}^n$. Similarly, fixing $|I|, |J| \leq n$ and using that $\mathbf{a}_{ij} = \sum_{|I|,|J|\leq 1} \lambda_{ij}^{IJ} e_I \sqcup\!\sqcup\, e_J$ for some $\lambda_{ij}^{IJ} \in \mathbb{R}$ we get that

$$
\begin{aligned}
\langle e_{I'}, \mathbb{Y}_s\rangle \sigma_{i_{|I|}}(Y_t)\big(\langle e_{J'}, \mathbb{Y}_s\rangle \sigma_{i_{|J|}}(Y_t)\big)^\top &= \langle e_{I'}, \mathbb{Y}_s\rangle\langle e_{J'}, \mathbb{Y}_s\rangle\langle \mathbf{a}_{i_{|I|} j_{|J|}}, \mathbb{Y}_s\rangle \\
&= \sum_{|H_1|,|H_2|\leq 1} \lambda_{i_{|I|} j_{|J|}}^{H_1 H_2} \langle e_{I'} \sqcup\!\sqcup\, e_{H_1}, \mathbb{Y}_s\rangle\langle e_{J'} \sqcup\!\sqcup\, e_{H_2}, \mathbb{Y}_s\rangle.
\end{aligned}
$$

This shows that the components of the corresponding diffusion matrix are polynomials of degree 2 in $\mathbb{Y}^n$. The polynomial property follows from Lemma 2.2 in [19]. □

As the linear operator L maps the finite-dimensional vector space $T^{(n)}(\mathbb{R}^d)$ into itself, it can be represented by a matrix.

Definition 2.4 We call the operator L defined in (6) *dual operator corresponding to* $\mathbb{Y}$ and denote by G the *d_n-dimensional matrix representative of* L. Explicitly, for each $|I| \leq n$ we consider coefficients $\eta_{IJ} \in \mathbb{R}$ such that

$$
L e_I = \sum_{|J|\leq n} \eta_{IJ} e_J,
$$

and fix a labelling injective function $\mathscr{L} : \{I : |I| \leq n\} \to \{1, \ldots, d_n\}$. The matrix $G \in \mathbb{R}^{d_n \times d_n}$ is then given by

$$G_{\mathscr{L}(I)\mathscr{L}(J)} := \eta_{IJ}. \tag{7}$$

Note that using the notation of (4), for each $\mathbf{u} \in T^{(n)}(\mathbb{R}^d)$ it holds

$$\mathbf{vec}(L\mathbf{u}) = G\mathbf{vec}(\mathbf{u}).$$

The results of the following theorem have been implemented and the code is available at the repository github.com/sarasvaluto/AffPolySig presented in [13].

Theorem 2.5 *Let $(Y_t)_{t\geq 0}$ be a polynomial process satisfying (5), G be the d_n-dimensional matrix representative of the dual operator corresponding to $\mathbb{Y}$, and $(\mathcal{F}_t)_{t\geq 0}$ be the filtration generated by $(Y_t)_{t\geq 0}$. Then for each $T, t \geq 0$ and each $|I| \leq n$ it holds*

$$\mathbb{E}[\langle e_I, \mathbb{Y}^n_{T+t}\rangle | \mathcal{F}_T] = \sum_{|J|\leq n} (e^{tG^\top})_{\mathscr{L}(I)\mathscr{L}(J)} \langle e_J, \mathbb{Y}^n_T\rangle,$$

or equivalently,

$$\mathbb{E}[\mathbf{vec}(\mathbb{Y}^n_{T+t}) | \mathcal{F}_T] = e^{tG^\top}\mathbf{vec}(\mathbb{Y}^n_T),$$

where $e^{(\cdot)}$ denotes the matrix exponential.

Proof Lemma 2.3 yields that $\mathbf{vec}(\mathbb{Y}^n)$ is a polynomial process and the claim follows by Theorem 3.1 in [19] for polynomials of degree 1. □

In the special case given by $Y = W$ the matrix G is nilpotent and we obtain a more explicit representation. Loosely speaking, the signature element $\langle e_I, \widehat{\mathbb{W}}_t\rangle$ has nonzero expectation only if all the indices $i_j \neq 0$ in I are grouped in blocks of even length.

Corollary 2.6 *Let W be a vector of d correlated Brownian motions with correlation matrix ρ. Consider a multi-index I admitting the representation*

$$e_I = e_0^{\otimes k_0} \otimes e_{J_1} \otimes e_0^{\otimes k_1} \otimes e_{J_2} \otimes \cdots \otimes e_0^{\otimes k_m}, \tag{8}$$

for some $J_i \in \{1, \ldots, d\}^{2h_i}$ and $h_i, k_i \in \mathbb{N}_0$. Then,

$$\mathbb{E}[\langle e_I, \widehat{\mathbb{W}}_t\rangle] = \frac{t^{\sum_{i=0}^m k_i + \sum_{i=1}^m h_i}}{(\sum_{i=0}^m k_i + \sum_{i=1}^m h_i)!} \left(\frac{1}{2}\right)^{\sum_{i=1}^m h_i} \prod_{i=1}^m \rho(J_i), \tag{9}$$

where $\rho(J) := \prod_{k=1}^{|J|/2} \rho_{j_{2k-1},j_{2k}}$. If I does not admit representation (8) then $\mathbb{E}[\langle e_I, \widehat{\mathbb{W}}_t\rangle] = 0$. Additionally, for each $s, t \geq 0$ it holds

$$\mathbb{E}[\langle e_I, \widehat{\mathbb{W}}_{s+t}\rangle|\mathcal{F}_s] = \sum_{e_{I_1}\otimes e_{I_2}=e_I} \langle e_{I_1}, \widehat{\mathbb{W}}_s\rangle \mathbb{E}[\langle e_{I_2}, \widehat{\mathbb{W}}_t\rangle].$$

Note that letting I_0 be the number of 0s in I and setting $I_{>0} := |I| - I_0$ we can rewrite (9) as

$$\mathbb{E}[\langle e_I, \widehat{\mathbb{W}}_t\rangle] = \frac{t^{I_0+I_{>0}/2}}{(I_0 + I_{>0}/2)!}\left(\frac{1}{2}\right)^{\frac{I_{>0}}{2}} \prod_{i=1}^{m} \rho(J_i).$$

Example 2.7 Note that in the special case where W is a d-dimensional (standard) Brownian motion we get that $\rho_{ij} = 1_{\{i=j\}}$. Corollary 2.6 then yields

$$\mathbb{E}[\langle e_I, \widehat{\mathbb{W}}_t\rangle] = \frac{t^{I_0+I_{>0}/2}}{(I_0 + I_{>0}/2)!}\left(\frac{1}{2}\right)^{\frac{I_{>0}}{2}},$$

for each I of the form $e_I = e_{j_0}^{\otimes k_0} \otimes \cdots \otimes e_{j_m}^{\otimes k_m}$ with k_i even whenever $j_i > 0$.

In order to better understand Theorem 2.5 we propose two examples. The first concerns a vector of correlated Brownian motions and constitutes the core of the proof of Corollary 2.6. The second one leads to the formulas that will be applied later on in the chapter.

Example 2.8 Let W be a d-dimensional Brownian motion and a process $(X_t)_{t\geq 0}$ given by

$$\mathrm{d}X_t = \sqrt{a(X_t)}\mathrm{d}W_t, \qquad X_0 = x_0,$$

for some matrix a of the form $a_{ij}(X_t) = \sigma^i\sigma^j\rho_{ij}$ and some constant $\sigma^i > 0$ and $\rho_{ij} \in [-1, 1]$. Observe that $\widehat{X}$ satisfies (5) in $d + 1$ dimensions for

$$b_j(\widehat{X}_t) = 1_{\{j=0\}} \qquad \text{and} \qquad a_{ij}(\widehat{X}_t) = \sigma^i\sigma^j\rho_{ij}1_{\{i,j\neq 0\}}.$$

The corresponding $\mathbf{b}$ and $\mathbf{a}$ are given by $\mathbf{b}_j = e_\emptyset 1_{\{j=0\}}$ and $\mathbf{a}_{ij} = e_\emptyset\sigma^i\sigma^j\rho_{ij}1_{\{i,j\neq 0\}}$ and we thus get

$$Le_I = e_{I'}1_{\{i_{|I|}=0\}} + \frac{1}{2}e_{I''}\sigma^{i_{|I|-1}}\sigma^{i_{|I|}}\rho_{i_{|I|-1}i_{|I|}}1_{\{i_{|I|-1},i_{|I|}\neq 0\}}.$$

This for instance implies that $L(e_1) = 0$, $L(e_I \otimes e_0) = e_I \sqcup\!\sqcup\, \mathbf{b}_0 + \frac{1}{2}e_{I'} \sqcup\!\sqcup\, \mathbf{a}_{i_{|I|}0} = e_I$, and $L(e_0 \otimes e_1 \otimes e_2) = \frac{1}{2}e_0\sigma^1\sigma^2\rho_{12}$. Letting $(\mathcal{F}_t)_{t\geq 0}$ be the filtration generated by $(\widehat{X}_t)_{t\geq 0}$, by Theorem 2.5 we can thus conclude that

$$\mathbb{E}[\mathbf{vec}(\widehat{\mathbb{X}}^n_{T+t})|\mathcal{F}_T] = e^{tG^\top}\mathbf{vec}(\widehat{\mathbb{X}}^n_T), \tag{10}$$

or equivalently,

$$\mathbb{E}[\langle e_I, \widehat{\mathbb{X}}^n_{T+t}\rangle | \mathcal{F}_T] = \sum_{|J|\leq n} (e^{tG^\top})_{\mathcal{L}(I)\mathcal{L}(J)} \langle e_J, \widehat{\mathbb{X}}^n_T \rangle, \tag{11}$$

where G denotes the $(d+1)_n$-dimensional matrix representative of L.

Example 2.9 In a similar setting as of the previous example let W be a d-dimensional Brownian motion and set $Z_t := (X_t, B_t)$ for an another one-dimensional Brownian motion $(B_t)_{t\geq 0}$. Suppose that

$$\mathrm{d}X_t = \mathrm{diag}(\kappa)(\theta - X_t)\mathrm{d}t + \sqrt{a(X_t)}\mathrm{d}W_t, \qquad X_0 = x_0,$$

where $\kappa, \theta \in \mathbb{R}^d$ and $\mathrm{diag}(\kappa)$ denotes a diagonal matrix consisting of the components of κ and a is as before. Observe that $\widehat{Z}$ satisfies (5) in $d+2$ dimensions for

$$b_j(\widehat{Z}_t) = 1_{\{j=0\}} + \kappa^j(\theta^j - \widehat{Z}^j_t)1_{\{j\neq 0\}} \qquad \text{and} \qquad a_{ij}(\widehat{Z}_t) = \sigma^i\sigma^j\rho_{ij}1_{\{i,j\neq 0\}},$$

where $\kappa^{d+1} := 0$, $\sigma^{d+1} := 1$, and $\rho_{j(d+1)}$ is the correlation between X^j and B. The corresponding $\mathbf{b}$ is then given by $\mathbf{b}_j = e_\emptyset(1_{\{j=0\}} + \kappa^j(\theta^j - \widehat{Z}^j_0)1_{\{j\neq 0\}}) - e_j\kappa^j 1_{\{j\neq 0\}}$ and we thus get

$$\begin{aligned} Le_I = {} & e_{I'}(1_{\{i_{|I|}=0\}} + \kappa^{i_{|I|}}(\theta^{i_{|I|}} - \widehat{Z}^{i_{|I|}}_0)1_{\{i_{|I|}\neq 0\}}) - (e_{I'} \sqcup\!\sqcup\, e_{i_{|I|}})\kappa^{i_{|I|}}1_{\{i_{|I|}\neq 0\}} \\ & + \frac{1}{2}e_{I''}\sigma^{i_{|I|-1}}\sigma^{i_{|I|}}\rho_{i_{|I|-1}i_{|I|}}1_{\{i_{|I|-1},i_{|I|}\neq 0\}}. \end{aligned}$$

3 Linear Signature Models

Let us now introduce a framework for signature-based asset price models. We fix a time horizon $T > 0$, consider a d-dimensional continuous semimartingale $X := (X^1, \ldots, X^d)$ and its matrix-valued quadratic covariation $[X]$. We suppose that X encodes all the information to represent the market asset S. Recall that for notational convenience we assume that S is one-dimensional.

Furthermore, we suppose that X has some tractability properties which are made precise below and which are for instance satisfied by a d-dimensional Brownian motion, but more generally also all polynomial diffusion processes.

3.1 Definition and First Properties

We regard the continuous semimartingale X as the *primary process*, being the main modeling building block. Its time extension is denoted by

$$\widehat{X}_t := (t, X_t).$$

We use e_0 for the component of $\widehat{X}$ corresponding to time and e_k for its component corresponding to X^k. Its signature is denoted by $\widehat{\mathbb{X}}_t$.

Throughout the section we then make the following standing assumption about diffusion matrix of X.

For all $i, j \in \{1, \ldots, d\}$ it holds

$$\mathrm{d}[X^i, X^j]_t = \sum_{|I| \leq m} a_{ij}^I \langle e_I, \widehat{\mathbb{X}}_t \rangle \mathrm{d}t \tag{12}$$

for some $m \in \mathbb{N}$. Observe that this assumption is always satisfied when X is given by a polynomial process.

Remark 3.1 It could be of interest to consider the extension $\widehat{X}_t := (t, X_t, [X]_t)$, which includes the d^2-dimensional process given by the quadratic covariation of the primary process X. In this case the components of the diffusion matrix do not need to be linear functions of the time extended signature, but could for instance be general path-dependent functionals. Moreover, the map $t \mapsto [X]_t$ does not need to be absolutely continuous with respect to the Lebesgue measure, hence X does not need to be an Itô-semimartingale. For ease of exposition we focus here on the simple time extension and refer to [11] for the slightly more general setup.

Our goal consists in approximating the dynamics of S with a *signature model.*

Definition 3.2 (Signature Model) A *signature model* is a stochastic process of the form

$$S_n(\ell)_t := \ell_\emptyset + \sum_{0 < |I| \leq n} \ell_I \langle e_I, \widehat{\mathbb{X}}_t \rangle, \tag{13}$$

where $n \in \mathbb{N}$ and $\ell := \{\ell_\emptyset, \ell_I : 0 < |I| \leq n\}$.

Remark 3.3 By Proposition 3.4 below the class of Sig-SDEs models considered in [37] can be embedded in our framework by choosing a properly extended one-dimensional Brownian motion as primary process.

In the following we list several important properties which make signature models a tractable framework for stochastic finance.

- For each $t \in [0, T]$, $S_n(\ell)_t$ is linear in $\widehat{\mathbb{X}}_t$. This in particular implies that having precomputed $\widehat{\mathbb{X}}$ an update of the parameters ℓ boils down to evaluating the scalar product in (13).
- The quadratic variation of processes of form (13) is again of the form (13).
- Representation (13) remains invariant under polynomial transformations.
- Itô-integrals of processes of form (13) with respect to processes of form (13) are again processes of form (13). This includes in particular the signature $\widehat{\mathbb{S}}_n(\ell)$ of $\widehat{S}_n(\ell)_t := (t, S_n(\ell)_t)$ or expressions of the form $\int_0^{\cdot} S_n(\ell)_s \mathrm{d}X_s^i$.
- The latter point implies that the expected signature of $\widehat{S}_n(\ell)$ can be expressed as

$$\mathbb{E}[\langle e_J, \widehat{\mathbb{S}}_n(\ell)_t\rangle] = P_J(\ell, \mathbb{E}[\widehat{\mathbb{X}}_t]),$$

  for some P_J such that $P_J(\cdot, \mathbb{E}[\widehat{\mathbb{X}}_t])$ is a polynomial of degree $|J|$ and $P_J(\ell, \cdot)$ is a linear map for each ℓ (see Theorem 3.9 and Remark 3.11 for more details).
- Due to the universal approximation theorem (see e.g. [7, 15]) this yields approximations

$$\mathbb{E}\big[f\big((\widehat{\mathbb{S}}_n^2(\ell)_t)_{t\in[0,T]}\big)\big] \approx P_f(\ell, \mathbb{E}[\widehat{\mathbb{X}}_T])$$

  for each map f, which is continuous in a suitable sense, where P_f is given by a finite linear combination of maps P_J as above and with $\widehat{\mathbb{S}}^2$ we denote here the second order lift. This includes representations for

$$\mathbb{E}\big[\tilde{f}(\widehat{S}_n(\ell)_T)\big] \qquad \text{and} \qquad \mathbb{E}\left[\tilde{f}\left(\int_0^T \widehat{S}_n(\ell)_t \mathrm{d}t\right)\right]$$

  for maps $\tilde{f}$ being payoff functions and where the expectation is taken with respect to a risk-neutral measure.

We now prove several representation results for the signature model of form (13).

Proposition 3.4 *Fix $n \in \mathbb{N}$. Then, there is a one to one correspondence between representations of the form* (13) *and representations of the form*

$$S_n(\ell)_t = \ell_\emptyset + \int_0^t \Big(\ell_\emptyset^0 + \sum_{0<|I|\le n-1} \ell_I^0 \langle e_I, \widehat{\mathbb{X}}_s\rangle\Big)\mathrm{d}s$$
$$+ \sum_{k=1}^d \int_0^t \Big(\ell_\emptyset^k + \sum_{0<|I|\le n-1} \ell_I^k \langle e_I, \widehat{\mathbb{X}}_s\rangle\Big)\mathrm{d}X_s^k,$$

for $\ell := \{\ell_\emptyset, \ell_I^k : |I| \le n-1$ and $k \in \{0, \dots, d\}\}$.

Before presenting the proof of let us formulate the following technical lemma.

Lemma 3.5 *For each $|I| > 0$ it holds*

$$\int_0^t \langle e_I, \widehat{\mathbb{X}}_s \rangle \mathrm{d}s = \langle e_I \otimes e_0, \widehat{\mathbb{X}}_t \rangle, \qquad \textit{and} \qquad \int_0^t \langle e_I, \widehat{\mathbb{X}}_s \rangle \mathrm{d}X_s^k = \langle \tilde{e}_I^k, \widehat{\mathbb{X}}_t \rangle,$$

where

$$\tilde{e}_I^k := e_I \otimes e_k - \sum_{|J| \le m} \frac{a^J_{i_{|I|}k}}{2} (e_{I'} \sqcup\mathchoice{\mkern-7mu}{\mkern-7mu}{\mkern-3.2mu}{\mkern-3.8mu}\sqcup e_J) \otimes e_0.$$

Proof The representations of $\int_0^t \langle e_I, \widehat{\mathbb{X}}_s \rangle \mathrm{d}s$ follows by the definition of signature. We proceed with the proof of the representation of $\int_0^t \langle e_I, \widehat{\mathbb{X}}_s \rangle \mathrm{d}X_s^k$. For $I = \emptyset$ the claim follows. By the definition of the Stratonovich integral,

$$\begin{aligned}
\int_0^t \langle e_I, \widehat{\mathbb{X}}_s \rangle \mathrm{d}X_s^k &= \int_0^t \langle e_I, \widehat{\mathbb{X}}_s \rangle \circ \mathrm{d}X_s^k - \frac{1}{2} [\langle e_I, \widehat{\mathbb{X}} \rangle, X^k]_t \\
&= \langle e_I \otimes e_k, \widehat{\mathbb{X}}_t \rangle - \frac{1}{2} \int_0^t \langle e_{I'}, \widehat{\mathbb{X}}_s \rangle \mathrm{d}[X^{i_{|I|}}, X^k]_s.
\end{aligned}$$

The shuffle property and the definition of the signature yield then

$$\begin{aligned}
\int_0^t \langle e_I, \widehat{\mathbb{X}}_s \rangle \mathrm{d}[X^{k_1}, X^{k_2}]_s &= \int_0^t \sum_{|J| \le m} a^J_{k_1 k_2} \langle e_I, \widehat{\mathbb{X}}_s \rangle \langle e_J, \widehat{\mathbb{X}}_s \rangle \mathrm{d}s \\
&= \sum_{|J| \le m} a^J_{k_1 k_2} \langle (e_I \sqcup\mathchoice{\mkern-7mu}{\mkern-7mu}{\mkern-3.2mu}{\mkern-3.8mu}\sqcup e_J) \otimes e_0, \widehat{\mathbb{X}}_t \rangle,
\end{aligned}$$

and the claim follows. □

Suppose that X is a vector of correlated Brownian motions with correlation matrix ρ. Then, the transformation introduced in Lemma 3.5 reads

$$\tilde{e}_\emptyset^k = e_k \qquad \text{and} \qquad \tilde{e}_I^k = e_I \otimes e_k - \frac{\rho_{i_{|I|},k}}{2} \mathbf{1}_{\{i_{|I|} \neq 0\}} e_{I'} \otimes e_0,$$

for each $|I| > 0$.

We are now ready to provide the proof of Proposition 3.4.

Proof of Proposition 3.4 Let $S_n(\ell)$ be as in the statement of the proposition. By Lemma 3.5 it holds

$$\begin{aligned}
S_n(\ell)_t = \ell_\emptyset &+ \Big(\ell_\emptyset^0 \langle e_0, \widehat{\mathbb{X}}_t \rangle + \sum_{0 < |I| \le n-1} \ell_I^0 \langle e_I \otimes e_0, \widehat{\mathbb{X}}_t \rangle \Big) \\
&+ \sum_{k=1}^d \Big(\ell_\emptyset^k \langle \tilde{e}_\emptyset^k, \widehat{\mathbb{X}}_t \rangle + \sum_{0 < |I| \le n-1} \ell_I^k \langle \tilde{e}_I^k, \widehat{\mathbb{X}}_t \rangle \Big),
\end{aligned}$$

which is of the form given by (13). Conversely, by Lemma 3.5 we also have

$$\langle e_I \otimes e_0, \widehat{\mathbb{X}}_t \rangle = \int_0^t \langle e_I, \widehat{\mathbb{X}}_s \rangle \mathrm{d}s,$$

$$\langle e_I \otimes e_k, \widehat{\mathbb{X}}_t \rangle = \int_0^t \langle e_I, \widehat{\mathbb{X}}_s \rangle \mathrm{d}X_s^k + \frac{1}{2} \sum_{|J| \leq m} \int_0^t \langle a^J_{i_{|I|}k}(e_{I'} \sqcup\!\sqcup e_J), \widehat{\mathbb{X}}_s \rangle \mathrm{d}s,$$

yielding the claim. □

3.2 Absence of Arbitrage and Universality Properties

The results obtained in the representation above will enable us to establish conditions ensuring absence of arbitrage. More precisely, we shall make the assumption that the principle of "no free lunch with vanishing risk" [16] holds. This assumption is equivalent to the existence of an equivalent local martingale measure $\mathbb{Q}$ as $S_n(\ell)$ has continuous sample paths. Throughout this section, we will assume zero interest rates and that the asset has already been discounted. To formulate a precise condition for the absence of arbitrage, we rely on the following corollary, which is a direct consequence of Proposition 3.4 and Lemma 3.5.

Corollary 3.6 *Suppose that X is a local martingale. Then $S_n(\ell)$ is a local martingale if and only if it admits a representation of the form*

$$S_n(\ell)_t = \ell_\emptyset + \sum_{k=1}^{D} \left(\ell_\emptyset^k \langle e_k, \widehat{\mathbb{X}}_t \rangle + \sum_{0<|I|\leq n-1} \ell_I^k \langle \tilde{e}_I^k, \widehat{\mathbb{X}}_t \rangle \right), \tag{14}$$

for some $D \in \{1, \ldots, d\}$ and $\ell := \{\ell_\emptyset, \ell_I^k : 0 \leq |I| \leq n-1 \text{ and } k \in \{1, \ldots, d\}\}$.

Proof Since X is a local martingale, $S_n(\ell)$ in the representation of Proposition 3.4 is a local martingale if and only if all integrals with respect to time and the quadratic variation process vanish. This means that $S_n(\ell)$ is of form

$$S_n(\ell) = \ell_\emptyset + \sum_{k=1}^{d} \int_0^t \left(\ell_\emptyset^k + \sum_{0<|I|\leq n-1} \ell_I^k \langle e_I, \widehat{\mathbb{X}}_s \rangle \right) \mathrm{d}X_s^k \tag{15}$$

and Lemma 3.5 yields the assertion. □

Let us now formulate sufficient no-arbitrage conditions.

Corollary 3.7 *Suppose that there is an equivalent measure $\mathbb{Q} \sim \mathbb{P}$ such that X is a local $\mathbb{Q}$-martingale. Then the following holds.*

(i) The model $S_n(\ell)$ is free of arbitrage if it admits a representation as of (14).

(ii) If $\mathbb{Q}$ is an equivalent local martingale measure for $S_n(\ell)$, then $S_n(\ell)$ is necessarily of form (14).

Proof The first assertion is a direct consequence of Corollary 3.6, as form (14) implies that $S_n(\ell)$ is a local $\mathbb{Q}$-martingale and thus $\mathbb{Q}$ is an equivalent local martingale measure. The assumption of the second assertion implies that $S_n(\ell)$ is a local martingale under $\mathbb{Q}$, whence by Corollary 3.6 it has to be of form (14). □

Remark 3.8

(i) It is important to recognize that $S_n(\ell)$ could be free of arbitrage without being of the form in (14). This scenario arises when the primary process X is not a local martingale under any of the equivalent local martingale measures.
(ii) Additionally, note that if $S_n(\ell)$ is of the form (14) and a local martingale under $\mathbb{Q}$, it does not necessarily imply that X is a local $\mathbb{Q}$-martingale. This situation arises when the drift terms in (15) cancel each other out. It is worth mentioning that in the one-dimensional case ($d = 1$), such a scenario is impossible, and consequently, X is inevitably a local $\mathbb{Q}$-martingale. Moreover, $\mathbb{Q}$ is unique and the model thus complete.

3.3 The Expected Signature of $S_n(\ell)$

For the pricing of so-called *sig-payoffs* discussed in Sect. 3.4, it is crucial to be able to compute the expected signature of $\widehat{S}_n(\ell)_t = (t, S_n(\ell)_t)$. In this section, we present formulas that link this computation to the calculation of the expected signature of $\widehat{X}_t$, which can often be explicitly computed. Specifically, when X is a Brownian motion, this is a well-known result (refer to, e.g., [18]). For the case where X is a polynomial process, the computation can be carried out using polynomial processes techniques (see Sect. 2). In a general setting, it can be computed by solving an infinite-dimensional system of linear PDEs corresponding to the Kolmogorov forward equation of the signature process (see [36]). For a comprehensive treatment of signature cumulants, i.e., the logarithm of the expected signature, we direct the readers to [21].

Theorem 3.9 *Fix $n \in \mathbb{N}$, a multi-index J, and $D \in \{1, \dots, d\}$ and denote by $\widehat{\mathbb{S}}_n(\ell)_t$ the signature of $\widehat{S}_n(\ell)_t$. Let e_0 be the component of $\widehat{S}_n(\ell)$ corresponding to time and e_1 its component corresponding to $S_n(\ell)$. Define $e(\emptyset, \ell) := \tilde{e}(\emptyset, \ell) := e_\emptyset$ and*

$$e(J, \ell) = \widetilde{\sqcup\!\sqcup}_{i=1}^{|J|} \Big(e_0 \mathbf{1}_{\{j_i = 0\}} + \Big(\sum_{0<|I|\leq n} \ell_I e_I \Big) \mathbf{1}_{\{j_i = 1\}} \Big),$$

$$\tilde{e}(J, \ell) = \widetilde{\sqcup\!\sqcup}_{i=1}^{|J|} \Big(e_0 \mathbf{1}_{\{j_i = 0\}} + \Big(\sum_{k=1}^{D} \Big(\ell_\emptyset^k e_k + \sum_{0<|I|\leq n-1} \ell_I^k \tilde{e}_I^k \Big) \Big) \mathbf{1}_{\{j_i = 1\}} \Big),$$

for $|J| > 0$ with $\tilde{ш}$ the half-shuffle being introduced in Definition 1.1. Then the following representations hold.

- $\langle e_J, \widehat{\mathbb{S}}_n(\ell)_t\rangle = \langle e(J,\ell), \widehat{\mathbb{X}}_t\rangle$, *if $S_n(\ell)$ is given by* (13), *and*
- $\langle e_J, \widehat{\mathbb{S}}_n(\ell)_t\rangle = \langle \tilde{e}(J,\ell), \widehat{\mathbb{X}}_t\rangle$, *if $S_n(\ell)$ is given as in Corollary 3.6.*

Proof We proceed by induction to prove the claim. Fix $S_n(\ell)$ as in (13). For $J = \emptyset$ the claim follows by the definition of signature. Suppose the claim holds true for each J such that $|J| = m - 1$, and fix J with $|J| \leq m$. Then

$$\begin{aligned}
\langle e_J, \widehat{\mathbb{S}}_n(\ell)_t\rangle &= \int_0^t \langle e_{J'}, \widehat{\mathbb{S}}_n(\ell)_s\rangle \circ \mathrm{d}\langle e_{j_m}, \widehat{S}_n(\ell)_s\rangle \\
&= \int_0^t \langle e_{J'}, \widehat{\mathbb{S}}_n(\ell)_s\rangle \circ \mathrm{d}\langle e_0 \mathbf{1}_{\{j_m=0\}} + \Big(\sum_{0<|I|\leq n} \ell_I e_I\Big)\mathbf{1}_{\{j_m=1\}}, \widehat{\mathbb{X}}_s\rangle \\
&= \mathbf{1}_{\{j_m=0\}} \int_0^t \langle e_{J'}, \widehat{\mathbb{S}}_n(\ell)_s\rangle \circ \mathrm{d}\langle e_0, \widehat{\mathbb{X}}_s\rangle \\
&\quad + \mathbf{1}_{\{j_m=1\}} \sum_{0<|I|\leq n} \ell_I \int_0^t \langle e_{J'}, \widehat{\mathbb{S}}_n(\ell)_s\rangle \circ \mathrm{d}\langle e_I, \widehat{\mathbb{X}}_s\rangle.
\end{aligned}$$

The induction hypothesis and Remark 1.2 yield the first claim and the second one is analogous. □

Remark 3.10 The linear combinations of multiindices given by $e(J,\ell)$ might look very abstract and we thus provide some additional details. The intuition behind its construction is as follows. For each 0 in J we are integrating with respect to time and we thus add $\tilde{ш} e_0$ to the current linear combination of multiindices. For each 1 in J we are instead integrating with respect to $S_n(\ell)$ and we thus need to add $\tilde{ш}(\sum_{0<|I|\leq n} \ell_I e_I)$. Since integrating with respect to $S_n(\ell)$ corresponds to integrating with respect to $S_n(\ell) - S_n(\ell)_0$ the term $\ell_\emptyset e_\emptyset$ can be omitted. Choosing $J = (0,1)$ and $J = (1,1)$ we would for instance get

$$\begin{aligned}
e((0,1),\ell) &= e_0 \tilde{ш}\Big(\sum_{0<|I|\leq n} \ell_I e_I\Big) = \sum_{0<|I|\leq n} \ell_I e_0 \tilde{ш} e_I, \\
e((1,1),\ell) &= \Big(\sum_{0<|I|\leq n} \ell_I e_I\Big) \tilde{ш}\Big(\sum_{0<|I|\leq n} \ell_I e_I\Big) = \sum_{0<|I_1|,|I_2|\leq n} \ell_{I_1}\ell_{I_2} e_{I_1} \tilde{ш} e_{I_2}.
\end{aligned}$$

The intuition behind $\tilde{e}(J,\ell)$ is similar.

Remark 3.11

(i) Let $S_n(\ell)$ be as in (13) and observe that

$$e(J,\ell) = \widetilde{\sqcup\!\sqcup}_{i=1}^{|J|} \sum_{0<|I|\le n} \Big(e_I \mathbf{1}_{\{j_i=0\}} \mathbf{1}_{\{I=(0)\}} + \ell_I e_I \mathbf{1}_{\{j_i=1\}} \Big).$$

Setting $c(j, I, \ell) := \mathbf{1}_{\{j=0\}}\mathbf{1}_{\{I=(0)\}} + \ell_I \mathbf{1}_{\{j=1\}}$ we thus obtain that

$$\begin{aligned} \mathbb{E}[\langle e_J, \widehat{\mathbb{S}}_n(\ell)_t\rangle] &= \mathbb{E}[\langle e(J,\ell), \widehat{\mathbb{X}}_t\rangle] \\ &= \sum_{|I_1|,\ldots,|I_{|J|}|=1}^{n} \mathbb{E}[\langle \widetilde{\sqcup\!\sqcup}_{i=1}^{|J|} e_{I_i}, \widehat{\mathbb{X}}_t\rangle] \prod_{i=1}^{|J|} c(j_i, I_i, \ell). \end{aligned}$$

Although this representation may seem intricate at first glance, it is, in fact, quite convenient as the expectations $\mathbb{E}[\langle \widetilde{\sqcup\!\sqcup}_{i=1}^{|J|} e_{I_i}, \widehat{\mathbb{X}}_t\rangle]$ can be computed just once in advance. Since $c(j, I, \cdot)$ is affine, we also immediately obtain that the map

$$(\ell, \mathbb{E}[\widehat{\mathbb{X}}_t]) \mapsto P_J(\ell, \mathbb{E}[\widehat{\mathbb{X}}_t]) := \mathbb{E}[\langle e_J, \widehat{\mathbb{S}}_n(\ell)_t\rangle]$$

is polynomial of degree $|J|$ in its first argument and linear in the second one.

(ii) Similarly, let $S_n(\ell)$ be as in Corollary 3.6, set $\tilde{e}^k_\emptyset := e_k$, and observe that

$$\tilde{e}(J,\ell) = \widetilde{\sqcup\!\sqcup}_{i=1}^{|J|} \sum_{k=1}^{D} \sum_{0\le|I|\le n-1} \Big(\tilde{e}^k_I \mathbf{1}_{\{j_i=0\}} \mathbf{1}_{\{I=(0)\}} \mathbf{1}_{\{k=1\}} + \ell^k_I \tilde{e}^k_I \mathbf{1}_{\{j_i=1\}} \Big).$$

Setting $\tilde{c}(j, I, k, \ell) := \mathbf{1}_{\{j=0\}}\mathbf{1}_{\{I=(0)\}}\mathbf{1}_{\{k=1\}} + \ell^k_I \mathbf{1}_{\{j=1\}}$ yields

$$\mathbb{E}[\langle e_J, \widehat{\mathbb{S}}_n(\ell)_t\rangle] = \sum_{k_1,\ldots,k_{|J|}=1}^{D} \sum_{|I_1|,\ldots,|I_{|J|}|=0}^{n-1} \mathbb{E}[\langle \widetilde{\sqcup\!\sqcup}_{i=1}^{|J|} \tilde{e}^{k_i}_{I_i}, \widehat{\mathbb{X}}_t\rangle] \prod_{i=1}^{|J|} \tilde{c}(j_i, I_i, k_i, \ell).$$

We obtain that the map

$$(\ell, \mathbb{E}[\widehat{\mathbb{X}}_t]) \mapsto \widetilde{P}_J(\ell, \mathbb{E}[\widehat{\mathbb{X}}_t]) := \mathbb{E}[\langle e_J, \widehat{\mathbb{S}}_n(\ell)_t\rangle]$$

is polynomial of degree $|J|$ in its first argument and linear in the second one.

The expressions above also provide a formula of the variance of $S_n(\ell)$.

Corollary 3.12 *Let $S_n(\ell)$ be as in Corollary 3.6 and assume that it is a true martingale. Then*

$$\mathrm{Var}(S_n(\ell)_t) = 2 \sum_{k_1,k_2=1}^{D} \sum_{|I_1|,|I_2|=1}^{n-1} \mathbb{E}[\langle \tilde{e}^{k_1}_{I_1} \sqcup\!\sqcup\, \tilde{e}^{k_2}_{I_2}, \widehat{\mathbb{X}}_t\rangle] \ell_{I_1}\ell_{I_2}. \tag{16}$$

Proof The martingale property guarantees that $\mathbb{E}[S_n(\ell)_t] = S_n(\ell)_0$ and hence, by the shuffle product, $\mathrm{Var}(S_n(\ell)_t) = 2\mathbb{E}[\langle e_1 \otimes e_1, \widehat{\mathbb{S}}_n(\ell)_t\rangle]$. By Remark 3.11 we can conclude that

$$\mathrm{Var}(S_n(\ell)_t) = 2 \sum_{k_1,k_2=1}^{D} \sum_{|I_1|,|I_2|=1}^{n-1} \mathbb{E}[\langle \tilde{e}_{I_1}^{k_1} \sqcup\!\sqcup \tilde{e}_{I_2}^{k_2}, \widehat{\mathbb{X}}_t\rangle]\tilde{c}(1, I_1, k_1, \ell)\tilde{c}(1, I_2, k_2, \ell)$$

and the claim follows. □

3.4 Pricing of Sig-Payoffs

We recall here the notion of *sig-payoffs* as introduced in [34].

Definition 3.13 (Sig-Payoff) Suppose that the price process S is given by a continuous semimartingale. A payoff $F : \Omega \to \mathbb{R}$ is said to be a sig-payoff if there exists $m \in \mathbb{N}$, and $f := \{f_\emptyset, f_J : 0 < |J| \leq m\}$, such that

$$F := f_\emptyset + \sum_{0<|J|\leq m} f_J \langle e_J, \widehat{\mathbb{S}}_T\rangle,$$

where $\widehat{\mathbb{S}}$ denotes the signature of $\widehat{S}_t = (t, S_t)$.

Example 3.14 Let $K > 0$ be a strike price and $T > 0$ a maturity time. Then, Asian forwards written on S are payoffs of the form

$$\frac{1}{T}\int_0^T S_t \mathrm{d}t - K = \frac{1}{T}\int_0^T (S_t - S_0)\mathrm{d}t - K + S_0 = \frac{1}{T}\langle e_1 \otimes e_0, \widehat{\mathbb{S}}_T\rangle + (K - S_0)\langle e_\emptyset, \widehat{\mathbb{S}}_T\rangle,$$

and are thus sig-payoffs.

While standard vanilla derivatives like call and put options do not fall under the category of sig-payoffs, approximate sig-payoffs can still serve as efficient control variates in Monte Carlo pricing, as explained in Sect. 3.5.2.

In this section, we consistently denote $\widehat{S}_t$ as (t, S_t) and focus on pricing sig-payoffs when S follows a signature model, as specified in the subsequent corollary.

Corollary 3.15 *Let the dynamics of $S_n(\ell)$ under a local martingale measure $\mathbb{Q}$ be specified as in Corollary 3.6. Consider a sig-payoff*

$$F = f_\emptyset + \sum_{0<|J|\leq m} f_J \langle e_J, \widehat{\mathbb{S}}_n(\ell)_T\rangle.$$

Then, using the notation of Remark 3.11 we can write the corresponding price as

$$\begin{aligned}\mathbb{E}_{\mathbb{Q}}[F] &= f_{\emptyset} + \sum_{0<|J|\leq m} f_J \widetilde{P}_J(\ell, \mathbb{E}_{\mathbb{Q}}[\widehat{\mathbb{X}}_T]) \\ &= f_{\emptyset} + \sum_{|J|=1}^{m} \sum_{k_1,\dots,k_{|J|}=1}^{D} \sum_{|I_1|,\dots,|I_{|J|}|=0}^{n} \mathbb{E}_{\mathbb{Q}}[\langle \widetilde{\sqcup\!\sqcup}_{i=1}^{|J|} \tilde{e}_{I_i}^{k_i}, \widehat{\mathbb{X}}_T\rangle] f_J \prod_{i=1}^{|J|} \tilde{c}(j_i, I_i, k_i, \ell),\end{aligned} \tag{17}$$

for $\tilde{c}(j, I, k, \ell) := \mathbf{1}_{\{j=0\}}\mathbf{1}_{\{I=(0)\}}\mathbf{1}_{\{k=1\}} + \ell_I^k \mathbf{1}_{\{j=1\}}$.

Proof This is a direct consequence of Theorem 3.9 and Remark 3.11. □

Remark 3.16

(i) Expression (17) admits also a second representation that turns out to be useful for coding:

$$\begin{aligned}\mathbb{E}_{\mathbb{Q}}[F] = f_{\emptyset} + \sum_{|J|=1}^{m} \sum_{k_1,\dots,k_{|J|}=1}^{D} \sum_{I_1,\dots,I_{|J|}\in\mathcal{I}} \mathbb{E}_{\mathbb{Q}}[\langle \widetilde{\sqcup\!\sqcup}_{i=1}^{|J|} \tilde{e}_{I_i}^{k_i}, \widehat{\mathbb{X}}_T\rangle] \\ \times\ f_{(\mathbf{1}_{\{I_1\neq I^t, k_1\neq 0\}},\dots,\mathbf{1}_{\{I_{|J|}\neq I^t, k_{|J|}\neq 0\}})} \prod_{i=1}^{|J|} \Big(\mathbf{1}_{\{I_i=I^t, k_i=0\}} + \ell_{I_i}^{k} \mathbf{1}_{\{I_i\neq I^t, k_i\neq 0\}}\Big).\end{aligned}$$

(ii) With a similar procedure it is also possible to provide a representation of $\mathrm{Var}_{\mathbb{Q}}(F)$. By the shuffle product we know that

$$(F - f_{\emptyset})^2 = \sum_{|J_1|,|J_2|=1}^{m} f_{J_1} f_{J_2} \langle e_{J_1} \sqcup\!\sqcup\, e_{J_2}, \widehat{\mathbb{S}}_n(\ell)_T\rangle.$$

Setting $\widetilde{P}_{I_1 \sqcup\!\sqcup I_2} := \sum_{i=1}^{K} \widetilde{P}_{J_i}$ for J_i and K satisfying $e_{I_1} \sqcup\!\sqcup\, e_{I_2} = \sum_{i=1}^{K} e_{J_i}$ we thus obtain

$$\begin{aligned}\mathrm{Var}_{\mathbb{Q}}(F) &= \mathbb{E}_{\mathbb{Q}}[(F - f_{\emptyset})^2] - \mathbb{E}_{\mathbb{Q}}[F - f_{\emptyset}]^2 \\ &= \sum_{|J_1|,|J_2|=1}^{m} f_{J_1} f_{J_2} \Big(\widetilde{P}_{J_1\sqcup\!\sqcup J_2}(\ell, \mathbb{E}_{\mathbb{Q}}[\widehat{\mathbb{X}}_T]) - \widetilde{P}_{J_1}(\ell, \mathbb{E}_{\mathbb{Q}}[\widehat{\mathbb{X}}_T]) \widetilde{P}_{J_2}(\ell, \mathbb{E}_{\mathbb{Q}}[\widehat{\mathbb{X}}_T])\Big).\end{aligned}$$

(iii) From this representations we can see that the maps $P_F(f, \ell, \mathbb{E}_{\mathbb{Q}}[\widehat{\mathbb{X}}_T]) := \mathbb{E}_{\mathbb{Q}}[F]$ and $P_{\mathrm{Var}(F)}(f, \ell, \mathbb{E}_{\mathbb{Q}}[\widehat{\mathbb{X}}_T]) := \mathrm{Var}_{\mathbb{Q}}(F)$ inherit good properties from $\widetilde{P}_J$. In particular, P_F is linear in f and polynomial of degree m in ℓ, and $P_{\mathrm{Var}(F)}$ is quadratic in f and polynomial of degree $2m$ in ℓ. Both maps are linear in $\mathbb{E}[\widehat{\mathbb{X}}_T]$.

3.5 Calibration to Option Prices

In this section our objective is to calibrate models to call (and put) option prices, aiming to closely match their implied volatilities.

Throughout this section, we assume the existence of a pricing measure $\mathbb{Q}$ and that the primary process X is a vector of correlated $\mathbb{Q}$-Brownian motions. The essence of this approach lies in selecting the parameters ℓ to reproduce the market prices of options on S. Given the prices of N call options $\pi^*(T_1, K_1), \ldots, \pi^*(T_N, K_N)$ characterized by maturities T_i and strikes K_i, our goal is to determine these parameters for an optimal fit. Then, in spirit of [6], we translate the current calibration task into finding ℓ such that

$$\ell \in \operatorname{argmin}_\ell \ L_{\text{option}}(\ell),$$

where

$$L_{\text{options}}(\ell) = \sum_{i=1}^{N} \gamma_i \Big(\pi^*(T_i, K_i) - \pi^{\text{model}}(\ell, T_i, K_i)\Big)^2, \tag{18}$$

with γ_i Vega weights and $\pi^{\text{model}}(\ell, T_i, K_i)$ denoting the price of the option under the signature model with parameters ℓ, maturity T_i and strike K_i.

Remark 3.17 The loss function described in (18) can be adapted depending on the available data. A common approach is for instance to weigh the options by their bid-ask spreads, see for instance [6]. This reflects the relative importance of reproducing different option prices precisely. In the present work we shall employ Vega weights since bid-ask spreads are not always provided in the data and since they are well-suited to match the implied volatility surface from option prices, see also Remark 5.8 in [13].

For notational convenience, let us here briefly recall the definition of implied volatility, which serves as our criterion for goodness of fit. Notice that for liquidity reasons we only calibrate to vanilla options.

Definition 3.18 Let $\pi(T, K)$ be the price of a call option with maturity T and strike price K written on the asset S. The implied volatility of $\pi(T, K)$ is defined as the volatility $\sigma_{\text{IV}}(T, K)$ that solves the equation

$$\pi^{BS}(T, K, \sigma_{\text{IV}}(T, K)) = \pi(T, K), \tag{19}$$

where π^{BS} denotes the Black-Scholes option price. Then $(\sigma_{\text{IV}}(T, K))_{T,K}$ is called (implied) volatility surface, and $(\sigma_{\text{IV}}(T, K))_K$ is called (implied) volatility smile for each fixed maturity T.

Once the optimal parameters ℓ^* are found via (18) we then need to solve (19) numerically for $\pi(T, K) = \pi^{\text{model}}(\ell^*, T, K)$ to assess the goodness of fit. Indeed we measure it in terms of the absolute relative error (in percentage) between the implied volatilities from the market and the model. Another goodness of fit test is to check whether the calibrated implied volatilities lie in the market's bid-ask spread (if available), see Sect. 4.2.

3.5.1 Monte Carlo Pricing

We now turn our attention to the task of computing the price $\pi^{\text{model}}(\ell, T, K)$ for a given strike $K \in \mathbb{R}$ and maturity $T > 0$ under the signature model.

Before presenting our Monte Carlo-based method, it is worth recalling that Sect. 3.4 provides a closed-form formula for computing sig-payoffs without relying on Monte Carlo techniques. Additionally, thanks to the universal approximation theorem, observe that call options payoffs can be approximated arbitrarily well by sig-payoffs, or even just by polynomials, on compact sets. The combination of these two properties forms the basis of the approach employed by [37], which involves breaking down the computation of $\pi^{\text{model}}(\ell, T, K)$ into two components: an approximation of call (put) payoffs using sig-payoffs and the pricing of the latter through the closed-form formula derived in Sect. 3.4. However, achieving an acceptable error between the original call payoff and the sig-payoff requires selecting a sig-payoff with signature terms of sufficiently high order. This approximation must hold for all sets of coefficients ℓ involved in the optimization procedure, encompassing a large compact set K.

It is important to note that while sig-payoffs may be used as control variates in Monte Carlo pricing techniques, as we will discuss in Sect. 3.5.2, we opted for a Monte Carlo approach. The linearity of the model makes this approach particularly manageable, rendering the entire calibration computationally feasible within a reasonable time while delivering highly accurate results.

For the Monte Carlo price we thus fix a number of samples $N_{MC} > 0$ and approximate $\pi^{\text{model}}(\ell, T, K)$ via

$$\pi^{\text{model}}(\ell, T, K) \approx \frac{1}{N_{MC}} \sum_{i=1}^{N_{MC}} (S_n(\ell)_T(\omega_i) - K)^+.$$

We stress again that this can be computed fast. Indeed, by the linearity of the model simulating $(S_n(\ell)_T(\omega_i))_{i=1}^{N_{MC}}$ boils down to the following steps:

- simulate $(X_t(\omega_i))_{t\in[0,T]}$, which in the current setting are just trajectories for correlated Brownian motions, for each $i \in \{1, \dots, N_{MC}\}$;
- compute $\langle e_I, \widehat{\mathbb{X}}_T(\omega_i)\rangle$ for all $i = 1, \dots, N_{MC}$ and for all multi-indices I such that $|I| \leq n$;
- take linear combinations to compute $\langle \tilde{e}_I, \widehat{\mathbb{X}}_T(\omega_i)\rangle$ for all $i = 1, \dots, N_{MC}$ and for all multi-indices $|I| \leq n-1$ as described in Lemma 3.5;
- retrieve $(S_n(\ell)_T(\omega_i))_{i=1}^{N_{MC}}$ via (14).

It is worth noting that the parameters ℓ come into play only in the final step, enabling the precomputation and storage of all other quantities. This contrasts with other models where the calibration parameters are involved at each time point during the simulation steps, such as in an Euler scheme for classical Markovian models or more complex schemes employed in rough volatility models.

3.5.2 Variance Reduction with Sig-Payoffs

Even though Monte Carlo pricing is fast since all essential quantities can be precomputed as explained above, we here discuss variance reduction techniques (see e.g. [26]) that can speed up the procedure even further. The idea is to introduce a control variate, i.e., a random variable Φ^{cv} such that:

$$\mathbb{E}_{\mathbb{Q}}[\Phi^{cv}] = 0, \qquad \mathrm{Var}\big((S_n(\ell)_T - K)^+ - \Phi^{cv}\big) < \mathrm{Var}\big((S_n(\ell)_T - K)^+\big).$$

An example of control variates used for pricing and calibrating neural SDE models can be found in [10, 25], where Φ^{cv} is constructed from the Delta hedge. A possible other choice of control variates for signature models are sig-payoffs. Indeed, one can use the pricing formula derived in Sect. 3.4 to define:

$$\Phi^{cv}(\ell, T, K) := f_\emptyset + \sum_{0<|J|\leq m} f_J \langle e_J, \widehat{\mathbb{S}}_n(\ell)_T \rangle - \widetilde{P}_f(\ell, \mathbb{E}_{\mathbb{Q}}[\widehat{\mathbb{X}}_T]),$$

for some fixed $m > 0$ where

$$(S_n(\ell)_T - K)^+ \approx f_\emptyset + \sum_{0<|J|\leq m} f_J \langle e_J, \widehat{\mathbb{S}}_n(\ell)_T \rangle,$$

for a wide range of ℓ and with high probability. This can be done by performing a linear regression to obtain the coefficients $f = (f_J)_{|J|\leq m}$. Alternatively, a polynomial approximation of the payoff's function can also be employed.

The properties of Φ^{cv} then guarantee the accuracy of the approximation

$$\pi^{\text{model}}(\ell, T, K) \approx \frac{1}{N_{MC}} \Big(\sum_{i=1}^{N_{MC}} (S_n(\ell)_T(\omega_i) - K)^+ - \Phi^{cv}(\ell, T, K)(\omega_i) \Big),$$

already for smaller values of N_{MC}. In the following remark we report a numerical experiment in this regard.

Example 3.19 Fix $n, d = 2$ and consider the parameters $\ell^* \in \mathbb{R}^{13}$ calibrated to the first smile of the market data shown in Fig. 4 (left). Consider as example a call option with maturity $T = 30$ days and strike $K = 85\%$ of the spot price. Let p be a polynomial of order $m = 4$ approximating the function $f : x \mapsto (x - K)^+$ on the compact $[70\% S_0, 120\% S_0]$. Recall that $\mathbb{E}_{\mathbb{Q}}[p(S_n(\ell)_T)]$

can be computed analytically as polynomials are sig-payoffs. Using $\Phi^{cv}(\ell) := p(S_n(\ell)_T) - \mathbb{E}_{\mathbb{Q}}[p(S_n(\ell)_T)]$ as control variate we can reduce the sample variance of the Monte Carlo estimator from approximately $3.89 \cdot 10^{-5}$ to $6.5 \cdot 10^{-7}$. Observe that $p(S_n(\ell)_T)$ here coincides with the sig-payoff without time-augmentation i.e.,

$$p(S_n(\ell)_T) = \sum_{|I|\leq 4} f_I \langle e_I, \mathbb{S}_n(\ell)_T \rangle.$$

3.5.3 Model Performance

In the following we discuss the problem of minimizing the functional (18) using the Monte Carlo method as described in Sect. 3.5.1 to compute the model prices. We consider the model described in Corollary 3.6 for two correlated Brownian motions B and W with correlation coefficient $\rho = -0, 5$, and $D = 1$.

As a first example we consider synthetic data, where the implied volatility surface to fit is generated by a Heston model whose dynamics are given by

$$\begin{aligned} dS_t &= \mu S_t dt + S_t \sqrt{V_t} dB_t^{\mathbb{P}} \\ dV_t &= \kappa(\theta - V_t)dt + \sigma \sqrt{V_t} dW_t^{\mathbb{P}}, \end{aligned} \tag{20}$$

where in both cases $d[B^{\mathbb{P}}, W^{\mathbb{P}}]_t = \rho dt$ where $\rho \in [-1, 1]$.

We consider 7 maturities $(T_k)_{k=1}^7$ ranging from 30 days to 2 years and 13 strikes $(K_j)_{j=1}^{13}$ ranging from 80% to 120% of the spot price. The truncation parameter is fixed to $n = 3$ and the number of Monte Carlo samples to $N_{MC} = 10^6$. The results for the following two sets of parameters under a risk neutral measure

κ	θ	σ	ρ	V_0
0.2	0.3	0.5	−0.5	0.08

are displayed in the first and in the second row of Fig. 1, respectively.

We stress that these calibrations to Heston generated implied volatility surfaces can take between 4 and 15 minutes on a standard laptop. We consider now implied volatility data as of 17/03/2021 for call options written on the S&P 500 index. Our dataset provided by Bloomberg consists of 7 maturities $(T_k)_{k=1}^7$, ranging from 30 days to 2 years, and 9 strike prices $(K_j)_{j=1}^9$ for each maturity which vary between 80% and 120% of the spot price. Again, the truncation parameter is fixed to $n = 3$ and the Monte Carlo's parameter to $N_{MC} = 10^6$. The results are displayed in Fig. 2.

Figure 2 indicates that the volatility smiles for short maturities have not been captured adequately. Specifically, as with many continuous models, the challenge lies in fitting the shortest maturities, primarily due to the high at-the-money skew in the market (refer to Chapter 3 and Chapter 7 of [22]).

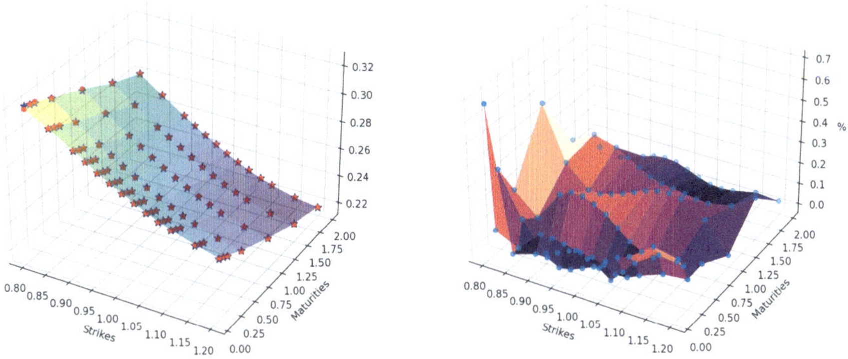

Fig. 1 On the left: blue stars correspond to the implied volatilities of the Heston models, red dots denote the calibrated implied volatilities of $S_n(\ell)$ with $n = 3$ (13 estimated parameters). On the right: absolute relative errors between the two surfaces are expressed in percentages

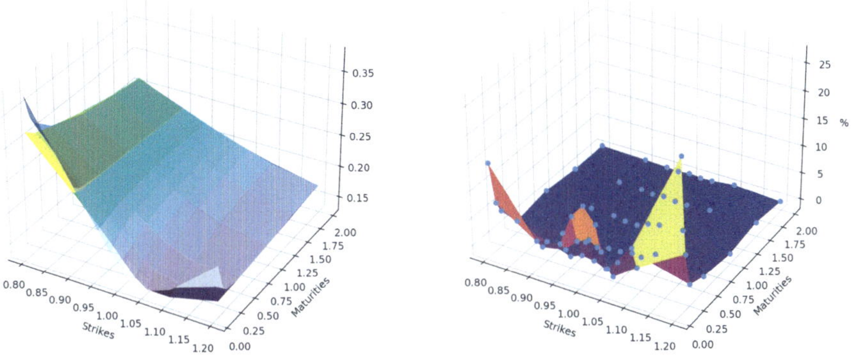

Fig. 2 On the left: the upper surface represents the implied volatility of the S&P 500 index as of 17-03-2021, the lower one is the calibrated implied volatility of $S_n(\ell^*)$ with $n = 3$ (13 parameters). On the right: absolute relative error between the two surfaces in percentages

To assess the model's ability to replicate short maturity smiles, we conduct calibrations using different loss functions that penalize outliers more severely. In the initial calibration procedure described earlier, we denote by w_i the absolute error between the target implied volatility and the approximated one for maturity T_i and strike K_i. Subsequently, inspired by generative-adversarial distances as considered in, e.g., [10], we define a new loss function

$$L_{p,\alpha}(\ell) = \sum_{i=1}^{N} (\gamma_i + \alpha w_i)|\pi^*(T_i, K_i) - \pi^{\text{model}}(\ell, T_i, K_i)|^p, \tag{21}$$

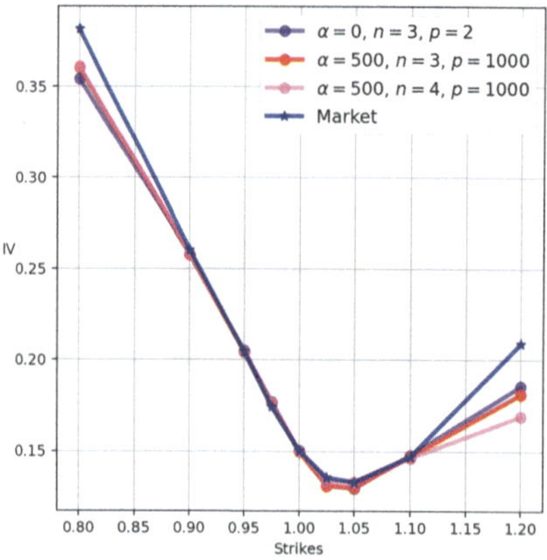

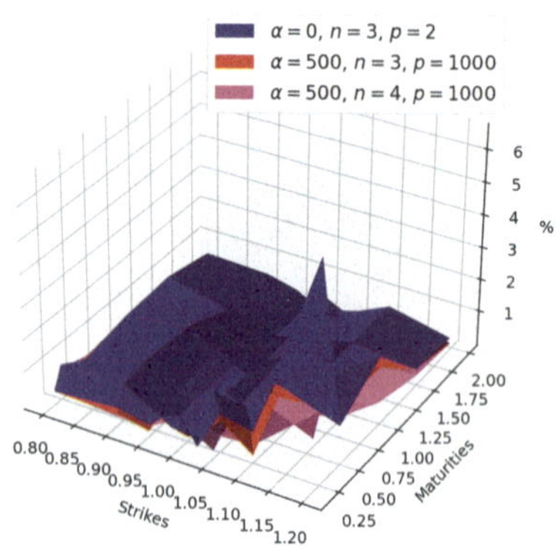

Fig. 3 Comparison between the calibrated implied volatility smiles for different parameters and the S&P 500 index as of 17-03-2021 smile (in blue) at maturity $T_1 = 30$ days (on the left) and for maturities ranging from 60 days to 2 years (on the right). Calibration has been performed using (21)

depending on parameters p and α that need to be chosen. By taking high values for p and α we can approximate the sup-distance between the two price surfaces, i.e.,

$$L_\infty(\ell) := \sup_{i=1,\dots,N} |\pi^*(T_i, K_i) - \pi^{\text{model}}(\ell, T_i, K_i)|,$$

without compromising differentiability with respect to ℓ. The result for different choices of the parameters α and p but also the truncation level n is displayed in Fig. 3. As can be guessed from the figure, although the maximal absolute relative error for maturities larger than 60 days is (almost) acceptable (5.6% for $n = 3$, $p = 1000$, and $\alpha = 500$ and 2.3 % for $n = 4$ and $p = 1000$, and $\alpha = 500$), the absolute relative error for the shortest maturity and the far in and out of the money strikes are still above 12 % and 18 %, respectively. Observe that the performance for $n = 4$ is the best for every maturity larger than 60 days as well as for the at-the-money region of the shortest maturity.

In a final experiment, we conduct two separate calibrations: one to the shortest maturity alone and another to every other maturity combined. The first calibration is performed for $T_1 = 30$ days, and the second calibration encompasses 6 maturities ranging from 60 days to 2 years. In both cases, we consider 9 strikes ranging from 80% to 120% of the spot price. For the first calibration, we fix the parameters at $n = 2$, $p = 2$, and $\alpha = 0$, while for the subsequent maturities, the parameters are set to $n = 4$, $p = 300$, and $\alpha = 500$. The results are displayed in Fig. 4. Notably, the fit for the first maturity is remarkably accurate, as well as for the entire implied volatility surface. The computational time required for the calibration of the first smile is approximately 2 minutes, while the calibration to the remaining surface might take longer, mainly due to the higher value of n and the calibration to all maturities.

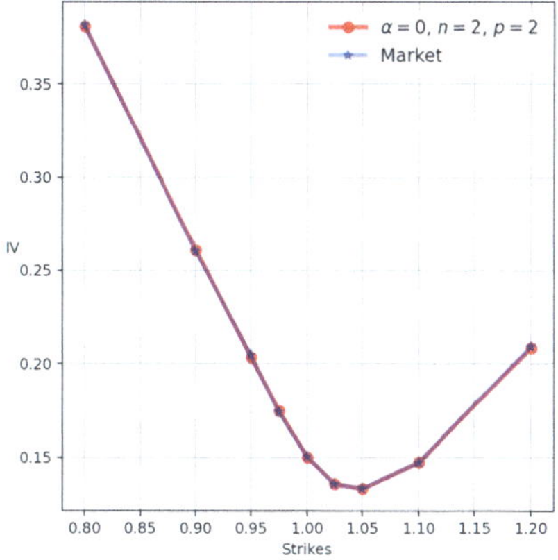

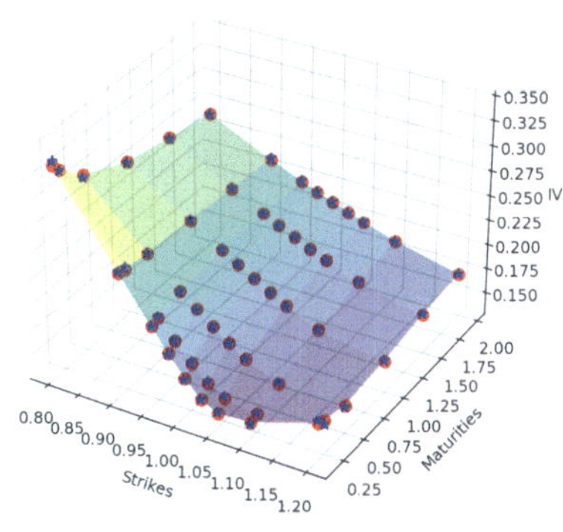

Fig. 4 Comparison between the calibrated implied volatility smiles and the SPX-500 Index as of 17-03-2021 smile (in blue) at maturity $T_1 = 30$ days (on the left) and for maturities ranging from 60 days to 2 years (on the right). Two different calibrations have been performed using (21)

These results suggest that introducing maturity-dependent parameters and performing a slice-wise calibration to the individual smiles as for instance in [10] or [25] can be of interest to obtain both, an excellent accuracy and a low computational time. For details in this regard and the out-of-sample performance see [11].

4 A Signature Model for Index Options and Volatility Derivatives

This section is dedicated to define the second model $(S_t)_{t\geq 0}$ for the S&P 500 index in detail. Under a risk-neutral probability measure $\mathbb{Q}$ the model's dynamics are given by:

$$\mathrm{d}S_t = S_t \sigma_t^S \mathrm{d}B_t. \tag{22}$$

Here, $S_0 \in \mathbb{R}^+$, $\sigma^S = (\sigma_t^S)_{t\geq 0}$ is the volatility process, and $B = (B_t)_{t\geq 0}$ is a one-dimensional Brownian motion that is correlated with the volatility process. Note that the instantaneous variance is given by $V_t := (\sigma_t^S)^2$ for every $t \geq 0$. Choosing a functional form of the volatility process is the crucial modelling choice. We set

$$\sigma_t^S(\ell) := \ell_\emptyset + \sum_{0<|I|\leq n} \ell_I \langle e_I, \widehat{\mathbb{X}}_t \rangle, \tag{23}$$

i.e. the volatility process is determined by a linear function of the (time-extended) signature of a primary process X. Moreover, we assume that X is a polynomial process (recall Definition 2.1) and that the model parameters are $\ell := \{\ell_I \in \mathbb{R} : |I| \leq n\} \in \mathbb{R}^{(d+1)_n}$.

For later convenience, we now introduce the process $(Z_t)_{t\geq 0}$ where $Z_t = (X_t, B_t)$ and denote by $(\widehat{\mathbb{Z}}_t)_{t\geq 0}$ its time-extended signature. The correlation of the components of $(Z_t)_{t\geq 0}$ is given by $\rho = (\rho_{ij})_{i,j=1,\dots,d+1}$, i.e.,

$$\rho_{ij} = \frac{[Z^i, Z^j]}{\sqrt{[Z^i]}\sqrt{[Z^j]}} \in [-1, 1],$$

Recall that we use $[\cdot, \cdot]$ for the quadratic variation. We will often write $(\sigma_t^S)_{t\geq 0} = (\sigma_t^S(\ell))_{t\geq 0}$ as in (22) to keep the notation light and only mention the dependence on ℓ when it is explicitly needed.

Remark 4.1 (Interest Rates and Dividends) To calibrate our model to option prices we will both need the discounted, dividend-adjusted and undiscounted, unadjusted prices. This is due to the fact that the VIX is defined via the discounted, dividend-adjusted prices by the CBOE but the claims on the S&P500 itself are written on the undiscounted, unadjusted prices. Recall that $(S_t)_{t\geq 0}$ given by (22) is the discounted, dividend-adjusted price process. Including an interest rate r and dividend q, the corresponding undiscounted, unadjusted price process is given by

$$\mathrm{d}\tilde{S}_t = (r - q)\tilde{S}_t \mathrm{d}t + \tilde{S}_t \sigma_t^S(\ell)\mathrm{d}B_t.$$

The price of a call option with time to maturity $T > 0$ and strike price $K \in \mathbb{R}$ written on the S&P 500 index, is therefore

$$C(T, K) = \mathbb{E}\left[e^{-rT}(\tilde{S}_T(\ell) - K)^+\right] = \mathbb{E}[e^{-rT}(e^{(r-q)T}S_T(\ell) - K)^+],$$

where we used that $\tilde{S}_t(\ell) = e^{(r-q)t}S_t(\ell)$.

Remark 4.2 Recall our assumption that $\widehat{X}$ is a polynomial process. This is important since then $\widehat{\mathbb{X}}^n$ is a finite-dimensional polynomial process in sense of [19] and [9] and therefore the expected signature of $\widehat{X}$ and conditional expected signature can be found by solving a finite-dimensional ODE, more specifically, it is given by a finite-dimensional matrix exponential as described in Sect. 2.

This assumption still leaves us with a broad class of admissible primary processes. Some prominent examples are correlated Brownian motions, Cox-Ingersoll-Ross (CIR) processes, geometric Brownian motions, Jacobi processes, OU processes, and all continuous affine processes.

Remark 4.3 Several stochastic volatility models are encapsulated in our modelling framework (23). Let us elaborate on the following:

- The Stein-Stein model, as introduced in [40], is obtained if we choose a one-dimensional OU process as our primary process $(X_t)_{t\geq 0}$ and set $n = 1$, $\ell_\emptyset = \ell_{(0)} = 0$ and $\ell_{(1)} \neq 0$.

- The SABR model, as introduced initially in [31] with $\beta = 1$. To this end we choose $(X_t)_{t\geq 0}$ to be a 1-dimensional geometric Brownian motion without drift and again let $n = 1$, with $\ell_\emptyset = \ell_{(0)} = 0$ and $\ell_{(1)} \neq 0$.
- Let $(X_t)_{t\geq 0}$ be a one-dimensional OU process with $n = 5$, $\ell_\emptyset, \ell_{(1)}, \ell_{(1,1,1)}$, $\ell_{(1,1,1,1,1)}$ non-zero and $\ell_I = 0$ otherwise. This results in the model described in [1], except for the fact that therein a deterministic input curve is added. Moreover, we can embed the entire set of Gaussian polynomial volatility models introduced in [1], if we allow X not to be a semimartingale and do not add a time-augmentation.

In the next section we discuss the pricing of VIX options with (23) as well as the nature of the log-price process.

4.1 Explicit Formulas for the VIX and the Log-Price

The CBOE Volatility Index (VIX) measures the market's expected volatility of the S&P 500 index. More specifically, its current value corresponds to the expected annualized change in the S&P 500 index over the 30 days ahead. Indeed, the index value is computed by

$$\mathrm{VIX}_T = \sqrt{\mathbb{E}\left[-\frac{2}{\Delta} \log\left(\frac{S_{T+\Delta}}{S_T}\right) \Big| \mathcal{F}_T \right]}, \tag{24}$$

where $\Delta = 30$ days, S_T denotes the price process and $\mathcal{F}_T$ the filtration at time $T > 0$. Note that for *any* stochastic volatility model of the form

$$dS_t = S_t \sqrt{V_t} dB_t,$$

where $(B_t)_{t\geq 0}$ is a Brownian motion and $(V_t)_{t\geq 0}$ satisfies

$$\mathbb{E}\left[\int_0^T V_s \mathrm{d}s\right] < \infty, \tag{25}$$

the VIX index at $T > 0$ is given by

$$\mathrm{VIX}_T = \sqrt{\frac{1}{\Delta}\mathbb{E}\left[\int_T^{T+\Delta} V_t \mathrm{d}t \,|\mathcal{F}_T\right]}, \tag{26}$$

see e.g., [35]. Although there are of course put and call options written on the VIX, we from now on consider without loss of generality only call options and will simply refer to them as *VIX options*.

We shall now derive an analytical expression for the VIX index value (24) under our model, i.e. if S is defined as in (22) and (23) and for a polynomial process X. More precisely, we state in Theorem 4.4 that the value of the VIX is given by the square-root of a quadratic function in the parameters ℓ. This is achieved via polynomial technology and the computation of a matrix exponential, as we showed in Sect. 2.

Theorem 4.4 *Let the price process $S = (S_t)_{t\geq 0}$ be given by*

$$\mathrm{d}S_t = S_t\sigma_t^S(\ell)\mathrm{d}B_t,$$

with volatility process $\sigma^S(\ell) = (\sigma_t^S(\ell))_{t\geq 0}$ and $B = (B_t)_{t\geq 0}$ denoting a one-dimensional Brownian motion. Recall our modeling assumption, namely that $\sigma^S(\ell)$ and X satisfy (23) *and note that under this* (25) *is fulfilled. As introduced in Sect. 2 we fix an injective labeling function $\mathscr{L} : \{I : |I| \leq n\} \to \{1, \ldots, (d+1)_{2n+1}\}$ and denote by G be the $(d+1)_{(2n+1)}$-dimensional matrix representative of the dual operator corresponding to $\widehat{\mathbb{X}}$. Then, the following expression for the VIX at time $T > 0$ holds*

$$VIX_T(\ell) = \sqrt{\frac{1}{\Delta}\ell^\top Q(T, \Delta)\ell}, \tag{27}$$

with

$$Q_{\mathscr{L}(I),\mathscr{L}(J)}(T, \Delta) = \mathbf{vec}((e_I \sqcup\!\sqcup\, e_J) \otimes e_0)^\top (e^{\Delta G^\top} - \mathrm{Id})\mathbf{vec}(\widehat{\mathbb{X}}_T^{2n+1}), \tag{28}$$

and $\mathrm{Id} \in \mathbb{R}^{(d+1)_{2n+1}\times(d+1)_{2n+1}}$ *denoting the identity matrix. Note that Q is positive semidefinite and symmetric and hence allows for a Cholesky decomposition.*

Proof Note that $V_t(\ell) = (\sigma^S(\ell))^2$ can be expressed as

$$V_t(\ell) = \left(\sum_{|I|\leq n} \ell_I \langle e_I, \widehat{\mathbb{X}}_t\rangle\right)^2 = \sum_{|I|,|J|\leq n} \ell_I\ell_J\langle e_I \sqcup\!\sqcup\, e_J, \widehat{\mathbb{X}}_t\rangle,$$

by the shuffle-property. Moreover, recall that continuous polynomials processes have finite moments of every degree and hence by Remark 1.2 it follows that (25) holds. Plugging-in the above equation for $V_t(\ell)$ the value of the VIX is for each $T > 0$ given by

$$\mathrm{VIX}_T^2(\ell) = \frac{1}{\Delta}\sum_{|I|,|J|\leq n} \ell_I\ell_J\mathbb{E}\left[\int_T^{T+\Delta} \langle e_I \sqcup\!\sqcup\, e_J, \widehat{\mathbb{X}}_t\rangle \mathrm{d}t|\mathcal{F}_T\right] = \frac{1}{\Delta}\ell^\top Q(T, \Delta)\ell,$$

where for each $T > 0$ the matrix Q is given by

$$
\begin{aligned}
Q_{\mathcal{L}(I)\mathcal{L}(J)}(T,\Delta) :&= \mathbb{E}\Big[\int_T^{T+\Delta} \langle e_I ⧢ e_J, \widehat{\mathbb{X}}_t\rangle \mathrm{d}t | \mathcal{F}_T\Big] \\
&= \mathbb{E}\Big[\int_0^{T+\Delta} \langle e_I ⧢ e_J, \widehat{\mathbb{X}}_t\rangle \mathrm{d}t - \int_0^{T} \langle e_I ⧢ e_J, \widehat{\mathbb{X}}_t\rangle \mathrm{d}t | \mathcal{F}_T\Big] \\
&= \mathbb{E}\big[\langle (e_I ⧢ e_J)\otimes e_0, \widehat{\mathbb{X}}_{T+\Delta}\rangle - \langle (e_I ⧢ e_J)\otimes e_0, \widehat{\mathbb{X}}_{T}\rangle | \mathcal{F}_T\big] \\
&= \mathbb{E}\big[\langle (e_I ⧢ e_J)\otimes e_0, \widehat{\mathbb{X}}_{T+\Delta}\rangle | \mathcal{F}_T\big] - \langle (e_I ⧢ e_J)\otimes e_0, \widehat{\mathbb{X}}_{T}\rangle.
\end{aligned}
$$

By Theorem 2.5 we can also express Q as

$$
\begin{aligned}
Q_{\mathcal{L}(I)\mathcal{L}(J)}(T,\Delta) &= \mathbf{vec}((e_I ⧢ e_J)\otimes e_0)^\top e^{\Delta G^\top} \mathbf{vec}(\widehat{\mathbb{X}}_T^{2n+1}) \\
&\quad - \mathbf{vec}((e_I ⧢ e_J)\otimes e_0)^\top \mathbf{vec}(\widehat{\mathbb{X}}_T^{2n+1}) \\
&= \mathbf{vec}((e_I ⧢ e_J)\otimes e_0)^\top (e^{\Delta G^\top} - \mathrm{Id})\mathbf{vec}(\widehat{\mathbb{X}}_T^{2n+1}),
\end{aligned}
$$

which proves our claim. The fact that Q is positive semidefinite and symmetric follows from the shuffle property. Under these properties, it is well-known that a Cholesky decomposition exists, meaning that there is an upper triangular matrix $U_T \in \mathbb{R}^{(d+1)_n \times (d+1)_n}$ such that

$$
\mathrm{VIX}_T(\ell) = \sqrt{\frac{1}{\Delta}\ell^\top U_T U_T^\top \ell} = \frac{1}{\sqrt{\Delta}}\sqrt{(U_T^\top \ell)^2} = \frac{1}{\sqrt{\Delta}}\|U_T^\top \ell\|.
$$

□

Moreover, an explicit expression of the log price in terms of a linear function of the signature of $\widehat{Z}$ can be obtained, which we present in the next proposition.

Proposition 4.5 *Let S be given by* (22) *with $S_0 = 1$, as well as σ^S and X satisfy* (23). *Under these assumptions the log-price at time $t \geq 0$ is given by,*

$$
\log(S_t(\ell)) = -\frac{1}{2}\ell^\top Q^0(t)\ell + \sum_{|I|\leq n} \ell_I \langle \tilde{e}_I^B, \widehat{\mathbb{Z}}_t\rangle, \tag{29}
$$

with

$$
\tilde{e}_\emptyset^B := e_{d+1}, \qquad \tilde{e}_I^B := e_I \otimes e_{d+1} - \sum_{|J|\leq m} \frac{a^J_{i_{|I|}(d+1)}}{2}(e_{I'} ⧢ e_J)\otimes e_0,
$$

for each $|I| > 0$, and the matrix $Q^0(t) \in \mathbb{R}^{(d+1)_n \times (d+1)_n}$ has components

$$
Q^0_{\mathcal{L}(I)\mathcal{L}(J)}(t) = \langle (e_I ⧢ e_J)\otimes e_0, \widehat{\mathbb{X}}_t\rangle,
$$

for an arbitrary but fixed labeling function $\mathcal{L} : \{I : |I| \leq n\} \to \{1, \dots, (d+1)_n\}$.

Proof Using Itô's lemma and the form or S, σ^S and X under our model, we obtain

$$\begin{aligned}
\log(S_t(\ell)) &= -\frac{1}{2}\int_0^t V_s(\ell)\mathrm{d}s + \int_0^t \sigma_s^S(\ell)\mathrm{d}B_s \\
&= -\frac{1}{2}\sum_{|I|,|J|\leq n} \ell_I \ell_J \int_0^t \langle e_I \sqcup\!\sqcup\, e_J, \widehat{\mathbb{X}}_s\rangle \mathrm{d}s + \sum_{|I|\leq n} \ell_I \int_0^t \langle e_I, \widehat{\mathbb{X}}_s\rangle \mathrm{d}B_s \\
&= -\frac{1}{2}\sum_{|I|,|J|\leq n} \ell_I \ell_J \langle (e_I \sqcup\!\sqcup\, e_J) \otimes e_0, \widehat{\mathbb{X}}_t\rangle + \sum_{|I|\leq n} \ell_I \langle \tilde{e}_I^B, \widehat{\mathbb{Z}}_t\rangle \\
&= -\frac{1}{2}\ell^\top Q^0(t)\ell + \sum_{|I|\leq n} \ell_I \langle \tilde{e}_I^B, \widehat{\mathbb{Z}}_t\rangle,
\end{aligned}$$

where we used that $\int_0^t \langle e_I, \widehat{\mathbb{X}}_s\rangle \mathrm{d}B_s = \langle \tilde{e}_I^B, \widehat{\mathbb{Z}}_t\rangle$ (Lemma 3.5) in the second to last equality. □

4.2 *Joint Calibration of SPX and VIX Options*

We point out again that we only work with call options in the following, however this choice was made merely for simplicity and it is straightforward to include any other liquid options. The time-to-maturities of options available on the market, as well as strikes-prices available on the market may differ for VIX and SPX. Accordingly, we denote by $\mathcal{T}^{\mathrm{SPX}}$, $\mathcal{T}^{\mathrm{VIX}}$ the set of maturities and by $\mathcal{K}^{\mathrm{SPX}}$, $\mathcal{K}^{\mathrm{VIX}}$ the set of strikes for SPX and VIX options respectively.

Proposition 4.5 implies that the SPX call option payoff is given by

$$e^{-rT}(\tilde{S}_T(\ell)-K)^+ = e^{-rT}\left(\exp\left\{(r-q)T - \frac{1}{2}\ell^\top Q^0(t)\ell + \sum_{|I|\leq n} \ell_I \langle \tilde{e}_I^B, \widehat{\mathbb{Z}}_T\rangle\right\} - K\right)^+,$$

for a given maturity $T > 0$ and a strike price $K > 0$. Recall from Remark 4.1 that $\tilde{S}$ stands for the undiscounted, unadjusted price process, r is the interest rate and q the dividend. Similarly, according to Theorem 4.4, the pay-off for the VIX reads

$$\begin{aligned}
e^{-rT}(\mathrm{VIX}_T(\ell) - K)^+ &= e^{-rT}\left(\sqrt{\frac{1}{\Delta}\ell^\top Q(T,\Delta)\ell} - K\right)^+ \\
&= e^{-rT}\left(\frac{1}{\sqrt{\Delta}}\|U_T^\top \ell\| - K\right)^+,
\end{aligned}$$

where we applied the Cholesky decomposition to $Q(T, \Delta) = U_T U_T^\top$. To evaluate the expectations in the pricing formula for the above options as well as the VIX's futures, we apply a Monte-Carlo procedure. However, we want to point out that we do not need an additional Monte-Carlo simulation to evaluate the conditional expectation in the definition of the VIX. This is a significant computational benefit of our model thanks to the polynomial technology.

The loss function to be minimized in the joint calibration is simply a convex combination of the individual loss functions of VIX and SPX respectively:

$$L_{\text{joint}}(\ell, \lambda) := \lambda L_{\text{SPX}}(\ell) + (1-\lambda) L_{\text{VIX}}(\ell), \tag{30}$$

for $\lambda \in (0, 1)$. In particular, for some loss function $\mathcal{L}^\beta$ to be specified below,

- the VIX' loss function is given by

$$L_{\text{VIX}}(\ell) := \sum_{\substack{T \in \mathcal{T}^{VIX} \\ K \in \mathcal{K}^{VIX}}} \mathcal{L}^\beta\Big(\pi_{\text{VIX}}^{\text{model}}(\ell, T, K), \pi_{\text{VIX}}^{b,a}(T, K), \sigma_{\text{VIX}}^{b,a}(T, K), F_{\text{VIX}}^{\text{model}}(\ell, T), F_{\text{VIX}}^{mkt}(T)\Big),$$

  where π_{VIX}^{model} and F_{VIX}^{model} are the Monte Carlo option- and futures-prices under our model (i.e. for $\text{VIX}_T(\ell, \omega_i)$ defined as in (27));
- the SPX' loss function is given by

$$L_{\text{SPX}}(\ell) := \sum_{T \in \mathcal{T}^{\text{SPX}}, K \in \mathcal{K}^{\text{SPX}}} \mathcal{L}^\beta(\pi_{\text{SPX}}^{\text{model}}(\ell, T, K), \pi_{\text{SPX}}^{b,a}(T, K), \sigma_{\text{SPX}}^{b,a}(T, K)).$$

  Note that in the case of the SPX no future prices need to be calibrated (because the index value can directly be used as spot price) and we therefore do not add them as inputs to $\mathcal{L}^\beta$ by slight abuse of notation.

For some $\beta \in (0, 1)$, we define $\mathcal{L}^\beta$ to be the loss function

$$\mathcal{L}^\beta(\pi, \pi^{mkt,b,a}, \sigma^{mkt,b,a}, F, F^{mkt}) = \tag{31}$$
$$\left(\frac{\left(\beta \tilde{1}_{\{\pi \notin \pi^{mkt,b,a}\}} + (1-\beta)\right)\left|\pi - (\pi^{mkt,a} + \pi^{mkt,b})/2\right| + \left|\delta^{mkt} e^{-rT}(F - F^{mkt})\right|}{\upsilon^{mkt}(\sigma^{mkt,a} - \sigma^{mkt,b})}\right)^2,$$

with

- υ^{mkt}, δ^{mkt} denoting the Vega and Delta of the option under the Black-Scholes model (but note that they depend on both maturity and strike price);
- $\pi^{mkt,b,a} = [\pi^{mkt,b}, \pi^{mkt,a}]$,$\sigma^{mkt,b,a} = [\sigma^{mkt,b}, \sigma^{mkt,a}]$, with $\pi^{mkt,b}$, $\pi^{mkt,a}$, $\sigma^{mkt,b}$ and $\sigma^{mkt,a}$ denoting the market bid and ask prices and implied volatilities respectively.

- F and F^{mkt} standing for the models and markets futures respectively (with maturity T);
- $\tilde{1}_{\{x \notin [y^b, y^a]\}} := s(y^b - x) + s(x - y^a)$ for $s(x) := \frac{1}{2}\tanh(100x) + \frac{1}{2}$ a smooth approximation of the indicator function, which penalizes implied volatilities that lie out of the bid-ask spread.

4.3 Numerical Results

Before presenting the numerical results concerning the joint calibration problem within our framework, it is worth to mention that two main approaches in the literature are taken into account when choosing the maturities to fit.

1. One approach is to choose $T_1^{\mathrm{SPX}} = T_1^{\mathrm{VIX}}$ the first maturity to be the same for VIX and SPX and then for higher maturities to set $j \geq 2$, $T_j^{\mathrm{SPX}} = T_{j-1}^{\mathrm{VIX}} + \Delta$, see for instance [24, 27, 29, 30]. Sometimes, one may also choose the first maturity to differ by a few days between SPX and VIX, if the same maturity is not available for both on the market.
2. Another approach consists of taking the same set of maturities for SPX and VIX $\mathcal{T}^{\mathrm{SPX}} = \mathcal{T}^{\mathrm{VIX}}$ (or close together) as it was done for example by [4, 23, 39].

In the following we will adhere to the first approach and refer the reader to [13] for results regarding the second approach.

The trading day we consider for our calibration is 02/06/2021, the same as was used in [24, 30] and we use call options for both VIX and SPX. Maturities are reported in the following tables with the corresponding range of strikes (in percentage) with respect to the spot and the market's futures prices.

In the table below, we report the maturities used in the calibrations (where $T = 1 \equiv 365.25$ days) and report the corresponding moneyness range (i.e. strike price normalized by spot price or markets futures price) within which we calibrated our model.

$T_1^{\mathrm{VIX}} = 0.0383$	$T_2^{\mathrm{VIX}} = 0.0767$	$T_1^{\mathrm{SPX}} = 0.0383$	$T_2^{\mathrm{SPX}} = 0.1205$	$T_3^{\mathrm{SPX}} = 0.1588$
(90%,220%)	(90%,220%)	(92%,105%)	(70%,105%)	(80%,120%)

The maturities in days are $\mathcal{T}^{SPX} = (14, 44, 58)$ an $\mathcal{T}^{VIX} = (14, 28)$. Let us also point out the high moneyness region (up to 220%) for VIX options is known to be challenging to fit. In our model, we need to choose both a primary process and a level of truncation of the signature a priori. Indeed, we treat these choices as *hyperparameters* and do not train them. As primary process we choose a three-dimensional OU process (see Example 2.9) with randomly chosen parameters

$$\kappa = (0.1, 25, 10)^\top, \qquad \theta = (0.1, 4, 0.08)^\top, \qquad \sigma = (0.7, 10, 5)^\top,$$

$$\rho = \begin{pmatrix} 1 & 0.213 & -0.576 & 0.329 \\ \cdot & 1 & -0.044 & -0.549 \\ \cdot & \cdot & 1 & -0.539 \\ \cdot & \cdot & \cdot & 1 \end{pmatrix}, \qquad X_0 = (1, 0.08, 2)^\top.$$

and truncate the signature at $n = 3$. In this setting, our model has 85 trainable parameters, meaning that $\ell \in \mathbb{R}^{85}$. Note that choosing instead a two-dimensional Brownian motion as primary process does not lead to good fits at low truncation levels of the signature ($n = 3$), which is shown in [13, Appendix A]. The OU process is therefore a natural choice for a polynomial process which is tractable but exhibits richer dynamics than correlated Brownian motions alone. Moreover, OU processes have qualified for volatility modeling as shown in various articles, see e.g. [1, 2, 38, 40]. In terms of loss function, we choose the parameters $\lambda = 0.35$ and $\beta = 1$. For the evaluation of the pricing functional we simulate $N_{MC} = 80{,}000$ Monte Carlo samples.

The calibrated implied volatility smiles are depicted in Fig. 5 and the absolute value of the relative error between the models and markets future prices for each VIX-maturity are reported in the table below.

$T_1^{\text{VIX}} = 0.0383$	$T_2^{\text{VIX}} = 0.0767$
$\varepsilon_{T_1^{\text{VIX}}} = 9.8 \cdot 10^{-6}$	$\varepsilon_{T_2^{\text{VIX}}} = 6.6 \cdot 10^{-8}$

Indeed, the model does seem to fit the market implied volatilities of the VIX and SPX options at the given maturities rather well. As shown in Fig. 5, the model's implied volatilities lie within the market's bid-ask spreads and the calibrated future's prices coincide with the ones of the market. The latter is confirmed by the small relative errors reported in the table. However, it is also worth noting that this numerical experiment only consists of a small set of maturities and a single trading day. We refer the reader to [13] for a more thorough empirical study with calibration results for various sets of maturities and tests for parameter stability under re-calibration over multiple weeks.

Acknowledgments The first three authors gratefully acknowledge financial support through grant Y 1235 and grant I 3852 of the Austrian Science Fund. All authors acknowledge financial support through the OEAD WTZ project FR.

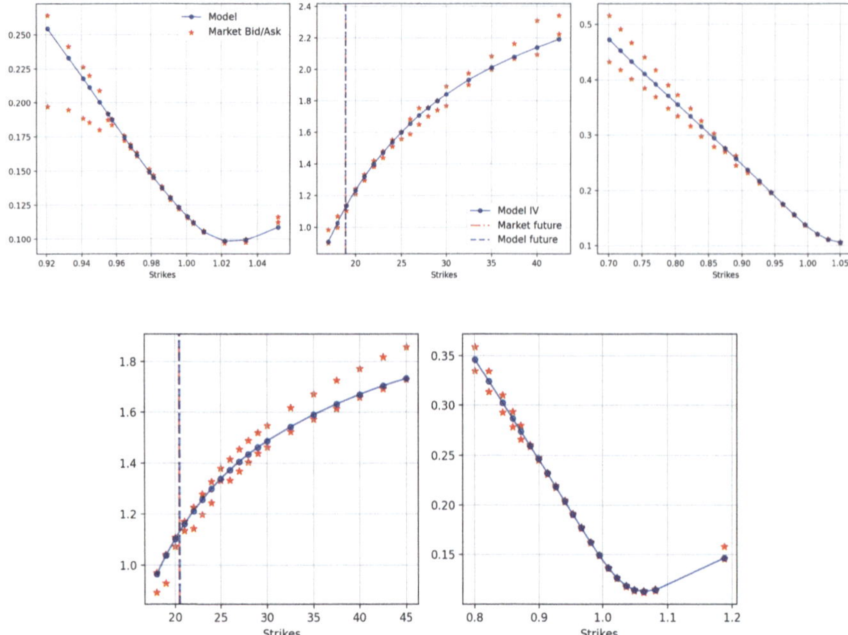

Fig. 5 The implied volatility smiles correspond to T_1^{SPX}, T_1^{VIX}, T_2^{SPX}, T_2^{VIX}, T_3^{SPX} (from top left to bottom right). The blue dots correspond to the values given by our calibrated model and the red stars are the markets bid-ask implied volatilities (from market call options). For the VIX, we additionally show the markets future price by the red dashed line and the one of the calibrated model by a blue dashed line

References

1. E. Abi Jaber, C. Illand, S. Li, Joint SPX-VIX calibration with Gaussian polynomial volatility models: deep pricing with quantization hints. Preprint. arXiv:2212.08297 (2022)
2. E. Abi Jaber, C. Illand, S. Li, The quintic Ornstein-Uhlenbeck volatility model that jointly calibrates SPX & VIX smiles. Preprint. arXiv:2212.10917 (2022)
3. H. Boedihardjo, J. Diehl, M. Mezzarobba, H. Ni, The expected signature of Brownian motion stopped on the boundary of a circle has finite radius of convergence. Bull. Lond. Math. Soc. **53**(1), 285–299 (2021)
4. A. Bondi, S. Pulido, S. Scotti, The rough Hawkes Heston stochastic volatility model. Math. Finance **1**, 1–45 (2024)
5. T. Cass, E. Ferrucci, On the wiener chaos expansion of the signature of a gaussian process, in *Probability Theory and Related Fields* (2024), pp. 1–39
6. R. Cont, S. Ben Hamida, Recovering volatility from option prices by evolutionary optimization. J. Comput. Finance **8**(4), 43–76 (2005)
7. C. Cuchiero, J. Möller, Signature methods in stochastic portfolio theory. Preprint. arXiv:2310.02322 (2023)
8. C. Cuchiero, S. Svaluto-Ferro, Infinite-dimensional polynomial processes. Finance Stoch. **25**(2), 383–426 (2021)

9. C. Cuchiero, M. Keller-Ressel, J. Teichmann, Polynomial processes and their applications to mathematical finance. Finance Stoch. **16**, 711–740 (2012)
10. C. Cuchiero, W. Khosrawi, J. Teichmann, A generative adversarial network approach to calibration of local stochastic volatility models. Risks **8**(4), 101 (2020)
11. C. Cuchiero, G. Gazzani, S. Svaluto-Ferro, Signature-based models: Theory and calibration. SIAM J. Financ. Math. **14**(3), 910–957 (2023)
12. C. Cuchiero, S. Svaluto-Ferro, J. Teichmann, Signature SDEs from an affine and polynomial perspective. Preprint. arXiv:2302.01362 (2023)
13. C. Cuchiero, G. Gazzani, J. Möeller, S. Svaluto-Ferro, Joint calibration to SPX and VIX options with signature-based models. Math. Finance **1**, 1–53 (2024)
14. C. Cuchiero, F. Guida, L. Di Persio, S. Svaluto-Ferro, Measure-valued affine and polynomial diffusions. Stoch. Process. Appl. **175**, 104392 (2024)
15. C. Cuchiero, F. Primavera, S. Svaluto-Ferro, Universal approximation theorems for continuous functions of càdlàg paths and Lévy-type signature models. Finance Stoch. (2025), to appear
16. F. Delbaen, W. Schachermayer, A general version of the fundamental theorem of asset pricing. Math. Ann. **300**(1), 463–520 (1994)
17. G. Di Nunno, K. Kubilius, Y. Mishura, A. Yurchenko-Tytarenko, From constant to rough: a survey of continuous volatility modeling. Mathematics **11**(19), 4201 (2023)
18. T. Fawcett, Problems in stochastic analysis. Connections between rough paths and non-commutative harmonic analysis. PhD Thesis, Univ. Oxford, 2003
19. D. Filipović, M. Larsson, Polynomial diffusions and applications in finance. Finance Stoch. **20**(4), 931–972 (2016)
20. P.K. Friz, A. Shekhar, General rough integration, Lévy rough paths and a Lévy–Kintchine-type formula. Ann. Probab. **45**(4), 2707–2765 (2017)
21. P.K. Friz, P.P. Hager, N. Tapia, Unified signature cumulants and generalized magnus expansions, in *Forum of Mathematics, Sigma*, vol. 10 (Cambridge University Press, Cambridge, 2022), p. e42
22. J. Gatheral, *The Volatility Surface: A Practitioner's Guide* (Wiley, 2011)
23. J. Gatheral, T. Jaisson, M. Rosenbaum, Volatility is rough. Quant. Finance **18**(6), 933–949 (2018)
24. G. Gazzani, J. Guyon, Pricing and calibration in the 4-factor path-dependent volatility model. Preprint. arXiv:2406.02319 (2024)
25. P. Gierjatowicz, M. Sabate-Vidales, D. Siska, L. Szpruch, Z. Zuric, Robust pricing and hedging via neural SDEs. J. Comput. Finance **3**(26), 1–32 (2022)
26. P. Glasserman, *Monte Carlo Methods in Financial Engineering*, vol. 53 (Springer, 2004)
27. I. Guo, G. Loeper, J. Obloj, S. Wang, Optimal transport for model calibration. Preprint. arXiv:2107.01978 (2021)
28. J. Guyon, The joint S&P 500/VIX smile calibration puzzle solved. Risk, April (2020)
29. J. Guyon, Dispersion-constrained martingale Schrödinger problems and the exact joint S&P 500/VIX smile calibration puzzle. Finance Stoch. **28**, 1–53 (2023)
30. J. Guyon, J. Lekeufack, Volatility is (mostly) path-dependent. Quant. Finance, **23**, 1–38 (2023)
31. P.S. Hagan, D. Kumar, A.S. Lesniewski, D. Woodward, Managing smile risk. Best Wilmott **1**, 249–296 (2002)
32. T. Lyons, H. Ni, Expected signature of Brownian motion up to the first exit time from a bounded domain. Ann. Probab. **43**(5), 2729–2762 (2015)
33. T. Lyons, N. Victoir, Cubature on Wiener space. Proc. R. Soc. Lond. Ser. A Math. Phys. Eng. Sci. **460**(2041), 169–198 (2004)
34. T. Lyons, S. Nejad, I. Perez Arribas, Non-parametric pricing and hedging of exotic derivatives. Appl. Math. Finance **27**(6), 457–494 (2020)
35. A. Neuberger, The log contract. J. Portfolio Manag. **20**(2), 74 (1994)
36. H. Ni, The expected signature of a stochastic process. Ph.D. thesis, University of Oxford, 2012
37. I. Perez Arribas, C. Salvi, L. Szpruch, Sig-SDEs model for quantitative finance, in *Proceedings of the First ACM International Conference on AI in Finance* (2020), pp. 1–8

38. S. Rømer, Empirical analysis of rough and classical stochastic volatility models to the SPX and VIX markets. Quant. Finance **22**, 1–34 (2022)
39. M. Rosenbaum, J. Zhang, Deep calibration of the quadratic rough Heston model. Preprint. arXiv:2107.01611 (2021)
40. E.M. Stein, J.C. Stein, Stock price distributions with stochastic volatility: an analytic approach. Rev. Financ. Stud. **4**(4), 727–752 (1991)

Signature Trading Strategies

Owen Futter and Magnus Wiese

Abstract In this chapter, the primary objective is to recall the concept of a trading strategy and to extend this concept to the setting of signatures. We discuss how one can extend classical methods of optimal trading by tackling stochastic control problems through the lens of signatures, in order to incorporate path-dependencies. First introduced in (Signatures in machine learning and finance. PhD thesis, University of Oxford, 2020), the notion of a signature trading strategy allows to solve path-dependent optimization problems in a simple and explicit way. We first outline the key concepts of signature trading and highlight its associated benefits. Additionally, we present a solution to the mean-variance criterion and illustrate the advantages of signature-based methods using intuitive examples. Finally, we explore how signature trading strategies can be applied to obtain approximate solutions to optimal execution problems.

1 Introduction

1.1 Background and Motivation

Financial Market Participants
There are many different types of participants in financial markets, each with different objectives and preferences, deploying various methods and strategies depending on their reasons for trading at any given time. Certain investors, such as mutual funds, pension funds, and endowments, typically engage in longer-term investments without the need for frequent rebalancing of their positions. These firms generally opt for more classical portfolio construction techniques. On the other end

O. Futter (✉)
Imperial College London, London, UK
e-mail: o.futter21@imperial.ac.uk

M. Wiese
University of Kaiserslautern, Kaiserslautern, Germany

C. Bayer et al. (eds.), *Signature Methods in Finance*, Springer Finance Lecture Notes, https://doi.org/10.1007/978-3-031-97239-3_8

of the scale are fast algorithmic traders, including hedge funds and proprietary trading firms, who utilise quantitative methods to capitalise on short-term market inefficiencies, with more frequent, dynamic rebalancing of their positions. These participants rely heavily on sophisticated algorithms, machine learning models, and vast computational resources to execute trades at speeds measured in microseconds. They also employ risk models capable of managing a variety of more complex risks dynamically. The dichotomy between these market participants illustrates the diverse approaches to financial optimization, where slower asset managers aim for long-term stability and growth, while fast algorithmic traders pursue rapid gains through advanced quantitative techniques. In this chapter of the book, we are more focused on the latter, traders who generally already have access to predictive signals and want to transform those signals into a trading strategy while exploiting any path dependencies and temporal structures that may exist.

Modelling Signal and Noise

In any market setting, the interplay between buyers and sellers dictates the current price of any asset, reflecting the transactions that have taken place in that asset over some past time horizon. This price is influenced by the information available to market participants, encompassing intrinsic asset value, reactions to order flow, news, and broader economic, political, or geographic events. This information then causes market participants to buy or sell in reaction to it, creating chain reactions and feedback loops within the price. For example, if there is a flurry of buyers within a market, other participants may believe that the market has positive information (that they do not have) and also decide to buy, driving the price even higher. True *signal* emanates from diverse sources, arriving at different speeds as market participants react in various ways. Additionally, market orders may be placed by institutions that are driven by factors such as capital rebalancing or option flow hedging, which occur randomly without necessarily implying an expectation of future price movements. These "noisy" orders in the market are made without access to any exogenous signal to inform when or how that asset should be bought or sold, adding extra layers of noise on top of any true signal that may be present in asset price movements. Information within the market is also noisy itself, since it can be subjective (either positive or negative based on one's opinion). Consequently, the relationship between signal and noise in finance is complex and takes a variety of non-linear forms and varying speeds, giving rise to quantitative traders tasked with slicing through the layers of noise by deploying mathematical tools to model the behavior of asset prices.

1.2 Trading Strategies

Transforming a Signal into a Strategy

The design and development of trading strategies is a critical element of finance and has been the subject of extensive research over recent decades. Creating a

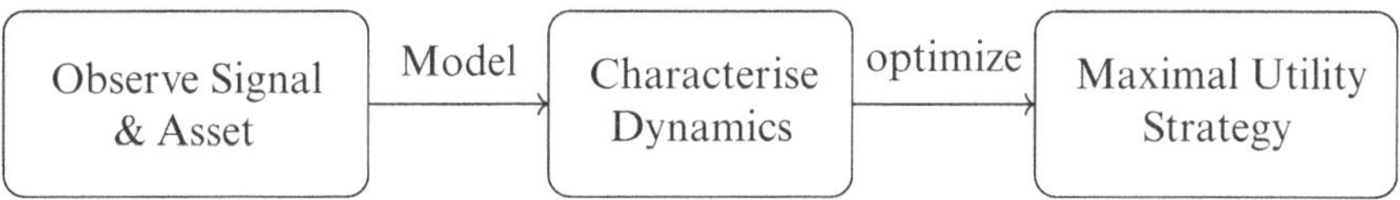

Fig. 1 Standard blueprint for obtaining an optimal trading strategy

trading strategy can often be divided into two areas: extracting *alpha* (expected excess returns) and managing the associated *execution* and *risk*. Methods of extracting alpha (or an edge) are done in a *modelling* phase, perhaps achieved through statistical techniques, in order to capture the signal within the noisy price dynamics mentioned previously. The exact modelling setup chosen is dependent on the underlying asset that is being traded, as well as the trade frequency and the investment horizon. Meanwhile, the choice of execution and allocation of the associated risk generally depends on your choice of model and objective criterion. Conventionally, choosing an optimal strategy with respect to some risk objective is done in an *optimization* phase, where various methods are used to transform the signal-asset dynamics in a way that maximizes the chosen utility function of the strategy's PnL, given the underlying model. A pioneering example of financial optimization is modern portfolio theory, introduced in Markowitz's seminal work [18] and then extended to continuous time in Merton's portfolio problem (both of which we discuss further in Sect. 1.3.1). These simple theories base portfolio allocation on investors' risk and return preferences, resulting in a well-diversified portfolio.

As illustrated in Fig. 1, the range of possible model choices is extensive. This includes various methods of feature engineering (selecting predictive exogenous variables), return prediction via statistical techniques, and the selection of probabilistic models. The choice of optimization technique is then tailored to the chosen utility function and model.

1.3 Classical Financial Optimization Problems

1.3.1 Portfolio Optimization

Portfolio optimization frameworks help investors decide how to allocate capital across a universe of assets, this is typically large and can be in the number of hundreds. Typically, such strategies are *self-financing*, meaning they operate with a fixed initial wealth, without additional capital injections over time. This assumption reflects the priority of growing the portfolio's value over the long term, rather than on capturing short-term market movements. In such self-financing portfolio strategies, the control (decision) variable is the portfolio's *weights*, representing the proportion of total capital allocated to each asset. These weights must sum to one, ensuring the portfolio remains fully invested without the need for external funds.

This framework is especially suitable for portfolios that do not require frequent rebalancing, as excessive trading incurs transaction costs that can quickly erode returns, especially for portfolios with a large number of assets. For rebalancing to be justified, any expected excess returns (or *alpha*) must be substantial enough to outweigh these costs, making infrequent adjustments generally more practical.

Markowitz Portfolio Framework (1952)
The classical Markowitz mean-variance problem [18] that led to Modern Portfolio Theory (MPT), is considered the foundational approach to portfolio optimization and provides a simple and useful benchmark. The Markowitz model is a *one-period* framework where an investor maintains a constant allocation $w = (w_1, \ldots, w_d)$ across d underlying assets at an initial time, say $t = 0$, until some time in future $t = T$. In this framework we assume that the returns of each asset are normally distributed, where the expected return of each asset from now until time T is given as $\boldsymbol{\mu} = (\mu_1, \ldots, \mu_d)^\top$ and the corresponding covariance of the returns is given in a covariance matrix Σ. The portfolio expected PnL (return), and variance of PnL at time T, is given as

$$\mathbb{E}(V_T) = w^\top \mu$$
$$\mathrm{Var}(V_T) = w^\top \Sigma w$$

The goal is to maximize expected *return* (change in wealth) for a specified level of *risk* (variance). If there are no weight constraints, the optimal solution is straightforward:

$$w^* = \Sigma^{-1}\mu.$$

When a weight constraint is imposed (e.g., weights summing to 1), the optimization problem can still be solved using quadratic programming, reducing to a linear system by applying a Lagrangian and first-order conditions. We can observe that this is among the simplest portfolio optimization models: it assumes normally distributed returns and uses *static* weights that remain fixed over the investment period. The objective function that we are considering is quadratic and hence convex, leaving a closed form, intuitive and easy to compute solution.

Merton's Portfolio Problem (1969)
Building on the success of Modern Portfolio Theory and utility maximization approaches like the Kelly criterion (1956), Merton extended portfolio optimization to continuous time, allowing for dynamic rebalancing over both finite and infinite time horizons. In Merton's framework, the price of the underlying asset follows the stochastic differential equation (SDE):

$$\frac{dX_t}{X_t} = \mu dt + \sigma dW_t \tag{1}$$

where $(W_t)_{t\geq 0}$ is a standard one-dimensional Brownian motion on the filtered probability space $(\Omega, \mathcal{F}, \mathbb{F} = (\mathcal{F}_t)_{t\geq 0}, \mathbb{P})$. We denote the *fraction* of wealth of the strategy time t as π_t, where the remainder of the investors wealth is invested in a risk-free asset, hence *self-financing*. Assuming we are trading over a short enough time horizon such that interest rates are negligible and there is no consumption of wealth, the PnL of the strategy at time T is given by:

$$V_T^{\xi} = \int_0^T \pi_t dX_t = \int_0^T X_t \pi_t \mu dt + \int_0^T X_t \pi_t \sigma dW_t$$

where the instantaneous gain/loss at time t is given as $dV_t = \pi_t dX_t$. We note here that the PnL V is simply the wealth of the process at time T minus the initial wealth at time 0, and this is the *control process* that we are able to influence with π. In stochastic control problems, our goal is to solve problems of the form

$$\sup_{\xi} \mathbb{E}\left[\int_0^T g\big(t, V_t^{\xi}\big)dt + F\big(T, V_T^{\xi}\big)\right] \tag{2}$$

where g is a running reward function and F is the terminal reward function, which are utility functions that are generally non-decreasing and concave. This extension allows for continuous rebalancing of the portfolio in order to maximize our given utility function.

Stochastic Portfolio Theory

Stochastic Portfolio Theory (SPT), introduced by Robert Fernholz [8], is a framework for analyzing and constructing portfolios in a continuous-time, market-relative setting. Unlike traditional approaches that focus on absolute returns, SPT models the dynamics of portfolio weights and asset market capitalizations over time, aiming to achieve robust portfolio growth relative to a benchmark. A key feature of SPT is that it does not rely on specific distributional assumptions about asset returns, and no requirement of *alpha* or expectation on the underlying drift. For a given universe of assets $X^1, \ldots, X^n$, each of which is assigned a *market weight* $\mu_i(t)$, satisfying $\sum_{i=1}^{d} \mu_i(t) = 1$, where the market weights will evolve over time as the asset prices change. The portfolio weights are denoted as $\pi(t) = (\pi_1(t), \ldots, \pi_d(t))$ and the portfolio *value* $V(t)$ evolves according to

$$dV(t) = V(t) \sum_{i=1}^{d} \pi_i(t) \frac{dX_i(t)}{X_i(t)}.$$

This theory is well-suited for longer term portfolios that aim to maximize long term *growth*, due to the multiplicative nature of the portfolio value, making it more analagous to the Kelly framework. Intuitively, if we assumed that our stock dynamics follow a geometric Brownian motion with static coefficients, then the optimal proportion of our wealth to invest should also be static in time. Therefore, when trading a martingale (and assuming you have no extra information), there is no temporal structure or *alpha* to exploit, and hence no path-dependent characteristics to model. However, the reality of financial markets is that volatility is not constant and is in fact stochastic and path-dependent, causing asset returns to be non-normal and fat-tailed over a fixed time horizon. Recently, an excellent use case of Stochastic Portfolio Theory was developed by Cuchiero and Möller in [6], in which the portfolio weights are parameterised as path-dependent functionals. These strategies are relatively general, and includes a particularly relevant case where the functional is modeled as a linear functional on the signature, an approach aligned with the topics of this chapter. This is particularly powerful in financial markets, since by using stochastic volatility and path-dependencies, you are able to better minimise risk and maximize growth over the longer term.

1.3.2 Optimal Trading and Execution

Alternatively, some traders and quantitative funds trade intraday at a higher frequency, and use market information (signals) to inform their decision making. These kind of financial market participant are generally not capital constrained, meaning the notion of being *fully invested* (weights summing to one) does not make sense here. Rather than having a fixed amount of cash to invest, they are instead constrained by *liquidity*, *volatility* and *transaction costs (market impact)*. It is common for trading strategies to be allocated a specific daily *dollar volatility*, meaning that they are constrained with how much their PnL can vary from day to day for a given average exposure of one dollar. These traders often only handle trading strategies with a much smaller universe of assets as opposed to those in the portfolio optimization case seen previously. With this, there are two main optimization tasks that are required to efficiently trade based on the information that they have. The first of which is finding the optimal amount of inventory (or *position*) to hold at a given time, such that it maximizes the expected PnL, subject to a terminal time variance constraint. Secondly, once the desired position in known, they are tasked with how to transition from the current position to the desired position, while minimizing the costs and risk associated with doing so. We discuss further the mean-variance optimization for the optimal *position* in Sect. 3 and the optimal execution case in Sect. 4.

1.4 Path-Dependent Optimization

In this chapter, we are concerned with finding optimal *systematic* and *dynamic* trading strategies such that the trader continuously updates their position as new information filters in. This is done with no discretion and is defined as a function of the market state, which continuously updates through time, leading to a new position for each time t. Throughout, we denote $T \in \mathbb{R}$ the terminal time. Let

$(\Omega, \mathcal{F}, (\mathcal{F}_t)_{t\in[0,T]}, \mathbb{P})$ be a filtered probability space. Furthermore, denote by $X = (X_t)_{t\in[0,T]}$ a non-negative $\mathbb{R}^d$-valued stochastic process satisfying $X_0^m = 1$ for $m \in \{1, \ldots, d\}$. We are interested in finding an optimal, predictable dynamic trading strategy $(\xi)_{t\in[0,T]}$ that maximizes the expected utility of the PnL

$$\max_{\substack{(\xi_t)_{t\in[0,T]} \\ \text{s.t constraints}}} \quad \mathbb{E}(U(V_T^{\xi})) \tag{3}$$

We can see that this optimization is within the stochastic control framework in (2), while in this chapter we do not consider a running reward function, although this is a natural extension. The optimal strategy ξ is a function of the market state (i.e. filtration $\mathcal{F}$ at time t) and its optimality is with respect to the constraints and utility function that the trader chooses. V_T^{ξ} is the terminal value of the trading strategy (i.e. the PnL),

$$V_T^{\xi} = \sum_{m=1}^{d} \int_0^T \xi_t^m dX_t^m. \tag{4}$$

In order to first fix the dynamics of the underlying asset price X, we usually need to construct a model. In the classical stochastic control frameworks, this may be done by modelling the asset price and the trader's wealth via a diffusion process such as (1) and using dynamic programming to solve the Hamilton-Jacobi-Bellman (HJB) equation. However for quantitative traders, it is most likely that they are trading under the presence of exogenous information f to enrich the filtration $\mathcal{F}$ (and hence the model of the dynamics of X) and so can estimate its coefficients through statistical techniques and forecasts. For example, in a one-period mean-variance optimization, we require estimates of the mean and variance of the underlying asset returns, which will need to be calibrated via some form of prediction or modelling. Therefore, a large number of models tend to fall into the *predict-then-optimize* framework, where assumptions are made on the asset returns/signals that are input into any prediction, leaving the final solution exposed to residual errors that are fat tailed and have non-trivial autocorrelation structure, eventually becoming compounded in the optimization phase, leading to large positions in assets.

Common objective criteria focus on a single trade-by-trade optimization basis, overlooking the potential path that the trading strategy will take. However, in practice signals and asset prices can have strong temporal dependencies and so subsequent trading strategy positions will inherit autocorrelation structure, which we can use to our advantage to exploit through increasing our expected returns, but also by recycling risks in a pathwise manner. As a motivating example in Fig. 2, we consider the comparison between a strategy that is mean-reverting (has autocorrelation) vs one that doesn't. Both strategies yield identical daily PnL distributions (and hence Sharpe ratio), but have different distributions at a future time due to the temporal structure within the strategy over time.

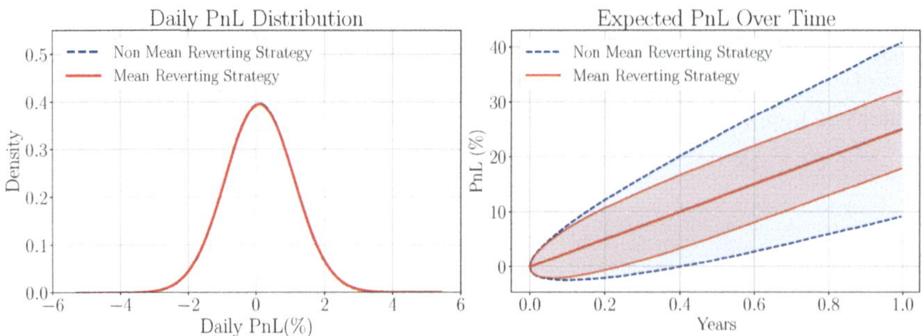

Fig. 2 Comparison of the PnL profile through time of a mean-reverting strategy vs a non mean-reverting strategy

The main difference from the stochastic control methods, is that we aim to work in a "model-free" sense, so we are not fixing a type of stochastic process of our underlying asset or any information (signal) that we may have. We instead choose a class of functions to model the trading strategy itself, this means we can bypass direct return prediction, resulting in an *end-to-end* optimization framework and thereby mitigating the accumulation of errors in the two-stage approach. Why would we do this? Since we want to optimize over a large class of stochastic processes, for example

$$dX_t = \mu(t, X_{0,t})dt + \sigma(t, X_{0,t})dW_t$$

where the SDE coefficients are dependent on the history of the path. For example if the drift is path-dependent, there is temporal structure that is exploitable as α and if there is path-dependent volatility, then we are able to manage our risk dynamically. The general form for a path-dependent trading strategy is

$$\xi_t = \phi(X_{0,t}).$$

Hence, the objective function that we are considering becomes

$$\mathbb{E}[U(V_T^{\xi})] = \mathbb{E}\left[U\left(\int_0^T \phi(X_{0,t})dX_t\right)\right] = J^{\phi}(X_{0,t}) \tag{5}$$

which is a path-dependent criterion, that is maximized by a path-dependent function ϕ^*. Like most function estimation/optimization, we require a class of functions to parameterise and then solve the optimization by finding the optimal parameters. In this chapter, we parameterise this class of functionals as linear functionals on the signature. In this chapter we also focus on optimising under a *dynamic* mean-variance criterion. The general dynamic mean-variance optimization can be framed as

$$\max_{(\xi_t)_{t\in[0,T]}} \mathbb{E}\left(V_T^\xi\right) - \frac{\lambda}{2}\operatorname{Var}\left(V_T^\xi\right), \tag{6}$$

where the quantity

$$V_T^\xi := \sum_{m=1}^{d} \int_0^T \xi_t^m \, dX_t^m$$

is the strategy PnL at some future time T and λ is a risk aversion parameter that can be solved for to obtain a specific level of variance. Many classical mean-variance techniques either require a one-period optimization, heavy modelling assumptions or require numerical techniques. However, by using signatures, we are able to recover a closed form solution when observing the class of signature trading strategies.

2 Signature Trading Strategies

First introduced in the work of Perez et al. in [1, 14, 17], *Signature Trading Strategies* were adopted to capture a large class of trading strategies by representing them as a linear functional applied to the signature of the past path. Naturally, this makes sense, since in Sect. 1.1 we discussed how signal takes complex path-dependent forms. We can think of a trading strategy as a decision made using the knowledge of the current state (e.g. the previous price path, plus some exogenous trading signal); in other words as a function from path space to some decision process. This inevitably raises the question—can the class of linear functionals on the signature be seen as a rich enough class of maps that represent such trading strategies? We will show that this is possible to due to signatures' capability to approximate continuous functions on paths. Also perhaps crucially, the Signature Trading framework does not impose the restriction that the underlying asset or signal be Markovian, allowing the Sig-Trader to capture auto-correlation and mean-reverting behaviours within the process that perhaps some more classical methods are not able to exploit. This enables us to incorporate path-dependencies into our trading decisions, while still obtaining a closed-form solution, that is easy to compute and simple to analyse. Throughout the remainder of this chapter there are some notations that we will frequently use and for the avoidance of doubt (since there is often different notation for the same mathematical objects), we list them below.

Notation

$T((\mathbb{R}^d))$	extended tensor algebra over $\mathbb{R}^d$
$T^{(N)}(\mathbb{R}^d)$	truncated tensor algebra over $\mathbb{R}^d$ of order $N \in \mathbb{N}$
$T((\mathbb{R}^d)^*)$	dual space of the tensor algebra
C^{p-var}	the space of paths with bounded p-variation
$G^{\lfloor p \rfloor}(\mathbb{R}^d)$	the space of geometric p-rough paths
X	continuous underlying tradeable asset process
$\hat{X}$	time-augmented process of X, (t, X)
$\hat{X}^{LL}$	Hoff lead-lag process of $\hat{X}$
$\mathbb{X}^n$	n-th term of the signature of X
$\mathbb{X}^{\leq M}$	signature of X truncated at order N
$\mathbb{X}^{<\infty}$	full signature of X
α	un-tradeable exogenous signal
$\hat{Z}$	market signal process (t, X_t, α_t)
$\mathcal{Z}^\alpha$	space of all market signal trajectories
$\hat{\mathbb{Z}}^{LL,<\infty}$	full signature of the Hoff lead-lag process $\hat{Z}^{LL}$
$\mathbf{w}$	a word corresponding to a multi-index of the tensor algebra
$\mathbf{f}(m)$	an index mapping from one word to another
$\mathbf{wv}$	concatenation of two words $\mathbf{w}$ and $\mathbf{v}$
$\mathcal{A}_d$	alphabet of $d \in \mathbb{N}$ letters
$\mathcal{W}(\mathcal{A}_d)$	vector space of all words with letters in the alphabet of $\mathcal{A}_d$
$\sqcup\!\sqcup$	shuffle product
$\mathcal{T}^Z$	space of continuous trading strategies with respect to $Z = (t, X, \alpha)$
$\mathcal{T}^Z_{\text{sig}}$	space of linear signature trading strategies with respect to Z
$\mathcal{S}^Z_{\text{sig}}$	space of linear signature trading speeds with respect to Z
ℓ	linear functional as an element in the dual space of the tensor algebra
ξ	an $\mathbb{R}^d$-valued trading strategy process
θ	an $\mathbb{R}$-valued trading speed

2.1 Signature Trading Preliminaries

Definition 2.1 (Hoff Lead-Lag Process, [9, 13]) Let $\hat{X} : [0, 1] \to \mathbb{R}^{d+1}$ be the continuous time-augmented process of X, discretely sampled at $t = t_0, \ldots, t_{2N} \in [0, 1]$. The *Hoff lead-lag transformed path* is defined as the piecewise linear interpolation $\hat{X}^{LL} : [0, 1] \to \mathbb{R}^{2(d+1)}$ such that

$$(\hat{X}^{LL}_{t_i})_{i=1}^{2N} = (\hat{X}^{\text{lag}}_{t_i}, \hat{X}^{\text{lead}}_{t_i})_{i=1}^{2N}$$

where

$$\hat{X}_t^{\text{lead}} = \begin{cases} \hat{X}_{t_{k+1}}, & \text{if } t \in [\frac{2k}{2N}, \frac{2k+1}{2N}] \\ \hat{X}_{t_{k+1}} + 2(t-(2k+1))(X_{t_{k+2}} - X_{t_{k+1}}), & \text{if } t \in [\frac{2k+1}{2N}, \frac{2k+3/2}{2N}] \\ \hat{X}_{t_{k+2}}, & \text{if } t \in [\frac{2k+3/2}{2N}, \frac{2k+2}{2N}], \end{cases}$$

$$\hat{X}_t^{\text{lag}} = \begin{cases} \hat{X}_{t_k}, & \text{if } t \in [\frac{2k}{2N}, \frac{2k+3/2}{2N}] \\ \hat{X}_{t_{k+1}} + 2(t-(2k+\frac{3}{2}))(X_{t_{k+1}} - X_{t_k}), & \text{if } t \in [\frac{2k+3/2}{2N}, \frac{2k+2}{2N}]. \end{cases}$$

Remark 2.2 There also exists a more intuitive and straightforward definition of a lead-lag process, however we decide not to opt for this version and instead consider the so-called *Hoff process* [13], due to its fundamental properties in the case when we want to calculate the PnL of our trading strategy (Theorem 2.14). This is due to the powerful, non-trivial result of Theorem 2.13, proven in [9] that states that the Itô integral of a function of a process X against itself, can be recovered via the components of the Hoff lead-lag transformation.

Definition 2.3 (Linear Functional on the Tensor Algebra) Note that there is a natural pairing between the extended tensor algebra $T((\mathbb{R}^d))$ and its dual space $T((\mathbb{R}^d)^*)$, by which we denote

$$\langle \cdot, \cdot \rangle : T((\mathbb{R}^d)^*) \times T((\mathbb{R}^d)) \to \mathbb{R}$$

and is given by

$$\langle \ell, \mathbb{X} \rangle = \sum_{\mathbf{w} \in \mathcal{W}(\mathcal{A}_d)} \ell_{\mathbf{w}} \mathbb{X}^{\mathbf{w}}$$

for $\ell \in T((\mathbb{R}^d)^*)$, $\mathbb{X} \in T((\mathbb{R}^d))$. We make clear that there exists a canonical isomorphism $(\mathbb{R}^d)^* \cong \mathbb{R}^d$ through the mapping $(\mathbb{R}^d)^* \ni \langle \ell, \cdot \rangle \mapsto \ell \in \mathbb{R}^d$.

Definition 2.4 (Shuffle Product Property) Let $\mathbb{X}_{0,T}^{<\infty} \in G^{\lfloor p \rfloor}(\mathbb{R}^d)$ be a geometric rough path and let $\ell_1, \ell_2 \in T((\mathbb{R}^d)^*)$ be elements of the dual space of the tensor algebra, then

$$\langle \ell_1, \mathbb{X}_{0,T}^{<\infty} \rangle \langle \ell_2, \mathbb{X}_{0,T}^{<\infty} \rangle = \langle \ell_1 \sqcup\!\sqcup\, \ell_2, \mathbb{X}_{0,T}^{<\infty} \rangle.$$

We can define the shuffle product $\sqcup\!\sqcup : \mathcal{W}(\mathcal{A}_d) \times \mathcal{W}(\mathcal{A}_d) \to \mathcal{W}(\mathcal{A}_d)$ inductively by

$$\mathbf{ua} \sqcup\!\sqcup\, \mathbf{vb} = (\mathbf{u} \sqcup\!\sqcup\, \mathbf{vb})\mathbf{a} + (\mathbf{ua} \sqcup\!\sqcup\, \mathbf{v})\mathbf{b}$$

$$\mathbf{w} \sqcup\!\sqcup\, \emptyset = \emptyset \sqcup\!\sqcup\, \mathbf{w} = \mathbf{w}$$

for all words $\mathbf{u}, \mathbf{v}$ and letters $\mathbf{a}, \mathbf{b} \in \mathcal{W}(\mathcal{A}_d)$.

Example 2.5 Let $\mathbf{w} = \mathbf{12}$, $\mathbf{v} = \mathbf{34}$, then $\mathbf{w} \sqcup\!\sqcup \mathbf{v}$ is given by

$$\mathbf{12} \sqcup\!\sqcup \mathbf{34} = \mathbf{1234} + \mathbf{1324} + \mathbf{1342} + \mathbf{3124} + \mathbf{3142} + \mathbf{3412}$$

Remark 2.6 We make extensive use of the fact that polynomials of linear functionals can be expressed as shuffle products of the linear functionals themselves.

Definition 2.7 (Trading Signal) We will treat α_t to be an un-tradable exogenous trading signal. This can be seen of as a prediction of a particular asset return (for a given future time-horizon) or perhaps a critical piece of data that we believe is informative about the future price movement of the asset that we are trading. We may often interchange the use of the words *signal, alpha* and *factor* all to mean the same thing.

Definition 2.8 (Market Signal Process) Let $X = (X_t)_{t\in[0,T]}$ be a d-dimensional tradable underlying asset process and $\{\alpha^i\}_{i=1}^N$ are N un-tradable exogenous trading signals. Then we define the (time-augmented) *market signal process* as the $(1+d+N)$-dimensional process

$$\hat{Z}_t := (t, X_t, \alpha_t)$$

that induces the natural filtered probability space $(\Omega, \mathcal{F}^Z, (\mathcal{F}_t^Z)_{t\in[0,T]}, \mathbb{P})$, where the filtration of X satisfies $\mathcal{F}^X \subseteq \mathcal{F}^Z$ such that X is driven by $\{\alpha^i\}_{i=1}^N$. Throughout the remainder of this paper, we maintain such assumptions on $\hat{Z}$. We define as the space of all *market signal trajectories* for given assets X and signals α,

$$\mathcal{Z}^f := \left\{\hat{Z}_t = (t, X_t, \alpha_t) \;\middle|\; X\colon [0, T] \to \mathbb{R}^d, \alpha : [0, T] \to \mathbb{R}^N \text{ and } Z_0 = (0, 1, 1)\right\}.$$

Definition 2.9 (Exogenous Signature Trading Strategy) Let $\hat{Z}$ be a market signal process as defined in Definition 2.8. Let $\xi = (\xi)_{t\in[0,T]}$ be an adapted, $\mathcal{F}_t^Z$-predictable and integrable strategy such that $\int_0^T \xi_s^2 ds < \infty$. If the strategy ξ is then a function of the market state, that is $\xi_t = \phi(\hat{Z}_{0,t})$, a continuous function of the previous *market signal trajectory* up to time t. We can extend ξ to be a *signature trading strategy*, such that

$$\xi_t^m = \langle \ell_m, \hat{\mathbb{Z}}_{0,t}\rangle \approx \phi^m(t, Z_{0,t}), \quad \forall m = 1, \ldots, d.$$

We define the space of all *exogenous signature trading strategies* with respect to exogenous signals α as

$$\mathcal{T}_{\text{sig}}^Z := \left\{(\xi_t)_{t\in[0,T]} \;\middle|\; \text{if } \exists\, \ell \in T((\mathbb{R}^{N+d+1})^*) \text{ s.t } \xi_t = \langle \ell, \hat{\mathbb{Z}}_{0,t}\rangle \text{ and } \int_0^T \xi_s^2 ds < \infty\right\}.$$

A natural question to ask is if this class of strategies $\mathcal{T}_{\text{sig}}^{Z}$ is reasonable or if in fact it is a large enough class of trading strategies to consider. In fact, we can show that the class of signature trading strategies is dense in the space of continuous trading strategies on compact sets of paths.

Lemma 2.10 *Let $K \subset C^{1-var}([0,T],\mathbb{R}^d) \subset \mathcal{Z}^f$ be a compact set of market signal trajectories. Then for any exogenous signature trading strategy $\xi = \phi(\hat{Z}_{0,T})$ that acts on paths in K and for every $\epsilon > 0$, there exists a linear functional $\ell \in T((\mathbb{R}^{N+d+1})^*)$, such that*

$$\sup_{Z\in K} \|\phi(\hat{Z}_{0,T}) - \langle \ell, \hat{\mathbb{Z}}_{0,T}\rangle\| < \epsilon.$$

where the choice of suitable candidate topology is discussed in [5].

Proof We can see that this result follows from the universal approximation of continuous functions on paths by linear functionals acting on the signature. □

Remark 2.11 If instead we wanted to trade *endogenously* without any trading signals, we would simply take the market signals to be the null-process $\emptyset$, such that $\hat{X} = \hat{Z}$, the results would still hold.

Now that we have defined the framework required in order to trade a signature trading strategy embedded with market signals, we can state a key result in order to combine the importance of the Hoff process in Theorem 2.13, that allows us to explicitly state the PnL of the exogenous sig-trader without the need for an integral at all.

2.2 Signature Trading Strategy PnL

Definition 2.12 (Trading Strategy PnL) Let $\hat{Z} = (t, X_t, \alpha_t)$ be a market signal process with X a tradable underlying process and α an untradable signal process. If $\xi \in \mathcal{T}_{\text{sig}}^{Z}$ is a linear signature trading strategy such that $\xi_t^m = \langle \ell_m, \hat{\mathbb{Z}}_{0,t}\rangle$, for each asset $m = 1, \ldots, d$, then we define the PnL of the signature trading strategy between time 0 and time T as

$$V_T = \sum_{m=1}^{d} \int_0^T \langle \ell_m, \hat{\mathbb{Z}}_{0,t}\rangle dX_t^m \tag{7}$$

where the integral is understood in the Itô sense.

It is crucial to distinguish the difference between the Itô integral used in this definition versus the Stratonovich integral in the definition of the signature. If that integral was in fact an Itô integral then we could describe the above definition of PnL directly in terms of the add-time signature $\hat{\mathbb{Z}}_{0,t}$, however this is unfortunately not the case! So in order to develop a more friendly version of V_T, we require the following theorem involving the *Hoff lead-lag process* as seen in Definition 2.1.

Theorem 2.13 (Recovery of Itô Integral Using the Hoff Process, (Theorem 5.1, [9])) *Let $X = (X_t)_{t\in[0,T]}$ be a stochastic process on the filtered probability space $(\Omega, \mathcal{F}, (\mathcal{F}_t)_{t\in[0,T]}, \mathbb{P})$. Suppose we observe piecewise smooth streams of X that are discretely sampled over a sequence of times $\{t_i\}_{i=0}^N$. Let $\hat{X}^{LL} = (\hat{X}^{\text{lag}}_{t_i}, \hat{X}^{\text{lead}}_{t_i})_{i=1}^{2N}$ be the associated observed Hoff lead-lag transform of $\hat{X}$ as defined in Definition 2.1. Let $\phi = (\phi^1, \dots, \phi^d)$ be continuous functions on paths as described in Lemma 2.10, then we have that*

$$\sum_{m=1}^{d} \int_0^T \phi^m(\hat{X}^{\text{lag}}_t) d\hat{X}^{m,\text{lead}}_t \to \int_0^T \phi(X_t) dX_t := \sum_{m=1}^{d} \int_0^T \phi^m(X_t) dX^m_t$$

$$\text{as} \max_{t_i, t_{i+1}} |t_{i+1} - t_i| \to 0,$$

in either probability or L^p-norm.

Proof The basic idea of the proof is that the areas between the *lead* and *lag* components of the Hoff lead-lag process (captured by $\hat{X}^{LL}$), introduce a correction factor in the stochastic integral limit, consequently allowing us to recover the Itô, not Stratonovich, integral. □

This result allows us to show that applying a function to the observed lagged path, and integrating against the observed leading path, we recover the true Itô integral against the stochastic process X as the mesh size goes to zero. In this sense, we see that we are able to approximate asymptotically the true Itô integral as instead an integral of the observed, Hoff lead-lag stream. A natural next step is to consider the case where our strategy ξ is in fact a *signature trading strategy*, i.e. $\xi^m_t = \langle \ell_m, \hat{\mathbb{Z}}_{0,t} \rangle$ for each asset $m = 1, \dots, d$. Moreover, how can we find an expression for the PnL V_T in (7), in the Itô integral sense, using Theorem 2.13?

Theorem 2.14 (PnL of a d-Asset Signature Trading Strategy Under Exogenous Signal, [11, Theorem 2.11]) *Let $\hat{Z} := (t, X_t, \alpha_t)_{t\in[0,T]}$ be the market signal process, where X is a d-dimensional tradable stochastic process and α is a N-dimensional un-tradable signal process. Let $\ell_1, \dots, \ell_d \in T((\mathbb{R}^{N+d+1})^*)$ and define our trading strategy as $\xi^m_t = \langle \ell_m, \hat{\mathbb{Z}}^{<\infty}_{0,t} \rangle$ for $m = 1, \dots, d$. Then, we have that the PnL of the strategy between time 0 and time T can be represented as*

$$V_T = \sum_{m=1}^{d} \int_0^T \langle \ell_m, \hat{\mathbb{Z}}^{<\infty}_{0,t} \rangle dX^m_t \approx \sum_{m=1}^{d} \int_0^T \langle \ell_m, \hat{\mathbb{Z}}^{\text{lag}, <\infty}_{0,t} \rangle dX^{m,\text{lead}}_t \tag{8}$$

$$= \sum_{m=1}^{d} \langle \ell_m \mathbf{f}(m), \hat{\mathbb{Z}}^{LL, <\infty}_{0,T} \rangle \tag{9}$$

where $\mathbf{f}(m) : \{1, \ldots, d\} \to \mathcal{W}(\mathcal{A}_{2(N+d+1)})$ *is a shift operator which is defined as* $\mathbf{f}(m) = \pi(e^*_{m+N+d+1})$ *where* $\pi : T((\mathbb{R}^{2(N+d+1)})^*) \to \mathcal{W}(\mathcal{A}_{2(N+d+1)})$ *the canonical isomorphism between the dual space* $T((\mathbb{R}^{2(N+d+1)})^*)$ *and the space of all words* $\mathcal{W}(\mathcal{A}_{2(N+d+1)})$.

Now, due to the linearity of expectation, we are able to represent the *expected PnL* at time T as the following:

$$\mathbb{E}(V_T) = \sum_{m=1}^{d} \langle \ell_m \mathbf{f}(m), \mathbb{E}(\hat{\mathbb{Z}}^{LL,<\infty}_{0,T}) \rangle. \tag{10}$$

where $\mathbb{E}(\hat{\mathbb{Z}}^{LL,<\infty}_{0,T})$ is the expected signature of the Hoff lead-lag transformation of $\hat{Z}$.

Remark 2.15 Note that $\mathbf{f}(m) : \{1, \ldots, d\} \to \mathcal{W}(\mathcal{A}_{2(N+d+1)})$ is simply just the shift operator that allows any multi-index $(j_1, \ldots, j_n) \in \{1, \ldots, d\}^n$ for the original signature terms to be transformed into a multi-index for the signature terms of the $2(N+d+1)$-dimensional time-augmented lead-lag process. Let us consider the case of when we have two assets $X = (X^1, X^2)$, one signal α, such that $d = 2$, $N = 1$. Then we define the 4-dimensional market signal process $\hat{Z} := (t, X^1, X^2, \alpha)$ and its corresponding Hoff lead-lag process $\hat{Z}^{LL}$ will have dimension $2(N + d + 1) = 8$. Given that the assets correspond to $m = 1, 2$, then we define the shift operator for each asset as

$$\mathbf{f}(1) = e^*_{1+N+d+1} = \mathbf{5}$$
$$\mathbf{f}(2) = e^*_{2+N+d+1} = \mathbf{6}$$

This simply means that the linear function ℓ is then applied to the correct expected lead-lag signature terms. By correct, we mean the dimensions of the lead-lag process that correspond to the *lead* part of the integrator $dX_t^{m,\text{lead}}$ in the Itô integral of (8). To show this further, the dimensions of the lead-lag process are as follows:

$$(t^{\text{lag}}, \underbrace{X^{1,\text{lag}}, X^{2,\text{lag}}}_{m=1,2}, \alpha^{1,\text{lag}}, t^{\text{lead}}, \underbrace{X^{1,\text{lead}}, X^{2,\text{lead}}}_{\mathbf{f}(1)=\mathbf{5}, \mathbf{f}(2)=\mathbf{6}}, \alpha^{1,\text{lead}})$$

More intuition on the shift operator will be provided in Example 3.3.

Proof The full proof is given in [11, Appendix B.1]. □

Remark 2.16 This result allows us to express any integral of a linear functional on the signature of a time-augmented market signal path as a newly defined linear functional on the time-augmented lead-lag market signal path instead.

Lemma 2.17 (Variance of the PnL of a Signature Trading Strategy Under Exogenous Signal, [11, Lemma 2.12]) *Let X be a d-dimensional tradable stochastic process and let α be a N-dimensional un-tradable signal process. Define $\hat{Z}_t := (t, X_t, \alpha_t)$ as the market signal process and $\ell_1, \ldots, \ell_d \in T((\mathbb{R}^{N+d+1})^*)$ and define our trading strategy as $\langle \ell_m, \hat{\mathbb{Z}}^{<\infty}_{0,s} \rangle$ for $m = 1, \ldots, d$. Let the expected PnL of the strategy between time 0 and time T be defined as in (10). Then the Variance of the PnL at time T is defined as*

$$\mathrm{Var}(V_T) = \sum_{m=1}^{d} \sum_{n=1}^{d} \langle \ell_m \mathbf{f}(m) \sqcup\!\sqcup\, \ell_n \mathbf{f}(n), \mathbb{E}(\hat{\mathbb{Z}}^{LL}_{0,T}) \rangle - \langle \ell_m \mathbf{f}(m), \mathbb{E}(\hat{\mathbb{Z}}^{LL}_{0,T}) \rangle \langle \ell_n \mathbf{f}(n), \mathbb{E}(\hat{\mathbb{Z}}^{LL}_{0,T}) \rangle.$$

where $\sqcup\!\sqcup$ is the shuffle product defined in Definition 2.4.

Proof This results follows simply from Theorem 2.14 and the fact that variance of a random variable is defined as $\mathrm{Var}(V_T) = \mathbb{E}(V_T^2) - (\mathbb{E}(V_T))^2$, where $\mathbb{E}(V_T)$ is defined in (10) and

$$\mathbb{E}(V_T^2) = \sum_{m=1}^{d} \sum_{n=1}^{d} \langle \ell_m \mathbf{f}(m) \sqcup\!\sqcup\, \ell_n \mathbf{f}(n), \mathbb{E}(\hat{\mathbb{Z}}^{LL}_{0,T}) \rangle \tag{11}$$

$$\mathbb{E}(V_T)^2 = \sum_{m=1}^{d} \sum_{n=1}^{d} \langle \ell_m \mathbf{f}(m), \mathbb{E}(\hat{\mathbb{Z}}^{LL}_{0,T}) \rangle \langle \ell_n \mathbf{f}(n), \mathbb{E}(\hat{\mathbb{Z}}^{LL}_{0,T}) \rangle. \tag{12}$$

□

3 Path-Dependent Mean-Variance Optimization

In this section we derive an explicit and concise expression for the optimal signature trading strategy in the presence of exogenous market signals considering a path-dependent extension of the classical mean-variance criterion, as seen in [11]. Using the trading strategy expected PnL and variance of PnL as defined in Sect. 2.2, we show that the path-dependent mean-variance optimization problem is convex in the weights of the linear functionals $\ell_1, \ldots, \ell_d$.

3.1 Mean-Variance Optimal Solution

Theorem 3.1 (Optimal Mean-Variance Signature Trading Strategy [11, Theorem 3.1]) *Denote $T \in \mathbb{N}$ the terminal time. Let X be a d-dimensional tradable* *stochastic process and let* α *be a* N*-dimensional un-tradable signal process. Define* $\hat{Z}_t := (t, X_t, \alpha_t)$ *as the market signal process. Define a signature trading strategy* ξ_t^m *through a linear functional on the signature of the market signal process, i.e.* $\xi_t^m = \langle \ell_m, \hat{\mathbb{Z}}_{0,t} \rangle$*. Then, for a given truncation level* M*, the mean-variance optimal signature trading strategy* $\ell^* = (\ell_1^*, \ldots, \ell_d^*)$*,* $\ell_m^* \in T^{(M)}((\mathbb{R}^{N+d+1})^*)$*, satisfies*

$$\ell_m^* := \underset{\substack{\ell_m \in T^{(M)}((\mathbb{R}^{N+d+1})^*) \\ Var(V_T) \leq \Delta}}{\operatorname{argmax}} \sum_{m=1}^{d} \left\langle \ell_m \mathbf{f}(m), \mathbb{E}(\hat{\mathbb{Z}}_{0,T}^{LL,<\infty}) \right\rangle, \quad \forall m \in \{1, \ldots, d\} \tag{13}$$

and is given by

$$\langle \ell_m^*, e_{\mathbf{w}} \rangle = \frac{((\Sigma^{\text{sig}})^{-1} \boldsymbol{\mu}^{\text{sig}})_{\mathbf{wf}(m)}}{2\lambda}, \qquad m \in \{1, \ldots, d\}, \mathbf{w} \in \mathcal{W}_{N+d+1}^M$$

where the variance-scaling parameter λ *is given by*

$$\lambda = \frac{1}{2\sqrt{\Delta}} \left(\sum_{m=1}^{d} \sum_{n=1}^{d} \sum_{\mathbf{w},\mathbf{v} \in \mathcal{W}_{N+d+1}^M} ((\Sigma^{\text{sig}})^{-1} \boldsymbol{\mu}^{\text{sig}})_{\mathbf{wf}(m)} ((\Sigma^{\text{sig}})^{-1} \mu^{\text{sig}})_{\mathbf{vf}(n)} \Sigma_{\mathbf{wf}(m),\mathbf{vf}(n)}^{\text{sig}} \right)^{\frac{1}{2}}.$$

We define the "Signature PnL attribution" as the $d \cdot |\mathcal{W}_{N+d+1}^M|$*-length vector* $\mu^{\text{sig}} = (\mu_1^{\text{sig}}, \ldots, \mu_d^{\text{sig}})^\top$ *as*

$$\mu_{\mathbf{wf}(m)}^{\text{sig}} = \left\langle \mathbf{wf}(m), \mathbb{E}(\hat{\mathbb{Z}}_{0,T}^{LL,<\infty}) \right\rangle, \quad \forall \mathbf{w} \in \mathcal{W}_{N+d+1}^M, m \in \{1, \ldots, d\} \tag{14}$$

and the "Signature PnL covariances" as the $d \cdot |\mathcal{W}_{N+d+1}^M| \times d \cdot |\mathcal{W}_{N+d+1}^M|$ *matrix* Σ^{sig} *as*

$$\Sigma_{\mathbf{wf}(m),\mathbf{vf}(n)}^{\text{sig}} = \left\langle \mathbf{wf}(m) \sqcup\!\sqcup \mathbf{vf}(n), \mathbb{E}(\hat{\mathbb{Z}}_{0,T}^{LL,<\infty}) \right\rangle \tag{15}$$

$$- \left\langle \mathbf{wf}(m), \mathbb{E}(\hat{\mathbb{Z}}_{0,T}^{LL,<\infty}) \right\rangle \left\langle \mathbf{vf}(n), \mathbb{E}(\hat{\mathbb{Z}}_{0,T}^{LL,<\infty}) \right\rangle \tag{16}$$

for all $\mathbf{w}, \mathbf{v} \in \mathcal{W}_{N+d+1}^M$ *and* $m, n \in \{1, \ldots, d\}$.

Proof In order to solve the constrained optimization problem (13), we can introduce the Lagrangian

$$\mathcal{L}(\ell_1, \ldots, \ell_d, \lambda) = \mathbb{E}(V_T) - \lambda(\text{Var}(V_T) - \Delta)$$

and we are interested in finding saddle points $\ell_1^*, \ldots, \ell_d^* \in T^{(M)}((\mathbb{R}^{N+d+1})^*)$, $\lambda^0 \in \mathbb{R}$ that satisfy $\nabla \mathcal{L}(\ell_1^*, \ldots, \ell_d^*, \lambda) = 0$. For this purpose recall Theorem 2.14, where the expected terminal PnL is given by (10). Furthermore, we can decompose the variance of the terminal PnL, given in Eqs. (11) and (12). Next, we can compute for each asset $m \in \{1, \ldots, d\}$ and each word $\mathbf{w} \in \mathcal{W}_{N+d+1}^M$, the gradients of (10), (11), and (12) with respect to $\langle \ell_m, e_\mathbf{w} \rangle$

$$\frac{\partial \mathbb{E}(V_T)}{\partial \langle \ell_m, e_\mathbf{w} \rangle} = \langle \mathbf{wf}(m), \mathbb{E}(\hat{\mathbb{Z}}_{0,T}^{LL,<\infty}) \rangle$$

$$\frac{\partial \mathbb{E}(V_T^2)}{\partial \langle \ell_m, e_\mathbf{w} \rangle} = 2 \sum_{n=1}^{d} \sum_{\mathbf{v} \in \mathcal{W}_{N+d+1}^M} \langle \ell_n, e_\mathbf{v} \rangle \langle \mathbf{wf}(m) \sqcup\!\sqcup \mathbf{vf}(n), \mathbb{E}(\hat{\mathbb{Z}}_{0,T}^{LL,<\infty}) \rangle$$

$$\frac{\partial \mathbb{E}(V_T)^2}{\partial \langle \ell_m, e_\mathbf{w} \rangle} = 2 \sum_{n=1}^{d} \sum_{\mathbf{v} \in \mathcal{W}_{N+d+1}^M} \langle \ell_n, e_\mathbf{v} \rangle \langle \mathbf{wf}(m), \mathbb{E}(\hat{\mathbb{Z}}_{0,T}^{LL,<\infty}) \rangle \langle \mathbf{vf}(n), \mathbb{E}(\hat{\mathbb{Z}}_{0,T}^{LL,<\infty}) \rangle$$

Using these computed gradients, we can compute the gradient of the Lagrangian with respect to $\langle \ell_m, e_\mathbf{w} \rangle$

$$\frac{\partial \mathcal{L}(\ell_1, \ldots, \ell_d, \lambda)}{\partial \langle \ell_m, e_\mathbf{w} \rangle} = \langle \mathbf{wf}(m), b \rangle - 2\lambda \sum_{n=1}^{d} \sum_{\mathbf{v} \in \mathcal{W}_{N+d+1}^M} \langle \ell_n, e_\mathbf{v} \rangle \Sigma_{\mathbf{wf}(m),\mathbf{vf}(n)}^{\text{sig}}$$

where we define $\mu^{\text{sig}} = (\mu_1^{\text{sig}}, \ldots, \mu_d^{\text{sig}})^\top$ as

$$\mu_{\mathbf{wf}(m)}^{\text{sig}} = \langle \mathbf{wf}(m), \mathbb{E}(\hat{\mathbb{Z}}_{0,T}^{LL,<\infty}) \rangle, \quad \forall \mathbf{w} \in \mathcal{W}_{N+d+1}^M, m \in \{1, \ldots, d\}$$

and the matrix Σ^{sig} as

$$\begin{aligned} \Sigma_{\mathbf{wf}(m),\mathbf{vf}(n)}^{\text{sig}} =& \langle \mathbf{wf}(m) \sqcup\!\sqcup \mathbf{vf}(n), \mathbb{E}(\hat{\mathbb{Z}}_{0,T}^{LL,<\infty}) \rangle \\ &- \langle \mathbf{wf}(m), \mathbb{E}(\hat{\mathbb{Z}}_{0,T}^{LL,<\infty}) \rangle \langle \mathbf{vf}(n), \mathbb{E}(\hat{\mathbb{Z}}_{0,T}^{LL,<\infty}) \rangle \end{aligned}$$

for all $\mathbf{w}, \mathbf{v} \in \mathcal{W}_{N+d+1}^M$ and $m, n \in \{1, \ldots, d\}$.

Using the first order conditions, observe that we obtain the system of linear equations

$$\mu_{\mathbf{wf}(m)}^{\text{sig}} = 2\lambda \sum_{n=1}^{d} \sum_{\mathbf{v} \in \mathcal{W}_{N+d+1}^M} \langle \ell_n, e_\mathbf{v} \rangle \Sigma_{\mathbf{wf}(m),\mathbf{vf}(n)}^{\text{sig}}, m \in \{1, \ldots, d\}, \mathbf{w} \in \mathcal{W}_{N+d+1}^M. \tag{17}$$

Under truncation, we find that (17) is a system of

$$\begin{aligned} d_M &= |I| \cdot |\mathcal{W}_{N+d+1}^M| \\ &= d \sum_{k=0}^{M} (N+d+1)^k \\ &= (N+d+1)^{M+1} - 1 \end{aligned}$$

equations and d_M unknowns. Hence, assuming that Σ^{sig} is invertible, we can solve (17) and obtain for $m \in \{1, \ldots, d\}$, $\mathbf{w} \in \mathcal{W}_{N+d+1}^M$

$$\langle \ell_m^*, e_{\mathbf{w}} \rangle = \frac{((\Sigma^{\text{sig}})^{-1} \mu^{\text{sig}})_{\mathbf{wf}(m)}}{2\lambda} \tag{18}$$

where we assumed by complementary slackness (KKT) that $\lambda \neq 0$. We can then substitute (18) into the variance constraint to obtain two solutions for λ

$$\begin{aligned} \lambda_{\pm} &= \pm \frac{1}{2\sqrt{\Delta}} \left(\sum_{m=1}^{d} \sum_{n=1}^{d} \sum_{\mathbf{w},\mathbf{v} \in \mathcal{W}_{N+d+1}^M} \langle \ell_m, e_{\mathbf{w}} \rangle \langle \ell_n, e_{\mathbf{v}} \rangle \Sigma_{\mathbf{wf}(m),\mathbf{vf}(n)}^{\text{sig}} \right)^{\frac{1}{2}} \\ &= \pm \frac{1}{2\sqrt{\Delta}} \left(\sum_{m=1}^{d} \sum_{n=1}^{d} \sum_{\mathbf{w},\mathbf{v} \in \mathcal{W}_{N+d+1}^M} ((\Sigma^{\text{sig}})^{-1} \mu^{\text{sig}})_{\mathbf{wf}(m)} ((\Sigma^{\text{sig}})^{-1} \mu^{\text{sig}})_{\mathbf{vf}(n)} \right. \\ &\qquad \left. \times \Sigma_{\mathbf{wf}(m),\mathbf{vf}(n)}^{\text{sig}} \right)^{\frac{1}{2}} \end{aligned}$$

Hence, we obtain the solution

$$\langle \ell_m^*, e_{\mathbf{w}} \rangle = \frac{(\Delta)^{\frac{1}{2}} ((\Sigma^{\text{sig}})^{-1} \mu^{\text{sig}})_{\mathbf{wf}(m)}}{\left(\sum_{\substack{m,n \in \\ \{1,\ldots,d\}}} \sum_{\mathbf{w},\mathbf{v} \in W_{N+d+1}^N} ((\Sigma^{\text{sig}})^{-1} \mu^{\text{sig}})_{\mathbf{wf}(m)} ((\Sigma^{\text{sig}})^{-1} \mu^{\text{sig}})_{\mathbf{vf}(n)} \Sigma_{\mathbf{wf}(m),\mathbf{vf}(n)}^{\text{sig}} \right)^{\frac{1}{2}}}$$

for $m \in \{1, \ldots, d\}$, $\mathbf{w} \in \mathcal{W}_{N+d+1}^M$. □

Remark 3.2 The matrix Σ^{sig} and vector μ^{sig} are just placeholders for different terms and combinations of $\mathbb{E}(\hat{\mathbb{Z}}_{0,T}^{LL,<\infty})$. Hence, all we need in order to know what our optimal functional is, is the expected Hoff lead-lag signature, and this is straightforward to compute for reasonable orders of truncation and number of assets and signals. The following example aims to provide some practical intuition behind how the optimal strategy is calculated and the relation between the signature and words of the tensor algebra.

Example 3.3 Let us consider the case of when we have two assets $X = (X^1, X^2)$, one signal α, such that $d = 2$, $N = 1$. Then we define the 4-dimensional market signal process $\hat{Z} := (t, X^1, X^2, \alpha)$. Let us fix the truncation level to be $M = 2$. The 2nd level truncated signature $\mathbb{Z}_{0,t}^{\leq 2}$ has $|\mathcal{W}_4^2| = 21$ terms, namely the associated words are:

$$\mathcal{W}_4^2 = \{\mathbf{\emptyset}, \mathbf{0}, \mathbf{1}, \mathbf{2}, \mathbf{3}, \mathbf{00}, \mathbf{01}, \mathbf{02}, \mathbf{03}, \mathbf{10}, \mathbf{11}, \mathbf{12}, \mathbf{13}, \mathbf{20}, \mathbf{21}, \mathbf{22}, \mathbf{23}, \mathbf{30}, \mathbf{31}, \mathbf{32}, \mathbf{33}\} \tag{19}$$

Hence, the optimal linear signature trading strategy will correspond to two linear functionals (one for each asset) ℓ_1, ℓ_2, each of length 21, defined as

$$\xi_t^1 = \langle \ell_1, \hat{\mathbb{Z}}_{0,t}^{\leq 2} \rangle$$
$$\xi_t^2 = \langle \ell_2, \hat{\mathbb{Z}}_{0,t}^{\leq 2} \rangle$$

In the above theorem, we can see the vector $\mu^{\mathrm{sig}} = (\mu_1^{\mathrm{sig}}, \mu_2^{\mathrm{sig}})$ will be defined as follows:

$$\mu^{\mathrm{sig}} := \left\{ \mathbb{E}\left(\hat{\mathbb{Z}}_{\mathbf{wf}(m)}^{LL, \leq 3} \right) \right\}_{\mathbf{w} \in \mathcal{W}_4^2, m=1,2}.$$

Hence, μ^{sig} will be a $d \cdot |\mathcal{W}_{N+d+1}^M| = 2 \times 21 = 42$ length vector, containing elements of the expected lead-lag signature of order 3.

Recall that the shift operator $\mathbf{f}(m)$ is defined for each asset $m = 1, 2$ as $\mathbf{f}(1) = \mathbf{5}, \mathbf{f}(2) = \mathbf{6}$, which correspond to the 5th and 6th dimensions of the lead-lag process. Hence, we see that μ_1^{sig} contains the expected lead-lag signature terms corresponding to the index of words

$$I_1 := \{\mathbf{5}, \mathbf{05}, \mathbf{15}, \mathbf{25}, \mathbf{35}, \mathbf{005}, \mathbf{015}, \mathbf{025}, \mathbf{035}, \mathbf{105}, \mathbf{115}, \mathbf{125}, \mathbf{135}, \mathbf{205}, \mathbf{215}, \mathbf{225}, \mathbf{235}, \mathbf{305}, \mathbf{315}, \mathbf{325}, \mathbf{335}\}$$

and μ_2^{sig} contains the expected lead-lag signature terms corresponding to the index of words

$$I_2 := \{\mathbf{6}, \mathbf{06}, \mathbf{16}, \mathbf{26}, \mathbf{36}, \mathbf{006}, \mathbf{016}, \mathbf{026}, \mathbf{036}, \mathbf{106}, \mathbf{116}, \mathbf{126}, \mathbf{136}, \mathbf{206}, \mathbf{216}, \mathbf{226}, \mathbf{236}, \mathbf{306}, \mathbf{316}, \mathbf{326}, \mathbf{336}\},$$

such that we have:

$$\mu^{\text{sig}} = \begin{bmatrix} \left\langle \mathbf{5}, \mathbb{E}\left(\hat{\mathbb{Z}}^{LL,\leq 3}\right)\right\rangle \\ \vdots \\ \left\langle \mathbf{335}, \mathbb{E}\left(\hat{\mathbb{Z}}^{LL,\leq 3}\right)\right\rangle \\ \left\langle \mathbf{6}, \mathbb{E}\left(\hat{\mathbb{Z}}^{LL,\leq 3}\right)\right\rangle \\ \vdots \\ \left\langle \mathbf{336}, \mathbb{E}\left(\hat{\mathbb{Z}}^{LL,\leq 3}\right)\right\rangle \end{bmatrix} \Big\} \text{ 42 elements}$$

Now, we consider the 42×42 matrix Σ^{sig}, defined element-wise as

$$\begin{aligned} \Sigma^{\text{sig}}_{\mathbf{wf}(m),\mathbf{vf}(n)} &= \left\langle \mathbf{wf}(m) \sqcup\!\sqcup \mathbf{vf}(n), \mathbb{E}(\hat{\mathbb{Z}}^{LL,<\infty}_{0,T})\right\rangle \\ &\quad - \left\langle \mathbf{wf}(m), \mathbb{E}(\hat{\mathbb{Z}}^{LL,<\infty}_{0,T})\right\rangle \left\langle \mathbf{vf}(n), \mathbb{E}(\hat{\mathbb{Z}}^{LL,<\infty}_{0,T})\right\rangle \end{aligned}$$

for all $\mathbf{w}, \mathbf{v} \in \mathcal{W}_4^2$ and $m, n = 1, 2$.

Each element of this matrix represents a covariance term between the PnL *attributions* that we discussed previously. For example, let us observe an arbitrary element of Σ^{sig}. Let $\mathbf{w} = \mathbf{01}$, $\mathbf{v} = \mathbf{23}$, $m = 1$, $n = 2$. Then $\mathbf{wf}(m) = \mathbf{015}$, $\mathbf{vf}(n) = \mathbf{236}$ and the corresponding element in the matrix is given as

$$\Sigma^{\text{sig}}_{\mathbf{015},\mathbf{236}} = \left\langle \mathbf{015} \sqcup\!\sqcup \mathbf{236}, \mathbb{E}(\hat{\mathbb{Z}}^{LL,\leq 6}_{0,T})\right\rangle - \left\langle \mathbf{015}, \mathbb{E}(\hat{\mathbb{Z}}^{LL,\leq 3}_{0,T})\right\rangle \left\langle \mathbf{236}, \mathbb{E}(\hat{\mathbb{Z}}^{LL,\leq 3}_{0,T})\right\rangle$$

where $\mathbf{015} \sqcup\!\sqcup \mathbf{236}$ is a sum of 20 different words in $\mathcal{W}_4^6$. We refer the reader also to Example 2.5 for another example of the shuffle product. It is evident that, while our linear functional is only applied to the second order truncated signature, we require the sixth order signature in order to compute the covariance matrix, which can cause a computational bottleneck in practice.

3.2 *Examples*

3.2.1 Lead-Lag and Pairs Trading

The true advantage that data-driven methods have over classical parametric frameworks is that they can exploit the inefficiencies present in financial time series data, without specifying the explicit dynamics that they are trying to capture. In this example, we try to understand the multivariate relationship between two assets that *track* each other, perhaps due to the fact that the intrinsic value of these two assets are extremely similar. There are many such examples in finance, perhaps they are two adjacent interest rates or futures contracts, or perhaps they are just two stocks that happen to be very similar to one another. The fundamental idea is that the fair value of these two assets should be similar or the same as one another, and hence they will move in the same direction. If the asset prices move apart from one another for whatever reason, then this can represent a *mispricing*, and one that a trader would like to exploit, causing the prices to converge back to one another. If one of the assets always moves before the other, then we could see this as a *lead-lag* relationship, otherwise if there is no clear leader or lagger then this can be seen as a general *pairs trading* strategy, where one aims to capitalise on divergences in the two asset prices. The sentiment is that while both assets have their own dynamics, the difference (spread) between the prices of the two assets should hold some predictability on future (co-)movements. Classical literature focuses on modelling this relationship using a mean-reverting process such as an Ornstein-Uhlenbeck process. The strategy should then contain a *buy signal* when the spread falls below some threshold and a *sell signal* when the spread is above the threshold, in the anticipation that this spread should converge back to the threshold. In this experiment we take two assets such that

$$\begin{aligned} dX_t &= \sigma^X dW_t^X \\ dY_t &= \kappa(X_t - Y_t)dt + \sigma^Y dW_t^Y \end{aligned}$$

where X is a standard Brownian motion with zero drift and volatility σ^X. Y however is modelled as a mean-reverting process where its drift is proportional to the spread between X and Y. That is, X is the *leading* process and Y is a lagging process. We can imagine that we are a mid-frequency quantitative trader who wants to exploit this relationship and trades/rebalances every 30 minutes, and wants to maximize their expected PnL and minimise the variance of the PnL at the end of the day (perhaps due to margin/capital constraints being defined daily), just like in our objective in (3). In a 9-hour trading day, we would see this essentially gives us 16 trading opportunities. We can clearly see in this toy example that the only exploitable *alpha* within the framework is the temporal dependence through mean-reversion in Y since there is no long term drift in X or Y.

If a trader was to deploy a static buy and hold strategy here, the PnL would be zero on average, however, for higher order Sig-Traders, it is possible to exploit the mean-reversion dynamics. In fact, this example demonstrates how the above-mentioned *alpha* is self-contained within the 1st level of the signature (which is the increment of the path, or the 'drift') and so orders of truncation of $M \geq 2$ do not contain any more predictive power on excess returns than that of $M = 1$. However, on the right hand side (RHS) of Fig. 3, the signature efficient frontier illustrates that when trading with a daily look-forward horizon, the ratio of return to variance of daily PnL is greater as we increase the order of the Sig-Trader. This is due to the higher levels of the signature capturing non-linear path-dependencies that can help reduce variance in the PnL distribution at the end of the day. We relate back

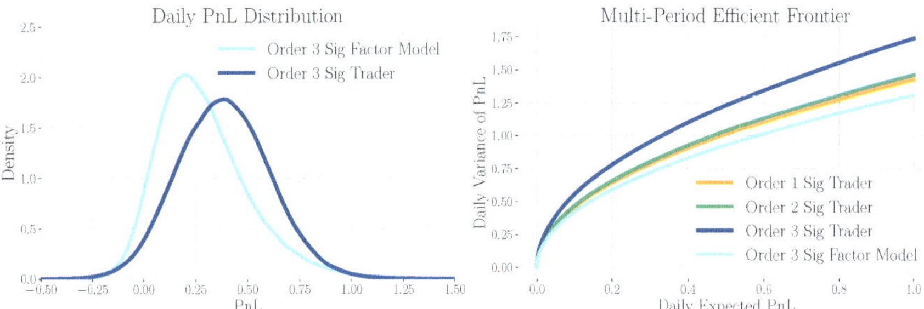

Fig. 3 Weekly PnL distribution (LHS) and the signature efficient frontier comparing different orders of truncation (RHS)

to Fig. 2 to demonstrate that in fact higher order Sig-Traders are able to construct mean-reverting strategies that naturally limit drawdowns, which are an inherently path-dependent characteristic.

Comparison to Signature Factor Model

In this synthetic example, we compare the Sig-Trader strategy to that of the original factor model. The generic factor model is set up as a supervised linear regression on future returns, as a function of the current signature,

$$\mu_{t+1} = \mathbb{E}[r_{t+1}|\mathcal{F}_t] = B\hat{\mathbb{Z}}_{0,t} + \varepsilon_{t+1}$$

where r is the 2-dimensional asset returns, B is a $2 \times N$ matrix of factor coefficients, $\hat{\mathbb{Z}}_{0,t}$ is a vector of the signature values of the market signal process and ε is a vector of the 2 assets' (unexplained) residuals returns. We can see that this model uses the same input as the Sig-Trader (the signature up to time t), with the same tools (applying a linear function). The key distinguishment however is that the factor model is optimal for maximising the mean-variance PnL profile of *30-minute* portfolio returns, which ignores any pathwise dynamics of the strategy itself, throughout the course of the day. This is very important, since in fact, volatility/volume profiles are not constant over the course of a trading day, and hence risks are very different depending on the time of day. Figure 4 demonstrates the difference in nature between the Sig-Trader and the factor model. In the order 1 case, as described above, the exploitable alpha in terms of returns are captured by the 1st level of the signature and so we see a similar position profile for the Sig-Trader and the factor model. However, when the factor model tends to go more short (or long), the higher order Sig-Traders, e.g. order 3, tend to reduce its position, since this will be more beneficial to minimising the PnL at the end of the trading day. We can think of this behaviour as being similar to applying a sigmoid function to your position in search of robustness, or applying a stop-loss to avoid becoming too leveraged - which are common practices. By incorporating path-dependencies in the strategy, we have access to a more robust and intuitive extension to classic factor models.

Fig. 4 Comparison between the Markowitz Order 3 factor model positions and the corresponding Sig Trader positions, for orders 1,2,3

4 Alternative Optimization Criteria

As discussed in Sect. 1, obtaining an *optimal* trading strategy is always with respect to the market scenario we are trading within. While mean-variance optimization is regarded as perhaps the most common and intuitive of utility preferences, various other objectives and circumstances may influence trading decisions, contingent upon personal utility and preferences. In addition to individual preferences, it's crucial to consider characteristics of market microstructure when trading, particularly for large investors or when dealing with illiquid assets, since the market becomes sensitive to any trade we make. This means that by actively trading, we are able to fundamentally alter the price process that we are trying to model, incurring a feedback loop. We explore such market microstructure problems in this section and present the results of Kalsi et al. in [14], in which they obtain approximate solutions to the problem of optimal execution within the class of signature trading strategies.

4.1 Optimal Execution Problem

The classical optimal execution problem concerns a trader who wants to transition from one position (or level of inventory) Q_0 to another Q_T over a specified time horizon. If the trader were to immediately post a market order in the limit order book (LOB), insufficient liquidity within the book might result in execution at a less favorable price, often referred to as *market impact*. To mitigate this impact, it may be prudent to instead split the *parent order* up into smaller *child orders* and execute them gradually since they will have much less market impact, however this approach exposes the trader to potential adverse market movements during the execution period. Consequently, the trader faces the dilemma of trading as swiftly as possible to minimize uncertainty regarding future market movements while simultaneously minimizing market impact. As seen in the blueprint of Fig. 1, in order to solve this type of optimal trading problem, we require a model for the asset midprice dynamics as well as a way of modelling market impact, including the execution price. Once the modeling phase is complete, the next step is to formulate an objective function aligned with the trader's preferences. The goal is typically to maximize total wealth at the terminal time, which depends on the prices executed between time 0 and T, as well as the final execution at the terminal time. Optionally, a penalty may be imposed on any remaining inventory held at time T as this is generally undesirable.

Signature Trading Speeds
In previous sections we modelled the position (holding) of the underlying asset at time t by ξ_t, as a *signature trading strategy*. In execution problems we often model our strategy as a *trading speed*, which is defined as the change in our position (inventory) per unit time.

Definition 4.1 (Signature Trading Speed) Let $\hat{Z}$ be a market signal process as defined in Definition 2.8 where $d = 1$ and we have access to N exogenous signals. We denote $\mathcal{S}^Z$ as the space of *trading speeds* with respect to market signal process Z, defined as the space of all continuous adapted processes with respect to the filtration generated by $\hat{Z}$. Similarly to Definition 2.9, we define the space of *signature trading speeds* $\mathcal{S}^Z_{\text{sig}}$ as

$$\mathcal{S}^Z_{\text{sig}} := \left\{(\theta_t)_{t\in[0,T]} \mid \text{ if } \exists\, \ell \in T((\mathbb{R}^{N+2})^*) \text{ s.t } \theta_t = \langle \ell, \hat{\mathbb{Z}}^{<\infty}_{0,t} \rangle\right\},$$

where the *inventory* held at time t after trading with speed θ is defined as

$$Q^\theta_t := Q_0 - \int_0^t \theta_s ds.$$

Modelling Market Impact
Modelling price impact from a large market order is a well-studied topic of mathematical finance and there exists many formulations in the literature. Generally the execution price P^θ_t of the strategy θ at time t is modelled as

$$P^\theta_t := X_t - I(\theta_{0,t})$$

where X is the unperturbed midprice if the trader does not trade the underlying asset. We observe that the market impact $I(\theta_{0,t})$ is a function of the path of past trading speeds, which in our framework is a linear function of the signature of the past path of the market signal process, that is to say

$$\begin{aligned} P^\theta_t &= X_t - I\left(\left\langle \ell, \hat{\mathbb{Z}}^{<\infty}_{0,t}\right\rangle\right) \\ &= X_t - \left\langle g^\ell, \hat{\mathbb{Z}}^{<\infty}_{0,t}\right\rangle \end{aligned} \tag{20}$$

where g^ℓ is a linear functional that models the market impact I as a function of the past market signal trajectory. There are several ways discussed in the literature to model market impact, derived from empirical observations, which include but not limited to:

1. *Temporary market impact.* We let $g^\ell = \lambda\ell$, with $\lambda > 0$, then

$$I(\theta_{0,t}) = \left\langle g^\ell, \hat{\mathbb{Z}}^{<\infty}_{0,t} \right\rangle = \lambda\theta_t,$$

   which is a linear temporary market impact studied in [2, 4, 16].
2. *Permanent market impact.* In [2–4], a permanent market impact given by $\int_0^t \theta_s ds$ is considered. If we set $g^\ell = \ell\mathbf{0}$ as a concatenation of the linear functional ℓ with the word $\mathbf{0}$ which is the designated time index of the process (t, X_t, f_t). Therefore the market impact is defined as

$$\left\langle g^\ell, \hat{\mathbb{Z}}^{<\infty}_{0,t} \right\rangle = \int_0^t \theta_s ds = \int_0^t \left\langle \ell, \hat{\mathbb{Z}}^{<\infty}_{0,t} \right\rangle ds = \left\langle \ell\mathbf{0}, \hat{\mathbb{Z}}^{<\infty}_{0,t} \right\rangle.$$

3. *Transient market impact.* In [7, 12], the authors considered a transient market impact that is given by $\int_0^t K(t-s)\theta_s ds$ where $K(x) := \exp(-\rho x)$ for $\rho > 0$ is an exponential time kernel. We can therefore find a linear functional g^ℓ such that the market impact is well approximated, i.e.

$$\int_0^t K(t-s)\theta_s ds \approx \left\langle g^\ell, \hat{\mathbb{Z}}^{<\infty}_{0,t} \right\rangle$$

   to an arbitrary accuracy due to the universal approximation theorem.

More generally, we can model the execution price as

$$P_t^\theta := \left\langle f^\theta, \hat{\mathbb{Z}}^{<\infty}_{0,t} \right\rangle. \tag{21}$$

In the previous example of (20), we had specifically separated the market impact from the midprice but since the midprice is one of the channels of the market signal process, it becomes a specific case of (21) where $f^\theta = \mathbf{1} + \boldsymbol{\emptyset} - g^\theta$.

Optimal Execution Formulation

Suppose a trader at time 0 is in the presence of Q_0 inventory and is hoping to liquidate all of their position by time T. The optimal execution task aims to strike an equilibrium between trading fast enough so as to avoid uncertain future price moves, while minimising their price impact, caused by market orders that are too large relative to the limit order book. The objective criterion is as follows:

$$\mathcal{V}(\theta) = \underbrace{W_T^\theta}_{\text{accumulated cash}} + \underbrace{Q_T^\theta(P_T^\theta - \alpha Q_T^\theta)}_{\text{terminal execution}} - \underbrace{\phi\int_0^T (Q_s^\theta)^2 ds}_{\text{running inventory penalty}} \tag{22}$$

where the *wealth accumulated at time t corresponding to the trading speed* θ, *is defined by*

$$W_t^\theta = \int_0^t P_s \theta_s ds$$

and the *remaining inventory at time t is defined by*

$$Q_t = Q_0 - \int_0^t \theta_s ds.$$

The first term of the objective function is clearly our accumulated wealth from liquidating our assets, which is influenced by our trade order size plus market movements, hence the balance that we are trying to strike. The second term refers to the terminal execution if we arrive at time T with remaining inventory, here $\alpha \geq 0$ is the terminal liquidation penalty parameter (the larger α then the larger the penalty for holding inventory at the terminal time since this would result in a greater market impact). The final term is a running inventory penalty, since we want to trade as quick as possible in order to transition into the new position. Therefore, the goal of the optimization is to find the *signature trading speed*, θ^*, that maximizes the objective criterion $V(\theta)$ defined in (22), i.e. we aim to solve

$$\sup_{\theta \in \mathcal{S}_{\text{sig}}^Z} \mathbb{E}\left[V(\theta)\right]. \tag{23}$$

4.2 Optimal Execution Solution with Signatures

We will show that, while the optimization problem in (23) is nonlinear in the underlying midprice process, since we are parameterising P, X, and θ as linear functions on the signature, then so too do W and Q. Due to the shuffle product property of the signature (Definition 2.4), we are able to re-write the value functions as a new linear functional on the signature of the market signal process. In (23) there are 4 quantities of interest (W, Q, $\int_0^t Q_s^2 ds$, $Q(P - \alpha Q)$) that we want to compute and then combine, in order to represent the objective as one main linear functional.

Lemma 4.2 *Let* $\theta \in \mathcal{S}_{\text{sig}}^Z$ *be a signature trading speed given by* $\langle \ell, \hat{\mathbb{Z}}_{0,t}^{<\infty} \rangle$, *hence* ℓ *is the linear functional that represents our trading speed that we are solving for. Recall that the word associated to the constant linear functional is* $\mathbf{\emptyset}$, *the word associated to the time dimension is* $\mathbf{0}$ *and the word associated with the midprice process is* $\mathbf{1}$. *Then the following statements hold:*

1. *The wealth process if we trade according to* ℓ*:*

$$\begin{aligned} W_T^\ell &= \int_0^T P_s^\ell \langle \ell, \hat{\mathbb{Z}}_{0,s}^{<\infty} \rangle ds = \int_0^t (X_s - \langle g^\ell, \hat{\mathbb{Z}}_{0,s}^{<\infty} \rangle) \langle \ell, \hat{\mathbb{Z}}_{0,s}^{<\infty} \rangle ds \\ &= \int_0^T \left\langle (\mathbf{1} + \boldsymbol{\emptyset} - g^\ell) \sqcup\!\sqcup\, \ell, \hat{\mathbb{Z}}_{0,s}^{<\infty} \right\rangle ds \\ &= \left\langle \left[(\mathbf{1} + \boldsymbol{\emptyset} - g^\ell) \sqcup\!\sqcup\, \ell \right] \mathbf{0}, \hat{\mathbb{Z}}_{0,T}^{<\infty} \right\rangle \end{aligned}$$

2. *The inventory process if we trade according to* ℓ*:*

$$\begin{aligned} Q_T^\ell &= Q_0 - \int_0^T \theta_s ds = Q_0 - \int_0^T \langle \ell, \hat{\mathbb{Z}}_{0,s}^{<\infty} \rangle ds \\ &= \langle Q_0 \boldsymbol{\emptyset} - \ell \mathbf{0}, \hat{\mathbb{Z}}_{0,T}^{<\infty} \rangle \end{aligned}$$

3. *The running inventory penalty if we trade according to* ℓ*:*

$$\begin{aligned} \int_0^T (Q_s^\ell)^2 ds &= \int_0^T \left\langle (Q_0 \boldsymbol{\emptyset} - \ell \mathbf{0})^{\sqcup\!\sqcup 2}, \hat{\mathbb{Z}}_{0,s}^{<\infty} \right\rangle ds \\ &= \left\langle \left((Q_0 \boldsymbol{\emptyset} - \ell \mathbf{0})^{\sqcup\!\sqcup 2} \right) \mathbf{0}, \hat{\mathbb{Z}}_{0,T}^{<\infty} \right\rangle \end{aligned}$$

4. *The terminal execution if we trade according to* ℓ*:*

$$\begin{aligned} Q_T^\ell (P_T^\ell - \alpha Q_T^\ell) &= \langle Q_0 \boldsymbol{\emptyset} - \ell \mathbf{0}, \hat{\mathbb{Z}}_{0,T}^{<\infty} \rangle \langle \mathbf{1} + \boldsymbol{\emptyset} - g^\ell - \alpha (Q_0 \boldsymbol{\emptyset} - \ell \mathbf{0}), \hat{\mathbb{Z}}_{0,T}^{<\infty} \rangle \\ &= \left\langle (Q_0 \boldsymbol{\emptyset} - \ell \mathbf{0}) \sqcup\!\sqcup\, (\mathbf{1} + \boldsymbol{\emptyset} - g^\ell) - \alpha (Q_0 \boldsymbol{\emptyset} - \ell \mathbf{0})^{\sqcup\!\sqcup 2}, \hat{\mathbb{Z}}_{0,T}^{<\infty} \right\rangle. \end{aligned}$$

Therefore, we can now state the overall optimal execution problem under the class of signature trading speeds.

Theorem 4.3 *Let* $\theta \in \mathcal{S}_{\text{sig}}^Z$ *be a signature trading speed given by* $\langle \ell, \hat{\mathbb{Z}}_{0,t}^{<\infty} \rangle$*, then the value function in (22) can be written as*

$$\begin{aligned} \mathcal{V}(\theta) = &\left\langle \left[(\mathbf{1} + \boldsymbol{\emptyset} - g^\ell) \sqcup\!\sqcup\, \ell \right] \mathbf{0} - \phi \left((Q_0 \boldsymbol{\emptyset} - \ell \mathbf{0})^{\sqcup\!\sqcup 2} \right) \mathbf{0} - \alpha (Q_0 \boldsymbol{\emptyset} - \ell \mathbf{0})^{\sqcup\!\sqcup 2}, \right. \\ &\left. + (Q_0 \boldsymbol{\emptyset} - \ell \mathbf{0}) \sqcup\!\sqcup\, (\mathbf{1} + \boldsymbol{\emptyset} - g^\ell), \hat{\mathbb{Z}}_{0,T}^{<\infty} \right\rangle. \end{aligned}$$

Hence, the optimal execution problem that we are trying to solve for θ^**, is reduced to finding a linear functional* ℓ^* *that satisfies*

$$\begin{aligned} \sup_{\ell \in T((\mathbb{R}^{N+2})^*)} &\left\langle \left[(\mathbf{1} + \boldsymbol{\emptyset} - g^\ell) \sqcup\!\sqcup\, \ell \right] \mathbf{0} - \phi \left((Q_0 \boldsymbol{\emptyset} - \ell \mathbf{0})^{\sqcup\!\sqcup 2} \right) \mathbf{0} - \alpha (Q_0 \boldsymbol{\emptyset} - \ell \mathbf{0})^{\sqcup\!\sqcup 2}, \right. \\ &\left. + (Q_0 \boldsymbol{\emptyset} - \ell \mathbf{0}) \sqcup\!\sqcup\, (\mathbf{1} + \boldsymbol{\emptyset} - g^\ell), \mathbb{E} \left[\hat{\mathbb{Z}}_{0,T}^{<\infty} \right] \right\rangle. \end{aligned}$$

Recall, that in the more general case, when we are under the presence of exogenous signals, we can incorporate the market impact function g^ℓ directly within a new linear functional that is dependent on the exogenous signals. In this specific case we have $f^\ell = \mathbf{1} + \emptyset - g^\ell$, however we can in fact model this as any linear functional f^ℓ, that is we solve the optimization

$$\sup_{\ell \in T((\mathbb{R}^{N+2})^*)} \Big\langle (f^\ell \sqcup\!\sqcup \ell)\mathbf{0} - \phi\left((Q_0\emptyset - \ell\mathbf{0})^{\sqcup\!\sqcup 2}\right)\mathbf{0} - \alpha(Q_0\emptyset - \ell\mathbf{0})^{\sqcup\!\sqcup 2}, \tag{24}$$
$$+(Q_0\emptyset - \ell\mathbf{0}) \sqcup\!\sqcup f^\ell, \mathbb{E}\left[\hat{\mathbb{Z}}^{<\infty}_{0,T}\right]\Big\rangle.$$

Numerically Solving the Optimal Execution Problem

While in Sect. 3 we were able to obtain a closed form solution for ℓ^*, in this problem we require numerical techniques to solve the optimization. We also previously required a notion of the Itô integral in our definition of the PnL of a trading strategy in order to capture the predictability of such a trading strategy. However, when working with trading speeds, we define our objective slightly differently since we are not directly to model every price movement, but instead focus on modelling the market impact and hence executable price, P. This leads us to instead only require the estimation of the signature of the market signal process and not the Hoff lead-lag process for the fitting procedure, which significantly reduces the bottleneck in computation. The main building blocks for numerically solving the problem are:

1. Pre-determine via a well-studied market impact model or empirically fit for a linear functional f^θ that satisfies $P^\theta_t = \langle f^\theta, \hat{\mathbb{Z}}^{<\infty}_{0,t}\rangle$.
2. Estimate the expected signature of the market signal process, $\mathbb{E}\left[\hat{\mathbb{Z}}^{<\infty}_{0,T}\right]$.
3. Numerically solve the optimization problem (24). If we consider linear permanent or temporary market impact, the value function becomes quadratic in the weights of ℓ and we can simply find the unique global maximum.

4.3 Numerical Examples

Geometric Brownian Motion

If the underlying asset paths behave as a geometric Brownian motion, the optimal execution solution can be obtained in closed form. The optimal execution strategy in this setting is a *time-weighted average price* (TWAP) strategy. When the signature trading model is fitted using geometric Brownian motion paths, we observe the same closed form solution, as demonstrated in Fig. 5, showing that the signature trading framework agrees with classical methods. Kalsi, Lyons and Perez in [14] also conduct experiments that incorporate exogenous information, such as order-flow, into the midprice dynamics, allowing to estimate the optimal trading speed when the closed form solution is unknown, outperforming the Almgren-Chriss benchmark. This is particularly useful since the execution algorithm is also dependent on how

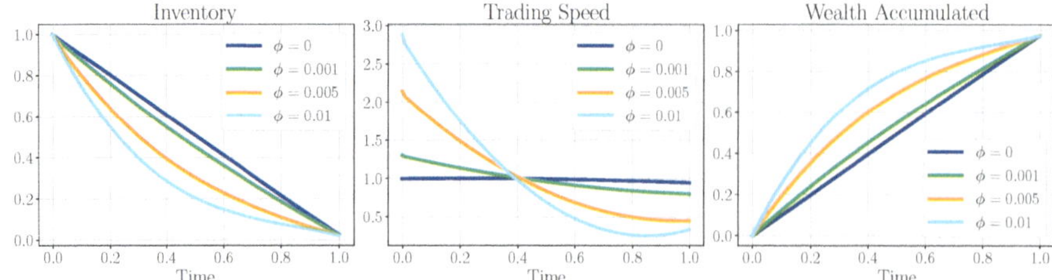

Fig. 5 Optimal inventory, trading speed and wealth trajectories for the optimal execution program when we are trading a Markovian Brownian motion, for different *urgency* parameters ϕ

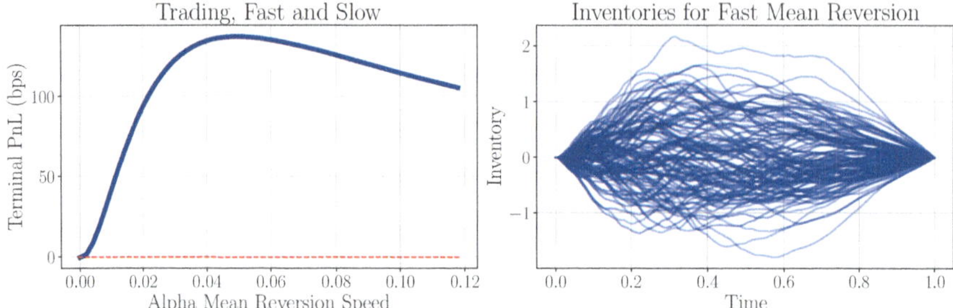

Fig. 6 PnL of a trading strategy, depending on the speed of mean-reversion (LHS). An example of the change in inventory when trading with a mean-reverting alpha signal (RHS)

much liquidity is in the market, which leads to solutions such as a Volume Weighted Average Price (VWAP) strategy, rather than time weighted. Further discussion and details of other variants including VWAP can be found in [3]. Figure 5 also demonstrates how increasing the parameter ϕ delivers a more *urgent* execution program. The choice of ϕ is dependent on many things, including the speed of your signal. If you believe the price is about to decrease in the next second, then you need to sell out of your position much quicker than if the price won't decrease for another 30 seconds into the future, hence this is a parameter that may be tuned at the discretion of the trader.

Ornstein-Uhlenbeck Process

As we have seen in Sect. 3, quantitative traders often have alpha that they can exploit, perhaps through the form of mean-reversion. However, there is a trade off since if we trade too quickly, we will incur a large market impact, but if we trade too slowly, we will be exposed to the risk of missing the price change that we were trying to position for. In Fig. 6, we see the PnL (including market impact/trading costs) of a trading strategy, depends on the speed of mean-reversion. As the mean-reversion speed increases, this is good since the signal becomes less noisy, meaning our signal to noise ratio is higher. However, there becomes a point at which the mean-reversion can be *too fast* since we are unable to trade it without incurring too large costs. This implies that there exists a sweet spot, in this case a mean-reversion of $\alpha = 0.05$.

Implementation
Throughout this chapter, we used the package `signatory` [15], alongside `PyTorch` for calculating and performing functionality related to tensors and the signature transform. We also performed some computations with `iisignature` integrated in `numpy`. The **GitHub** repository for this work can be obtained in [10]. Here, we provide an extensive overview and implementation of Signature Trading, both the mean-variance solution and optimal execution solution. We provide examples and functionality in both signatory/PyTorch, as well as iisignature/numpy, for whichever the user prefers.

The signature trading method can be completely self-contained in the sense that it does not require inputted assumptions such as expected returns or covariances and that they are inferred along with characteristics of the whole process within the algorithm itself. For both the mean-variance solution and the optimal execution strategy, once a linear functional ℓ is obtained from past data samples it is straightforward to unravel this into an implementable trading strategy characterised by the number of units to buy/sell at each time point $t \in [0, T]$. Since the strategy is dynamic, as new data arrives the Sig-Trader will continuously compute the signature and update their position accordingly.

Whilst Sig-Trading is data driven and does not require direct probabilistic assumptions on the underlying model, just like most frameworks, there is versatility when deploying Sig-Trading in practice as there does still remain quantities that need reliably estimating in the fitting procedure. In the mean-variance framework, the expected lead-lag signature can be noisy when calibrated through time and so we leave any questions of robustness (with respect to the fitting procedure) to future ongoing extensions of the Signature Trading project, as there remains question marks over how well defined such a solution is, especially in the context of rapidly changing regimes such as in financial markets. The most straightforward method to estimate the expected signature is a Monte Carlo approach by calculating the empirical expected lead-lag signature. However, there remains a multitude of alternative ways to approach this for some given sample/fitting data, in order to ensure robustness out of sample. This chapter does not directly discuss these approaches but we do point out that any trader's favourite statistical estimation and training procedures can work in this setting, for example it is possible to weight more to recent fitting data which may be incorporated via a rolling fitted model. Hence, we leave such statistical procedures directly to the discretion of the user and instead provide a framework in which such techniques can be ensembled.

5 Conclusion

Path dependencies, such as non-Markovian data structures or time series exhibiting temporal correlation, are common phenomena in financial data. Momentum and mean-reversion (of assets or signals) are two of the most fundamental features of time series data that a trader can exploit, and these features are inherently dependent on the entire path. However, many traditional techniques used for portfolio optimization are too inflexible to handle such structural complexities in the data and signals. Signature trading strategies, on the other hand, offer a versatile representation of investment strategies and have proven to be powerful tools in pricing, hedging, and optimal execution.

We observe that the advantages of signature trading strategies can, in fact, be extended to more general portfolio optimization problems, as they encompass many common trading styles used in practice (including the aforementioned momentum and mean-reversion). Specifically, in this chapter, we obtained solutions to two of the most fundamental stochastic control problems in finance: the mean-variance criterion and the optimal execution problem. In summary, signature trading provides an alternative to machine learning methods in its ability to handle path-dependent, nonlinear dynamics and signals, while also extending classical solutions in financial stochastic control problems. Overall, these solutions offer more intuition than ML-based methods and provide greater discretion in the fitting process than classical methods, offering a flexible framework that effectively addresses many challenges encountered when working with financial data.

References

1. I.P. Arribas, Signatures in machine learning and finance. PhD thesis, University of Oxford, 2020
2. Á. Cartea, S. Jaimungal, Optimal execution with limit and market orders. Quant. Finance **15**, 1279–1291 (2015)
3. A. Cartea, S. Jaimungal, A closed-form execution strategy to target volume weighted average price. SIAM J. Financ. Math. **7**, 760–785 (2016). https://doi.org/10.1137/16M1058406
4. Á. Cartea, S. Jaimungal, Incorporating order-flow into optimal execution. Math. Financ. Econ. **10**, 339–364 (2016)
5. T. Cass, W.F. Turner, Topologies on unparameterised path space. J. Funct. Anal. **286**(4), 110261 (2024)
6. C. Cuchiero, J. Möller, Signature methods in stochastic portfolio theory. Preprint arXiv:2310.02322 (2024)
7. G. Curato, J. Gatheral, F. Lillo, Optimal execution with non-linear transient market impact. Quant. Finance **17**, 41–54 (2017)
8. E.R. Fernholz, *Stochastic Portfolio Theory* (Springer, New York, 2002)
9. G. Flint, B. Hambly, T. Lyons, Discretely sampled signals and the rough Hoff process. Stoch. Process. Appl. **126**, 2593–2614 (2016)
10. O. Futter, Signature trading, GitHub repository (accessed August 2024). https://github.com/owenfutter/signature_trading

11. O. Futter, B. Horvath, M. Wiese, Signature trading: A path-dependent extension of the mean-variance framework with exogenous signals. Preprint arXiv:2308.15135 (2023)
12. J. Gatheral, A. Schied, A. Slynko, Transient linear price impact and Fredholm integral equations. Math. Finance **22**, 445–474 (2012)
13. B. Hoff, The Brownian frame process as a rough path. PhD thesis, University of Oxford, 2006
14. J. Kalsi, T. Lyons, I.P. Arribas, Optimal execution with rough path signatures. SIAM J. Financ. Math. **11**, 470 (2020)
15. P. Kidger, T. Lyons, Signatory: differentiable computations of the signature and logsignature transforms, on both CPU and GPU, in *ICLR 2021 - 9th International Conference on Learning Representations* (2021)
16. C.A. Lehalle, E. Neuman, Incorporating signals into optimal trading. Finance Stoch. **23**, 275–311 (2019)
17. T. Lyons, S. Nejad, I.P. Arribas. Non-parametric pricing and hedging of exotic derivatives. Appl. Math. Finance **27**, 457–494 (2019)
18. H. Markowitz, Portfolio selection. J. Finance **7**, 77–91 (1952)

Optimal Stopping for Non-Markovian Asset Price Processes

Christian Bayer, Paul P. Hager, and Sebastian Riedel

Abstract Some of the most liquidly traded options in equity markets are American and Bermudan options, whose owner may choose the option's exercise date—within a certain range. Hence, these options are *optimal stopping problems* from a mathematical perspective. There is a huge literature on *solving* optimal stopping problems, and most of the prevalent methods (e.g., solving the Hamilton-Jacobi-Bellman PDE, least squares Monte Carlo, dual martingale methods) strongly rely on the Markov property for the underlying dynamics, to avoid the curse of dimensionality. In this chapter, we will show how the signature can be used to adopt classical, Markovian numerical methods for the non-Markovian case.

1 Optimal Stopping

Optimal stopping problems form an important class of problems within the wider realm of *stochastic optimal control*. A prime example in finance is the pricing of *American options*. Unlike European options, American options may be exercised (by the owner) at any time t before a terminal time T, in which case the counterparty has to pay the owner a payoff $\phi(S_t)$ given in terms of the current price S_t of the underlying. For example, in the case of an American put option, we have $\phi(S_t) = (K - S_t)^+ = \max(K - S_t, 0)$, where K is a fixed threshold value, the *strike price*.

C. Bayer (✉)
Weierstrass Institute, Berlin, Germany
e-mail: christian.bayer@wias-berlin.de

P. P. Hager
Department of Statistics and Operations Research, University of Vienna, Wien, Austria
e-mail: paul.peter.hager@univie.ac.at

S. Riedel
FernUniversität in Hagen, Gebäude 3, Fakultät für Mathematik und Informatik, Lehrgebiet Angewandte Stochastik, Hagen, Germany
e-mail: sebastian.riedel@fernuni-hagen.de

C. Bayer et al. (eds.), *Signature Methods in Finance*, Springer Finance Lecture Notes, https://doi.org/10.1007/978-3-031-97239-3_9

When we mathematically model an American option, we obviously need to allow the owner to stop at a random time $t = t(\omega)$, to accommodate reasonable strategies such as "stop at the first time when S is lower than a given level L". At the same time, our model must not allow arbitrary random exercise times, to reflect that the owner's decision to exercise at time t has to be taken at (or before) time t—i.e., the owner cannot "look into the future". This, in particular, excludes trivial "solutions", such as exercising at time $\arg\max_{t\in[0,T]} \phi(S_t)$—the pathwise optimal time.

"Not looking into the future" is a statement about the *flow of information*, which is modeled by *filtrations* in probability theory. Given a filtration, exercise times conforming to the above-mentioned restrictions are then modeled as *stopping times*. In what follows, we will in parallel consider stopping times taking values in a set $\mathcal{T}$ of possible times, where we restrict ourselves to either of the following important cases:

1. Continuous time, i.e., $\mathcal{T} = [0, T]$ for some $T > 0$;
2. Discrete time, i.e., $\mathcal{T} = \{0 = t_0 < t_1 < \cdots < t_N = T\}$ for some $T > 0$, $N \geq 1$.

Definition 1.1 Given a probability space $(\Omega, \mathcal{F}, \mathbb{P})$ and a filtration $(\mathcal{F}_t)_{t\in\mathcal{T}}$, a *stopping time* is a random variable $\tau : \Omega \to \mathcal{T} \cup \{\infty\}$ such that for all $t \in \mathcal{T}$: $\{\tau \leq t\} \in \mathcal{F}_t$. We denote the set of all stopping times by $\mathcal{S}$.

We can now define the optimal stopping problem, where we, for simplicity, assume that interest rates are $r = 0$, or, equivalently, all cash-flows are already properly discounted.

Definition 1.2 In the context of Definition 1.1, let $(Y_t)_{t\in\mathcal{T}}$ be an adapted (i.e., for all $t \in \mathcal{T}$, Y_t is $\mathcal{F}_t$-measurable) *reward process* taking values in $\mathbb{R}_{\geq 0}$, which is bounded in the sense that $\mathbb{E}\left[\sup_{t\in\mathcal{T}} Y_t\right] < \infty$. Then the *optimal stopping problem* is the problem of computing

$$\sup_{\tau\in\mathcal{S}} \mathbb{E}\left[Y_{\tau\wedge T}\right].$$

Example 1.3 An especially important example of the optimal stopping problem is given when the reward process satisfies $Y_t = \phi(S_t)$ in terms of an underlying asset price process S and a payoff function ϕ. The value of the optimal stopping problem (under a risk-neutral probability measure) then corresponds to the value of an *American option* if $\mathcal{T} = [0, T]$ and of a *Bermudan option* if $\mathcal{T} = \{0 = t_0 < t_1 < \cdots < t_N = T\}$. The financial interpretation of $\mathcal{T}$ is as the set of all possible exercise times.

For concreteness' sake, we will usually consider the filtration to be generated by a (continuous) stochastic process $X = (X_t)_{t\in\mathcal{T}}$ taking values in $\mathbb{R}^d$. A natural class of stopping times are *first hitting times* of a process w.r.t. a set.

Example 1.4 Given a (for simplicity) continuous, adapted process $(S_t)_{t\in\mathcal{T}}$ taking values in $\mathbb{R}$ and a measurable subset $A \subset \mathbb{R}$, the first hitting time

$$\tau_A := \inf\{t \in \mathcal{T} | S_t \in A\}$$

is a stopping time.

As we shall see later, for optimal stopping we can essentially restrict ourselves to stopping times, which are hitting times.

How can we compute the value $\sup_{\tau \in \mathcal{S}} \mathbb{E}[Y_{\tau \wedge T}]$ of the optimal stopping problem as well as an optimal stopping time τ^*? Several possibilities come to mind. We could, for instance, hope that optimal stopping times are, indeed, hitting times of the underlying driving process X w.r.t. sets A, parameterize a sufficiently large class of such sets A_θ, and optimize over $\mathbb{E}[Y_{\tau_{A_\theta} \wedge T}]$. This approach forms the basis of the first signature-based method suggested in Sect. 3.

There are two important systematic approaches for solving optimal stopping problems, sometimes called *primal* and *dual* methods.

On the primal side, numerical methods are usually based on the *dynamic programming principle*, which we shall formulate in discrete time for simplicity. Let us first introduce the *value process* (also known as the *Snell envelope*)

$$V_t := \operatorname*{ess\,sup}_{\tau \in \mathcal{S}_t} \mathbb{E}[Y_{\tau \wedge T} \mid \mathcal{F}_t], \quad \mathcal{S}_t := \{\tau \in \mathcal{S} | \tau \geq t\}, \ t \in \mathcal{T}, \tag{1}$$

which serves as the backbone for most of the theory of the optimal stopping problem. In particular, V is a super-martingale, and is, in fact, the smallest super-martingale dominating the reward process Y. In discrete time, it is easy to see that it satisfies a semi-explicit backward recursive formula. Indeed, we have

$$V_{t_n} = \begin{cases} Y_T, & n = N, \\ \max\left(Y_{t_n}, \mathbb{E}[V_{t_{n+1}} \mid \mathcal{F}_{t_n}]\right), & 0 \leq n < N, \end{cases} \tag{2}$$

see, for instance, [15, Section 1.1]. Note that (2) implies an explicit formula of the optimal stopping time (as a first hitting time):

$$\tau^* := \inf\{t \in \mathcal{T} | V_t = Y_t\}. \tag{3}$$

Note that we can solve (2), provided we can compute the conditional expectation, i.e., solve a *regression problem*. We will come back to this approach in the context of signatures in Sect. 3. Note that the process V is a supermartingale.

From a computational point of view, it is very convenient to have primal and dual formulations (without duality gap) of the same problem, since this enables us to *sandwich* the unknown value V_0.

For the optimal stopping problem, it turns out that the dual problem presented in [16] is a minimization problem over martingales. More precisely, we have

$$V_0 = \inf_{M \in \mathcal{M}_0} \mathbb{E}\left[\sup_{t \in \mathcal{T}} (Y_t - M_t)\right], \tag{4}$$

where $\mathcal{M}_0$ contains all (say, square integrable) martingales starting in $M_0 = 0$.

Remark 1.5 Suppose that τ is an approximation of (3), i.e., a genuine, but sub-optimal stopping time. If we now compute $\mathbb{E}[Y_{\tau \wedge T}]$—using independent samples from the samples used to construct τ in the first place—we obtain a *lower bound* for the (unknown) true value V_0. Conversely, given an approximation M of the optimal martingale in (4), i.e., a genuine, but sub-optimal martingale. Plugging M into the representation—again using independent samples to preserve the martingale structure, we obtain an upper bound for V_0, i.e.,

$$\mathbb{E}[Y_{\tau \wedge T}] \leq V_0 \leq \mathbb{E}\left[\sup_{t \in \mathcal{T}} (Y_t - M_t)\right].$$

Hence, the error in either approximation can be bounded by the difference between our (computed) lower and upper bounds.

Finally, let us comment on the role of the Markov property for both the dynamic programming problem and the dual problem. As already pointed out above, the main computational challenge in the dynamic programming principle is the computation of the *continuation value*, i.e., the conditional expectation $\mathbb{E}[V_{t_{n+1}} \mid \mathcal{F}_{t_n}]$. In the Markovian case, this can be expressed as $\mathbb{E}[V_{t_{n+1}} \mid X_{t_n}]$, and we can make an *ansatz*

$$\mathbb{E}[V_{t_{n+1}} \mid X_{t_n}] = \sum_i c_i \varphi_i(X_{t_n}),$$

reducing the problem to a standard least-squares problem. If the Markov property does not hold, we can still re-express

$$\mathbb{E}[V_{t_{n+1}} \mid \mathcal{F}_{t_n}] = \mathbb{E}[V_{t_{n+1}} \mid X_{t_0}, X_{t_1}, \dots, X_{t_n}],$$

but this corresponds to an (infeasible) regression problem in $\mathbb{R}^{d(n+1)}$.

In the dual approach, we can often prove the existence of an optimal martingale expressible in terms of some function of X under the Markov property—e.g., in the Brownian setting, some optimal martingale is given by

$$M_t = \int_0^t f(s, X_s) \mathrm{d}W_s.$$

Again, we are left with a (low-dimensional) functional approximation problem.

Code Notebook
The complete and documented code for this chapter can be found in a Jupyter Notebook in the following GitHub repository:
https://github.com/hagerpa/signature_stopping.git

2 The Space of Stopped Rough Paths

The optimal stopping problem consists of finding a rule that tells us at every time instance whether it is favourable to stop a stochastic process now or to let it still run to maximize a certain gain. The information available to us is the whole past of the process until the present time. If the process satisfies the Markov property, the problem simplifies a lot since the past simply does not matter here. Consequently, most of the work that was done in the field of optimal stopping concentrates on the Markovian case, cf. e.g. the classical text [18]. However, one can argue that in many situations, assuming the Markov property leads to an oversimplification of the problem. In the general path-dependent case, the history of the process can be encoded using the signature, and it should not be too surprising that the optimal stopping problem can be reformulated in terms of the signature, too. This will lead to a computational convenient way to solve the optimal stopping problem in a very general form.

In this chapter, we will consider exclusively the continuous-time case $\mathcal{T} = [0, T]$. We will always assume that $\mathcal{F} = (\mathcal{F}_t)_{t\in\mathcal{T}}$ is a filtration generated by a continuous stochastic process $X = (X_t)_{t\in\mathcal{T}}$, i.e. $\mathcal{F}_t = \sigma(X_s \,|\, s \in \mathcal{T}, s \leq t)$. Let τ be an $\mathcal{F}$-stopping time. Note that for any continuous $\mathbb{R}$-valued process Y, we have the identity

$$Y_{\tau\wedge T} = Y_0 + \int_0^T \mathbb{1}_{[0,\tau\wedge T)}(t)\,\mathrm{d}Y_t.$$

For any $\mathcal{F}$-adapted stochastic process U and any $z > 0$, the random variable

$$\tilde{\tau} := \inf\left\{t \geq 0 \,|\, \int_0^t U_s^2\,\mathrm{d}s \geq z\right\} \wedge T$$

defines a stopping time, and thus

$$Y_{\tilde{\tau}\wedge T} = Y_0 + \int_0^T \mathbb{1}_{[0,z)}\left(\int_0^t U_s^2\,\mathrm{d}s\right)\mathrm{d}Y_t.$$

The idea is now to approximate

$$\sup_{\tau\in S} \mathbb{E}[Y_{\tau\wedge T}] \tag{5}$$

by

$$\sup_U \mathbb{E}\left[Y_0 + \int_0^T \mathbb{1}_{[0,z)}\left(\int_0^t U_s^2\,\mathrm{d}s\right)\mathrm{d}Y_t\right] \tag{6}$$

where the supremum runs over a sufficiently rich class of $\mathcal{F}$-adapted stochastic processes U. To see that (5) and (6) can indeed be equal, we can informally define for any given stopping time τ an adapted process U by setting

$$U_t := \begin{cases} 0 & \text{for } t \leq \tau \wedge T \\ \infty & \text{for } t > \tau \wedge T \end{cases}$$

that clearly satisfies

$$Y_{\tau \wedge T} = Y_0 + \int_0^T \mathbb{1}_{[0,z)} \left(\int_0^t U_s^2 \, \mathrm{d}s \right) \mathrm{d}Y_t.$$

Using the signature, we will be able to parametrize a subclass of adapted processes that will still be large enough to attain the supremum in (5).

One key point for using the signature method in optimal stochastic control theory is to model the notion of adaptedness via a suitable factorization over a space of path segments or *stopped paths*. To illustrate the idea, recall that if U is a random variable that is $\sigma(Z)$-measurable for another random variable Z, there is a measurable map f such that $U = f(Z)$. Therefore, if $(U_t)_{t \in \mathcal{T}}$ is $\mathcal{F}$-adapted, we know (all technicalities aside) that for every $t \in \mathcal{T}$, there is a measurable map f_t such that $U_t = f_t(X|_{[0,t]})$. Here, the maps f_t are defined on suitable path spaces where the paths are defined on the interval $[0, t]$. We would now like to remove the dependency of the map f on the time point t. To do this, we need to find a suitable space that contains paths that are defined on all different time intervals $[0, t]$, $t \in \mathcal{T}$, i.e. on path segments. From now, we assume that X is a continuous stochastic process taking values in $\mathbb{R}^d$. The following definition first appeared in Dupire's work on functional Itō-calculus [8].

Definition 2.1 Set $C_t := C([0, t], \mathbb{R}^d)$ and define the set

$$\mathscr{C}_T := \bigcup_{t \in [0,T]} C_t.$$

For $X, Y \in \mathscr{C}_T$, $X : [0, t] \to \mathbb{R}^d$ and $Y : [0, s] \to \mathbb{R}^d$, we define

$$d_{\mathscr{C}_T}(X, Y) := \sup_{u \in [0,T]} |X_{u \wedge t} - Y_{u \wedge s}| + |t - s|.$$

It is not hard to show that $(\mathscr{C}_T, d_{\mathscr{C}_T})$ is a metric space. With a bit more work, one can show that $\mathscr{C}_T$ is complete and separable, i.e. a Polish space. The following proposition shows that the space $\mathscr{C}_T$ indeed serves as a suitable space for the desired factorization.

Proposition 2.2 *Let $U = (U_t)_{t \in [0,T]}$ be an $\mathbb{R}$-valued, $\mathcal{F}$-adapted and left continuous stochastic process. Then there exists a Borel-measurable map $\theta : \mathscr{C}_T \to \mathbb{R}$ such that*

$$U_t(\omega) = \theta(X(\omega)|_{[0,t]})$$

for every $t \in [0, T]$ and $\omega \in \Omega$.

We shifted the proof of this Proposition to the appendix. Let us note that the left-continuity assumption can actually be dropped in the case when U is progressively measurable. A proof of this more general result can be found in the forthcoming work [1].

Proposition 2.2 will not be useful for us since not every continuous path has a well-defined signature. Instead, we will first consider the smaller space of bounded variation paths in the next section for which the signature always exists.

2.1 Signature Factorization for Bounded Variation Paths

In the last section, we saw that for every left-continuous $\mathcal{F}$-adapted process U, there exists a measurable map $\theta\colon \mathscr{C}_T \to \mathbb{R}$ such that $U_t = \theta(X|_{[0,t]})$. In this section, we will assume that the paths of X are of bounded variation such that for every $0 \leq s \leq t \leq T$, the signature

$$\mathbb{X}^{<\infty}_{s,t} = \left(1, X_t - X_s, \int_{s<t_1<t_2<t} \mathrm{d}X_{t_1} \otimes \mathrm{d}X_{t_2}, \ldots\right)$$

exists as an element in the *extended tensor algebra*

$$T((\mathbb{R}^d)) := \prod_{m=0}^{\infty} (\mathbb{R}^d)^{\otimes m},$$

cf. Sect. 2.3.3 of chapter "A Primer on the Signature Method in Machine Learning" and Sects. 2 and 3 of chapter "The Signature Kernel". The projection of $T((\mathbb{R}^d))$ on the n-th tensor power $(\mathbb{R}^d)^{\otimes n}$ is denoted by $\pi_n\colon T((\mathbb{R}^d)) \to (\mathbb{R}^d)^{\otimes n}$. $T((\mathbb{R}^d))$ is the algebraic dual of the *tensor algebra*

$$T((\mathbb{R}^d)^*) := \bigoplus_{m=0}^{\infty} ((\mathbb{R}^d)^*)^{\otimes m}$$

where the pairing is denoted by $\langle \cdot, \cdot \rangle\colon T((\mathbb{R}^d)^*) \times T((\mathbb{R}^d)) \to \mathbb{R}$, cf. Sect. 4 of chapter "The Signature Kernel". The shuffle product $ш$ that was introduced in Definition 2.12 of chapter "A Primer on the Signature Method in Machine Learning" extends to a product on $T((\mathbb{R}^d)^*)$ and the shuffle identity from Theorem 2.14 of chapter "A Primer on the Signature Method in Machine Learning" can be reformulated as

$$\langle l_1, \mathbb{X}_{0,T}^{<\infty}\rangle\langle l_2, \mathbb{X}_{0,T}^{<\infty}\rangle = \langle l_1 \sqcup\!\sqcup l_2 \mathbb{X}_{0,T}^{<\infty}\rangle$$

for every $l_1, l_2 \in T((\mathbb{R}^d)^*)$ (see also Eq. (8) of chapter "The Signature Kernel").

Our goal will be to identify a class of functionals Θ such that for any given map θ, we can find a $\vartheta \in \Theta$ such that

$$\theta(X|_{[0,\cdot]}) \approx \vartheta(\mathbb{X}_{0,\cdot}^{<\infty})$$

where $\mathbb{X}^{<\infty}$ denotes the signature of X. It is clear that this can only work if X can be recovered by $\mathbb{X}^{<\infty}$. We therefore consider the time-augmented process $\widehat{X}_t := (t, X_t)$ which is a stochastic process with values in $\mathbb{R}^{1+d}$. Its signature is denoted by $\widehat{\mathbb{X}}^{<\infty}$. From Proposition 2.34 of chapter "A Primer on the Signature Method in Machine Learning", we know that X can indeed by recovered by $\widehat{\mathbb{X}}^{<\infty}$.

Inspired by Dupire, we make the following definition:

Definition 2.3 Let $\widehat{\mathcal{X}}_t$ denote the set of paths $X\colon [0, t] \to \mathbb{R}^{1+d}$ having bounded variation and for which the first component is time, i.e. $X_s = (s, X_s^1, \ldots, X_s^d)$ for every $s \in [0, t]$. Set

$$\widehat{\mathscr{X}_T} := \bigcup_{t\in[0,T]} \widehat{\mathcal{X}}_t.$$

We equip $\widehat{\mathscr{X}_T}$ with the following metric: For $X, Y \in \widehat{\mathscr{X}_T}$ with $X\colon [0, t] \to \mathbb{R}^{1+d}$, $Y\colon [0, s] \to \mathbb{R}^{1+d}$, we set

$$d_{\widehat{\mathscr{X}_T}}(X, Y) := |X_0 - Y_0| + \int_0^T |\dot{X}_{u\wedge t} - \dot{Y}_{u\wedge s}|\, du + |t - s|.$$

We can now prove that continuous functionals defined on $\widehat{\mathscr{X}_T}$ can be arbitrarily well approximated by linear functionals of the signature on compact sets. The following proposition is a generalization of Proposition 2.35 in chapter "A Primer on the Signature Method in Machine Learning".

Proposition 2.4 *Let $\theta\colon \widehat{\mathscr{X}_T} \to \mathbb{R}$ be continuous and $K \subset \widehat{\mathcal{X}}_T$ be a compact set. Then for every $\varepsilon > 0$ there exists and element $l \in T((\mathbb{R}^{1+d})^*)$ such that*

$$\sup_{X|_{[0,t]}\in K} |\theta(X|_{[0,t]}) - \langle l, \mathbb{X}_{0,t}^{<\infty}\rangle| \leq \varepsilon.$$

Proof One first checks that for every $l \in T((\mathbb{R}^{1+d})^*)$, the map

$$X|_{[0,t]} \mapsto \langle l, \mathbb{X}_{0,t}^{<\infty}\rangle$$

defined from $\widehat{\mathscr{X}_T}$ to $\mathbb{R}$ is continuous. Set

$$\mathcal{W} := \{\theta \in C(\widehat{\mathscr{X}_T}, \mathbb{R}) \mid \exists l \in T((\mathbb{R}^{1+d})^*) \text{ s.t. } \theta(X|_{[0,t]}) = \langle l, \mathbb{X}_{0,t}^{<\infty} \rangle \; \forall X|_{[0,t]} \in \widehat{\mathscr{X}_T}\}.$$

We will prove that $\mathcal{W}$ satisfies the conditions of the Stone-Weierstrass theorem, see Theorem 2.33 in chapter "A Primer on the Signature Method in Machine Learning". The first two properties of that theorem are immediate. Let $\theta_1, \theta_2 \in \mathcal{W}$ with corresponding $l_1, l_2 \in T((\mathbb{R}^{1+d})^*)$. From the shuffle product identity, it holds that for every $X|_{[0,t]} \in \widehat{\mathscr{X}_T}$,

$$\begin{aligned} \theta_1(X|_{[0,t]})\theta_2(X|_{[0,t]}) &= \langle l_1, \mathbb{X}_{0,t}^{<\infty} \rangle \langle l_2, \mathbb{X}_{0,t}^{<\infty} \rangle \\ &= \langle l_1 \sqcup\!\sqcup l_2, \mathbb{X}_{0,t}^{<\infty} \rangle \\ &= \theta(X|_{[0,t]}) \end{aligned}$$

where $\theta \in \mathcal{W}$ is given by $\theta(X|_{[0,t]}) = \langle l_1 \sqcup\!\sqcup l_2, \mathbb{X}_{0,t}^{<\infty} \rangle$ for all $X|_{[0,t]} \in \widehat{\mathscr{X}_T}$. Thus $\mathcal{W}$ is indeed an algebra. To prove the point-separating property, let X and Y be two distinct paths. By Proposition 2.34 in chapter "A Primer on the Signature Method in Machine Learning", their signatures $\mathbb{X}_{0,T}^{<\infty}$ and $\mathbb{Y}_{0,T}^{<\infty}$ are distinct, too. Therefore, there is an element $l \in T((\mathbb{R}^{1+d})^*)$ such that

$$\langle l, \mathbb{X}_{0,T}^{<\infty} \rangle \neq \langle l, \mathbb{Y}_{0,T}^{<\infty} \rangle$$

which proves the last condition of the Stone-Weierstrass theorem. The claim of our proposition thus follows. □

Remark 2.5 It is possible to prove that $(\widehat{\mathscr{X}_T}, d_{\widehat{\mathscr{X}_T}})$ is a complete metric space. However, since the space of bounded variation paths is not separable, $\widehat{\mathscr{X}_T}$ is not separable either. As a remedy, one can instead consider the spaces $\widehat{\mathcal{X}}_t^0$ that are defined as the closure of smooth paths $X\colon [0,t] \to \mathbb{R}^{1+d}$ of the form $X_s = (s, X_s^1, \ldots, X_s^d)$, $s \in [0,t]$, with respect to the bounded variation metric. It is then possible to prove that the corresponding space

$$\widehat{\mathscr{X}_T^0} = \bigcup_{t \in [0,T]} \widehat{\mathcal{X}}_t^0,$$

is Polish.

Recall that our goal was to approximate $\mathcal{F}$-adapted processes U by functionals of the signature of the process X that generates $\mathcal{F}$. Looking back at the results of Propositions 2.2 and 2.4, it remains to prove that measurable functions on a metric space can be approximated by continuous functions. The following proposition is well-known. For completeness, we decided to provide a proof in the appendix.

Proposition 2.6 *Let $(\mathcal{M}, d)$ be a metric space and let μ be a probability measure defined on the Borel sets of $\mathcal{M}$. If $f\colon M \to \mathbb{R}$ is measurable, we can find a sequence of continuous maps (f_n) such that*

$$f_n \to f$$

μ-*almost surely as* $n \to \infty$.

Finally, we can state the following:

Proposition 2.7 *Let* $X\colon [0,T] \to \mathbb{R}^d$ *be a stochastic process. Assume that the process* $\widehat{X}_t = (t, X_t)$ *has sample paths in* $\widehat{\mathcal{X}}^0_T$ *almost surely. Let* U *be a left-continuous process adapted to the filtration generated by* X. *Then there is a sequence of elements* $l_n \in T((\mathbb{R}^{1+d})^*)$ *such that*

$$\langle l_n, \widehat{\mathbb{X}}^{<\infty}_{0,t}(\omega)\rangle \to U_t(\omega)$$

$(\mathbb{P} \otimes \mathrm{Leb})$-*almost surely as* $n \to \infty$.

Proof Let μ denote the law of $\widehat{X}$ that is a probability measure on $\widehat{\mathcal{X}}^0_T$. Let ν be the probability measure on the Borel sets of $\widehat{\mathscr{X}^0_T}$ that is defined as the image measure of $\mu \otimes \mathrm{Leb}$ under the map

$$(\widehat{X}, t) \mapsto \widehat{X}|_{[0,t]}.$$

Similar as in Proposition 2.2, we can find a measurable map $\theta\colon \widehat{\mathscr{X}^0_T} \to \mathbb{R}$ such that $U_t(\omega) = \theta(\widehat{X}(\omega)|_{[0,t]})$. From Proposition 2.6, we can find a sequence of continuous functions $\theta_n\colon \widehat{\mathscr{X}^0_T} \to \mathbb{R}$ such that $\theta_n \to \theta$ ν-almost surely as $n \to \infty$. Next, let $\tilde{\theta}\colon \widehat{\mathscr{X}^0_T} \to \mathbb{R}$ be any continuous function. We claim that for arbitrary $\varepsilon > 0$ and $\delta > 0$, we can find an element $l \in T((\mathbb{R}^{1+d})^*)$ such that

$$\nu(|\tilde{\theta}(\widehat{X}|_{[0,t]}) - \langle l, \widehat{\mathbb{X}}^{<\infty}_{0,t}\rangle| \geq \varepsilon) < \delta. \tag{7}$$

Since $\widehat{\mathscr{X}^0_T}$ is Polish, we can find a compact set $K \subset \widehat{\mathscr{X}^0_T}$ such that $\nu(K^c) < \delta$. On the set K, we can use Proposition 2.4 to see that there is an element $l \in T((\mathbb{R}^{1+d})^*)$ such that

$$\sup_{X|_{[0,t]} \in K} |\tilde{\theta}(\widehat{X}|_{[0,t]}) - \langle l, \widehat{\mathbb{X}}^{<\infty}_{0,t}\rangle| < \varepsilon.$$

and (7) follows. From these results, the assertion of the proposition can be deduced. □

2.2 Signature Factorization for Rough Paths

The most important stochastic processes, like the Brownian motion, do not have sample paths of bounded variation. In this section, we want to explain how to define the signature for those paths.

One way to measure roughness of a path $X\colon [0,T] \to \mathbb{R}^d$ is to consider its *p-variation* for some $p \in [1,\infty)$, i.e.

$$\|X\|_p := \sup_{\mathcal{D}\subset[0,T]} \left(\sum_{[t_i,t_{i+1}]\in\mathcal{D}} |X_{t_{i+1}} - X_{t_i}|^p \right)^{\frac{1}{p}}$$

where the supremum ranges over all finite partitions of the interval $[0,T]$. It is not hard to see that $\|\cdot\|_p$ is a seminorm on the space of paths and that for $p=1$ it coincides with the bounded variation seminorm. It is also not hard to check that α-Hölder paths have finite $\frac{1}{\alpha}$-variation (however, the converse is not true: the square root function $t \mapsto \sqrt{t}$ is of bounded variation but not Lipschitz continuous).

Let us assume that a path X is not of bounded variation. Examples from stochastic analysis teach us that, in general, there is no canonical choice of what the iterated integrals of X should be. In fact, if X denotes the Brownian motion, we can make sense of iterated integrals as Itō- or Stratonovich integrals, both satisfying basic properties of an integral, but being different objects. One of Lyons' key insights was that if finitely many iterated integrals up to a certain order (depending on the regularity parameter p) are given, the remaining iterated integrals, and hence the signature, are uniquely determined. This leads us to the notion of a *rough path* that we are going to define now.

Set $\Delta_T := \{s,t \in [0,T] \,|\, s \le t\}$. For $p \in [1,\infty)$, choose $N \in \mathbb{N}$ such that $p-1 < N \le p$. The *truncated tensor algebra* is defined as

$$T^N(\mathbb{R}^d) := \bigoplus_{m=0}^{N} (\mathbb{R}^d)^{\otimes m}.$$

For paths $\mathbf{X}, \mathbf{Y}\colon \Delta_T \to T^N(\mathbb{R}^d)$, we define

$$\|\mathbf{X}\|_p := \max_{k=1,\dots,N} \sup_{\mathcal{D}\subset[0,T]} \left(\sum_{[t_i,t_{i+1}]\in\mathcal{D}} |\pi_k(\mathbf{X}_{t_i,t_{i+1}})|^{kp} \right)^{\frac{1}{kp}}$$

and

$$d_p(\mathbf{X},\mathbf{Y}) := \max_{k=1,\dots,N} \sup_{\mathcal{D}\subset[0,T]} \left(\sum_{[t_i,t_{i+1}]\in\mathcal{D}} |\pi_k(\mathbf{X}_{t_i,t_{i+1}}) - \pi_k(\mathbf{Y}_{t_i,t_{i+1}})|^{kp} \right)^{\frac{1}{kp}}.$$

Definition 2.8 Let $p \in [1,\infty)$ and $p-1 < N \le p$.

(i) For a continuous path $X\colon [0,T] \to \mathbb{R}^d$ of bounded variation, we define its *canonical lift to a p-rough path* $\mathbf{X}\colon \Delta_T \to T^N(\mathbb{R}^d)$ by setting

$$\mathbf{X}_{s,t} := \left(1, X_t - X_s, \int_{s<t_1<t_2<t} \mathrm{d}X_{t_1} \otimes \mathrm{d}X_{t_2}, \dots, \right.$$
$$\left. \int_{s<t_1<\dots<t_N<t} \mathrm{d}X_{t_1} \otimes \dots \otimes \mathrm{d}X_{t_N} \right).$$

(ii) A continuous path $\mathbf{X}\colon \Delta_T \to T^N(\mathbb{R}^d)$ is called a *p-rough path* if there exists a sequence of smooth paths $X(n)\colon [0, T] \to \mathbb{R}^d$ such that their canonical lifts $\mathbf{X}(n)$ satisfy

$$d_p(\mathbf{X}, \mathbf{X}(n)) \to 0$$

as $n \to \infty$. We denote the space of all p-rough paths by Ω_T^p and equip it with the metric d_p.

(iii) A continuous path $\mathbf{X}\colon \Delta_T \to T^N(\mathbb{R}^{1+d})$ is called a *time-augmented p-rough path* if there exists a sequence of smooth paths $X(n)\colon [0, T] \to \mathbb{R}^{1+d}$ of the form $X_t(n) = (t, X_t^1(n), \dots, X_t^d(n))$ such that their canonical lifts $\mathbf{X}(n)$ satisfy

$$d_p(\mathbf{X}, \mathbf{X}(n)) \to 0$$

as $n \to \infty$. The space of all time-augmented p-rough paths is equipped with the metric d_p, too, and will be denoted by $\widehat{\Omega}_T^p$

For every p-rough path $\mathbf{X}$, one can uniquely define its time-augmented rough path $\widehat{\mathbf{X}}$ (this follows from the more general result [11, Theorem 9.26]). Moreover, if $X(n)\colon [0, T] \to \mathbb{R}^d$ is a sequence of smooth paths such that their canonical lifts $\mathbf{X}(n)$ satisfy $d_p(\mathbf{X}, \mathbf{X}(n)) \to 0$ as $n \to \infty$, the canonical lifts $\widehat{\mathbf{X}}(n)$ of the time-augmented smooth paths $\widehat{X}_t = (t, X_t)$ satisfy $d_p(\widehat{\mathbf{X}}, \widehat{\mathbf{X}}(n)) \to 0$ as $n \to \infty$ [11, Theorem 9.30].

Example 2.9 For $p = 1$, one has $N = 1$ and a 1-rough path $\mathbf{X}\colon \Delta_T \to T^1(\mathbb{R}^d)$ is of the form $\mathbf{X}_{s,t} = (1, X_{s,t})$. Let $X(n)$ be a sequence of smooth paths such that $d_1(\mathbf{X}(n), \mathbf{X}) \to 0$ as $n \to \infty$. From the identity $X_t(n) - X_s(n) = X_t(n) - X_u(n) + X_u(n) - X_s(n)$, it follows that $X_{s,t} = X_{s,u} + X_{u,t}$. Setting $X_t := X_{0,t}$, it follows that $X_t - X_s = X_{s,t}$. Therefore, the space Ω_T^1 can be identified with the closure of the space of smooth paths under the bounded variation metric. Moreover, $\widehat{\Omega}_T^1$ can be identified with the space $\widehat{\mathcal{X}}_T^0$.

For us, the most important point about rough paths is that their signature can be uniquely defined.

Theorem 2.10 (Lyons' Extension Theorem) *Let $\mathbf{X} \in \Omega_T^p$. Then there is a unique map $\mathbb{X}^{<\infty}\colon \Delta_T \to T((\mathbb{R}^d))$ with the following properties:*

(i) $\mathbb{X}^{<\infty}$ extends $\mathbf{X}$, i.e. $\pi_n(\mathbb{X}_{s,t}^{<\infty}) = \pi_n(\mathbf{X}_{s,t})$ for every $s, t \in \Delta_T$ and every $n = 0, \dots, N$.

(ii) $\mathbb{X}^{<\infty}$ *satisfies* Chen's identity, *i.e. for every* $s \leq u \leq t$, *one has*

$$\mathbb{X}^{<\infty}_{s,t} = \mathbb{X}^{<\infty}_{s,u} \otimes \mathbb{X}^{<\infty}_{u,t}.$$

(iii) $\mathbb{X}^{<\infty}$ *satisfies the* shuffle identity, *i.e. for every* $s \leq t$, *one has*

$$\langle l_1, \mathbb{X}^{<\infty}_{s,t} \rangle \langle l_2, \mathbb{X}^{<\infty}_{s,t} \rangle = \langle l_1 \sqcup\!\sqcup l_2, \mathbb{X}^{<\infty}_{s,t} \rangle$$

for every $l_1, l_2 \in T((\mathbb{R}^d)^*)$.

(iv) *For every* $M \in \mathbb{N}$,

$$\max_{k=1,\dots,M} \sup_{\mathcal{D} \subset [0,T]} \left(\sum_{[t_i, t_{i+1}] \in \mathcal{D}} |\pi_k(\mathbb{X}^{<\infty}_{t_i, t_{i+1}})|^{kp} \right)^{\frac{1}{kp}} < \infty.$$

Proof [14, Theorem 3.7]. □

The signature $\mathbb{X}^{<\infty}_{0,T}$ of a rough path $\mathbf{X}$ determines the path under suitable conditions. In fact, one can prove the following:

Theorem 2.11 *Suppose that* $\mathbf{X}, \mathbf{Y} \in \widehat{\Omega}^p_T$ *and that* $\mathbb{X}^{<\infty}_{0,T} = \mathbb{Y}^{<\infty}_{0,T}$. *Then* $\mathbf{X} = \mathbf{Y}$.

Proof Follows from the results obtained in [5]. □

Analogous to the bounded variation case, we can now define the space of stopped rough paths:

Definition 2.12 The space of stopped p-rough paths is defined as

$$\Lambda^p_T := \bigcup_{t \in [0,T]} \widehat{\Omega}^p_t.$$

It is equipped with the following metric: If $\mathbf{X} \in \widehat{\Omega}^p_t$ and $\mathbf{Y} \in \widehat{\Omega}^p_s$, we define $\tilde{\mathbf{X}} \in \widehat{\Omega}^p_T$ and $\tilde{\mathbf{Y}} \in \widehat{\Omega}^p_T$ by $\tilde{\mathbf{X}}_{u,v} := \mathbf{X}_{u \wedge t, v \wedge t}$ resp. $\tilde{\mathbf{Y}}_{u,v} := \mathbf{Y}_{u \wedge s, v \wedge s}$. We then set

$$d_{\Lambda^p_T}(\mathbf{X}, \mathbf{Y}) := d_p(\tilde{\mathbf{X}}, \tilde{\mathbf{Y}}) + |t - s|.$$

Using exactly the same arguments as in Proposition 2.4, we can prove the following:

Proposition 2.13 *Let* $\theta \colon \Lambda^p_T \to \mathbb{R}$ *be continuous and* $K \subset \Lambda^p_T$ *be a compact set. Then for every* $\varepsilon > 0$ *there exists and element* $l \in T((\mathbb{R}^{1+d})^*)$ *such that*

$$\sup_{\mathbf{X}|_{[0,t]} \in K} |\theta(\mathbf{X}|_{[0,t]}) - \langle l, \mathbb{X}^{<\infty}_{0,t} \rangle| \leq \varepsilon.$$

This allows us to finally prove the following result:

Proposition 2.14 *Let* $\mathbf{X}$ *be a stochastic process with sample paths in a rough path space* Ω_T^p. *Consider the filtration* $\mathcal{F} = (\mathcal{F}_t)$ *generated by* $\mathbf{X}$, *i.e.*

$$\mathcal{F}_t = \sigma(\mathbf{X}_{u,v} \,|\, (u, v) \in \Delta_t).$$

Let $\widehat{\mathbf{X}}$ *denote the time-augmented process having sample paths in* $\widehat{\Omega}_T^p$ *and let* $\widehat{\mathbb{X}}^{<\infty}$ *denote its signature. Let* U *be a left-continuous process adapted to the filtration* $(\mathcal{F}_t)$. *Then there is a sequence of elements* $l_n \in T((\mathbb{R}^{1+d})^*)$ *such that*

$$\langle l_n, \widehat{\mathbb{X}}_{0,t}^{<\infty}(\omega)\rangle \to U_t(\omega)$$

$(\mathbb{P} \otimes \mathrm{Leb})$-*almost surely as* $n \to \infty$.

Example 2.15 Let $B\colon [0, T] \to \mathbb{R}^d$ be a Brownian motion. It is known that the sample paths of B are α-Hölder continuous for every $\alpha < \frac{1}{2}$. Therefore, the trajectories of B have finite p-variation for every $p > 2$. It can be shown that B can be lifted to a process $\mathbf{B}$ having trajectories in the rough path space Ω_T^p for every $p > 2$ [10, Chapter 3]. In fact, $\pi_2(\mathbf{B})$ is given by the iterated Stratonovich integrals of B:

$$\pi_2(\mathbf{B}_{s,t}) = \int_s^t (B_u - B_s) \otimes \circ B_u.$$

The time-augmented process $\widehat{\mathbf{B}}$ is given on the second level by

$$\pi_2(\widehat{\mathbf{B}}_{s,t}) = \begin{pmatrix} \int_s^t (u - s)\, \mathrm{d}u & \int_s^t (u - s)\, \mathrm{d}B_u \\ \int_s^t (B_u - B_s)\, \mathrm{d}u & \int_s^t (B_u - B_s) \otimes \circ B_u \end{pmatrix}.$$

The signature $\widehat{\mathbb{B}}^{<\infty}$ consists of all iterated Stratonovich integrals of the process $\widehat{B}_t = (t, B_t)$. It is clear that the filtration generated by B coincides with the one generated by $\mathbf{B}$, i.e.

$$\mathcal{F}_t = \sigma(\mathbf{B}_{u,v} \,|\, (u, v) \in \Delta_t) = \sigma(B_u \,|\, u \in [0, t]).$$

From Proposition 2.14, it then follows that every left-continuous process that is adapted to the Brownian filtration can be approximated by linear functionals of the signature of the time-augmented Brownian motion, i.e. by linear combinations of iterated Stratonovich integrals of the process $\widehat{B}_t = (t, B_t)$.

3 Approximation Methods

We now give a more detailed look at three different approximation methods for the optimal stopping problem.

3.1 *Parametrization of Stopping Times*

In the first approach, we parameterize stopping times. As motivated in Sect. 1, the general idea is to encode stopping times as *first hitting times* of sets, think of *optimal exercise regions*. Under Markovian assumptions, it would make sense to consider hitting times of the process X itself, i.e., consider first hitting times w.r.t. subsets of $[0, T] \times \mathbb{R}^d$. In order to generalize to the case with memory, we will instead consider hitting times w.r.t. subsets of $T((\mathbb{R}^{1+d}))$, i.e., consider

$$\tau_A := \inf\{t \in [0, T] | \widehat{\mathbb{X}}^{<\infty}_{0,t} \in A\}, \quad A \subset T((\mathbb{R}^{1+d})), \tag{8}$$

recalling that $\widehat{\mathbb{X}}^{<\infty}_{0,t}$ denotes the signature of the time-extended path (t, X_t). Given that $T((\mathbb{R}^{1+d}))$ is infinite-dimensional, this leaves us with the question of finding a tractable family of subsets $A \subset T((\mathbb{R}^{1+d}))$ rich enough to obtain good approximations to optimal stopping times.

In [2], the parameterization finally chosen was

$$\tau_\ell := \inf\{t \in [0, T] | \int_0^t \left\langle \ell , \widehat{\mathbb{X}}^{<\infty}_{0,s} \right\rangle^2 \mathrm{d}s \geq 1\}, \quad \ell \in T((\mathbb{R}^{1+d})^*). \tag{9}$$

The integral in (9) is increasing in t, which simplifies computing the minimizer in t. Obviously, we have $\left\langle \ell , \widehat{\mathbb{X}}^{<\infty}_{0,s} \right\rangle^2 = \left\langle \ell \sqcup\!\sqcup \ell , \widehat{\mathbb{X}}^{<\infty}_{0,s} \right\rangle$. Additionally, taking into account that running time t was added as first component of the process $\widehat{X}_t = (t, X_t)$ as well as the distributative property of the tensor product, we obtain

$$\int_0^t \left\langle \ell , \widehat{\mathbb{X}}^{<\infty}_{0,s} \right\rangle^2 \mathrm{d}s = \int_0^t \left\langle \ell \sqcup\!\sqcup \ell , \widehat{\mathbb{X}}^{<\infty}_{0,s} \right\rangle \mathrm{d}s = \left\langle (\ell \sqcup\!\sqcup \ell) 1 , \widehat{\mathbb{X}}^{<\infty}_{0,t} \right\rangle.$$

Hence, τ_ℓ is the first hitting time of the signature process at an *affine hyperplane* in $T((\mathbb{R}^{1+d}))$. Given the *shuffle identity*, we expect that such hitting times are a very expressive class of stopping times. Note that hitting times are, generally, discontinuous functions of the path, so that we would need to appeal to a version of Prop. 2.14 to get a proper universal approximation theorem.

However, from a numerical perspective, a promising alternative might be to first *regularize* the function instead. It turns out that such regularization can be achieved by *randomization*, see [2]. Indeed, let us adjust (9) by stopping when we first hit a random level Z, i.e.,

$$\tau_\ell^r := \inf\{t \in [0,T] | \int_0^t \langle \ell , \widehat{\mathbb{X}}_{0,s}^{<\infty} \rangle^2 \mathrm{d}s \geq Z\}, \quad \ell \in T((\mathbb{R}^{1+d})^*), \tag{10}$$

where, for simplicity, we fix $Z \sim \mathrm{Exp}(1)$, independent from X. After just having introduced the auxiliary random variable Z, we immediately integrate against it, giving

$$\mathbb{E}\left[Y_{\tau_\ell^r \wedge T} \mid X_\cdot \right] = Y_0 + \int_0^T \exp\left(- \int_0^t \langle \ell , \widehat{\mathbb{X}}_{0,s}^{<\infty} \rangle \mathrm{d}s \right) \mathrm{d}Y_t, \tag{11}$$

see [2, Prop. 4.3] for details. Note that the right hand side in (11) is, in fact, *smooth* in ℓ, so that higher order numerical optimization methods can be applied for computing

$$V_0 = \sup_{\ell \in T((\mathbb{R}^{1+d})^*)} \mathbb{E}\left[Y_0 + \int_0^T \exp\left(- \int_0^t \langle \ell , \widehat{\mathbb{X}}_{0,s}^{<\infty} \rangle \mathrm{d}s \right) \mathrm{d}Y_t \right]. \tag{12}$$

Of course, in addition to linear signature stopping rules such as (12), we could also use non-linear ones, i.e.,

$$V_0 = \sup_{\theta \in \Theta} \mathbb{E}\left[Y_0 + \int_0^T \exp\left(- \int_0^t f_\theta(\widehat{\mathbb{X}}_{0,s}^{<\infty}) \mathrm{d}s \right) \mathrm{d}Y_t \right] \tag{13}$$

for some suitable rich class of functionals f_θ of the signature parameterized by $\theta \in \Theta$. In fact, (13) was proved in [2, Prop. 7.4] for a class of neural network functions in the signature—θ corresponding to the network weights.

In either way, several additional approximations are necessary in order to turn (12) or (13), respectively, into a fully implementable numerical method. Concentrating on (13) for concreteness, we obtain to following scheme:

1. For a given *truncation level* $I \in \mathbb{N}$ of the signature and a time grid $0 = t_0 < \cdots < t_N = T$, simulate samples $\widehat{\mathbb{X}}_{0,t_n}^{\leq I,(m)}$, $Y_{t_n}^{(m)}$, $n = 1, \ldots, N$, $m = 1, \ldots, M$, possibly based on a discretization of the iterated integrals forming the signature.
2. Using a suitable *optimization algorithm*, solve the optimization problem

$$\widehat{\theta} \in \arg\min_{\theta \in \Theta} \frac{1}{M} \sum_{m=1}^{M} \left(Y_0^{(m)} + \sum_{n=0}^{N-1} \exp\left(- \sum_{j=0}^{n-1} f_\theta(\widehat{\mathbb{X}}_{0,t_j}^{\leq I,(m)})(t_{j+1} - t_j) \right) \right.$$
$$\left. \times (Y_{t_{n+1}}^{(m)} - Y_{t_n}^{(m)}) \right).$$

Remark 3.1 The stopping time $\widehat{\tau}$ corresponding to the estimated parameter $\widehat{\theta}$ (or, alternatively, $\widehat{\ell}$ in the linear case) is a (sub-optimal) stopping time, and we have

$$\mathbb{E}[Y_{\widehat{\tau}}] \leq V_0 = \mathbb{E}[Y_{\tau^* \wedge T}].$$

Hence, a *low-biased* estimator for the optimal value can be found by generating new, independent samples $\widehat{\mathbb{X}}_{0,t_n}^{\leq I,(m)}$, $Y_{t_n}^{(m)}$, $n = 1, \ldots, N$, $m = M+1, \ldots, 2M$, say, and compute

$$\widehat{V}_0 = \frac{1}{M} \sum_{m=M+1}^{2M} Y_{\widehat{\tau}^{(m)}}^{(m)},$$

noting that $\widehat{\tau}^{(m)}$ depends on $\widehat{\mathbb{X}}_{0,t_n}^{\leq I,(m)}$, $Y_{t_n}^{(m)}$, $n = 1, \ldots, N$, $m = M+1, \ldots, 2M$. The sub-optimality of the stopping time $\widehat{\tau}$ implies that $\widehat{V}_0$ is biased low, i.e., that $\mathbb{E}\left[\widehat{V}_0\right] \leq V_0$. If we manage to also construct a high-biased estimator (see the next subsection), then this gives a powerful method to estimate errors: If the gap between low- and high-biased estimators is small, the accuracy is high.

3.2 *The Signature Longstaff–Schwartz Algorithm*

The celebrated *Longstaff–Schwartz algorithm* [13] is based on the dynamic programming principle (2). We assume that we are given a Bermudan option, which allows exercising at times $0 = t_0 < t_1 < \cdots < t_N = T$. Recall that the optimal value function satisfies the recursive formula

$$V_{t_n} = \max\left(Y_{t_n}, C_{t_n}\right), \quad 0 \leq n < N, \quad C_{t_n} := \mathbb{E}\left[V_{t_{n+1}} \mid \mathcal{F}_{t_n}\right]$$

with the terminal value $V_T = Y_T$. In particular, any approximation $\widehat{C}_{t_n} \approx C_{t_n}$, $n = 0, \ldots, N-1$, of the *continuation value* C_{t_n}, defines an approximation $\widehat{V}_{t_n}$ of the value V_{t_n} by plugging in the recursion. However, in practice, $\widehat{V}$ can give unreasonable value, especially for extreme paths. A more stable approximation can be found by recursively constructing stopping times τ_n taking values in $\{t_n, \ldots, t_N\}$—which should be thought as (approximate) solutions to the optimal stopping problems started at time $t = t_n$. For this, we start with $\tau_N := t_N$. Having constructed stopping times $\tau_{n+1}, \ldots, \tau_N$, we then define

$$\tau_n := \begin{cases} t_n, & Y_{t_n} \geq \mathbb{E}\left[Y_{\tau_{n+1}} \mid \mathcal{F}_{t_n}\right], \\ \tau_{n+1}, & \text{else}. \end{cases} \tag{14}$$

Note that, by construction, τ_n takes values in $\{t_n, \ldots, t_N\}$. Moreover, by the dynamic programming principle, we have $\mathbb{E}\left[Y_{\tau_n} \mid \mathcal{F}_{t_n}\right] = V_{t_n}$ a.s., i.e., (14) is an equivalent formulation of (2). Given any algorithm computing approximations $\widehat{\mathbb{E}}_{t_n}$ of the conditional expectation operator $\mathbb{E}[\cdot \mid \mathcal{F}_{t_n}]$, we obtain approximations of the optimal stopping times τ_n by setting $\widehat{\tau}_N := t_N$ and

$$\widehat{\tau}_n := \begin{cases} t_n, & Y_{t_n} \geq \widehat{\mathbb{E}}_{t_n}\left[Y_{\tau_{n+1}}\right], \\ \widehat{\tau}_{n+1}, & \text{else}. \end{cases} \tag{15}$$

If the approximate conditional expectation $\widehat{\mathbb{E}}_{t_n}$ is always $\mathcal{F}_{t_n}$-measurable, then $\widehat{\tau}_n$ is a proper stopping time, $n = 0, \ldots, N$.

Now we are only left with the task of finding approximations to the conditional expectation operator. In the Markovian situation, Longstaff and Schwartz [13] suggest to use *least squares Monte Carlo*. By the Markov property, we know that the conditional expectation $\mathbb{E}\left[Z \mid \mathcal{F}_{t_n}\right] = f_{t_n}(X_{t_n})$ for some measurable function f. In fact, by the very definition of the conditional expectation (for r.v.s with finite second moment), it solves the minimization problem

$$f_{t_n} \in \underset{g \text{ measurable}}{\arg\min}\ \mathbb{E}\left[\left(Z - g(X_{t_n})\right)^2\right].$$

This is of the form of a regression problem. Of course, the space of measurable functions to too big to handle properly, so the standard approach is to work in a finite-dimensional subspace instead. Identify K basis functions $\phi_1, \ldots, \phi_K$, and make the *ansatz* that $\mathbb{E}\left[Z \mid \mathcal{F}_{t_n}\right]$ can be written as a linear combination of $\phi_1(X_{t_n}), \ldots, \phi_K(X_{t_n})$. Then, the *coefficients* of the linear combination can be found by solving the least squares problem

$$\mathbb{E}\left[\left(Z - \sum_{k=1}^{K} c_k \phi_k(X_{t_n})\right)^2\right] \xrightarrow{c_1,\ldots,c_K} \min! \tag{16}$$

Finally, given M independent samples $Z^{(m)}$, $X_{t_n}^{(m)}$, $m = 1, \ldots, M$, from (the joint distribution of) Z and X_{t_n}, we define

$$\begin{aligned} \widehat{\mathbb{E}}_{t_n}^M[Z] &:= \sum_{k=1}^{K} \widehat{c}_k \phi_k(X_{t_n}), \quad (\widehat{c}_1, \ldots, \widehat{c}_K) \\ &= \underset{c_1,\ldots,c_K \in \mathbb{R}}{\arg\min} \frac{1}{M} \sum_{m=1}^{M} \left(Z^{(m)} - \sum_{k=1}^{K} c_k \phi_k(X_{t_n}^{(m)})\right)^2. \end{aligned} \tag{17}$$

Remark 3.2 We have not yet touched upon the issue of finding good basis functions $\phi_1, \ldots, \phi_K$. A popular choice is to choose polynomials up to some (mixed) degree together with some polynomials of the payoff function itself, we refer to [12] for more details.

If X is not a Markov process, then there is, in general, no function f_{t_n} s.t. $\mathbb{E}\left[Z \mid \mathcal{F}_{t_n}\right] = f_{t_n}(X_{t_n})$. Instead, the conditional expectation is a function of the whole path, $\mathbb{E}\left[Z \mid \mathcal{F}_{t_n}\right] = f_{t_n}(X|_{[0,t_n]})$. Unfortunately, the space of functions on paths $X|_{[0,t_n]}$ is much larger than the space of functions on X_{t_n}, so that the task of finding efficient basis functions $\phi_1, \dots, \phi_K$ on path-space might seem hopeless at first glance. However, as we know now, linear functions of the (truncated, normalized) signature do an excellent job for this problem.

This motivates the following approach put forth in [3]: Use (15) together with

$$\widehat{\mathbb{E}}_{t_n}^{I,M}[Z] := \langle \ell^{I,M}, \widehat{\mathbb{X}}_{0,t_n}^{\leq I} \rangle, \quad \ell^{I,M} := \arg\min_{\ell} \frac{1}{M} \sum_{m=1}^{M} \left(Z^{(m)} - \langle \ell, \widehat{\mathbb{X}}_{0,t_n}^{\leq I,(m)} \rangle \right)^2, \tag{18}$$

where $I \in \mathbb{N}$ is a truncation value, and $\widehat{\mathbb{X}}_{0,t_n}^{\leq I,(m)}$ denotes the truncated signature at level I associated with the sample $X^{(m)}$, $m = 1, \dots, M$.

Under some technical assumptions, [3, Prop. 3.3] shows convergence of the value associated with the stopping time $\widehat{\tau}_0$ constructed as above as $I, M \to \infty$.

Note that the Longstaff–Schwartz algorithm produces a low-biased estimator in the form of Remark 3.1.

3.3 *Dual Martingale Methods*

The optimal stopping problem has a dual problem, which comes in the form of a minimization problem over martingales, see [16]. Indeed, it turns out that

$$V_0 = \inf_{M \in \mathcal{M}_0} \mathbb{E}\left[\sup_{t \in \mathcal{T}} (Y_t - M_t) \right], \tag{19}$$

where $\mathcal{M}_0$ denotes the space of martingales M with $M_0 = 0$ and $\mathbb{E}[\sup_{t \in \mathcal{T}} M_t] < \infty$, and Y needs to satisfy a uniform integrability condition. In fact, the proof of this remarkable result is both simple and enlightening, so that we will present it here.

Sketch of Proof of (19) For concreteness, we give the proof in continuous time, $\mathcal{T} = [0, T]$, and assume that Y has continuous paths. In that case, the supermartingale V has a *Doob–Meyer decomposition*

$$V_t = V_0 + M_t^* - A_t^*$$

into a martingale $M^* \in \mathcal{M}_0$ and a predictable increasing process A^*. For concreteness, we shall require $M_0^* = A_0^* = 0$. For the proof, note that for any $M \in \mathcal{M}_0$, we have

$$V_0 = \sup_{\tau \in \mathcal{S}_0} \mathbb{E}[Y_{\tau \wedge T}] = \sup_{\tau \in \mathcal{S}_0} \mathbb{E}[Y_{\tau \wedge T} - M_{\tau \wedge T}] \leq \mathbb{E}\left[\sup_{0 \leq t \leq T} (Y_t - M_t) \right].$$

For the converse inequality, note that $Y \leq V = V_0 + M^* - A^*$. Hence,

$$\inf_{M \in \mathcal{M}_0} \mathbb{E}\left[\sup_{0 \le t \le T} (Y_t - M_t)\right] \le \mathbb{E}\left[\sup_{0 \le t \le T} \left(Y_t - M_t^*\right)\right]$$
$$\le \mathbb{E}\left[\sup_{0 \le t \le T} \left(V_t - M_t^*\right)\right] = \mathbb{E}\left[\sup_{0 \le t \le T} \left(V_0 - A_t^*\right)\right] = V_0,$$

using that A^* is increasing. □

Note that the above proof never relied on the Markov property, and, indeed, (19) does not require it. However, similarly to Sect. 3.2, the Markov property can strongly facilitate the actual computation. Indeed, similar to the Longstaff–Schwartz algorithm, an obvious numerical approach for computing the value of the optimal stopping problem based on the dual formulation is to introduce a parametric family of martingales (say, M^θ, indexed by a finite-dimensional parameter space Θ) and then solve the finite-dimensional minimization problem

$$V_0 \approx \inf_{\theta \in \Theta} \mathbb{E}\left[\sup_{t \in \mathcal{T}} \left(Y_t - M_t^\theta\right)\right]. \tag{20}$$

If the underlying state process X is a Markov process, we can actually describe the Doob–Meyer decomposition of the value process V. First, note that in the Markovian case, there is a function v s.t. $V_t = v(t, X_t)$. Hence, we can directly obtain an optimal martingale M^* by applying Itô's formula to the process $V_t = v(t, X_t)$. Indeed, suppose for concreteness' sake that X is a diffusion process (departing from our usual assumption of bounded variation), i.e.,

$$\mathrm{d}X_t = \mu(t, X_t)\mathrm{d}t + \sigma(t, X_t)\mathrm{d}W_t \tag{21}$$

for a Brownian motion W. Then

$$M_t^* = \int_0^t \sigma(s, X_s)\partial_x v(s, X_s)\mathrm{d}W_s.$$

Remark 3.3 The above equation has to be interpreted in the proper way. If both X and W are one-dimensional processes, then no ambiguity exists. If, say, both X and W are d-dimensional, than μ and σ take values in $\mathbb{R}^d$ and $\mathbb{R}^{d \times d}$, respectively, $\partial_x v(s, X_s)$ is a gradient, while M^* is still a scalar-valued martingale, i.e., the equation needs to be interpreted as

$$M_t^* = \int_0^t (\sigma(s, X_s)\partial_x v(s, X_s))^\top \mathrm{d}W_s.$$

When we are trying to solve the optimal stopping problem, we obviously do not know v yet: it is, after all, the goal of our computation. But the above formula gives us an important structural information: at least one optimal martingale is of the form

$$M_t^* = \int_0^t f(s, X_s)\mathrm{d}W_s$$

for some function f. Hence, a prime candidate for a parametric family of martingales can be constructed by, once again, choosing basis functions $\phi_k(t, x)$, and setting

$$M_t^\theta = \int_0^t \sum_{k=1}^K \theta_k \phi_k(s, X_s)\mathrm{d}W_s,$$

i.e., we make a similar functional approximation ansatz as for the primal problem in Sect. 3.2 in the Markovian case.

In the non-Markovian case, assuming that our filtration $(\mathcal{F}_t)_{t\in[0,T]}$ is the filtration of a Brownian motion W, the martingale representation theorem implies that any martingale (including any optimal one) is of the form

$$M_t = M_0 + \int_0^t f_s \mathrm{d}W_s$$

for some adapted process f. Hence, when looking for a good parametric family of martingales, we need to solve a functional approximation problem for adapted processes, i.e., functionals $f_t = f(t, X|_{[0,t]})$. As in Sect. 3.2, a natural parameterization is given by (linear) functionals of the signature, i.e., we consider

$$M_t^\ell := \int_0^t \langle \ell, \widehat{\mathbb{X}}_{0,s}^{\leq I}\rangle \mathrm{d}W_s = \langle \ell, \int_0^t \widehat{\mathbb{X}}_{0,s}^{\leq I}\mathrm{d}W_s\rangle, \quad \ell \in \Theta := \prod_{i=0}^I (\mathbb{R}^{1+d})^{\otimes i}, \tag{22}$$

for some given truncation level I.

In order to obtain an implementable method for (20) in conjunction with (22), we have to take care of more approximation steps:

1. Replace the expectation $\mathbb{E}[\cdot]$ by a finite sample mean based on M samples $Y_t^{(m)}$, $\widehat{\mathbb{X}}_{0,t}^{\leq I,(m)}$, $W_t^{(m)}$;
2. Discretize the problem based on N time steps for the sup inside the expectation and J timestep for the computation of the signature as well as the stochastic integral in (22).

Let $L^{I,J,(m)}$ denote the discretization of $\int_0^t \widehat{\mathbb{X}}_{0,s}^{\leq I,(m)}\mathrm{d}W_s^{(m)}$ along a uniform time-grid of $[0, T]$ of mesh T/J. Then it can be shown (see [3, Prop. 3.8]) that under weak assumptions of the underlying process, we have that $V^{N,J,I,M} \to V_0$ as $N, J, I, M \to \infty$, where

$$V_0^{N,J,I,M} := \inf_{\ell\in\Theta} \frac{1}{M}\sum_{m=1}^M \max_{0\leq n\leq N}\left(Y_{nT/N}^{(m)} - \langle \ell, L_{0,nT/N}^{I,J,(m)}\rangle\right). \tag{23}$$

4 Martingale Duality in Details

Having provided a cursory look at three approximation methods for general optimal stopping problems based on the path-signature, we will now give an in-depth exposition for one of these methods, namely the dual martingale method from [3].

As explained above, the strategy is to approximate generic adapted processes by linear functionals of the signatures. The universality results for the standard signature, typically are formulated in terms of uniform convergence on compacts subsets of (rough) path-space. However, as explained in chapter "The Signature Kernel" of this book, there is a *robust signature* obtained by normalizing the standard signature, which is, in fact, uniformly bounded w.r.t. the underlying paths, and, hence, allows *global* universality—in the set of bounded continuous functions of (rough) path space—when considering the correct topology. Using this machinery, we can provide full convergence results (Theorem 4.1 below), and are not limited to local results—i.e., convergence "on compacts" or "with high probability". In what follows, we consider a tensor normalization λ and the corresponding robust signature $\lambda(\widehat{\mathbb{X}}^{<\infty}_{s,t})$ in the sense of chapter "The Signature Kernel".

We start by presenting a key technical lemma required in the analysis, which is a consequence of Theorem 8.4 in chapter "The Signature Kernel". Denote the linear functionals of the robust signature by

$$L_{\text{sig}}(\Lambda_T^p) := \left\{ \Lambda_T^p \ni \mathbf{X}|_{[0,t]} \mapsto \left\langle \ell \, , \lambda(\widehat{\mathbb{X}}^{<\infty}_{0,t}) \right\rangle \,\middle|\, \ell \in \bigoplus_{i=0}^{I} (\mathbb{R}^{1+d})^{\otimes i}, \; I \geq 1 \right\}, \tag{24}$$

where λ denotes an appropriate signature normalization, see chapter "The Signature Kernel", and we recall the definition of the space of stopped p-rough paths, see Definition 2.12.

The following result generalizes the universal approximation property of the signature in L^p w.r.t. some measure μ, extending Proposition 2.14. Note that measures μ on $\widehat{\Omega}_T^p$ can be extended to measures on the space of stopped rough paths Λ_T^p by restriction. We will denote its extension by μ, as well. Conversely, every measure μ on Λ_T^p also induces a measure on $\widehat{\Omega}_T^p$, which we will again denote by μ.

Theorem 4.1 *Let μ denote a finite measure on Λ_T^p which is supported on a compactly embedded subspace $V \subset \widehat{\Omega}_T^p$. Then for any $f \in L^p(\Lambda_T^p, \mu)$, $1 \leq p < \infty$, there is a sequence $f_n \in L_{sig}(\Lambda_T)$ s.t. $\|f - f_n\|_{L^p} \xrightarrow{n\to\infty} 0$.*

Remark 4.2 Many measures μ encountered in practice are, in fact, naturally supported on compactly embedded subspaces of $\widehat{\Omega}_T^p$. For instance, take μ to be the (rough) *Wiener measure*, i.e., the distribution of Brownian motion on rough path space. By Example 2.15, we can choose any $p > 2$ and obtain a probability measure μ on $\widehat{\Omega}_T^p$. However, we could also have chosen $2 < q < p$, and it turns out that μ is supported by $\widehat{\Omega}_T^q \subset \widehat{\Omega}_T^p$. This embedding is compact, see [11]. Hence, the law of Brownian motion satisfies the condition of Theorem 4.1, as do (by similar arguments) the laws of fractional Brownian motion (for any Hurst index) and diffusion processes.

Proof of Theorem 4.1 We follow the proof of [3, Th. 2.6].

By dominated convergence, we can first approximate f with a bounded measurable function f^K bounded by some K s.t. $\|f - f^K\|_{L^p} \leq \epsilon$. By abstract results from measure theory (Lusin's theorem, Tietze's extension theorem, cf. the proof of Prop. 2.6), we can further find $\widetilde{f} \in C_b(\Lambda_T; \mathbb{R})$ s.t. $\|f^K - \widetilde{f}\|_{L^p} \leq \epsilon$. Hence, we can, without loss of generality, assume that our initial $f \in C_b(\Lambda_T^p; \mathbb{R})$.

By Theorem 8.4 in chapter "The Signature Kernel", $L_{\text{sig}}(\Lambda_T^p) \subset C_b(\Lambda_T^p; \mathbb{R})$ is dense w.r.t. the strict topology. This means that there is a sequence $f_n \in L_{\text{sig}}(\Lambda_T^p)$ s.t. for any $\psi : \Lambda_T^p \to \mathbb{R}$ vanishing at infinity, we have that

$$\sup_{\mathbf{x}\in\Lambda_T^p} \psi(\mathbf{x})\,|f(\mathbf{x}) - f_n(\mathbf{x})| \xrightarrow{n\to\infty} 0.$$

Suppose that we can find a function ψ vanishing at infinity such that $\frac{1}{\psi} \in L^p(\mu)$. Then we can conclude that

$$\begin{aligned}\|f - f_n\|_{L^p}^p &= \int_{\Lambda_T^p} \left(\frac{\psi(\mathbf{x})(f(\mathbf{x}) - f_n(\mathbf{x}))}{\psi(\mathbf{x})}\right)^p \mu(\mathrm{d}\mathbf{x}) \\ &\leq \underbrace{\left(\sup_{\mathbf{x}\in\Lambda_T} \psi(\mathbf{x})\,|f(\mathbf{x}) - f_n(\mathbf{x})|\right)^p}_{\to 0} \underbrace{\int_{\Lambda_T^p} \frac{1}{\psi(\mathbf{x})^p} \mu(\mathrm{d}\mathbf{x})}_{<\infty} \xrightarrow{n\to\infty} 0.\end{aligned}$$

Existence of such a function ψ follows from Lemma 4.4 with the normed vector space $W = \widehat{\Omega}_T^p$, which concludes the argument. □

Lemma 4.3 *Let ν be a finite measure on $(\mathbb{R}_+, \mathcal{B}(\mathbb{R}_+))$. Then there is an increasing function $\eta : \mathbb{R}_+ \to \mathbb{R}_+$ s.t. $\lim_{x\to\infty} \eta(x) = \infty$ and $\int_{\mathbb{R}_+} \eta(x)\nu(\mathrm{d}x) < \infty$.*

The proof is left to the readers as an exercise, see also [3, Lemma 2.8]. The following lemma is a slight generalization of [3, Lemma 2.9].

Lemma 4.4 *Let μ be a finite measure on a normed vector space W which is supported on a compactly embedded normed vector space $V \subset W$ with norm $\|\cdot\|_V$. Let η be the function from Lemma 4.3 for the push-forward measure of $\mu|_V$ by the function $\|\cdot\|_V : V \to \mathbb{R}_+$. Then*

$$\psi(x) := \mathbf{1}_V(x)\left(\frac{1}{1 + \eta(\|x\|_V)}\right)^{1/p}$$

vanishes at infinity and satisfies $\frac{1}{\psi} \in L^p(\mu)$.

Proof Note that $\frac{1}{\psi} \in L^p(\mu)$ follows immediately by construction, so we are left to prove that ψ vanishes at infinity. Since $\eta \nearrow \infty$, for any $\epsilon > 0$ there is an $R > 0$ s.t. $\|x\|_V > R$ implies that $\left(\frac{1}{1+\eta(\|x\|_V)}\right)^{1/p} < \epsilon$. The claim follows by noting that $\{x \in V | \|x\|_V \leq R\}$ is compact as a subset of W. □

As a corollary of Theorem 4.1, we will see that all (square integrable) martingales can be approximated by integrals of linear functionals of the signature against the driving Brownian motion provided that the filtration is generated by Brownian motion. More concretely, we assume that X satisfies (21) driven by a d-dimensional Brownian motion W s.t. the diffusion matrix σ is invertible everywhere, implying that the filtrations generated by X and W coincide.

Lemma 4.5 *Let M be a square integrable one-dimensional martingale with $M_0 = 0$ w.r.t. the filtration $\mathcal{F}$. Then there is a sequence $f_n = (f_n^1, \ldots, f_n^d)$, $f_n^i \in L_{sig}(\Lambda_T^p)$ s.t.*

$$\mathbb{E}\left[\sup_{0\leq t\leq T} \left|M_t - M_t^n\right|^2\right] \xrightarrow{n\to\infty} 0, \quad M_t^n := \sum_{i=1}^d \int_0^t f_n^i(X|_{[0,s]})\mathrm{d}W_s^i.$$

Proof By the martingale representation theorem, see, e.g., [19], there is an adapted, square integrable process α_s s.t.

$$M_t = \sum_{i=1}^d \int_0^t \alpha_s^i \mathrm{d}W_s^i.$$

By our usual interpretation of adapted processes as measurable functions on Λ_T^p, we can find a measurable f s.t. $\alpha_t = f(\mathbf{X}|_{[0,t]})$, for all $t \in [0, T]$. By Theorem 4.1, for every $i = 1, \ldots, d$, we can find a sequence of functions $f_n^i \in L_{\mathrm{sig}}(\Lambda_T^p)$ s.t.

$$\mathbb{E}\left[\int_0^T \left(f^i(\mathbf{X}|_{[0,t]}) - f_n^i(\mathbf{X}|_{[0,t]})^2\right) \mathrm{d}t\right] \xrightarrow{n\to\infty} 0.$$

Doob's inequality together with the Itô isometry directly imply the result. □

We can now apply this result to the dual martingale formulation of the optimal stopping problem, see (19). Indeed, Lemma 4.5 implies that for any $M \in \mathcal{M}_0^2$

$$\sup_{0\leq t\leq T} (Y_t - M_t) = \lim_{n\to\infty} \sup_{0\leq t\leq T} \left(Y_t - \sum_{i=1}^d \int_0^t f_n^i(\mathbf{X}|_{[0,s]})\mathrm{d}W_s^i\right),$$

where the limit is understood in the L^2 sense. We conclude that

$$\mathbb{E}\left[\sup_{0\le t\le T}(Y_t - M_t)\right] = \lim_{n\to\infty}\mathbb{E}\left[\sup_{0\le t\le T}\left(Y_t - \sum_{i=1}^{d}\int_0^t f_n^i(\mathbf{X}|_{[0,s]})\mathrm{d}W_s^i\right)\right].$$

A simple argument then yields

$$V_0 = \inf_{\ell^1,\dots,\ell^d\in\bigoplus_{i=0}^{\infty}(\mathbb{R}^{1+d})^{\otimes i}}\mathbb{E}\left[\sup_{0\le t\le T}\left(Y_t - \sum_{i=1}^{d}\int_0^t \langle \ell^i,\,\lambda(\widehat{\mathbb{X}}_{0,s}^{<\infty})\rangle \mathrm{d}W_s^i\right)\right]. \tag{25}$$

We are now left with turning the duality formula (25) into a numerical algorithm. Several issues immediately come to mind: (25) includes several integrals (the stochastic integrals, but also the signature itself) which require discretization. This also applies to computing sup over an infinite set of times. The expectation needs to be computed. Finally, we need to minimize over linear functionals of the signature.

We start with time discretization. Let us first introduce a grid $0 = t_0 < t_1 < \cdots < t_N = T$, and restrict the sup inside the expectation to a max over the grid, i.e.,

$$V_0^N := \inf_{\ell^1,\dots,\ell^d\in\bigoplus_{i=0}^{\infty}(\mathbb{R}^{1+d})^{\otimes i}}\mathbb{E}\left[\max_{n=0,\dots,N}\left(Y_{t_n} - \sum_{i=1}^{d}\int_0^{t_n} \langle \ell^i,\,\lambda(\widehat{\mathbb{X}}_{0,s}^{<\infty})\rangle \mathrm{d}W_s^i\right)\right]. \tag{26}$$

The martingale duality formula (19) for the discrete time stopping problem, provided that we still use the continuous-time filtration $(\mathcal{F}_{t_n})_{n=0}^N$—i.e., *not* the filtration generated by $(X_{t_n})_{n=0}^N$. Hence, the error $\left|V_0 - V_0^N\right|$ only depends on the underlying processes W, X, Y, but not on the method, and we shall assume that $\left|V_0 - V_0^N\right|$ as $N \to \infty$, see, for instance, [9].

Next, let us truncate the signature at level I, which turns an infinite dimensional minimization problem into a finite-dimensional one, i.e.,

$$V_0^{I,N} := \inf_{\ell^1,\dots,\ell^d\in\bigoplus_{i=0}^{I}(\mathbb{R}^{1+d})^{\otimes i}}\mathbb{E}\left[\max_{n=0,\dots,N}\left(Y_{t_n} - \sum_{i=1}^{d}\int_0^{t_n} \langle \ell^i,\,\lambda(\widehat{\mathbb{X}}_{0,s}^{\le I})\rangle \mathrm{d}W_s^i\right)\right]. \tag{27}$$

It is not difficult to see that $\lim_{I\to\infty} V_0^{I,N} = V_0^N$: Indeed, note that $V_0^{I,N}$ is increasing in I. Besides, for every $\epsilon > 0$ we can find I_ϵ such that an ϵ-optimal solution $(\ell^1,\dots,\ell^d)$ is contained in $\bigoplus_{i=0}^{I_\epsilon}(\mathbb{R}^{1+d})^{\otimes i}$. This, however, already implies convergence.

Note that the infimum in (27) is attained, i.e., we can find minimizers $\ell^{1,*},\dots,\ell^{d,*} \in \bigoplus_{i=0}^{I}(\mathbb{R}^{1+d})^{\otimes i}$. Indeed, we note that the functional

$$\ell \mapsto \mathbb{E}\left[\max_{n=0,\dots,N}\left(Y_{t_n} - \sum_{i=1}^{d}\int_0^{t_n} \langle \ell^i,\,\lambda(\widehat{\mathbb{X}}_{0,s}^{\le I})\rangle \mathrm{d}W_s^i\right)\right]$$

is convex as an expectation of the pointwise maximum of affine (and, hence, convex) functions, see, e.g., [6, Section 3.2.1 together with Section 3.2.3]. The functional also grows to infinity as $\left|\ell^i\right| \to \infty$ and is continuous by dominated convergence (note the use of the robust signature). This implies the existence of minimizers.

If we now replace the exact truncated signature $\widehat{\mathbb{X}}^{\le I}_{0,s}$ by a discretized version $\widehat{\mathbb{X}}^{\le I,J}_{0,s}$ based on a grid $0 = s_0 < \cdots < s_J = T$ (containing $t_1, \ldots, t_N$)—think of $J \ll N$—and similarly approximate the stochastic integrals in (27) by approximation based on the same grid, i.e.,

$$V_0^{I,N,J} := \inf_{\ell^1,\ldots,\ell^d \in \bigoplus_{i=0}^I (\mathbb{R}^{1+d})^{\otimes i}} \mathbb{E}\left[\max_{n=0,\ldots,N}\left(Y_{t_n} - \sum_{i=1}^d \sum_{s_j < t_n} \langle \ell^i, \lambda(\widehat{\mathbb{X}}^{\le I,J}_{0,s_j})\rangle \times (W^i_{s_{j+1}} - W^i_{s_j})\right)\right]. \tag{28}$$

A similar argument as above shows that $\lim_{J\to\infty} V_0^{I,N,J} = V_0^{I,N}$.

This leaves the final approximation necessary, namely replacing the expectation by a finite sample (Monte Carlo) approximation. We are finally given an approximation problem

$$V_0^{I,N,J,M} := \inf_{\ell^1,\ldots,\ell^d \in \bigoplus_{i=0}^I (\mathbb{R}^{1+d})^{\otimes i}} \frac{1}{M} \sum_{m=1}^M \left[\max_{n=0,\ldots,N}\left(Y^{(m)}_{t_n} - \sum_{i=1}^d \sum_{s_j < t_n} \langle \ell^i, \lambda(\widehat{\mathbb{X}}^{\le I,J,(m)}_{0,s_j})\rangle (W^{i,(m)}_{s_{j+1}} - W^{i,(m)}_{s_j})\right)\right], \tag{29}$$

where $Y^{(m)}$, $\widehat{\mathbb{X}}^{\le I,J,(m)}_{0,s_j}$, $W^{i,(m)}_{s_j}$ are i.i.d. samples from the respective random variables, $m = 1, \ldots, M$.

The problem (29) is an example of a *sample average approximation* to a stochastic optimization problem. In order to streamline the discussion, let us denote all the sources of randomness in (28) by $\xi = \left((Y_{t_n})_{n=0}^N, (\widehat{\mathbb{X}}^{\le I,J}_{0,s_j})_{j=0}^J, (W^i_{s_j})_{j=0,\ldots,J,i=1,\ldots,d}\right)$. The we have

$$V_0^{I,N,J} = \inf_{\ell \in \left(\bigoplus_{i=0}^I (\mathbb{R}^{1+d})^{\otimes i}\right)^d} \mathbb{E}\left[F(\ell, \xi)\right],$$

for a function F implicitly defined by (28), whereas (29) is its sample average approximation

$$V_0^{I,N,J,M} = \inf_{\ell \in \left(\bigoplus_{i=0}^{I} (\mathbb{R}^{1+d})^{\otimes i}\right)^d} \frac{1}{M} \sum_{m=1}^{M} F(\ell, \xi^{(m)}).$$

The convergence theory for such approximations is well understood. We cite a (slightly modified) result from [17, Chapter 6, Theorem 4].

Lemma 4.6 *Given a measurable function* $F : \mathbb{R}^K \times \mathbb{R}^L \to \mathbb{R}$ *s.t.* $\ell \mapsto F(\ell, \xi)$ *is lower semi-continuous for every* ξ *and* $\ell \mapsto \mathbb{E}[F(\ell, \xi)]$ *is also lower semi-continuous. Additionally assume that* $\ell \mapsto F(\ell, \xi)$ *is convex for almost every* ξ *and the set of minimizers of* $\min_\ell \mathbb{E}[F(\ell, \xi)]$ *is non-empty and bounded. Then* $\min_\ell \frac{1}{M} \sum_{m=1}^{M} F(\ell, \xi^{(m)}) \xrightarrow[a.s.]{M\to\infty} \min_\ell \mathbb{E}[F(\ell, \xi)]$.

Since all the conditions of Lemma 4.6 are easily verified for (28), we can conclude that $\lim_{M\to\infty} V_0^{I,N,J,M} = V_0^{I,N,J}$ a.s. Hence, we obtain (as claimed earlier in Sect. 3.3)

$$\lim_{N\to\infty} \lim_{I\to\infty} \lim_{J\to\infty} \lim_{M\to\infty} V_0^{I,N,J,M} = V_0 \text{ a.s.},$$

with the caveat that some restrictions between the parameters exist—i.e., we can only choose $J \geq N$.

5 Numerical Example

The aim of this section is to illustrate the performance of the signature stopping methodology in numerically approximating optimal stopping values. We therefore investigate the optimal stopping of fractional Brownian motions. This choice allows us to explore a wide range of underlying path regularities, including scenarios involving non-Markovianity, all within a unified model.

Code Notebook

The complete and documented code for this numerical example can be found in a Jupyter Notebook in the following GitHub repository:

https://github.com/hagerpa/signature_stopping.git

This notebook explains how to implement the algorithms from Sects. 3.1 and 3.2 in Python. It is complemented with several exercises designed to deepen the understanding of signatures and their usage for parametrizing stopping times.

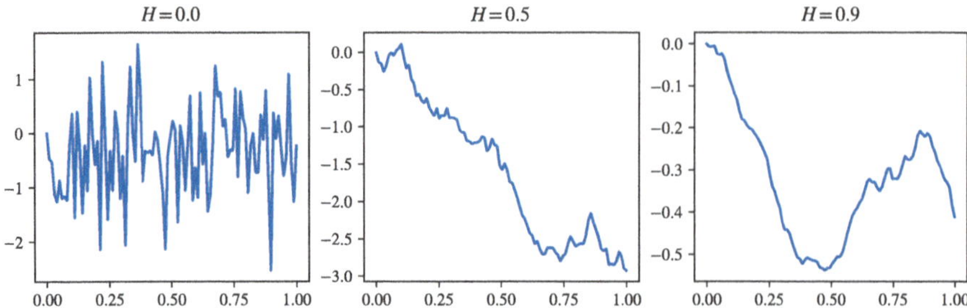

Fig. 1 Illustration of fractional Brownian Motion sample paths across different Hurst parameters H which is a strict upper bound on paths Hölder regularity. This case $H = 0.5$ corresponds to a standard Brownian motion

Table 1 Lower-biased approximated values to the optimal stopping problem of fractional Brownian motion using linear- and deep signature stopping rules. Evaluations were performed for several Hurst parameters H and truncation levels I using a time discretization of $N = 1000$ steps and around 2^{20} training and testing samples. The maximal Monte Carlo error was 0.001

	Linear					Deep				
H	$I = 1$	$I = 2$	$I = 3$	$I = 4$	$I = 5$	$I = 1$	$I = 2$	$I = 3$	$I = 4$	$I = 5$
0.1	0.350	0.625	0.840	0.995	1.094	1.247	1.346	1.352	1.358	1.359
0.3	0.131	0.213	0.266	0.297	0.307	0.392	0.418	0.417	0.422	0.418
0.5	−0.002	0.000	0.000	−0.001	0.000	0.002	0.000	−0.001	0.000	0.000
0.7	0.091	0.126	0.146	0.155	0.152	0.194	0.200	0.206	0.206	0.205
0.9	0.165	0.219	0.248	0.262	0.258	0.328	0.333	0.336	0.336	0.335

A one-dimensional fractional Brownian motion is a zero-mean Gaussian process X with a covariance function described by

$$\mathbb{E}(X_s X_t) = \frac{1}{2}\left(|s|^{2H} + |t|^{2H} - |t-s|^{2H}\right), \quad 0 \le s \le t, \tag{30}$$

where $H \in (0, 1]$ is referred to as the Hurst parameter. The sample paths of X are almost surely α-Hölder continuous for any $\alpha < H$. When $H = \frac{1}{2}$, X represents a standard Brownian motion. However, when $H \neq \frac{1}{2}$, X is neither a Markov process nor a semimartingale. Figure 1 illustrates the regularity of fractional Brownian sample paths for different Hurst parameters. As a one-dimensional process, X can be trivially lifted to a geometric rough path $\mathbb{X} \in \Omega_T^p(\mathbb{R}^d)$ for any $p > \frac{1}{H}$, defined as $\mathbb{X} = \pi_{\le \lfloor p \rfloor} \exp_\otimes(X_{s,t})$. In the particular example we consider, the stopped process and the underlying process are identical, i.e., $Y = X$ is the fractional Brownian motion itself. For all following experiments we fix a unit time horizon $T = 1.0$.

In Table 1, we present estimates of the optimal stopping values corresponding to various values of H. These estimates were obtained using the signature stopping time parametrizations introduced in Sect. 3.1. Specifically, we assess the performance of both linear and deep signature stopping rules, employing different signature truncation levels for each. The architecture of the deep neural networks

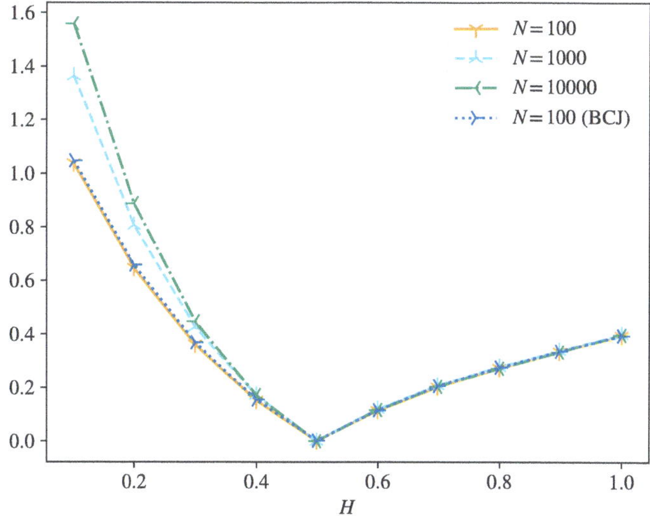

Fig. 2 An illustration of the estimated values (vertical axis) for the optimal stopping problem of fractional Brownian motion across a range of Hurst parameters (horizontal axis). These estimates were obtained using a deep signature stopping rule with a fixed signature truncation level of $I = 3$ on time discretization grids with varying numbers of steps ($N = 100, 1000, 10{,}000$). The values labeled by "BCJ" are those obtained in [4]

consists of 2 hidden layers with 30 additional neurons each, using ReLU activation functions. The observations indicate a significant improvement when employing higher truncation levels and transitioning from linear to deep signature strategies. Note that, as a consequence of Doob's optimal sampling theorem, the optimal stopping value for standard Brownian motion is 0. Therefore, the values obtained for $H = 0.5$ are consistent with this theoretical benchmark.

Figure 2 illustrates the estimates of optimal stopping values for a range of Hurst parameters using time discretizations with $N = 100, 1000, 10{,}000$ steps. These values were obtained using a deep signature stopping rule with a fixed signature truncation level of $I = 3$. For comparison, we have also included the benchmark values from [4]. In their work, the authors translated the problem into a Markovian setting by lifting the discretized fractional Brownian motion to a high-dimensional process, which includes its running history. We observe that when using the same time discretization ($n = 100$), our signature methods yield comparable results to [4]. It is worth noting, however, that for the signature methodology, increasing the number of time steps does not lead to an increase in the number of model parameters, as the input dimension of the network remains equal to the dimension of the truncated signatures. This contrasts starkly with the lifting approach used in [4]. Particularly for small Hurst parameters, we observe a significant improvement gained from choosing a finer discretization grid. This observation aligns with the theoretical insight from [2] stating that the optimal stopping value of fractional Brownian motion diverges as the Hurst parameter H approaches zero.

Table 2 Lower and upper biased approximations to the discrete-time optimal stopping of a fractional Brownian motion with $N = 100$ time points using the Signature Longstaff-Schwartz and dual martingale methods. These estimates were obtained with a signature truncation level of $I = 6$ and different choices for the time grid size J used for the signature approximation. In the right-most column, we provide values from [4] for comparison

	$J = 100$		$J = 500$		BCJ	
H	Lower	Upper	Lower	Upper	Lower	Upper
0.1	1.045	1.129	1.065	1.117	1.048	1.05
0.3	0.363	0.396	0.371	0.392	0.368	0.371
0.5	−0.001	0	−0.002	0	0	0.005
0.7	0.203	0.234	0.205	0.220	0.206	0.208
0.9	0.331	0.357	0.337	0.356	0.335	0.339

Let us now focus on the case of a Bermudan option with $N = 100$ equidistant exercise opportunities $0 = t_0 < t_1 < \cdots < t_N = T$, where the payoff process $Y = X$ is again a fractional Brownian motion. In this case, we can utilize the Longstaff-Schwartz method described in Sect. 3.2 and the dual martingale method described in Sect. 3.3 to compute lower and upper bounds. Note that, apart from the signature truncation level I, we have an additional parameter to consider: the number of time grid points, denoted as J, used to approximate the signature of the fractional Brownian motion. In Table 2, we present the results obtained for the choices $J = 100 = N$ and $J = 500 > N$ for various Hurst parameters. The values from [4] were also obtained on a grid with $N = 100$ points, providing comparable benchmarks for the discrete problem. We observe that the lower bounds obtained with the signature Longstaff-Schwartz method are comparable to those in [4] for $J = 100$, while for $J = 500$, we see a significant improvement. To obtain comparable upper bounds, it appears that choosing a finer grid for signature computation is necessary. We emphasize that even for discrete-time stopping problems with an underlying filtration generated by a non-Markovian continuous-time process, the ability to enhance results by increasing J, thereby incorporating more information about the underlying path without increasing the number of model parameters, is an inherent feature of the signature methodology. There seems to be no evident extension of methods such as those in [4] to incorporate such a feature.

6 Conclusions and Outlook

Signatures can serve as a natural set of basis functions for the numerical approximation of functions on path-space, analogous to polynomials on finite-dimensional spaces. In this chapter, we show how this conjecture can be used in the numerics of optimal stopping problems, when the underlying stochastic process is not a Markov process. More precisely, we derive non-Markovian extensions of some classical algorithms, namely the Longstaff–Schwartz algorithm and Rogers' dual algorithm.

In addition, we also present a direct algorithm based on parameterization of a rich class of stopping times defined in terms of the signature.

Hence, signature based methods for optimal stopping provide a very rich and efficient class of methods, as also supported by numerical examples. On the other hand, they can be challenging to implement, due to the large number of terms present in the signature, as well as the cost of computing the signature itself from time-series of the underlying process.

There are many open problems and largely unresolved questions in the context of the signature method for optimal stopping and, more generally, stochastic optimal control. Possibly the most prominent problem is to establish quantitative error bounds and rates of convergence for the proposed algorithms and other algorithms based on the signature. Note that the methods presented here require several numerical approximations (truncation of the signature, finite sample, discretization of the iterated integrals constituting the signature), and we would like to obtain sharp error estimates and rates for the total error in terms of these numerical parameters. Note that these approximations interact with each other, e.g., higher degree of truncation may lead to larger discretization errors or variances. Functional Taylor expansions as presented in chapter "Signature and the Functional Taylor Expansion" could serve as important tools for analyzing the approximation errors.

From a more applied point of view, it would be highly desirable to improve the performance of the signature methods if possible. To this end, dimension reduction techniques such as randomized signatures, see, for instance, [7], can be useful. Moreover, detailed case studies for important applications, such as pricing American options in rough volatility models, with an emphasis on obtaining tight upper and lower bounds would be desirable.

Acknowledgments C.B. gratefully acknowledges funding by the Deutsche Forschungsgemeinschaft (DFG, German Research Foundation) under Germany's Excellence Strategy (EXC-2046/1, project ID 390685689), as well as support from DFG CRC/TRR 388 "Rough Analysis, Stochastic Dynamics and Related Fields", Project B03.

Appendix

Proof of Proposition 2.2 Fix $t \in [0, T]$. Since U_t is $\mathcal{F}_t = \sigma(X_s \,|\, s \in [0, t])$-measurable, there is a measurable function $\theta_t : C_t \to \mathbb{R}$ such that

$$U_t(\omega) = \theta_t(X(\omega)|_{[0,t]})$$

for every $\omega \in \Omega$. Define $\phi \colon \mathscr{C}_T \to [0, T] \times C_T$ as follows: If $X \in C_t$, define $\tilde{X} \in C_T$ by setting $\tilde{X}_s := X_{s\wedge t}$ and we set $\phi(X) = (t, \tilde{X})$. Since ϕ is continuous, it is also measurable. Define $f \colon [0, T] \times C_T \to \mathbb{R}$ by $f(t, X) = \theta_t(X|_{[0,t]})$. For fixed t, $X \mapsto X_{[0,t]}$ as a map from C_T to C_t is continuous. Since θ_t is measurable, $X \mapsto f(t, X)$ is measurable. For $n \in \mathbb{N}$, define $I_k^n := [k/2^n T, (k+1)/2^n T)$ for $k = 0, \ldots, 2^n - 2$, $I_{2^n-1}^n := [(2^n - 1)/2^n T, T]$ and $t_k^n := k/2^n T$. Set

$$f_n(t, X) := \sum_{k=0}^{2^n-1} f(t_k^n, X) \mathbb{1}_{I_k^n}(t)$$

which is (jointly) measurable for every $n \in \mathbb{N}$. Set

$$\tilde{f}(t, X) := \limsup_{m\to\infty} \limsup_{n\to\infty} f_n(t + 1/m, X) \quad \text{and} \quad \theta(X|_{[0,t]}) := (\tilde{f} \circ \phi)(X|_{[0,t]}).$$

The map θ is thus measurable and satisfies

$$\theta(X(\omega)|_{[0,t]}) = \limsup_{m\to\infty} \limsup_{n\to\infty} \sum_{k=0}^{2^n-1} U_{t_k^n}(\omega) \mathbb{1}_{I_k^n}(t + 1/m) = U_t(\omega).$$

□

Proof of Proposition 2.6 It suffices to prove that for given $\varepsilon > 0$ and $\delta > 0$, there exists a continuous map $g \colon \mathcal{M} \to \mathbb{R}$ such that

$$\mu(|f - g| > \varepsilon) < \delta.$$

By Lusin's theorem, there is a closed set $A \subset \mathcal{M}$ such that $\mu(A^c) < \delta$ and $\tilde{g} := f|_A$ is continuous. Since A is closed, from Tietze's extension theorem, we can extend $\tilde{g}$ to a continuous map g on the whole space $\mathcal{M}$. The map g clearly satisfies

$$\{x \in \mathcal{M} : |f(x) - g(x)| > \varepsilon\} \subset A^c$$

for every $\varepsilon > 0$, thus the result follows. □

References

1. P. Bank, C. Bayer, P.P. Hager, S. Riedel, T. Nauen, Stochastic control with signatures. Preprint. arXiv:2406.01585 (2024)
2. C. Bayer, P.P. Hager, S. Riedel, J. Schoenmakers, Optimal stopping with signatures. Ann. Appl. Probab. **33**(1), 238–273 (2023)
3. C. Bayer, L. Pelizzari, J. Schoenmakers, Primal and dual optimal stopping with signatures. Preprint. arXiv:2312.03444 (2023)
4. S. Becker, P. Cheridito, A. Jentzen, Deep optimal stopping. J. Mach. Learn. Res. **20**(74), 1–25 (2019)
5. H. Boedihardjo, X. Geng, T. Lyons, D. Yang, The signature of a rough path: uniqueness. Adv. Math. **293**, 720–737 (2016)
6. S. Boyd, L. Vandenberghe, *Convex Optimization* (Cambridge University Press, 2004)
7. E.M. Compagnoni, A. Scampicchio, L. Biggio, A. Orvieto, T. Hofmann, J. Teichmann, On the effectiveness of randomized signatures as reservoir for learning rough dynamics, in *2023 International Joint Conference on Neural Networks (IJCNN)* (IEEE, 2023), pp. 1–8

8. B. Dupire, Functional Itô calculus. Quant. Finance **19**(5), 721–729 (2019)
9. E. Ekström, Properties of American option prices. Stoch. Process. Appl. **114**(2), 265–278 (2004)
10. P.K. Friz, M. Hairer, *A Course on Rough Paths* (Springer, Berlin, 2020)
11. P.K. Friz, N.B. Victoir, *Multidimensional Stochastic Processes as Rough Paths: Theory and Applications*, vol. 120 (Cambridge University Press, 2010)
12. P. Glasserman, Monte Carlo methods in financial engineering (Springer Science+Business Media, New York, 2003)
13. F.A. Longstaff, E.S. Schwartz, Valuing American options by simulation: a simple least-squares approach. Rev. Financ. Stud. **14**(1), 113–147 (2001)
14. T.J. Lyons, M. Caruana, T. Lévy, *Differential Equations Driven by Rough Paths* (Springer, 2007)
15. G. Peskir, A. Shiryaev, *Optimal Stopping and Free-boundary Problems* (Springer, Basel, 2006)
16. L.C. Rogers, Monte Carlo valuation of American options. Math. Finance **12**(3), 271–286 (2002)
17. A. Shapiro, Monte carlo sampling methods, in *Handbooks in Operations Research and Management Science*, vol. 10 (Elsevier, Amsterdam, 2003), pp. 353–425
18. A.N. Shiryaev, *Optimal Stopping Rules*, vol. 8 (Springer Science & Business Media, 2007)
19. S.E. Shreve, *Stochastic Calculus for Finance II: Continuous-time Models*, vol. 11 (Springer, New York, 2004)

Adapted Topologies and Higher-Rank Signatures

Chong Liu and Gudmund Pammer

Abstract Two stochastic processes can have similar laws but yield a vastly different outcome in applications such as optimal stopping or stochastic programming. The reason is that the usual weak topology does not account for the different available information in time that is stored in the filtration of the underlying process. To address the resulting discontinuities, Aldous introduced the extended weak topology, and subsequently, Hoover and Keisler showed that both, weak topology as well as extended weak topology, are just the first two topologies in a sequence of the so-called adapted weak topologies that get increasingly finer. In this short survey, we give a brief introduction to the recent advances of the applications of signature theory, in particular the higher-rank (expected) signatures, in adapted weak topologies and related fields, and highlight theoretical and computational properties.

1 Introduction to Adapted Topologies

Financial and economic agents frequently encounter decision making problems. Facing uncertainty, they have to make a sequence of choices in a complex environment which necessitates a systematic approach. Therefore, to enable the decision making process, mathematical models are employed, which assist in identifying reasonable courses of action. A common method of incorporating uncertainty into a model is through stochastic processes. The agent has to base their decision on the available information which is continually updated in time. This evolving information is formally represented by a filtration. Summarizing, the mathematical base of such a framework, where we consider throughout the article for ease of

C. Liu (✉)
ShanghaiTech University, Institute of Mathematical Sciences, Shanghai, China
e-mail: liuchong@shanghaitech.edu.cn

G. Pammer
Graz University of Technology, Institute of Statistics, Graz, Austria
e-mail: gudmund.pammer@tugraz.at

C. Bayer et al. (eds.), *Signature Methods in Finance*, Springer Finance Lecture Notes, https://doi.org/10.1007/978-3-031-97239-3_10

exposition a discrete-time framework with $T \in \mathbb{N}$ time steps and $I := \{1, \dots, T\}$, involves:

- a probability space $(\Omega, \mathcal{F}, \mathbb{P})$;
- a filtration $(\mathcal{F}_t)_{t \in I}$, i.e., an increasing sequence of σ-algebras with $\mathcal{F}_t \subseteq \mathcal{F}$; and
- a stochastic process $X = (X_t)_{t \in I}$ adapted to the filtration $(\mathcal{F}_t)_{t \in I}$, i.e., X_t is $\mathcal{F}_t$-measurable.

Informally speaking, as $\mathcal{F}_t$ reflects the information of the agent at time t, the adaptedness of X means that the agent has observed the path X up to time t.

To give an illustrative example of such a setup, we briefly consider the model of a simple financial market that can be found in classical mathematical finance literature, see for example [40]. There, the stochastic process $X = (X_t)_{t \in I}$ with $X_t \in \mathbb{R}^d$ may be used to model the uncertain evolution in time of d-asset prices. This means that each coordinate of $X_t = (X_t^i)_{i=1}^d$ represents the value of a single asset at a given point in time. Meanwhile, the filtration incorporates the information that is available to an investor, which they can use to base their decisions on. Therefore, the investor can only employ trading strategies that are adapted to this filtration. For example, let $(H_t)_{t \in I}$ be a self-financing dynamic trading strategy. When H_t^i denotes the amount of units of the i-th asset that the investor holds between time t and $t+1$ (that is, on $[t, t+1)$), then this amount is necessarily measurable with respect to $\mathcal{F}_t$. Moreover, the value of the self-financing portfolio at time t is given by

$$\sum_{s=1}^{t-1} H_s \cdot (X_{s+1} - X_s). \tag{1}$$

In more mathematical terms, we consider dynamic trading strategies that are predictable with respect to that filtration. Therefore, the collection of attainable claims of type (1) crucially depends on process and filtration, which we further stress by the example below. But beforehand, we introduce the following notion.

Definition 1.1 (Filtered Process) A filtered process $\mathbf{X}$ is a 5-tuplet

$$\mathbf{X} := \left(\Omega^{\mathbf{X}}, \mathcal{F}^{\mathbf{X}}, \mathbb{P}^{\mathbf{X}}, (\mathcal{F}_t^{\mathbf{X}})_{t \in I}, X\right),$$

consisting of a probability space $(\Omega^{\mathbf{X}}, \mathcal{F}^{\mathbf{X}}, \mathbb{P}^{\mathbf{X}})$, a filtration $(\mathcal{F}_t^{\mathbf{X}})_{t \in I}$ and an $(\mathcal{F}_t^{\mathbf{X}})_{t \in I}$-adapted stochastic process X. The class of all filtered processes is denoted by $\mathcal{FP}$.

As the herein studied framework can be condensed to such a 5-tuple, filtered processes can be seen as the mathematical base of models for many phenomena in finance and economics. In what follows we give an overview on past and recent developments in the systematic study of filtered processes, where the main motivation stems from model sensitivity.

We use $\mathbb{E}_{\mathbf{X}}$ to denote the expectation w.r.t. $\mathbb{P}^{\mathbf{X}}$ for some filtered process $\mathbf{X}$. When the probability space is unambiguous from the context, will omit the subscript and simple write $\mathbb{E}$.

Example 1.2 Let us consider the simple probability space $(\Omega, \mathcal{F}, \mathbb{P})$ where $\Omega := \{\omega_1, \omega_2\}$ has two elements, $\mathcal{F} := 2^{\Omega}$ is the power set of Ω, and $\mathbb{P}$ is the uniform distribution on Ω. We consider the process $X = (X_t)_{t=1}^2$ with

$$X_1(\omega) = 0, \quad X_2(\omega) := \begin{cases} 1 & \omega = \omega_1, \\ -1 & \omega = \omega_2, \end{cases}$$

for $\omega \in \Omega$. First assume that the investor's available information is given by the filtration $(\mathcal{F}_t)_{t=1}^2$ with $\mathcal{F}_1 := \{\emptyset, \Omega\}$ and $\mathcal{F}_2 := \mathcal{F}$, that is, we consider the filtered process $\mathbf{X} := (\Omega, \mathcal{F}, \mathbb{P}, (\mathcal{F}_t)_{t=1}^2, X)$. Then the set of attainable claims of type (1) is given by the collection

$$\{hX_2 : h \in \mathbb{R}\}.$$

By buying at time 1, h units of X, one can not make profit without the risk of losing money, since $h \cdot X_2 \geq 0$ $\mathbb{P}$-almost surely if and only if $h = 0$.

On the other hand, if the investor's available information is given by the filtration $(\mathcal{G}_t)_{t=1}^2$ with $\mathcal{G}_1 = \mathcal{G}_2 = \mathcal{F}$, that is, we have the filtered process $\mathbf{Y} := (\Omega, \mathcal{F}, \mathbb{P}, (\mathcal{G}_t)_{t=1}^2, X)$ as model, then the set of attainable claims is given by

$$\{HX_2 : H \text{ is } \mathcal{F}\text{-measurable}\}.$$

Since the investor knows at time $t = 1$ already the value of X_2, they can exploit this information by buying $H_1 = X_2$ units. This leads to profit without risk of losing money, since the value of the portfolio at time 2 is $HX_2 = 1$.

Even though Example 1.2 is admittedly a rather basic example, it still stresses the importance of studying stochastic processes closely in combination with filtration. Virtually any example of a sequential decision making problem exhibits similar properties, including the optimal stopping problem (cf. Example 1.3) that we will discuss in more detail below.

Before we move forward to the detailed discussions, for readers' convenience we first collect some important notations which will be used throughout the whole article and list them as follows:

Table of Notations

Symbol	Meaning	Page
Spaces		
$\mathcal{P}(A)$	The space of probability measures on a topological space A	337
$I \to A$	The space of all paths from I into a set A	339
$M^{(r)}$	The path space of rank r-prediction processes	343
$T((\mathbb{R}^d))$	The tensor algebra over $\mathbb{R}^d$	348
H	A Hilbert space in $T((\mathbb{R}^d))$ containing the image of signature	347
$H^{(r+1)}$	A Hilbert space in $T((\mathbb{R} \oplus H^{(r)}))$ containing the image of $\Phi_{\text{Sig}}^{(r)}$	353
$\boldsymbol{H}_{\text{Sig}}^{(r)}$	The RKHS induced by $\boldsymbol{k}_{\text{Sig}}^{(r)}$	355
Paths and processes		
$\text{pp}(\mathbf{X})$	The prediction process of a filtered process $\mathbf{X}$	339
$\text{pp}^{(r)}(\mathbf{X})$	The rank r-prediction process of a filtered process $\mathbf{X}$	343
$\boldsymbol{p}^{(r)}$	The paths in $M^{(r)}$	353
$\tilde{\boldsymbol{p}}^{(r)}$	The paths formed by taking the integral of $\Phi_{\text{Sig}}^{(r-1)}$ against $\boldsymbol{p}^{(r)}$	353
$\hat{x}^i$	The i-th sample path of process $\mathbf{X}$	360
$\hat{y}^j$	The j-th sample path of process $\mathbf{Y}$	360
$\widetilde{\text{pp}}^{(1)}(\mathbf{X})$	$\widetilde{\text{pp}}^{(1)}(\mathbf{X}) = \mathbb{E}_{\mathbf{X}}[\Phi_{\text{Sig}}(X) \mid \mathcal{F}_t^{\mathbf{X}}], t \in I$	361
$\widehat{\text{pp}}^{(1)}(\mathbf{X})$	The empirical estimator of $\widetilde{\text{pp}}^{(1)}(\mathbf{X})$	363
Topologies		
$\mathcal{T}^{(r)}$	The adapted weak topology of rank r	343
$\mathcal{T}$	The adapted weak topology	344
Signatures		
Φ_{Sig}	The signature on $I \to \mathbb{R}^d$	349
$\Phi_{\text{Sig}}^{(r)}$	The rank-r signature on $M^{(r)}$	352
$d_{\text{Sig}}^{(r)}$	The rank-r signature metric	354
$\boldsymbol{k}_{\text{Sig}}$	The signature kernel	355
$\boldsymbol{k}_{\text{Sig}}^{(r)}$	The rank-r signature kernel	355
$\mathcal{D}_{\text{Sig}}^{(r)}$	The rank-r signature kernel MMD	356
$\hat{\mathcal{D}}_{\text{Sig}}^{(1)}$	The empirical estimator of $\mathcal{D}_{\text{Sig}}^{(1)}$	363

1.1 Model Sensitivity

In the past 30 years, there was substantial progress in our understanding of model uncertainty in finance, starting with the seminal articles by Lyons [60], Avellaneda et al. [8], Soner et al. [75] on markets with uncertain volatilities, and Hobson [44] and Cont [32] on robust hedging and pricing. Given the numerous contributions to the field in recent years, we can not do justice to all, but refer readers to Beiglböck et al. [20], Galichon et al. [41] and Bouchard and Nutz [29], and the references therein. Parametric model sensitivity was studied, for example, by Kramkov and Sîrbu [54] and Kardaras and Žitković [50]. More recently, Backhoff et al. [13] shed new light on model sensitivity in finance by suggesting a non-parametric analysis based on the recent notion of adapted Wasserstein distance. We will loosely follow their motivation:

Models are based on data and observations that are naturally noisy as well as incomplete information. For this reason, a model can be at best a faithful approximation of the modeled scenario and optimal actions that are based on this model may not be optimal in reality. Consequently, it becomes crucial to gauge the sensitivity of decisions based on slightly erroneous models. In other words, one would like to get a grasp on the following question.

(Q) How well do decision strategies derived from a slightly erroneous model perform when applied to reality?

However, Question (Q) is not yet well-posed, because we have not clarified when two models should be considered as 'close'. Consequently, we need a suitable notion of proximity of models. In our setting, this translates to finding an appropriate notion of when two filtered processes (cf. Definition 1.1) $\mathbf{X}, \mathbf{Y}$ are close. However, before comprehending proximity, it is essential to understand when two filtered processes should be considered equal.

1.2 The Weak Topology

Two random variables X, Y taking values in $\mathbb{R}^d$, that are not necessarily supported on the same probability space, share the same probabilistic properties if their law coincides. At the same time, a reasonable notion of proximity is provided by the ubiquitous topology of weak convergence of measures. Let $C_b(\mathbb{R}^d)$ be the set of continuous bounded functions on $\mathbb{R}^d$. We say that X and Y have the same distribution if

$$\mathbb{E}[f(X)] = \mathbb{E}[f(Y)] \quad \text{for all } f \in C_b(\mathbb{R}^d). \tag{2}$$

A sequence $(X^n)_{n\in\mathbb{N}}$ of $\mathbb{R}^d$-valued random variables is said to converge weakly to a random variable X if

$$\mathbb{E}[f(X^n)] \to \mathbb{E}[f(X)] \quad \text{for all } f \in C_b(\mathbb{R}^d). \tag{3}$$

There are numerous ways of metrizing weak convergence of measures on the set of probability distributions on $\mathbb{R}^d$, $\mathcal{P}(\mathbb{R}^d)$, such as the Prokhorov metric, Wasserstein distances or maximum mean discrepancies. Throughout this article, we endow spaces of probability measure with weak convergence of measures.

In a first attempt, one could try to identify filtered processes that share the same law and say two filtered processes are close if their laws are similar. This is rather unsatisfactory, since we completely neglect any information that is encoded in the filtrations, which is inadequate for our purposes, cf. Example 1.2. But, even when considering only filtered processes $\mathbf{X}$ whose filtration $(\mathcal{F}^{\mathbf{X}})_{t\in I}$ is given by the natural filtration generated by the process X itself, proximity w.r.t. weak convergence of measures is not the correct notion in view of finding an answer to Question 1.1. In order to stress the shortcomings of the usual weak topology of measures when employed in our setting, we provide a simple example.

Example 1.3 The optimal stopping problem is perhaps one of the simplest information dependent optimization problems that is encountered on financial markets. Let $(\Omega, \mathcal{F}, \mathbb{P})$, $(\mathcal{F}_t)_{t=1}^2$, $(\mathcal{G}_t)_{t=1}^2$ and X be given as in Example 1.2. For $\varepsilon > 0$ we consider the simplistic scenarios

$$\mathbf{X} := (\Omega, \mathcal{F}, \mathbb{P}, (\mathcal{F}_t)_{t\in I}, X),$$
$$\mathbf{X}^\varepsilon := \left(\Omega, \mathcal{F}, \mathbb{P}, (\mathcal{G}_t)_{t\in I}, X^\varepsilon\right),$$

where the process

$$X_1^\varepsilon(\omega) := \begin{cases} \varepsilon & \omega = \omega_1, \\ -\varepsilon & \omega = \omega_2, \end{cases} \quad \text{and} \quad X_2^\varepsilon(\omega) := X_2(\omega),$$

see Fig. 1 below. Clearly, we have

$$\mathbb{E}[f(X^\varepsilon)] = \frac{1}{2}(f((-\varepsilon, -1)) + f((\varepsilon, 1))) \to \frac{1}{2}(f((0, -1) + f((0, 1))) = \mathbb{E}[f(X)],$$

while the optimal stopping values

$$\sup_{\tau\in \mathrm{ST}(\mathbf{X}^\varepsilon)} \mathbb{E}[X_\tau^\varepsilon] = \frac{1}{2}(1-\varepsilon) \to \frac{1}{2} > 0 = \sup_{\tau\in \mathrm{ST}(\mathbf{X})} \mathbb{E}[X_\tau],$$

for $\varepsilon \searrow 0$, where $\mathrm{ST}(\mathbf{X})$ denotes the set of $(\mathcal{F}_t^{\mathbf{X}})_{t\in I}$-stopping times.

Example 1.3 provides a very simple example that illustrates the core of the problem. Even though $X^\varepsilon \to X$ in distribution, the associated filtered processes $\mathbf{X}^\varepsilon$ are, from a probabilistic point of view, very different from $\mathbf{X}$. While X is a martingale on the filtered probability space of $\mathbf{X}$, the processes $(X^\varepsilon)_{\varepsilon\geq 0}$ are predictable on their respective filtered probability spaces. Of course, one could easily consider more sophisticated examples, also in continuous time. For example, it is possible to approximate a Brownian motion in law by "deterministic" processes, whose paths are completely known at initial time and observe the very same phenomenon, see for example [6, Chapter 4, Section 11].

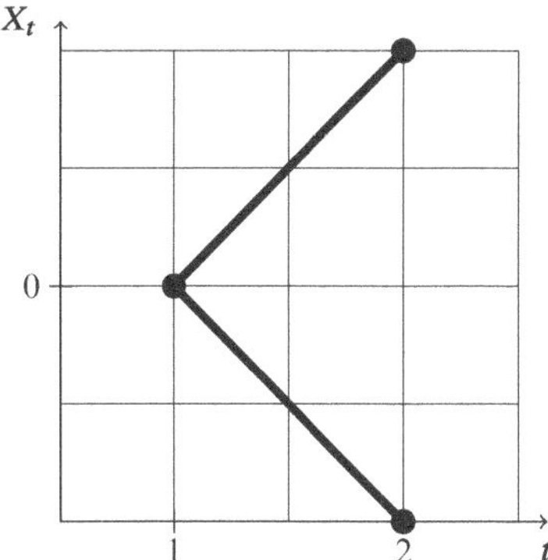

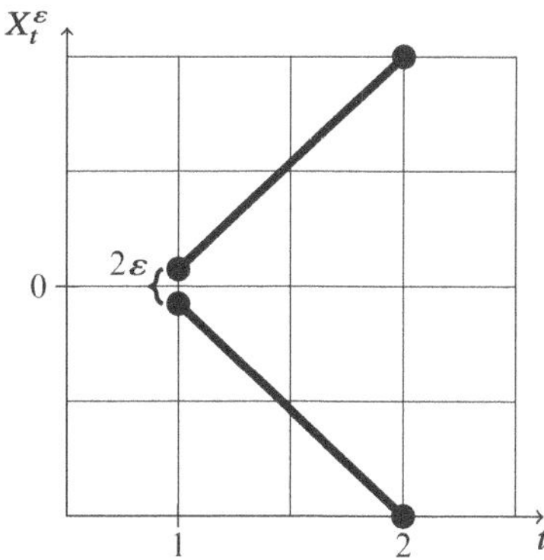

Fig. 1 Two processes with similar law but different probabilistic properties. A similar illustration can also be found in [13, Figure 1] and [28, Figure 1]

1.3 The Prediction Process

Multiple mathematicians have addressed the shortcomings presented in Sect. 1.2 and worked on ways of refining the weak topology in order to additionally incorporate the information that is encoded in the filtration. Notable contributions stem from the works of Aldous [6] and Hoover–Keisler [46]. For an in depth discussion on the multitude of related works we refer the interested reader to Sect. 4.

From now on, for $I = \{1, \dots, T\}$ and any set A, we will use the symbol $I \to A$ to denote the product space $A^T = A \times \dots \times A$ (T copies of A), which is the space of paths defined on I taking values in A.

Inspired by earlier work of Knight [53], Aldous defines the prediction process $\mathrm{pp}(\mathbf{X})$ of a filtered process $\mathbf{X}$ as the measure-valued process given by

$$\mathrm{pp}_t(\mathbf{X}) := \mathbb{P}^{\mathbf{X}}(X \in \cdot | \mathcal{F}_t^{\mathbf{X}}) \quad t \in I. \tag{4}$$

We have that $\mathrm{pp}(\mathbf{X})$ takes values in $I \to \mathcal{P}(I \to \mathbb{R}^d)$. As the name suggests, $\mathrm{pp}_t(\mathbf{X})$ is the predicted law of the process given the information at time t. This process can be used to define an equivalence relation that Aldous dubs *synonymity*. Two filtered processes $\mathbf{X}$ and $\mathbf{Y}$ are called synonymous if $\mathrm{pp}(\mathbf{X})$ and $\mathrm{pp}(\mathbf{Y})$ have the same distribution, i.e.,

$$\mathbb{E}_{\mathbf{X}}[f(\mathrm{pp}(\mathbf{X}))] = \mathbb{E}_{\mathbf{Y}}[f(\mathrm{pp}(\mathbf{Y}))] \quad \text{for all } C_b(I \to \mathcal{P}(I \to \mathbb{R}^d)). \tag{5}$$

A sequence $(\mathbf{X}^n)_{n \in \mathbb{N}}$ of filtered processes is said to converge in the *extended weak topology* to a filtered process $\mathbf{X}$ if

$$\mathbb{E}_{\mathbf{X^n}}[f(\mathrm{pp}(\mathbf{X}^n))] \to \mathbb{E}_{\mathbf{X}}[f(\mathrm{pp}(\mathbf{X}))] \quad \text{for all } f \in C_b(I \to \mathcal{P}(I \to \mathbb{R}^d)). \tag{6}$$

Remark 1.4 We have the following observations of topological nature:

1. It is straightforward to see that the extended weak topology is finer than convergence in law of $(X^n)_{n\in\mathbb{N}}$ to X. Indeed, any function $g \in C_b(I \to \mathbb{R}^d)$ has a representative $f \in C_b(I \to \mathcal{P}(I \to \mathbb{R}^d))$ such that

$$\mathbb{E}[g(X)] = \mathbb{E}[f(\mathrm{pp}(\mathbf{X})] \quad \text{for all filtered processes } \mathbf{X},$$

by simply setting $f(\rho) := \int g(x)\, d\rho_t(x)$ for every $\rho \in (I \to \mathcal{P}(I \to \mathbb{R}^d))$ and any $t \in I$, because we have by the tower property

$$\mathbb{E}[f(\mathrm{pp}(\mathbf{X}))] = \mathbb{E}\left[\int g(x)\, d\mathrm{pp}(\mathbf{X})_t(x)\right] = \mathbb{E}[\mathbb{E}[g(X)|\mathcal{F}_t]] = \mathbb{E}[g(X)].$$

2. Consider the prediction processes $(\mathbf{X}^\varepsilon)_{\varepsilon\geq 0}$ and $\mathbf{X}$ from Example 1.3. Then a simple computation shows that $\mathrm{pp}(\mathbf{X}) = (\mathrm{pp}_1(\mathbf{X}), \mathrm{pp}_2(\mathbf{X}))$ satisfies

$$\mathrm{pp}_1(\mathbf{X})(\omega_i) = \frac{1}{2}\left(\delta_{(0,-1)} + \delta_{(0,1)}\right) \text{ and } \mathrm{pp}_2(\mathbf{X})(\omega_i) = \delta_{X(\omega_i)}, \quad i = 1, 2;$$

where $\delta_{X(\omega_1)} = \delta_{(0,1)}$ and $\delta_{X(\omega_2)} = \delta_{(0,-1)}$. Similarly, as $\mathcal{G}_1 = \mathcal{G}_2 = \mathcal{F}$, we can show that $\mathrm{pp}(\mathbf{X}^\varepsilon) = (\mathrm{pp}_1(\mathbf{X}^\varepsilon), \mathrm{pp}_2(\mathbf{X}^\varepsilon))$ satisfies that

$$\mathrm{pp}_1(\mathbf{X}^\varepsilon)(\omega_i) = \mathrm{pp}_2(\mathbf{X}^\varepsilon)(\omega_i) = \delta_{X^\varepsilon(\omega_i)}, \quad i = 1, 2;$$

where $\delta_{X^\varepsilon(\omega_1)} = \delta_{(\varepsilon,1)}$ and $\delta_{X(\omega_2)} = \delta_{(-\varepsilon,-1)}$. Hence, $\mathrm{pp}(\mathbf{X}^\varepsilon) \to \mathrm{pp}(\mathbf{X}^0)$ in distribution while the laws of $\mathrm{pp}(\mathbf{X})$ and $\mathrm{pp}(\mathbf{X}^0)$ differ. Therefore, Example 1.3 provides an example of a sequence that converges weakly but not with respect to the extended weak topology.
3. Typical functions in $C_b(I \to \mathcal{P}(I \to \mathbb{R}^d))$ can be built as follows: For each $t \in I = \{1, \dots, T\}$, we choose a function $g_t \in C_b(I \to \mathbb{R}^d)$; then given a measure-valued path $\rho = (\rho_t)_{t\in I} \in I \to \mathcal{P}(I \to \mathbb{R}^d)$, we obtain a $\mathbb{R}^T$-valued vector by testing each measure $\rho_t \in \mathcal{P}(I \to \mathbb{R}^d)$ against the function $g_t \in C_b(I \to \mathbb{R}^d)$:

$$\left(\int g_1(x)d\rho_1(x), \dots, \int g_T(x)d\rho_T(x)\right) \in \mathbb{R}^T.$$

Finally, pick a continuous bounded function $h : \mathbb{R}^T \to \mathbb{R}$, we define

$$f(\rho) = h \circ \left(\int g_1(x)d\rho_1(x), \dots, \int g_T(x)d\rho_T(x)\right),$$

and get an element $f \in C_b(I \to \mathcal{P}(I \to \mathbb{R}^d))$. In fact these functions are used in [46] to study the adapted topologies.

As a concrete example, we assume $T = 2$ and $d = 1$, in this case $(I \to \mathbb{R}^d) = \mathbb{R}^2$, and $(I \to \mathcal{P}(I \to \mathbb{R}^d)) = \mathcal{P}(\mathbb{R}^2) \times \mathcal{P}(\mathbb{R}^2)$. We now take $g_1(x_1, x_2) = [(x_1 + x_2) \wedge 2] \vee (-2)$ and $g_2(x_1, x_2) = \sin(x_1 - x_2)$, and $h(x_1, x_2) = \exp(-|x_1| - x_2^2)$ to define a function $f \in C_b(I \to \mathcal{P}(I \to \mathbb{R}^d))$ as above. Then for any filtered process $\mathbf{X}$ on I taking values in $[-1, 1]$, we have (as $\mathrm{pp}(\mathbf{X}) \in \mathcal{P}(\mathbb{R}^2) \times \mathcal{P}(\mathbb{R}^2)$)

$$f(\mathrm{pp}(\mathbf{X})) = \exp(-|\mathbb{E}[X_1 + X_2|\mathcal{F}_1]| - \mathbb{E}[\sin(X_1 - X_2)|\mathcal{F}_2]^2),$$

and in this case

$$\mathbb{E}[f(\mathrm{pp}(\mathbf{X}))] = \mathbb{E}[-|\exp(\mathbb{E}[X_1 + X_2|\mathcal{F}_1]| - \mathbb{E}[\sin(X_1 - X_2)|\mathcal{F}_2]^2)].$$

We leave a simple exercise to curious readers to check that for filtered processes $\mathbf{X}$ and $\mathbf{X}^\varepsilon$ from Example 1.3, it holds that for the above function f,

$$\begin{aligned}\mathbb{E}[f(\mathrm{pp}(\mathbf{X}))] &= \mathbb{E}[\exp(-|\mathbb{E}[X_1 + X_2|\mathcal{F}_1]| - \mathbb{E}[\sin(X_1 - X_2)|\mathcal{F}_2]^2)] \\ &= \exp(-\sin^2(1))\end{aligned}$$

and

$$\begin{aligned}\mathbb{E}[f(\mathrm{pp}(\mathbf{X}^\varepsilon))] &= \mathbb{E}[\exp(-|\mathbb{E}[X_1^\varepsilon + X_2^\varepsilon|\mathcal{G}_1]| - \mathbb{E}[\sin(X_1^\varepsilon - X_2^\varepsilon)|\mathcal{G}_2]^2)] \\ &= \exp(-(1 + \varepsilon) - \sin^2(1 - \varepsilon)),\end{aligned}$$

so that $\lim_{\varepsilon \to 0} \mathbb{E}[f(\mathrm{pp}(\mathbf{X}^\varepsilon))] \neq \mathbb{E}[f(\mathrm{pp}(\mathbf{X}))]$.

Probabilistic properties such as being a martingale or being predictable, play an important role in stochastic analysis. Many of these properties are already captured by the prediction process.

Remark 1.5 Let $\mathbf{X}$ and $\mathbf{Y}$ be filtered processes that are synonymous, then we have:

(a) if X is a $(\mathcal{F}_t^{\mathbf{X}})_{t\in I}$-martingale, then Y is a $(\mathcal{F}_t^{\mathbf{Y}})_{t\in I}$-martingale;
(b) if X is $(\mathcal{F}_t^{\mathbf{X}})_{t\in I}$-predictable, then Y is $(\mathcal{F}_t^{\mathbf{Y}})_{t\in I}$-predictable;
(c) if X is a $(\mathcal{F}_t^{\mathbf{X}})_{t\in I}$-Markov process, then Y is a $(\mathcal{F}_t^{\mathbf{Y}})_{t\in I}$-Markov process.

To strengthen the intuition, we check the second claim as the other two properties can be shown with similar reasoning. Recall that X is $(\mathcal{F}_t^{\mathbf{X}})_{t\in I}$-predictable if X_{t+1} is $\mathcal{F}_t^{\mathbf{X}}$-measurable for all $t \in I$. Equivalently, we can express this property in terms of the prediction process:

$$\delta_{X_{t+1}} = \mathrm{law}(X_{t+1}|\mathcal{F}_t^{\mathbf{X}}) = (e_t)_\# \mathrm{law}(X|\mathcal{F}_t^{\mathbf{X}}) = (e_t)_\# \mathrm{pp}_t(\mathbf{X}), \tag{7}$$

where δ_x denotes a Dirac measure at $x \in \mathbb{R}^d$ and $(e_t)_\#$ is the pushforward of a measure with respect to the time-t evaluation map $e_t : (\mathbb{R}^d)^I \to \mathbb{R}^d$. Therefore,

X is $(\mathcal{F}_t^{\mathbf{X}})_{t\in I}$-predictable if and only if $(e_t)_\# \mathrm{pp}_t(\mathbf{X})$ takes values in the set of Dirac measures. But, the latter is a property that can be deduced by knowing just the law of $\mathrm{pp}(\mathbf{X})$. Since $\mathbf{X}$ and $\mathbf{Y}$ are synonymous, their prediction processes $\mathrm{pp}(\mathbf{X})$ and $\mathrm{pp}(\mathbf{Y})$ have the same distribution. It is immediate from (7) that Y also has to be $(\mathcal{F}_t^{\mathbf{Y}})_{t\in I}$-predictable.

In fact, these properties are even closed under convergence in the extended weak topology, except for the martingale property, that is Item (a). This holds true also for the latter if we additionally assume uniform integrability of the sequence of stochastic processes. We refer readers to [17] for detailed proofs.

Remark 1.5 suggests that synonymity may be the adequate notion of equivalence for filtered process. But, even though the prediction process is suitable to rectify many problems (see also the discussion in Sect. 4), it is still not complex enough to identify the value of sequential decision making problems, when admitting processes with general filtrations. We will show this point by means of optimal stopping and insist that similar examples can be constructed for most sequential decision making problems.

Example 1.6 Let U, V be independent fair coin flips such that $\mathbb{P}(U = 1) = \mathbb{P}(U = 0) = \mathbb{P}(V = 1) = \mathbb{P}(V = 0) = 1/2$, $I = \{1, 2, 3\}$ and X is given by

$$X := (X_t)_{t\in I} := (0.6, 0.4, V) \in \mathbb{R}^3.$$

Consider the two filtrations $(\mathcal{F}_t)_{t\in I}$ and $(\mathcal{G}_t)_{t\in I}$ where

$$\mathcal{F}_1 := \sigma(U),\ \ \mathcal{F}_2 := \sigma\left(U, \{U = 1, V = 0\}, \{U = 1, V = 1\}\right),\ \ \mathcal{F}_3 := \sigma(U, V),$$
$$\mathcal{G}_1 := \{\emptyset, \Omega\},\quad \mathcal{G}_2 := \mathcal{F}_2,\quad \mathcal{G}_3 := \mathcal{F}_3,$$

and define the filtered processes

$$\mathbf{X} := (\Omega, \mathcal{F}, \mathbb{P}, (\mathcal{F}_t)_{t\in I}, X) \quad \text{and} \quad \mathbf{Y} := (\Omega, \mathcal{F}, \mathbb{P}, (\mathcal{G}_t)_{t\in I}, X).$$

Observe that $\mathcal{F}_1$ encodes at time 1 whether or not we will know the value of X_3 at time 2, while $\mathcal{G}_1$ contains no information. The prediction process is ignorant to this kind of information, which is of "higher order". In fact, computing the prediction processes at time 1 we get

$$\mathrm{pp}_1(\mathbf{X}) = \mathrm{law}(X|\mathcal{F}_1) = \mathrm{law}(X|U) = \mathrm{law}(X) = \mathrm{law}(X|\mathcal{G}_1) = \mathrm{pp}_1(\mathbf{Y}),$$

where we used independence of U and V. In particular, we have shown that $\mathrm{pp}(\mathbf{X}) = \mathrm{pp}(\mathbf{Y})$. On the other hand, computing the optimal stopping values we find

$$\sup_{\tau\in\mathrm{ST}(\mathbf{X})} \mathbb{E}[X_\tau] = 0.65 > 0.6 = \sup_{\tau\in\mathrm{ST}(\mathbf{Y})} \mathbb{E}[X_\tau].$$

We leave the verification of the above inequality as an exercise to curious readers by checking that $\tau^{\mathbf{X}} := \mathbf{1}_{\{U=0\}} + 2\mathbf{1}_{\{U=1,V=0\}} + 3\mathbf{1}_{\{U=1,V=1\}}$ and $\tau^{\mathbf{Y}} := 2\mathbf{1}_{\{U=1,V=0\}} + 3\mathbf{1}_{\{U=1,V=1\}} + 3\mathbf{1}_{\{U=0\}}$ are optimal stopping times for $\sup_{\tau \in \mathrm{ST}(\mathbf{X})} \mathbb{E}[X_\tau]$ and $\sup_{\tau \in \mathrm{ST}(\mathbf{Y})} \mathbb{E}[X_\tau]$, respectively.

1.4 Higher-Rank Prediction Processes

We have seen that Aldous' notion of synonymity does not capture all probabilistic properties of a filtered process, in particular, it is agnostic to information of higher order as described in Example 1.6. Motivated by such an observation, Hoover–Keisler [46] define prediction processes of higher rank. For a rank $r \in \mathbb{N}$, the *rank r-prediction process* $\mathrm{pp}^{(r)}$ of a filtered process $\mathbf{X}$ is inductively defined by

$$\mathrm{pp}^{(0)}(\mathbf{X}) := X, \quad \text{and} \quad \mathrm{pp}_t^{(r)}(\mathbf{X}) := \mathrm{law}(\mathrm{pp}^{(r-1)}(\mathbf{X})|\mathcal{F}_t^{\mathbf{X}}) \quad \text{for } t \in I. \tag{8}$$

Schematically, we can think of the path space of the rank r prediction process as

$$\begin{aligned}
M^{(0)} &:= (I \to \mathbb{R}^d), \\
M^{(1)} &:= (I \to \mathcal{P}(I \to \mathbb{R}^d)), \\
&\vdots \\
M^{(r)} &:= (I \to \mathcal{P}\underbrace{(I \to \cdots \mathcal{P}(I \to \mathbb{R}^d))\cdots)}_{M^{(r-1)}}).
\end{aligned}$$

The space $M^{(r)}$ is constructed by alternately applying the following two operations:

- mapping a Polish space S to the set of Borel probability measures $\mathcal{P}(S)$ on S;
- taking the product $(I \to S) = S^I$ of a Polish space S.

As both operations preserve the property of being Polish,[1] it is immediate that $M^{(r)}$ is a Polish space, which mathematically justifies the inductive construction in (8). Two filtered processes $\mathbf{X}$ and $\mathbf{Y}$ have the same rank-r adapted distribution if

$$\mathbb{E}[f(\mathrm{pp}^{(r)}(\mathbf{X}))] = \mathbb{E}[f(\mathrm{pp}^{(r)}(\mathbf{Y}))] \quad \text{for all } f \in C_b(M^{(r)}). \tag{9}$$

A sequence $(\mathbf{X}^n)_{n \in \mathbb{N}}$ of filtered processes is said to converge in the *adapted weak topology of rank r* to a filtered process $\mathbf{X}$ if

[1] We equip $(I \to S)$ with the product topology and $\mathcal{P}(S)$ with weak convergence of measures.

$$\mathbb{E}[f(\mathrm{pp}^{(r)}(\mathbf{X}^n))] \to \mathbb{E}[f(\mathrm{pp}^{(r)}(\mathbf{X}))] \quad \text{for all } f \in C_b(M^{(r)}). \tag{10}$$

Finally, the *adapted distribution* of a filtered process $\mathbf{X}$ is the law of $\mathrm{pp}^{\infty}(\mathbf{X})$ given by

$$\mathrm{pp}_t^{(\infty)}(\mathbf{X}) := (\mathrm{pp}_t^{(0)}(\mathbf{X}), \mathrm{pp}_t^{(1)}(\mathbf{X}), \dots) \quad \text{for all } t \in I, \tag{11}$$

where $\mathrm{pp}^{(\infty)}(\mathbf{X}) = (\mathrm{pp}_t^{(\infty)}(\mathbf{X}))_{t\in I}$ takes values in

$$M^{(\infty)} := (I \to (\mathbb{R}^d \times \mathcal{P}(M^{(0)}) \times \mathcal{P}(M^{(1)}) \times \dots)).$$

Before proceeding, we want to remark that in the present finite discrete-time setting with T time steps, it suffices to simply consider the prediction process of rank $T-1$ instead of the whole adapted distribution. This interesting property is discussed in more detail in Remark 1.7 down below. Therefore, the reader may choose to ignore the case $r = \infty$ and read it instead as $r = T - 1$. We say that a sequence $(\mathbf{X}^n)_{n\in\mathbb{N}}$ of filtered processes converges in the adapted weak topology to $\mathbf{X}$ if (10) holds for $r = \infty$. Hence, convergence in the adapted weak topology is precisely convergence in distribution of the corresponding adapted distributions. Clearly, all filtered processes $\mathcal{FP}$ are a proper class containing a lot of redundancy. As a next step, we use the equivalence relation induced by the adapted distribution $\sim_\infty$ to put filtered processes with the "same probabilistic properties" into a single equivalence class. We define the factor space of filtered processes

$$\mathrm{FP} := \mathcal{FP}/_{\sim_\infty}. \tag{12}$$

After factorization we end up with the set FP, on which we can define the adapted weak topology of rank $r \in \mathbb{N}\cup\{0,\infty\}$, denoted by $\mathcal{T}^{(r)}$, whose mode of convergence is described by (10). In analogy to when working with L_p-spaces, we will for the most part not distinguish between single filtered process and their $\sim_\infty$-equivalence class. Moreover, we will write $\mathcal{T} := \mathcal{T}^{(\infty)}$ for the adapted weak topology.

Remark 1.7 We compile some facts on $\mathcal{T}^{(r)}$, the adapted weak topology of rank r:

(1) Observe that $\mathcal{T}^{(0)}$ corresponds to the usual weak topology, while $\mathcal{T}^{(1)}$ is precisely Aldous' extended weak topology and $\mathcal{T} = \mathcal{T}^{(\infty)}$ is the adapted weak topology.
(2) The topologies $\mathcal{T}^{(r)}$ become increasingly finer. We have

$$\mathcal{T}^{(0)} \subsetneq \mathcal{T}^{(1)} \subsetneq \dots \subsetneq \mathcal{T}^{(T-1)} = \dots = \mathcal{T}^{(\infty)}.$$

Indeed, this can be seen as follows: For any filtered process $\mathbf{X}$ we have

$$(e_t)_\#\mathrm{pp}_t^{(r+1)}(\mathbf{X}) = (e_t)_\#\mathrm{law}(\mathrm{pp}^{(r)}(\mathbf{X})|\mathcal{F}_t^{\mathbf{X}}) = \mathrm{law}(\mathrm{pp}_t^{(r)}(\mathbf{X})|\mathcal{F}_t^{\mathbf{X}}) = \delta_{\mathrm{pp}_t^{(r)}(\mathbf{X})},$$

where e_t is the time-t evaluation map and δ_x denotes a Dirac measure at $x \in M^{(r)}$. The second equality is by definition of the push-forward operation and the last is due to $\mathrm{pp}^{(r)}(\mathbf{X})$ being $(\mathcal{F}_t^{\mathbf{X}})_{t\in I}$-adapted. Consequently, one can build a continuous map

$$\iota : M^{(r)} \to M^{(r-1)} \tag{13}$$

that for all $\mathbf{X} \in \mathrm{FP}$, $\iota(\mathrm{pp}^{(r+1)}(\mathbf{X})) = \mathrm{pp}^{(r)}(\mathbf{X})$ almost surely, which readily provides the inclusion $\mathcal{T}^{(r)} \subseteq \mathcal{T}^{(r+1)}$. For additional details, we refer to [17, Theorem 4.11 and Theorem 7.1].

(3) As we consider the setting where $I = \{1, \dots, T\}$ consists of $T \in \mathbb{N}$ times, it is not necessary to consider prediction processes of arbitrary rank. In fact, we have by [17, Lemma 4.7 and Lemma 4.10] that, for every $r \in \mathbb{N}$, there exists a continuous map

$$F : M^{(T-1)} \to M^{(r)}$$

with the property that $F(\mathrm{pp}^{(T-1)}(\mathbf{X})) = \mathrm{pp}^{(r)}(\mathbf{X})$ almost surely, for all $\mathbf{X} \in \mathrm{FP}$. It is straightforward to see that $\mathcal{T}^{(r)} = \mathcal{T}^{(T-1)} = \mathcal{T}$, which means that all the necessary information is contained in the rank $(T-1)$-prediction process.

(4) In fact, as long as $r < T - 1$ the inclusion $\mathcal{T}^{(r)} \subsetneq \mathcal{T}^{(r+1)}$ is strict. This can be achieved with the construction shown in Example 1.6, see also [46, Example 3.2] and [17, Proposition 7.2].

(5) For more detailed descriptions and examples of $\mathrm{pp}^{(r)}(\mathbf{X})$, $M^{(r)}$ and $\mathcal{T}^{(r)}$ we refer readers to [46] and [17]. We also recommend readers to focus on the case $r = 1$ at the first reading.

Theorem 1.8 *The topological space $(FP, \mathcal{T})$ is Polish, that is, separable and completely metrizable.*

A proof of Theorem 1.8 can be found in [17, Theorem 1.3] and [22, Theorem 1.5]. It is worth mentioning that, even though all filtered processes are a proper class, $(\mathrm{FP}, \mathcal{T})$ is tractable in the sense that it is a Polish space and simple processes (for example, Markov processes living on finite probability spaces, cf. [17, Theorem 5.4]) are dense. The main purpose of the adapted weak topology is to finally have a topology that is sufficiently strong to have continuity of sequential decision making problems, but which is not unnecessarily strong. We refer to Sect. 4 below for a more detailed discussion. To get back at our introductory example, that is Example 1.3, we have continuity of optimal stopping.

Lemma 1.9 (Continuity of Optimal Stopping) *Let $c : (I \to \mathbb{R}^d) \times I \to \mathbb{R}$ be continuous, bounded and non-anticipative.*[2] *Then we have continuity of the value function*

[2] A function $c : (I \to \mathbb{R}^d) \times I \to \mathbb{R}$ is non-anticipative if for each $t \in I$, the value of $c(x, t)$ only depends on $x_1, \dots, x_t$, i.e., the trajectory of the path x before time t.

$$v_c : (FP, \mathcal{T}) \to \mathbb{R} : v_c(\mathbf{X}) := \sup_{\tau \in \mathrm{ST}(\mathbf{X})} \mathbb{E}[c(X, \tau)].$$

For illustrative purposes, we include a self-contained proof of this result.

Proof of Lemma 1.9, $T = 2$ When there are only two time points $I = \{1, 2\}$, optimal stopping boils down to deciding at time $t = 1$ whether or not to stop. At time $t = 1$, the gain to immediately stop is $c(X, 1) = \mathbb{E}_{\mathbf{X}}[c(X, 1)|\mathcal{F}_1^{\mathbf{X}}]$ while the expected profit of stopping at $t = 2$ is $\mathbb{E}_{\mathbf{X}}[c(X, 2)|\mathcal{F}_1^{\mathbf{X}}]$. Consequently, the value of the optimal stopping problem is given by

$$v_c(\mathbf{X}) = \mathbb{E}_{\mathbf{X}}[c(X, 1) \vee \mathbb{E}_{\mathbf{X}}[c(X, 2)|\mathcal{F}_1^{\mathbf{X}}]] = \mathbb{E}_{\mathbf{X}}[F(\mathrm{pp}^{(1)}(\mathbf{X}))], \tag{14}$$

where $F : M^{(1)} \to R$ is given by

$$F(\rho) = \int c(x, 1)\, d\rho_1(x) \vee \int c(y, 2)\, d\rho_1(y),$$

for $\rho = (\rho_1, \rho_2) \in M^{(1)}$. As c is continuous and bounded, the same is true for F. Using the latter, the claim follows directly from the definition of $\mathcal{T}$ and the representation given in (14). □

Proof of Lemma 1.9, $T \in \mathbb{N}$ The proof for general $T \in \mathbb{N}$ is essentially the same as in the case $T = 2$ but with an additional induction over the number of time steps. The value $v_c(\mathbf{X})$ is given by the associated Snell envelope $(U_t^{\mathbf{X}})_{t \in I \cup \{0\}}$ at time 0. This means that we have

$$v_c(\mathbf{X}) = U_0,$$

where the Snell envelope is given by the recursive scheme

$$U_T^{\mathbf{X}} := c(X, T) \quad \text{and} \quad U_t^{\mathbf{X}} := c(X, t) \vee \mathbb{E}_{\mathbf{X}}[U_{t+1}^{\mathbf{X}}|\mathcal{F}_t^{\mathbf{X}}],$$

for $t \in I$, and $U_0 = \mathbb{E}_{\mathbf{X}}[U_1^{\mathbf{X}}]$. Similarly, we define recursively

$$U_T^{\mathbf{X}} = \underbrace{c(\mathrm{pp}^{(0)}(\mathbf{X}), T)}_{:= F_T(\mathrm{pp}^{(0)}(\mathbf{X}))} \quad \text{and} \quad U_t^{\mathbf{X}} = \underbrace{c(\iota_t(\mathrm{pp}^{(t)}(\mathbf{X})), t) \vee (F_{T-t+1})_\# \mathrm{pp}^{(t)}(\mathbf{X})}_{:= F_{T-t}(\mathrm{pp}^{(t)}(\mathbf{X}))},$$

where $\iota_t : M^{(t)} \to M^{(0)}$, defined analogously to ι in Remark 1.7 ((2)), is a continuous map with $\iota_t(\mathrm{pp}^{(t)}(\mathbf{X})) = \mathrm{pp}^{(0)}(\mathbf{X}) = X$. Hence, $F_{T-t} : M^{(t)} \to \mathbb{R}$ is continuous. In particular, we can construct a continuous map $F_0 : \mathcal{P}(M^{(T-1)}) \to \mathbb{R}$ with

$$F_0(\mathrm{law}(\mathrm{pp}^{(T-1)}(\mathbf{X}))) := \mathbb{E}_{\mathbf{X}}[F_1(\mathrm{pp}^{(T-1)}(\mathbf{X}))] = U_0^{\mathbf{X}} = v_c(\mathbf{X}).$$

We conclude from the definition of convergence in $\mathcal{T}$, cf. (10), continuity of v_c. □

2 Higher-Rank (Expected) Signatures

In the introduction we established that the higher-rank adapted weak topology $\mathcal{T}^{(r)}$, $r \geq 1$, on the space of filtered processes FP is finer than the weak topology (the latter corresponds to $r = 0$), and is essential to addressing continuity for multi-stage optimization problems which are in general discontinuous with respect to the weak topology, such as optimal stopping problems as discussed in Example 1.2. Therefore, it should come as no surprise that many theoretical tools which are valid for the weak topology do not work any more for the higher-rank cases. For instance, the expectation of the classical signature feature, which is able to characterize the laws of stochastic processes as moment generating function for path-valued random variables, does not distinguish two filtered processes in the rank-1 adapted weak topology. Motivated by these observations, we want to accordingly modify the classical signature feature to introduce a family of so-called *higher-rank signatures*, that adequately capture higher-rank adapted weak topologies whilst retaining the advantageous properties of the classical signature, both in theoretical and computational aspects. For technical simplicity, throughout the whole section, we will implicitly assume that

(A1) All filtered processes take values in a compact subset $\mathcal{K} \subset \mathbb{R}^d$.

2.1 *Properties of Higher-Rank (Expected) Signatures*

In this subsection (which is based on [28, Sections 3,4]), we introduce the notion of higher-rank signatures, whose expectations can be used to distinguish filtered processes with respect to the higher-rank adapted weak topologies.

To provide the reader with sufficient intuition for these increasingly complex concepts, let us first disregard the time components and discuss the static case.

The Static Case

Since $\mathcal{K} \subset \mathbb{R}^d$ is a compact subset, the Stone–Weierstrass theorem asserts that the space of all polynomials on $\mathcal{K}$ is dense in $C_b(\mathcal{K})$ in the topology of uniform convergence. Furthermore, a fundamental result in probability theory states that two $\mathcal{K}$-valued random variables X, Y have the same law if and only if, for all $k \in \mathbb{N}$, they have the same k-th moment. Moreover, a sequence $(X^n)_{n\in\mathbb{N}}$ of $\mathcal{K}$-valued random variables converges to X in distribution if and only if, for all $k \in \mathbb{N}$, their k-th moments converge. These well–known results can be reformulated in terms of the notations used in chapter "The Signature Kernel" as follows:

Consider the Hilbert space

$$H := \left\{(a_0, a_1, a_2, \dots) \in \prod_{k=0}^{\infty} (\mathbb{R}^d)^{\otimes k} : \left(\sum_{k=0}^{\infty} \|a_k\|^2\right)^{1/2} < \infty\right\}, \tag{15}$$

which is embedded in the tensor algebra $T((\mathbb{R}^d)) := \prod_{k=0}^{\infty}(\mathbb{R}^d)^{\otimes k}$ and whose inner product is given by

$$\langle (a_0, a_1, a_2, \ldots,), (b_0, b_1, b_2, \ldots,) \rangle_H = \sum_{k=0}^{\infty} a_k \cdot b_k,$$

where $a_k \cdot b_k$ denotes the usual inner product of a_k and b_k in $(\mathbb{R}^d)^{\otimes k}$ which is obtained from the Euclidean inner product on $\mathbb{R}^d$. Next, define the *monomial* map $\Phi_{\text{Mon}} : \mathbb{R}^d \to H$ by

$$\Phi_{\text{Mon}}(x) := \Big(1, x, \frac{x^{\otimes 2}}{2!}, \frac{x^{\otimes 3}}{3!}, \ldots, \frac{x^{\otimes n}}{n!}, \ldots\Big),$$

where $x^{\otimes k} := (x_1^{m_1} \ldots x_d^{m_d})_{m_1+\ldots+m_d=k, m_i \in \mathbb{N}_0} \in (\mathbb{R}^d)^{\otimes k}$ for a vector $x = (x_1, \ldots, x_d) \in \mathbb{R}^d$. Thanks to the Stone–Weierstrass theorem we have

$$\forall \varepsilon > 0,\ \forall f \in C_b(\mathcal{K}),\ \exists \ell \in H : \sup_{x \in \mathcal{K}} |f(x) - \langle \ell, \Phi_{\text{Mon}}(x) \rangle_H| \leq \varepsilon,$$

$$\text{law}(X) = \text{law}(Y) \iff \mathbb{E}[\Phi_{\text{Mon}}(X)] = \mathbb{E}[\Phi_{\text{Mon}}(Y)],$$

$$\Big(\forall f \in C_b(\mathcal{K}), \lim_{n\to\infty} \mathbb{E}[f(X_n)] = \mathbb{E}[f(X)]\Big) \iff \lim_{n\to\infty} \mathbb{E}[\Phi_{\text{Mon}}(X_n)] = \mathbb{E}[\Phi_{\text{Mon}}(X)].$$

Writing $\Phi^n_{\text{Mon}}(x) := \frac{x^{\otimes n}}{n!}$, $n \in \mathbb{N}_0$ for the n-th component of $\Phi_{\text{Mon}}(x)$, notice that $\mathbb{E}[\Phi^n_{\text{Mon}}(X)] = \mathbb{E}[\frac{X^{\otimes n}}{n!}]$ is the n-th moment of X (up to the constant $1/n!$).

Remark 2.1 In fact, the preceding results regarding polynomials and re-scaled moments can be generalized further by relaxing the compactness assumption. To this end, we need to normalize the monomials map Φ_{Mon} in order to guarantee integrability. According to [31, Section 3.2], the renormalized version of Φ_{Mon} is called the *robust monomials*, and can be represented by $\Lambda \circ \Phi_{\text{Mon}}$ for a so-called tensor normalization $\Lambda : H \to \{(a_0, a_1, \ldots) \in H : \|(a_0, a_1, \ldots)\|_H \leq 1\}$.

Now, let us move forward to the case of filtered processes with respect to the weak topology, i.e., when the rank $r = 0$, cf. Sect. 1.4. In this setting, it is well-known that the notions of monomials on a compact subset $\mathcal{K} \subset \mathbb{R}^d$ admit natural generalizations to the path spaces $M^{(0)} = (I \to \mathcal{K})$, which are called signatures and robust signatures, respectively.

The Rank $r = 0$ Case

The classical signature theory tells us that the notions of monomials Φ_{Mon} on $\mathcal{K} \subset \mathbb{R}^d$, namely the static case, have natural generalizations to $\mathcal{K}$-valued paths, which are called the *signature*. To elaborate, an $\mathcal{K}$-valued (discrete-time) path $\boldsymbol{x} = (\boldsymbol{x}_1, \ldots, \boldsymbol{x}_T) \in M^{(0)} = (I \to \mathcal{K})$ is identified with a piecewise-linear path $\boldsymbol{x} : [0, T] \to \mathbb{R}^d$ via

$$\boldsymbol{x}_t := (i-t)\boldsymbol{x}_{i-1} + (t-(i-1))\boldsymbol{x}_i, \quad t \in [i-1, i],$$

and we write $\boldsymbol{x}_t = (x_t^1, \dots, x_t^d)$ for each $t \in [0, T]$. In other words, we embed the discrete-time path space $M^{(0)}$ into the space of continuous paths with bounded variation defined on $[0, T]$, which we denote by $(\mathbb{R}^d)_{1\text{-var}}$ as in chapter "The Signature Kernel". Then the *signature* of the path $\boldsymbol{x} \in (I \to \mathbb{R}^d)$ is defined by

$$\Phi_{\text{Sig}}(\boldsymbol{x}) := \left(1, \int d\boldsymbol{x}, \int d\boldsymbol{x}^{\otimes 2}, \dots\right) \in H,$$

where

$$\int d\boldsymbol{x}^{\otimes k} = \left(\int_{0<t_1<\dots<t_k<T} dx_{t_1}^{i_1} \dots dx_{t_k}^{i_k}\right)_{i_1,\dots,i_k \in \{1,\dots,d\}} \in (\mathbb{R}^d)^{\otimes k}$$

is the iterated Riemann integral for $k \geq 1$, and H is given as in (15). Using the same notation as in chapter "The Signature Kernel", for each $k \in \mathbb{N}$ we use $\Phi_{\text{Sig},k}(\boldsymbol{x})$ to denote the k-th iterated integrals of $\boldsymbol{x}$, namely $\Phi_{\text{Sig},k}(\boldsymbol{x}) = \int d\boldsymbol{x}^{\otimes k} \in (\mathbb{R}^d)^{\otimes k}$.

Thanks to the algebraic and analytic properties of iterated integrals, the following theorem holds true and implies that the signature feature Φ_{Sig} is the sequence of monomials on the path space $M^{(0)}$, respectively: Their linear combinations are dense in the space of continuous bounded functions on the path space, the expectation of $\Phi_{\text{Sig}}(X)$ (called the *expected signature*) can be used to characterize the weak convergence of the laws of stochastic processes (viewed as path-valued random variables). We summarize these facts into the following theorem, which can also be found in chapter "The Signature Kernel".

Theorem 2.2 *Let $\mathcal{K} \subset \mathbb{R}^d$ be compact. Let $M^{(0)} = (I \to \mathcal{K})$ denote the set of paths taking values in $\mathcal{K}$. For every $\boldsymbol{x} \in M^{(0)}$, we define the time-augmented version of $\boldsymbol{x}$ as $\bar{\boldsymbol{x}}(t) = (t, \boldsymbol{x}_t) \in \mathbb{R}^{d+1}$. Then the time-augmented signature feature*[3]

$$\bar{\Phi}_{Sig} : M^{(0)} \to H \subset T((\mathbb{R}^{d+1})), \quad \boldsymbol{x} \mapsto \Phi_{Sig}(\bar{\boldsymbol{x}})$$

has the following properties:

(i) $\bar{\Phi}_{Sig}$ is characteristic to the space of signed measures on $M^{(0)}$, i.e., for signed measures μ, ν on $M^{(0)}$ we have

$$\mu = \nu \iff \mathbb{E}_{X\sim\mu}[\bar{\Phi}_{Sig}(X)] = \mathbb{E}_{Y\sim\nu}[\bar{\Phi}_{Sig}(Y)].$$

(ii) For a sequence of $\mathcal{K}$-valued stochastic processes $X, X^1, X^2, \dots$, it holds that

$$\left(\forall f \in C_b(M^{(0)}),\ \lim_{n\to\infty} \mathbb{E}[f(X^n)] = \mathbb{E}[f(X)]\right) \iff \lim_{n\to\infty} \mathbb{E}[\bar{\Phi}_{Sig}(X^n)] = \mathbb{E}[\bar{\Phi}_{Sig}(X)].$$

[3] Here the Hilbert space H is defined as in (15) by replacing $\mathbb{R}^d$ with $\mathbb{R}^{d+1}$.

In other words, the (time-augmented) signature metric $d_{Sig}^{(0)} : \mathcal{P}(M^{(0)}) \times \mathcal{P}(M^{(0)}) \to \mathbb{R}_+$ *given by*

$$d_{Sig}^{(0)}(\mu, \nu) := \|\mathbb{E}_{X\sim\mu}[\bar{\Phi}_{Sig}(X)] - \mathbb{E}_{Y\sim\nu}[\bar{\Phi}_{Sig}(Y)]\|_H,$$

metrizes the weak topology on $\mathcal{P}(M^{(0)})$.

Remark 2.3

- In order to ensure the above characteristicness, it is necessary to consider the signature of the time-augmented path $\bar{\boldsymbol{x}}_t = (t, \boldsymbol{x}_t) \in \mathbb{R}^{d+1}$, see e.g. [31] for a detailed discussion on this issue. From now on, to keep the notations as simple as possible, we will not distinguish the signature feature from their time-augmented version, and will use the notations Φ_{Sig} instead of $\bar{\Phi}_{\mathrm{Sig}}$ for time-augmented signatures.
- As in the static case, one can use any tensor normalizing Λ to define a *robust signature* by $\Lambda \circ \Phi_{\mathrm{Sig}}$. Using this robust signature we can show that all of the above results hold for general $\mathbb{R}^d$-valued paths and drop the compact assumption, see [31]. To avoid unnecessary technicalities, we will stick to signature features for paths taking values in the compact subset $\mathcal{K}$ of $\mathbb{R}^d$ in the present article.

Now, let us think of an interesting question:

(Q) Can we adopt the signature methodology to define the moment generating function and a signature-type metric as in Theorem 2.2 to characterize the rank-r adapted weak topology on the space of filtered processes FP, for any rank $r \in \mathbb{N}_0$?

The Rank $r = 1$ Case

As the rank-0 adapted weak topology on FP is induced by the weak topology on $\mathcal{P}(M^{(0)})$, Theorem 2.2 shows that the signature feature Φ_{Sig} and the signature metric d_{Sig} achieve this goal when $r = 0$. Now, let us proceed with $r = 1$. According to the definition, the rank-1 adapted weak topology $\mathcal{T}^{(1)}$ is the initial topology induced by the law of the rank 1-prediction process $\mathrm{pp}^{(1)}(\mathbf{X}) = \mathrm{law}(X|\mathcal{F}_t^{\mathbf{X}})_{t\in I} \in M^{(1)}$ with respect to the weak topology on $\mathcal{P}(M^{(1)})$ where $M^{(1)} = (I \to \mathcal{P}(M^{(0)}))$. Therefore, to establish a signature-type metric for the rank-1 adapted weak topology, a natural idea is to define the signature feature for the rank 1-prediction process, which can be viewed as an $M^{(1)}$-valued random variable. The only technical issue here is that the paths $\boldsymbol{p} = (\boldsymbol{p}_t)_{t\in I}$ in $M^{(1)}$ are *measure-valued paths*, and it is not clear how to define iterated integrals for paths with values in a non-linear space (here, the non-linear space is $\mathcal{P}(M^{(0)})$).[4]

---

[4] Although $\mathcal{P}(M^{(0)})$ is a subset of the linear space of all signed measures on $M^{(0)}$, it is quite hard to compute the resulting signature explicitly if one uses the associated linear structure to define the iterated integrals of measure-valued paths.

One natural way to circumvent this difficulty is to transform a measure-valued path into a path taking values in a linear space, then use the iterated integrals to define the signature feature for the latter one, and use it as the signature for the original measure-valued path. In our situation, we can use the signature feature for $\mathbb{R}^d$-valued paths to achieve such a transformation. More precisely, for a given measure-valued path $\boldsymbol{p} = (\boldsymbol{p}_t)_{t\in I} \in M^{(1)}$, for each $t \in I$ we define $\tilde{\boldsymbol{p}}_t$ as the Bochner integral of the H-valued mapping Φ_{Sig}:

$$\tilde{\boldsymbol{p}}_t = \int_{\boldsymbol{x}\in M_0} \Phi_{\text{Sig}}(\boldsymbol{x})\boldsymbol{p}_t(d\boldsymbol{x}) \in H.$$

By a slight abuse of notation, we will still use $\tilde{\boldsymbol{p}}$ to denote the time-augmented version of the path $\tilde{\boldsymbol{p}}$, i.e., $\tilde{\boldsymbol{p}}_t = (t, \int_{\boldsymbol{x}\in M_0} \Phi_{\text{Sig}}(\boldsymbol{x})\boldsymbol{p}_t(d\boldsymbol{x})) \in \mathbb{R} \oplus H$ for each $t \in I$. Then, similar to $\mathbb{R}^d$-valued paths, we identify this discrete-time H-valued path with the piecewise-linear H-valued path $\tilde{\boldsymbol{p}}_t = (i-t)\tilde{\boldsymbol{p}}_{i-1} + (t-(i-1))\tilde{\boldsymbol{p}}_i, \quad t \in [i-1, i]$, which belongs to the space of H-valued path on $[0, T]$ with bounded variation. Subsequently, we can define iterated integrals and form a (time-augmented) signature for $\tilde{\boldsymbol{p}} = (\tilde{\boldsymbol{p}}_t)_{t\in I}$ by

$$\Phi_{\text{Sig}}(\tilde{\boldsymbol{p}}) = \left(1, \int d\tilde{\boldsymbol{p}}, \int d\tilde{\boldsymbol{p}}^{\otimes 2}, \dots\right) \in H^{(2)},$$

where $H^{(2)} \subset T((\mathbb{R} \oplus H)) = \prod_{k=0}^{\infty}(\mathbb{R} \oplus H)^{\otimes k}$ is a Hilbert space embedded in the tensor algebra $T((\mathbb{R} \oplus H))$ over the Hilbert space $\mathbb{R} \oplus H$, whose inner product $\langle \cdot, \cdot \rangle_{H^{(2)}}$ is induced by the inner product $\langle \cdot, \cdot \rangle_{\mathbb{R}\oplus H}$ on $\mathbb{R} \oplus H$. For a complete description of $H^{(2)}$ we refer to [47, Appendix A] or [28, Appendix B]. Here we remark that the superscript (2) of $H^{(2)}$ means that $H^{(2)}$ is formed by taking the tensor algebra over vector space twice, i.e., $H^{(2)}$ is contained in the tensor algebra over another tensor algebra $H \subset T((\mathbb{R}^d))$.

Definition 2.4 The $H^{(2)}$-valued mapping

$$\Phi_{\text{Sig}}^{(1)}(\boldsymbol{p}) := \left(1, \int d\tilde{\boldsymbol{p}}, \int d\tilde{\boldsymbol{p}}^{\otimes 2}, \dots\right) \in H^{(2)}$$

is called the *rank-*1 *signature* of the measure-valued path $\boldsymbol{p} = (\boldsymbol{p}_t)_{t\in I} \in M^{(1)}$.

Now, let $\mathbf{X} = \left(\Omega^{\mathbf{X}}, \mathcal{F}^{\mathbf{X}}, \mathbb{P}^{\mathbf{X}}, (\mathcal{F}_t^{\mathbf{X}})_{t\in I}, X\right) \in \text{FP}$ be a filtered process. Its rank 1-prediction process $\text{pp}^{(1)}(\mathbf{X}) = \text{law}(X|\mathcal{F}_t^{\mathbf{X}})$, $t \in I$, is a $M^{(1)}$-valued random variable, and it is easy to see that

$$\widetilde{\text{pp}}^{(1)}(\mathbf{X})_t := \mathbb{E}_{\mathbf{X}}[\Phi_{\text{Sig}}(X)|\mathcal{F}_t^{\mathbf{X}}], \quad t \in I$$

is an H-valued martingale, so that $\Phi_{\text{Sig}}^{(1)}(\text{pp}^{(1)}(\mathbf{X})) = \Phi_{\text{Sig}}(t \mapsto \widetilde{\text{pp}}^{(1)}(\mathbf{X})_t)$.

The following theorem justifies the above construction, since the so-defined rank-1 signature for measure-valued paths satisfies all properties (e.g., universality and characteristicness) which hold for the signature for $\mathbb{R}^d$-valued paths. For a detailed proof, we refer readers to [28, Theorem 2].

Theorem 2.5 *The rank-1 signature $\Phi^{(1)}_{Sig}$ has the following properties:*

(i) $\Phi^{(1)}_{Sig}$ is universal to $C_b(M^{(1)})$ equipped with the uniform topology, i.e., for every $f \in C_b(M^{(1)})$ and $\varepsilon > 0$, there is $\ell \in H^{(2)}$ with

$$\sup_{p \in M^{(1)}} |f(p) - \langle \ell, \Phi^{(1)}_{Sig}(p)\rangle_{H^{(2)}}| \leq \varepsilon.$$

(ii) $\Phi^{(1)}_{Sig}$ is characteristic to the space of signed measures on $M^{(1)}$. In particular, two filtered processes $\mathbf{X}$ and $\mathbf{Y}$ have the same rank-1 adapted distribution if and only if

$$\mathbb{E}_{\mathbf{X}}[\Phi^{(1)}_{Sig}(\mathrm{pp}^{(1)}(\mathbf{X}))] = \mathbb{E}_{\mathbf{Y}}[\Phi^{(1)}_{Sig}(\mathrm{pp}^{(1)}(\mathbf{Y}))].$$

(iii) The rank-1 signature metric $d^{(1)}_{Sig} : FP \times FP \to \mathbb{R}_+$ *given by*

$$d^{(1)}_{Sig}(\mathbf{X}, \mathbf{Y}) := \|\mathbb{E}_{\mathbf{X}}[\Phi^{(1)}_{Sig}(\mathrm{pp}^{(1)}(\mathbf{X}))] - \mathbb{E}_{\mathbf{Y}}[\Phi^{(1)}_{Sig}(\mathrm{pp}^{(1)}(\mathbf{Y}))]\|_{H^{(2)}},$$

metrizes the rank-1 adapted weak topology $\mathcal{T}^{(1)}$ on FP.

The General Higher Rank Case

Thanks to the hierarchical structure of the higher-rank prediction processes, we can repeat the above construction to obtain the rank-r signature, for any $r \in \mathbb{N}_0$, which then can be used to characterize the rank-r adapted weak topology as in Theorem 2.5. Roughly speaking, as the rank-$(r+1)$ adapted weak topology is induced by the rank $(r+1)$-prediction process, which is constructed as the prediction process of the rank r-prediction process, the corresponding rank-$(r+1)$ signature $\Phi^{(r+1)}_{\mathrm{Sig}}$ is obtained by taking the signature feature of the paths formed via (taking expectations of) the rank-r signature.

The concrete procedure is described as follows, which is also reflected in the diagram below:

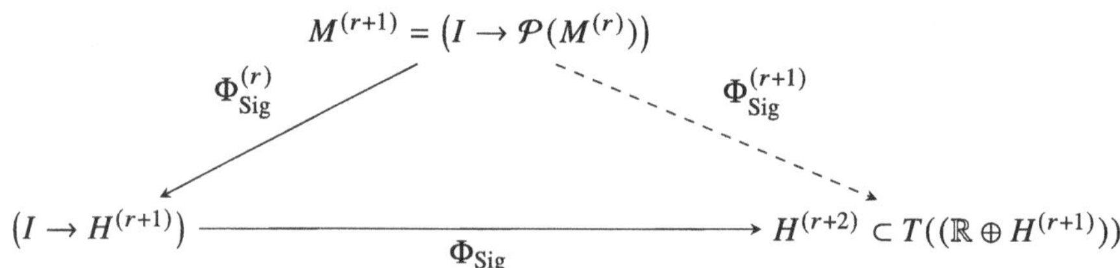

1. Assume that for $r \in \mathbb{N}_0$ we have defined the rank-r signature $\Phi^{(r)}_{\text{Sig}} : M^{(r)} \to H^{(r+1)}$ for the measure-valued path $\boldsymbol{p}^{(r)} = (\boldsymbol{p}^{(r)}_t)_{t \in I} \in M^{(r)}$, where $H^{(r+1)}$ is a Hilbert space . By convention we set $H^{(0)} := \mathbb{R}^d$ and $H^{(1)} := H$, and $\Phi^{(0)}_{\text{Sig}} = \Phi_{\text{Sig}}$ is the usual signature on $M^{(0)}$ defined as in Theorem 2.2.
2. Consider the measure-valued path space $M^{(r+1)} = (I \to \mathcal{P}(M^{(r)}))$. For a path $\boldsymbol{p}^{r+1} = (\boldsymbol{p}^{r+1}_t)_{t \in I} \in M^{(r+1)}$ and for each $t \in I$, $\boldsymbol{p}^{r+1}_t \in \mathcal{P}(M^{(r)})$ is a probability measure on the path space $M^{(r)}$. Hence, the following $H^{(r+1)}$-valued Bochner integral is well–defined :

$$\tilde{\boldsymbol{p}}^{(r+1)}_t = \int_{\boldsymbol{p}^{(r)} \in M^{(r)}} \Phi^{(r)}_{\text{Sig}}(\boldsymbol{p}^{(r)}) d\boldsymbol{p}^{r+1}_t(\boldsymbol{p}^{(r)}) \in H^{(r+1)}.$$

In particular, now we obtain a $H^{(r+1)}$-valued path $t \mapsto \tilde{\boldsymbol{p}}^{(r+1)}_t$. Again, by an abuse of notation, we will still use $\tilde{\boldsymbol{p}}^{(r+1)}$ to denote the time-augmented version of the path $\tilde{\boldsymbol{p}}^{(r+1)}$. Note that now the codomain of $\tilde{\boldsymbol{p}}^{(r+1)}_t$ is $\mathbb{R} \oplus H^{(r+1)}$.
3. Let $H^{(r+2)}$ be a Hilbert space embedded in the tensor algebra $T((\mathbb{R} \oplus H^{(r+1)})) = \prod_{k=0}^{\infty}(\mathbb{R} \oplus H^{(r+1)})^{\otimes k}$ whose inner product $\langle \cdot, \cdot \rangle_{H^{(r+2)}}$ is induced by the inner product $\langle \cdot, \cdot \rangle_{\mathbb{R} \oplus H^{(r+1)}}$. Then, use the iterated integrals of $\mathbb{R} \oplus H^{(r+1)}$-valued path $t \mapsto \tilde{\boldsymbol{p}}^{(r+1)}_t$[5] to define

$$\Phi^{(r+1)}_{\text{Sig}}(\boldsymbol{p}^{(r+1)}) = \Phi_{\text{Sig}}(\tilde{\boldsymbol{p}}^{(r+1)}) = \left(1, \int d\tilde{\boldsymbol{p}}^{(r+1)}, \int d(\tilde{\boldsymbol{p}}^{(r+1)})^{\otimes 2}, \ldots\right) \in H^{(r+2)}$$

as the rank-$(r+1)$ signature for paths in $M^{(r+1)}$.

Using the above construction, Theorem 2.5 can be further generalized to any rank $r \in \mathbb{N}_0$. It suffices to notice that for filtered process $\mathbf{X} \in \text{FP}$, the rank r-prediction process $\text{pp}^{(r)}(\mathbf{X})$ is a $M^{(r)}$-valued random variable, and the rank-r adapted weak topology on FP is the initial topology induced by the rank r-prediction process $\text{pp}^{(r)} : \text{FP} \to \mathcal{P}(M^{(r)})$.

Theorem 2.6 *For every rank $r \in \mathbb{N}_0$ the* rank-r signature $\Phi^{(r)}_{Sig}$ *has the following properties:*

(i) $\Phi^{(r)}_{Sig}$ is universal to $C_b(M^{(r)})$ equipped with the uniform topology, i.e., for every $f \in C_b(M^{(r)})$ and $\varepsilon > 0$, there is $\ell \in H^{(r+1)}$ with

$$\sup_{\boldsymbol{p}^{(r)} \in M^{(r)}} |f(\boldsymbol{p}^{(r)}) - \langle \ell, \Phi^{(r)}_{Sig}(\boldsymbol{p}^{(r)}) \rangle_{H^{(r+1)}}| \leq \varepsilon.$$

[5] As before, we use the linear interpolation to extend the discrete-time path $\tilde{\boldsymbol{p}}^r$ to a piecewise path on $[0, T]$.

(ii) $\Phi_{Sig}^{(r)}$ *is characteristic to the space of signed measures on* $M^{(r)}$*. In particular, two filtered processes* $\mathbf{X}$ *and* $\mathbf{Y}$ *have the same rank-r adapted distribution if and only if*

$$\mathbb{E}_{\mathbf{X}}[\Phi_{Sig}^{(r)}(\mathrm{pp}^{(r)}(\mathbf{X}))] = \mathbb{E}_{\mathbf{Y}}[\Phi_{Sig}^{(r)}(\mathrm{pp}^{(r)}(\mathbf{Y}))],$$

that is, they have the same rank r-expected signature.

(iii) The rank-r signature metric $d_{Sig}^{(r)} : FP \times FP \to \mathbb{R}_+$ *given by*

$$d_{Sig}^{(r)}(\mathbf{X}, \mathbf{Y}) := \|\mathbb{E}_{\mathbf{X}}[\Phi_{Sig}^{(r)}(\mathrm{pp}^{(r)}(\mathbf{X}))] - \mathbb{E}_{\mathbf{Y}}[\Phi_{Sig}^{(r)}(\mathrm{pp}^{(r)}(\mathbf{Y}))]\|_{H^{(r+1)}},$$

metrizes the rank-r adapted weak topology $\mathcal{T}^{(r)}$ *on FP.*

See [28, Theorem 4 and Remark 7] for a detailed proof of Theorem 2.6.

Remark 2.7 Again, by composing the rank-r signature with a tensor normalization Λ we can generalize the above results to general filtered processes taking values in $\mathbb{R}^d$ and drop the compactness assumption. For more details, we refer to [28, Theorem 2, Remark 7].

2.2 Higher-Rank (Expected) Signatures and Kernel Learning

From the construction of higher-rank signatures in the last section, we see that the rank-r signature $\Phi_{\mathrm{Sig}}^{(r)}$ takes values in a Hilbert space $H^{(r+1)}$, for $r \in \mathbb{N}_0$. The inner product structure on $H^{(r+1)}$ naturally allows us to use a kernel learning approach to handle $\Phi_{\mathrm{Sig}}^{(r)}$ for any $r \geq 0$, exactly as the classical signature kernel learning method (corresponding to the case $r = 0$) discussed in [57, Section 7]; in particular, one can apply kernel learning techniques to compute the higher-rank signature metrics, which will be reported in the next subsection.

Note that in the sequel we implicitly assume that all paths are time-augmented when taking signatures of them.

First let us have a quick review of the connection between the usual signature Φ_{Sig} and kernel learning, which is discussed in chapter "The Signature Kernel" systematically. As before, let $\Phi_{\mathrm{Sig}} : M^{(0)} \to H$ denote the robust signature feature mapping $\mathbb{R}^d$-valued path to a vector in the Hilbert space H. Then we can obtain an associated Reproducing Kernel Hilbert Space[6] (RKHS) over $M^{(0)}$ induced by Φ_{Sig} and $\langle \cdot, \cdot \rangle_H$, which will be denoted by $\boldsymbol{H}_{\mathrm{Sig}}$, together with its kernel

$$\boldsymbol{k}_{\mathrm{Sig}}(\boldsymbol{x}, \boldsymbol{y}) = \langle \Phi_{\mathrm{Sig}}(\boldsymbol{x}), \Phi_{\mathrm{Sig}}(\boldsymbol{y}) \rangle_H, \quad \boldsymbol{x}, \boldsymbol{y} \in M^{(0)}.$$

[6] For the definition of RKHS, see chapter "The Signature Kernel".

In this case, the reproducing kernel induced by Φ_{Sig} is simply given by

$$\boldsymbol{k}_{\text{Sig}} : M^{(0)} \to \boldsymbol{H}_{\text{Sig}}, \quad \boldsymbol{x} \mapsto \boldsymbol{k}_{\text{Sig}}(\boldsymbol{x}, \cdot)$$

and the inner product on the RKHS $\boldsymbol{H}_{\text{Sig}}$ satisfies that

$$\langle \boldsymbol{k}_{\text{Sig}}(\boldsymbol{x}), \boldsymbol{k}_{\text{Sig}}(\boldsymbol{y}) \rangle_{\boldsymbol{H}_{\text{Sig}}} = \boldsymbol{k}_{\text{Sig}}(\boldsymbol{x}, \boldsymbol{y}) = \langle \Phi_{\text{Sig}}(\boldsymbol{x}), \Phi_{\text{Sig}}(\boldsymbol{y}) \rangle_H.$$

Moreover, we have $\boldsymbol{H}_{\text{Sig}} = \{f : M^{(0)} \to \mathbb{R} : f(\boldsymbol{x}) = \langle h, \Phi_{\text{Sig}}(\boldsymbol{x}) \rangle_H, h \in H\}$. In fact, it is not hard to check that these two Hilbert spaces H and $\boldsymbol{H}_{\text{Sig}}$ are isometric by considering the mapping $h \in H \mapsto f(\cdot) = \langle h, \Phi_{\text{Sig}}(\cdot) \rangle_H \in \boldsymbol{H}_{\text{Sig}}$; in other words, the function space $\boldsymbol{H}_{\text{Sig}}$ just provides an alternative way of thinking about the Hilbert space H embedded in the tensor algebra.

Now, for any $r \geq 0$, we can define the so-called *rank-r signature kernel* $\boldsymbol{k}^{(r)}_{\text{Sig}} : M^{(r)} \to \boldsymbol{H}^{(r)}_{\text{Sig}}$ for some RKHS $\boldsymbol{H}^{(r)}_{\text{Sig}}$ (where we set $\boldsymbol{k}^{(0)}_{\text{Sig}} = \boldsymbol{k}_{\text{Sig}}$ and $\boldsymbol{H}^{(0)}_{\text{Sig}} = \boldsymbol{H}_{\text{Sig}}$) in the same spirit as we did before: invoking that the rank-r signature $\Phi^{(r)}_{\text{Sig}}$ on $M^{(r)}$ takes values in a Hilbert space $H^{(r+1)}$, We therefore obtain the associated reproducing kernel

$$\boldsymbol{k}^{(r)}_{\text{Sig}} : M^{(r)} \to \boldsymbol{H}^{(r)}_{\text{Sig}}, \quad \boldsymbol{p}^{(r)} \mapsto \langle \Phi^{(r)}_{\text{Sig}}(\boldsymbol{p}^{(r)}), \Phi^{(r)}_{\text{Sig}}(\cdot) \rangle_{H^{(r+1)}}$$

as the rank-r signature kernel, and note that

$$\boldsymbol{k}^{(r)}_{\text{Sig}}(\boldsymbol{p}^{(r)}, \boldsymbol{q}^{(r)}) = \langle \Phi^{(r)}_{\text{Sig}}(\boldsymbol{p}^{(r)}), \Phi^{(r)}_{\text{Sig}}(\boldsymbol{q}^{(r)}) \rangle_{H^{(r+1)}}$$

for any $\boldsymbol{p}^{(r)}, \boldsymbol{q}^{(r)} \in M^{(r)}$.

Recall that (see e.g. [31, Section 6]) a natural distance for probability measures on any topological space $\mathcal{X}$ is given by fixing a class of functions $\mathcal{G} \subset \mathbb{R}^{\mathcal{X}}$ and define the integral probability metric between $\mu, \nu \in \mathcal{P}(\mathcal{X})$ as

$$d(\mu, \nu) = \sup_{f \in \mathcal{G}} |\mathbb{E}_{X \sim \mu}[f(X)] - \mathbb{E}_{Y \sim \nu}[f(Y)]|.$$

For instance, if $\mathcal{G}$ denotes the space of Lipschitz functions on $\mathcal{X}$ with Lipschitz constant 1, then the above integral probability metric is exactly the first order Wasserstein distance. An insight from the theory of kernel learning is that if $\mathcal{G}$ is the unit ball of a RKHS $(\boldsymbol{H}, \boldsymbol{k})$, then the above integral probability metric is called a Maximum Mean Discrepancy (MMD), and the reproducing property implies that

$$d^2(\mu, \nu) = \mathbb{E}[\boldsymbol{k}(X, X')] - 2\mathbb{E}[\boldsymbol{k}(X, Y)] + \mathbb{E}[\boldsymbol{k}(Y, Y')],$$

where X, X', Y, Y' are independent with $X, X' \sim \mu$ and $Y, Y' \sim \nu$.

The next proposition shows that higher-rank signature metrics $d_{\mathrm{Sig}}^{(r)}(\cdot,\cdot)$ are MMDs induced by the higher-rank signature kernels $\boldsymbol{k}_{\mathrm{Sig}}^{(r)}(\cdot,\cdot)$ defined as above, which is straight forward to prove, and we leave it as an exercise for those curious readers.

Proposition 2.8 *For any $r \in \mathbb{N}_0$, the rank-r signature kernel MMD $\mathcal{D}_{Sig}^{(r)}$ on FP defined by*

$$\mathcal{D}_{Sig}^{(r)}(\mathbf{X},\mathbf{Y}) := \sup_{f \in \boldsymbol{H}_{Sig}^{(r)} : \|f\|_{\boldsymbol{H}_{Sig}^{(r)}}=1} |\mathbb{E}_{\mathbf{X}}[f(\mathrm{pp}^{(r)}(\mathbf{X}))] - \mathbb{E}_{\mathbf{Y}}[f(\mathrm{pp}^{(r)}(\mathbf{Y}))]|$$

satisfies $\mathcal{D}_{Sig}^{(r)}(\mathbf{X},\mathbf{Y}) = d_{Sig}^{(r)}(\mathbf{X},\mathbf{Y})$ and

$$\mathcal{D}_{Sig}^{(r)}(\mathbf{X},\mathbf{Y})^2 = \mathbb{E}[\boldsymbol{k}_{Sig}^{(r)}(Z,Z')] - 2\mathbb{E}[\boldsymbol{k}_{Sig}^{(r)}(Z,\hat{Z})] + \mathbb{E}[\boldsymbol{k}_{Sig}^{(r)}(\hat{Z},\hat{Z}')],$$

where $Z, Z', \hat{Z}, \hat{Z}'$ are independent with $Z, Z' \sim \mathrm{pp}^{(r)}(\mathbf{X})$ and $\hat{Z}, \hat{Z}' \sim \mathrm{pp}^{(r)}(\mathbf{Y})$.

In the next subsection we will show that the above expansion of $\mathcal{D}_{\mathrm{Sig}}^{(r)}(\mathbf{X},\mathbf{Y})^2$ is simply estimated from finite samples of law(X) and law(Y) as the higher-rank signature kernel $\boldsymbol{k}_{\mathrm{Sig}}^{(r)}(\mathbf{X},\mathbf{Y})$ is relatively cheap to evaluate, provided all $\mathbf{X}$ and $\mathbf{Y}$ are endowed with their natural filtration, respectively.

2.3 Algorithms for Computing Higher-Rank (Expected) Signature

In the last subsection we have seen that the higher-rank signature metric (equivalently speaking, the higher-rank signature kernel MMD) metrizes the higher-rank adapted weak topology $\mathcal{T}^{(r)}$ on the space of filtered processes FP. Although these theoretical results look satisfactory, a numerical computation of these distances is admittedly a highly challenging task, as the Hilbert spaces $H^{(r)}$ and $\boldsymbol{H}_{\mathrm{Sig}}^{(r)}$ are infinite dimensional with very complicated algebraic structure. In this subsection we will introduce two possible algorithms for computing these higher-rank signature distances, which have been verified in some numerical experiments, albeit one can indeed improve them further.

To simplify our notations and give a clearer description of the algorithm, let us make throughout the following assumptions:

(A2) All filtered processes are equipped with the natural filtration, that is, we focus on the subspace $\mathrm{FP}_{\mathrm{plain}}$ of FP given by

$$\mathrm{FP}_{\mathrm{plain}} = \left\{ \mathbf{X} = \left(\boldsymbol{\Omega}^{\mathbf{X}}, \mathcal{F}^{\mathbf{X}}, \mathbb{P}^{\mathbf{X}}, (\mathcal{F}_t^{\mathbf{X}})_{t \in I}, X \right) \in \mathrm{FP} : \mathcal{F}_t^{\mathbf{X}} = \sigma(X_s, s \leq t) \right\};$$

(A3) $r = 1$.

Method 1: A Backward Recursion

We will apply a backward recursion method to derive algorithms that efficiently compute the signature $\Phi_{\mathrm{Sig}}(\mathbf{X})$ and the rank-1 signature $\Phi_{\mathrm{Sig}}^{(1)}(\mathbf{X})$, provided $\mathbf{X} \in \mathrm{FP}_{\mathrm{plain}}$ is a nowhere recombining (that is, its value at time t uniquely determines $X_1, \ldots, X_{t-1}$) Markov chain (with finite states). We refer readers to [28, Section 5] for a detailed description of this algorithm and here we only summarize the essential idea and main steps.

Recall that higher-rank signatures take values in some tensor algebras. Therefore, to compute them we need the following lemma concerning dynamic programming in an algebra (see [28, Lemma 3]).

Lemma 2.9 *Let A be an (Banach) algebra and $Z = (Z_t)_{t \in I}$ be a nowhere recombining Markov chain taking values in A with finite states. The function*

$$u_t(a) = \mathbb{E}[Z_{t+1} \ldots Z_T | Z_t = a]$$

satisfies for every $t \in I$ and $a \in A$ the recursion

$$u_t(a) = \sum_{b \in A} \mathbb{P}(Z_{t+1} \ldots Z_T | Z_t = a) b u_{t+1}(b),$$

where $Z_{t+1} \ldots Z_T$ and $bu_{t+1}(b)$ are calculated by using the multiplication defined on the algebra A.

If $\mathbf{X} \in \mathrm{FP}_{\mathrm{plain}}$ is a nowhere recombining Markov chain, then the signature process $Z_t = \Phi_{\mathrm{Sig}}(\mathbf{X}|_{[0,t]})$ can also be viewed as a nowhere recombining Markov chain taking values in the tensor algebra $T((\mathbb{R}^d))$. As $I = \{1, \ldots, T\}$ is discrete, it is well–known that

$$\Phi_{\mathrm{Sig}}(X) = \exp(\Delta_1 X) \otimes \ldots \otimes \exp(\Delta_T X)$$

where $\otimes$ is the tensor product on $T((\mathbb{R}^d))$, exp is the tensor exponential on $T((\mathbb{R}^d))$ such that $\exp(x) = \sum_{n=0}^{\infty} \frac{x^{\otimes n}}{n!}$ and $\Delta_t X = X_t - X_{t-1}$. Hence, Lemma 2.9 hints at an efficient way to compute the expected signature $\mathbb{E}[\Phi_{\mathrm{Sig}}(\mathbf{X})] := u_0(X_0)$ (we set $X_0 := 0$) since the value function u satisfies the recursion

$$u_t(x) = \sum_{y \in \mathbb{R}^d} \mathbb{P}(X_{t+1} = y | X_t = x) \exp(y - x) \otimes u_{t+1}(y),$$

Algorithm 1 Pseudo-code for $\mathbb{E}[\Phi_{\text{Sig}}(\mathbf{X})]$

```
1:  Input: A Markov chain X represented as a rooted tree with root r
2:  procedure ExpSig_0(a)
3:      if a.children is empty then
4:          return 1
5:      end if
6:      sum ← 0
7:      for b in a.children do
8:          sum ← sum+p(a, b) exp{b − a} ⊗ ExpSig_0(b)
9:      end for
10:     return sum
11: end procedure
12: Output: ExpSig_0(r) = E[Φ_Sig(X)]
```

The computation of the rank-1 signature $\Phi_{\text{Sig}}^{(1)}(\mathbf{X})$ follows along the same lines. We now apply Lemma 2.9 with $A = T((H))$ (recall that H denotes the target space of $\Phi_{\text{Sig}}(X)$) and $Z_t = \exp(\Delta_t \widetilde{\text{pp}}^{(1)}(\mathbf{X})) = \exp(\mathbb{E}[\Phi_{\text{Sig}}(X)|X_t] - \mathbb{E}[\Phi_{\text{Sig}}(X)|X_{t-1}])$. Here, we note that the value of $Z_t = a$ depends on $X_t = x$ and $X_{t-1} = x'$; and, since X is nowhere recombining by assumption, the path segment $(X_s)_{s=1}^t$ such that $X_t = x$ is uniquely determined, which implies that the value of $X_{t-1} = x'$ is actually uniquely determined by $X_t = x$. Based on this observation, we can see that the function $v_t(a) := \mathbb{E}[Z_{t+1} \cdots Z_T | Z_t = a]$ in Lemma 2.9 can actually be viewed as a function of $X_t = x$ and therefore satisfies the recursion

$$v_t(x) = \sum_{y \in \mathbb{R}^d} \mathbb{P}(X_{t+1} = y | X_t = x) \cdot \exp\Big\{\mathbb{E}[\Phi_{\text{Sig}}(X)|X_{t+1} = y] - \mathbb{E}[\Phi_{\text{Sig}}(X)|X_t = x]\Big\} * v_{t+1}(y),$$

where we used the symbol $*$ to denote the tensor product on $T((H))$. Of course, the exponential appearing in the above expression is taken with respect to the operation $*$, that is, for $h \in H$, $\exp(h) = \sum_{n=0}^{\infty} \frac{h^{*n}}{n!}$. One can readily check that $v_0(X_0) = \mathbb{E}[\Phi_{\text{Sig}}^{(1)}(\mathbf{X})]$ (with $X_0 := 0$ as usual) is the quantity we want to compute.

This recursion is more involved as it requires two evaluations of $\mathbb{E}[\Phi_{\text{Sig}}(X)|X_t]$ at every step. However, if we assume that the process $\mathbf{X}$ is nowhere recombining we can rewrite this as the following system

$$v_t(x) = \sum_{y \in \mathbb{R}^d} \mathbb{P}(X_{t+1} = y | X_t = x) \cdot \exp\big\{\Phi_{\text{Sig}}(X|X_t{=}y) \otimes u_{t+1}(y) - \Phi_{\text{Sig}}(X|X_{t-1}{=}x) \otimes u_t(x)\big\} * v_{t+1}(y),$$

$$u_t(x) = \sum_{y \in \mathbb{R}^d} \mathbb{P}(X_{t+1} = y | X_t = x) \exp\big\{y - x\big\} \otimes u_{t+1}(y),$$

where $\Phi_{\text{Sig}}(X|X_t = y)$ denotes the signature of the path segment $(X_s)_{s=1}^t$ such that $X_t = y$, which is well defined since X is nowhere recombining by assumption. Unfortunately $v_t(x)$ depends on both u_t and u_{t+1} and since multiplication in $T((H))$ is non-commutative, there is no way to separate the two dependencies. Because of this $u_t(x)$ needs to be computed before $v_t(x)$ and the recursion is best solved by caching the relevant values at every function call. This approach is outlined in Algorithm 2.

Algorithm 2 Pseudo-code for $\mathbb{E}[\Phi^{(1)}_{\text{Sig}}(\mathbf{X})]$

1: **Input:** A Markov chain X represented as a rooted tree with root r, $s(a)$ denotes the signature of the sample path of X that ends at a.
2: **procedure** $\text{ExpSig}_1(a)$
3: $\quad a_1 \leftarrow 0$
4: $\quad a_2 \leftarrow 0$
5: $\quad$ **for** b **in** $a.children$ **do**
6: $\quad\quad b_1, b_2 \leftarrow \text{ExpSig}_1(b)$
7: $\quad\quad a_1 \leftarrow a_1 + p(a, b) \exp\{b - a\} \otimes b_1$
8: $\quad$ **end for**
9: $\quad$ **for** b **in** $a.children$ **do**
10: $\quad\quad b_1, b_2 \leftarrow \text{ExpSig}_1(b)$
11: $\quad\quad a_2 \leftarrow a_2 + p(a, b) \exp\{s(b) \otimes b_1 - s(a) \otimes a_1\} * b_2$
12: $\quad$ **end for**
13: $\quad$ **return** a_1, a_2
14: **end procedure**
15: **Output:** $\text{ExpSig}_1(r) = \mathbb{E}[\Phi^{(1)}_{\text{Sig}}(\mathbf{X})]$

The above algorithm based on the backward recursion method provides a practical way to compute the higher-rank signature numerically. However, in this algorithm one needs to *truncate* the signature features Φ_{Sig} in every step. However, it is a notorious property for truncated signatures that the dimension grows exponentially as the truncation level increases: $\Phi_{\text{Sig},k}(\boldsymbol{x}) = (1, \int \boldsymbol{x}, \ldots, \int d\boldsymbol{x}^{\otimes k}) \in (\mathbb{R}^d)^{1+d+\ldots+d^k}$, and the situation gets worse if one truncates the higher-rank signatures $\Phi^{(1)}_{\text{Sig}}$. On the other hand, in many applications (see Sect. 3) we are more interested in the higher-rank signature distance between two filtered processes instead of the higher-rank robust signature itself, therefore we can actually take use of the inner product structure of the target spaces of higher-rank robust signatures to design a more efficient algorithm to compute these distances, which are $\mathbb{R}$-valued objects, as the next paragraph will show.

Method 2: Conditional KME and Signature Kernel PDE Trick

The aim of this algorithm is to compute the rank-1 signature kernel MMD $\mathcal{D}^{(1)}_{\text{Sig}}(\mathbf{X}, \mathbf{Y})$ for $\mathbf{X}, \mathbf{Y} \in \text{FP}_{\text{plain}}$. This algorithm can be naturally extended to compute higher-rank signature MMD. In contrast to the first algorithm, we will employ here techniques from kernel learning extensively. The statements below are mainly borrowed from [73] and [47].

The starting point of this algorithm is the property from Proposition 2.8 that the square of $\mathcal{D}^{(1)}_{\text{Sig}}(\mathbf{X}, \mathbf{Y})$ admits the representation

$$\mathcal{D}^{(1)}_{\text{Sig}}(\mathbf{X}, \mathbf{Y})^2 = \mathbb{E}[\boldsymbol{k}^{(1)}_{\text{Sig}}(Z, Z')] - 2\mathbb{E}[\boldsymbol{k}^{(1)}_{\text{Sig}}(Z, \hat{Z})] + \mathbb{E}[\boldsymbol{k}^{(1)}_{\text{Sig}}(\hat{Z}, \hat{Z}')],$$

where $Z, Z', \hat{Z}, \hat{Z}'$ are independent with $Z, Z' \sim \text{pp}^{(1)}(\mathbf{X})$ and $\hat{Z}, \hat{Z}' \sim \text{pp}^{(1)}(\mathbf{Y})$.

Thus, to compute $\mathcal{D}^{(1)}_{\text{Sig}}(\mathbf{X}, \mathbf{Y})$ it suffices to compute the three expectations of inner product on the right hand side of the above equation separately. To avoid notational excess, we assume from now on without loss of generality that $\mathbf{X}$ and $\mathbf{Y}$ live on a common probability space such that $Z = \text{pp}^{(1)}(\mathbf{X})$ and $\hat{Z} = \text{pp}^{(1)}(\mathbf{Y})$ are independent. Let us compute $\mathbb{E}[\boldsymbol{k}^{(1)}_{\text{Sig}}(\text{pp}^{(1)}(\mathbf{X}), \text{pp}^{(1)}(\mathbf{Y}))]$ as an example. We decompose the procedure of the numerical solution into the following four steps. For more details on this algorithm, we refer readers to [73] or [47]. Now let us summarize the main ideas of this algorithm in the following informal style.

Informal Description

Recall that what we need to compute is the expectation of the rank-1 signature kernel on prediction processes $\text{pp}^{(1)}(\mathbf{X})$ and $\text{pp}^{(1)}(\mathbf{Y})$, namely $\mathbb{E}[\boldsymbol{k}^{(1)}_{\text{Sig}}(\text{pp}^{(1)}(\mathbf{X}), \text{pp}^{(1)}(\mathbf{Y}))]$. Suppose samples $\hat{x}^i \sim \text{law}(\mathbf{X})$, $i = 1, \ldots, M$ and $\hat{y}^j \sim \text{law}(\mathbf{Y})$, $j = 1, \ldots, N$ are given, we can indeed use the empirical measures $\mu^M := \frac{1}{M}\sum_{i=1}^{M} \delta_{\hat{x}^i}$ and $\nu^N := \frac{1}{N}\sum_{j=1}^{N} \delta_{\hat{y}^j}$ to construct an empirical estimator

$$\mathbb{E}_{\mu^M \otimes \nu^N}[\boldsymbol{k}^{(1)}_{\text{Sig}}(\text{pp}^{(1)}(\mathbf{X}), \text{pp}^{(1)}(\mathbf{Y}))] = \frac{1}{NM}\sum_{i=1}^{M}\sum_{j=1}^{N} \boldsymbol{k}^{(1)}_{\text{Sig}}(\text{pp}^{(1)}(\mathbf{X})(\hat{x}^i), \text{pp}^{(1)}(\mathbf{Y})(\hat{y}^j))$$

for $\mathbb{E}[\boldsymbol{k}^{(1)}_{\text{Sig}}(\text{pp}^{(1)}(\mathbf{X}), \text{pp}^{(1)}(\mathbf{Y}))]$. Based on this observation, our task boils down to the numerical computation of the inner product $\boldsymbol{k}^{(1)}_{\text{Sig}}(\text{pp}^{(1)}(\mathbf{X})(\hat{x}^i), \text{pp}^{(1)}(\mathbf{Y})(\hat{y}^j))$ for each pair (i, j). A well-known result is that for $\mathbb{R}^d$-valued piecewise smooth paths x and y defined on $[0, T]$ (recall that in our case all paths are discrete on $0, 1, \ldots, T$, and we view them as piecewise-linear paths on $[0, T]$ by linear interpolation), the signature kernel $\boldsymbol{k}_{\text{Sig}}(x, y)$ is the terminal value $v^{x,y}(T, T)$ of the unique solution $v^{x,y}(\cdot, \cdot)$ to the Goursat problem

$$\frac{\partial^2 v^{x,y}(s, t)}{\partial s \partial t} = \left\langle \frac{\partial}{\partial t} x_t, \frac{\partial}{\partial s} y_s \right\rangle_{\mathbb{R}^d} v^{x,y}(s, t),$$

see e.g. [30] or [57, Section 7.2, PDE Solver]. Actually, by the very definition of the rank-1 signature kernel $\boldsymbol{k}^{(1)}_{\text{Sig}}(\cdot, \cdot)$, which is essentially induced by the signature of an Hilbert space H-valued paths, using the same computation as for the above classical case, it is not hard to check that $\boldsymbol{k}^{(1)}_{\text{Sig}}(\text{pp}^{(1)}(\mathbf{X})(\hat{x}^i), \text{pp}^{(1)}(\mathbf{Y})(\hat{y}^j))$ is the terminal value $u^{i,j}(T, T)$ of the unique solution $u^{i,j}$ to the Goursat problem

$$\frac{\partial^2 u^{i,j}(s,t)}{\partial s \partial t} = \big\langle \frac{\partial}{\partial t}\widetilde{\mathrm{pp}}^{(1)}(\mathbf{X})(\hat{x}^i), \frac{\partial}{\partial s}\widetilde{\mathrm{pp}}^{(1)}(\mathbf{Y})(\hat{y}^j)\big\rangle_H u^{i,j}(s,t),$$

where $\widetilde{\mathrm{pp}}^{(1)}(\mathbf{X}) = \mathbb{E}_{\mathbf{X}}[\Phi_{\mathrm{Sig}}(X)|\mathcal{F}_t^{\mathbf{X}}]$ and $\widetilde{\mathrm{pp}}^{(1)}(\mathbf{Y}) = \mathbb{E}_{\mathbf{Y}}[\Phi_{\mathrm{Sig}}(Y)|\mathcal{F}_t^{\mathbf{Y}}]$ are conditional expectations of H-valued random variables $\Phi_{\mathrm{Sig}}(X)$ and $\Phi_{\mathrm{Sig}}(Y)$ respectively, which can be viewed as functions on the path space $(I \to \mathbb{R}^d)$, as the filtrations $\mathcal{F}^{\mathbf{X}}$ and $\mathcal{F}^{\mathbf{Y}}$ are natural. Clearly, if we can compute the inner product $\big\langle \frac{\partial}{\partial t}\widetilde{\mathrm{pp}}^{(1)}(\mathbf{X})(\hat{x}^i), \frac{\partial}{\partial s}\widetilde{\mathrm{pp}}^{(1)}(\mathbf{Y})(\hat{y}^j)\big\rangle_H$, then we would be able to solve the above Goursat type PDE, and then get the desired quantity $\boldsymbol{k}^{(1)}_{\mathrm{Sig}}(\mathrm{pp}^{(1)}(\mathbf{X})(\hat{x}^i), \mathrm{pp}^{(1)}(\mathbf{Y})(\hat{y}^j))$. Hence, our goal is to numerically approximate $\big\langle \frac{\partial}{\partial t}\widetilde{\mathrm{pp}}^{(1)}(\mathbf{X})(\hat{x}^i), \frac{\partial}{\partial s}\widetilde{\mathrm{pp}}^{(1)}(\mathbf{Y})(\hat{y}^j)\big\rangle_H$. To do this, as $\widetilde{\mathrm{pp}}_t^{(1)}(\mathbf{X}) = \mathbb{E}_{\mathbf{X}}[\Phi_{\mathrm{Sig}}(X)|\mathcal{F}_t^{\mathbf{X}}]$ and $\widetilde{\mathrm{pp}}_t^{(1)}(\mathbf{Y}) = \mathbb{E}_{\mathbf{Y}}[\Phi_{\mathrm{Sig}}(Y)|\mathcal{F}_t^{\mathbf{Y}}]$, we should of course find a consistent estimator of the above conditional expectations in terms of data $(\hat{x}^i)_{1\le i\le M}$ and $(\hat{y}^j)_{1\le j\le N}$.

From the above discussions, we can conclude that the algorithm of computing the rank-1 signature kernel MMD $\mathcal{D}^{(1)}_{\mathrm{Sig}}(\mathbf{X}, \mathbf{Y})$ consists of the following subtasks:

1. Use samples $(\hat{x}^i)_{1\le i\le M} \sim \mathrm{law}(\mathbf{X})$ and $(\hat{y}^j)_{1\le j\le N} \sim \mathrm{law}(\mathbf{Y})$ to construct empirical estimators for conditional kernel means $\widetilde{\mathrm{pp}}_t^{(1)}(\mathbf{X}) = \mathbb{E}_{\mathbf{X}}[\Phi_{\mathrm{Sig}}(X)|\mathcal{F}_t^{\mathbf{X}}]$ and $\widetilde{\mathrm{pp}}_t^{(1)}(\mathbf{Y}) = \mathbb{E}_{\mathbf{Y}}[\Phi_{\mathrm{Sig}}(Y)|\mathcal{F}_t^{\mathbf{Y}}]$. See **Step 1** and **Step 2** below.
2. Apply the empirical estimators constructed in the previous steps to approximate the inner product $\big\langle \frac{\partial}{\partial t}\widetilde{\mathrm{pp}}^{(1)}(\mathbf{X})(\hat{x}^i), \frac{\partial}{\partial s}\widetilde{\mathrm{pp}}^{(1)}(\mathbf{Y})(\hat{y}^j)\big\rangle_H$, so that one can find a numerical solution of the Goursat problem

$$\frac{\partial^2 u^{i,j}(s,t)}{\partial s \partial t} = \big\langle \frac{\partial}{\partial t}\widetilde{\mathrm{pp}}^{(1)}(\mathbf{X})(\hat{x}^i), \frac{\partial}{\partial s}\widetilde{\mathrm{pp}}^{(1)}(\mathbf{Y})(\hat{y}^j)\big\rangle_H u^{i,j}(s,t)$$

   for each pair of samples $(\hat{x}^i, \hat{y}^j)$. The resulting quantity $u^{i,j}(T,T)$ is an approximation of the rank-1 signature kernel $\boldsymbol{k}^{(1)}_{\mathrm{Sig}}(\mathrm{pp}^{(1)}(\mathbf{X})(\hat{x}^i), \mathrm{pp}^{(1)}(\mathbf{Y})(\hat{y}^j))$. From this we can build an empirical estimator $\hat{\mathcal{D}}^{(1)}_{\mathrm{Sig}}(\mathbf{X}, \mathbf{Y}) = \frac{1}{NM}\sum_{i=1}^M \sum_{j=1}^N u^{i,j}(T,T)$ for $\mathcal{D}^{(1)}_{\mathrm{Sig}}(\mathbf{X}, \mathbf{Y})$. See **Step 3** and **Step 4** below.
3. It is interesting to notice that thanks to the concrete form of the estimators of $\widetilde{\mathrm{pp}}^{(1)}(\mathbf{X}) = \mathbb{E}_{\mathbf{X}}[\Phi_{\mathrm{Sig}}(X)|\mathcal{F}_t^{\mathbf{X}}]$ and $\widetilde{\mathrm{pp}}^{(1)}(\mathbf{Y}) = \mathbb{E}_{\mathbf{Y}}[\Phi_{\mathrm{Sig}}(Y)|\mathcal{F}_t^{\mathbf{Y}}]$, the computation of the inner product $\big\langle \frac{\partial}{\partial t}\widetilde{\mathrm{pp}}^{(1)}(\mathbf{X})(\hat{x}^i), \frac{\partial}{\partial s}\widetilde{\mathrm{pp}}^{(1)}(\mathbf{Y})(\hat{y}^j)\big\rangle_H$ on the Hilbert space H can be reduced to finding numerical solutions of Goursat PDEs on $\mathbb{R}^d$, namely solving

$$\frac{\partial^2 \tilde{u}_{i,j}(s,t)}{\partial s \partial t} = \big\langle \frac{\partial}{\partial t}\hat{x}_t^i, \frac{\partial}{\partial s}\hat{y}_s^j\big\rangle_{\mathbb{R}^d}\tilde{u}^{i,j}(s,t), \quad (s,t) \in [0,T]\times[0,T].$$

   Please see **Step 4** below for the details.

Step 1: Estimation of Conditional Kernel Mean is a Minimization Problem
First, note that we have by definition

$$\boldsymbol{k}_{\mathrm{Sig}}^{(1)}(\mathrm{pp}^{(1)}(\mathbf{X}), \mathrm{pp}^{(1)}(\mathbf{Y})) = \langle \Phi_{\mathrm{Sig}}(\widetilde{\mathrm{pp}}^{(1)}(\mathbf{X})), \Phi_{\mathrm{Sig}}(\widetilde{\mathrm{pp}}^{(1)}(\mathbf{Y}))\rangle_H,$$

where $\widetilde{\mathrm{pp}}^{(1)}(\mathbf{X}) = \mathbb{E}_{\mathbf{X}}[\Phi_{\mathrm{Sig}}(X)|\mathcal{F}_t^{\mathbf{X}}]$, $t \in I$. Hence, our initial task is to estimate the conditional kernel mean $\mathbb{E}_{\mathbf{X}}[\Phi_{\mathrm{Sig}}(X)|\mathcal{F}_t^{\mathbf{X}}] \in H$ for each $t \in I$. To do this, we will adopt the *conditional kernel mean embedding* scheme proposed in [65]. Roughly speaking, as the filtration $\mathcal{F}_t^{\mathbf{X}} = \sigma(X_s, s \leq t)$ is the natural filtration generated by X, there is a measurable function $F_t : (I_t \to \mathbb{R}^d) \to H$ such that

$$\mathbb{E}_{\mathbf{X}}[\Phi_{\mathrm{Sig}}(X)|\mathcal{F}_t^{\mathbf{X}}] = F_t \circ X|_{I_t}$$

where $I_t = \{1, \ldots, t\}$, and F_t is the minimizer of the loss function $\mathbb{E}_{\mathbf{X}}[\|\Phi_{\mathrm{Sig}}(X) - F(X|_{I_t})\|_H^2]$ among all measurable functions $F \in L^2((I_t \to \mathbb{R}^d), H, \mathbb{P}^{\mathbf{X}} \circ (X|_{I_t})^{-1})$.

Step 2: Solve Minimization Problem via Empirical Data
Usually, people use the unique minimizer of the so-called regularized empirical surrogate loss with regularization parameter $\lambda_M > 0$ given by

$$\hat{\mathcal{E}}_{X|_{I_t}, M, \lambda_M}(F) := \frac{1}{M} \sum_{i=1}^{M} \|\Phi_{\mathrm{Sig}}(\hat{x}^i) - F(\hat{x}^i|_{I_t})\|_H^2 + \lambda_M \|F\|_{\mathcal{G}^t}^2$$

as a good estimator for F_t, where $\hat{x}^i$, $i = 1, \ldots, M$ are i.i.d. sample paths of law(X), and $\mathcal{G}^t \subset L^2((I_t \to \mathbb{R}^d), H, \mathbb{P}^{\mathbf{X}} \circ (X|_{I_t})^{-1})$ is a properly chosen Hilbert space which is dense in $L^2((I_t \to \mathbb{R}^d), H, \mathbb{P}^{\mathbf{X}} \circ (X|_{I_t})^{-1})$. Here we remark that such a special construction of $\mathcal{G}^t$ will significantly simplify the minimization problem of the functional $\hat{\mathcal{E}}_{X|_{I_t}, M, \lambda_M}(F)$, and the concrete construction of $\mathcal{G}^t$ is completely based on the universality of the signature.

As the next step, it can be shown that (see [65]) given samples $\hat{x}^i$, $i = 1, \ldots, M$, the unique minimizer $\hat{F}_t^{(\hat{x}^i)_{i=1,\ldots,M}} \in G^t$ (for simplicity, we write it as $\hat{F}_t$ in below) of the regularized empirical surrogate loss with regularization parameter λ_M is of the form that

$$\hat{F}_t(\cdot) = \boldsymbol{k}_{\mathrm{Sig}}^{X,t}(\cdot)(\mathbf{K}_{t,t}^X + M\lambda_M \mathrm{I}_M)^{-1} \Phi_{\mathrm{Sig}}^{X,T} \tag{16}$$

where $\boldsymbol{k}_{\mathrm{Sig}}^{X,t}(\cdot) = (\boldsymbol{k}_{\mathrm{Sig}}(\hat{x}^i|_{I_t}, \cdot))_{i=1,\ldots,M} : (I_t \to \mathbb{R}^d) \to \mathbb{R}^M$ is an $\mathbb{R}^M$-valued (continuous) function on the path space $(I_t \to \mathbb{R}^d)$, $\mathbf{K}_{t,t}^X = ([\mathbf{K}_{t,t}^X]_{i,j})_{i,j=1,\ldots,M}$ is an $M \times M$ matrix with entries $[\mathbf{K}_{t,t}^X]_{i,j} = \boldsymbol{k}_{\mathrm{Sig}}(\hat{x}^i|_{I_t}, \hat{x}^j|_{I_t}) = \langle \Phi_{\mathrm{Sig}}(\hat{x}^i), \Phi_{\mathrm{Sig}}(\hat{x}^j)\rangle_H$, $\Phi_{\mathrm{Sig}}^{X,T} = (\Phi_{\mathrm{Sig}}(\hat{x}^i))_{i=1,\ldots,M}$ is a vector in H^M and I_M is the $M \times M$-identity matrix. Clearly, for a given $x|_{I_t} \in (I_t \to \mathbb{R}^d)$, we first evaluate the function $\boldsymbol{k}_{\mathrm{Sig}}^{X,t}(\cdot)$ against it and obtain an $\mathbb{R}^M$ row vector $(\langle \Phi_{\mathrm{Sig}}(\hat{x}^i|_{I_t}), \Phi_{\mathrm{Sig}}(x|_{I_t})\rangle_H)_{i=1,\ldots,M}$, and then multiply it with the $M \times M$-matrix $(\mathbf{K}^X_{t,t} + M\lambda_M \mathrm{I}_M)^{-1}$ to get another $\mathbb{R}^M$ row vector $\boldsymbol{k}^{X,t}_{\mathrm{Sig}}(x|_{I_t})(\mathbf{K}^X_{t,t} + M\lambda_M \mathrm{I}_M)^{-1} := (v_1, \ldots, v_M)$; finally, we multiply the latter vector $(v_1, \ldots, v_M)$ with the H^M-vector $\Phi^{X,T}_{\mathrm{Sig}} = (\Phi_{\mathrm{Sig}}(\hat{x}^i))_{i=1,\ldots,M}$ to get a vector $\sum_{i=1}^M v_i \Phi_{\mathrm{Sig}}(\hat{x}^i)$ in the Hilbert space H. The above explains why $\hat{F}_t(\cdot)$ defined in (16) is an H-valued function on $(I_t \to \mathbb{R}^d)$.

Step 3: Construction of Empirical Estimator for Higher-Rank Signature MMD
Given i.i.d. sample paths $(\hat{x}^i)_{i\le M}$ of law(X), from Step 2 we can define the H-valued stochastic process $\widehat{\mathrm{pp}}^{(1)}(\mathbf{X})_t := \hat{F}_t^{(\hat{x}^i)_{i=1,\ldots,M}}(\cdot), t \in I$, which serves as an empirical estimator for $\widetilde{\mathrm{pp}}^{(1)}(\mathbf{X})$. Similarly, for a given i.i.d. sample paths $(\hat{y}^j)_{j\le N}$ of law(Y), we define the H-valued stochastic process $\widehat{\mathrm{pp}}^{(1)}(\mathbf{Y})_t = \hat{F}_t^{(\hat{y}_j)_{j=1,\ldots,N}}(\cdot), t \in I$, which is the empirical estimator for $\widehat{\mathrm{pp}}^{(1)}(\mathbf{Y})$. Then, for empirical measures $\hat{\mu}^M := \frac{1}{M}\sum_{i=1}^M \delta_{\hat{x}^i}$ and $\hat{\nu}^N := \frac{1}{N}\sum_{j=1}^N \delta_{\hat{y}^j}$, we define

$$\hat{\mathcal{D}}^{(1)}_{\mathrm{Sig}}(\mathbf{X}, \mathbf{Y})^2 := \mathbb{E}_{\hat{\mu}^M \otimes \hat{\nu}^N}[\|\Phi_{\mathrm{Sig}}(\widehat{\mathrm{pp}}^{(1)}(\mathbf{X})) - \Phi_{\mathrm{Sig}}(\widehat{\mathrm{pp}}^{(1)}(\mathbf{Y}))\|^2_{H^{(2)}}],$$

(note that here $\Phi_{\mathrm{Sig}}(\widehat{\mathrm{pp}}^{(1)}(\mathbf{X}))$ denotes the signature of H-valued process $\widehat{\mathrm{pp}}^{(1)}(\mathbf{X})$) and use it as an empirical estimator for the rank-1 signature kernel MMD $\mathcal{D}^{(1)}_{\mathrm{Sig}}(\mathbf{X}, \mathbf{Y})$, when we have samples $(\hat{x}^i)_{i=1,\ldots,M}$ and $(\hat{y}^j)_{j=1,\ldots,N}$. he next theorem justifies the validity of such estimator: we may use it to obtain an accurate estimation of $\mathcal{D}^{(1)}_{\mathrm{Sig}}(\mathbf{X}, \mathbf{Y})$ provided the size of samples M and N is very large. The corresponding proof can be found in [73, Appendix, Theorem 7].

Theorem 2.10 *For* $\mathbf{X}, \mathbf{Y} \in FP_{plain}$, $\hat{\mathcal{D}}^{(1)}_{Sig}(\mathbf{X}, \mathbf{Y})$ *is a consistent estimator of the rank* 1*-robust signature kernel MMD, i.e.,*

$$\hat{\mathcal{D}}^{(1)}_{Sig}(\mathbf{X}, \mathbf{Y}) \to \mathcal{D}^{(1)}_{Sig}(\mathbf{X}, \mathbf{Y})$$

in probability as $M, N \to \infty$*, provided that the regularization parameters* λ_M, λ_N*, cf.* (16)*, decay to* 0 *slower than* $O(M^{-1/2})$ *and* $O(N^{-1/2})$*, respectively.*

Step 4: Evaluation of Higher-Rank Signature Kernels: A PDE Trick
Thanks to Theorem 2.10, the approximation of the rank-1 signature MMD $\mathcal{D}^{(1)}_{\mathrm{Sig}}(\mathbf{X}, \mathbf{Y})$ boils down to computing its empirical counterpart $\hat{\mathcal{D}}^{(1)}_{\mathrm{Sig}}(\mathbf{X}, \mathbf{Y})$. Starting from the expression

$$\hat{\mathcal{D}}^{(1)}_{\mathrm{Sig}}(\mathbf{X}, \mathbf{Y})^2 = \mathbb{E}_{\hat{\mu}^M \otimes \hat{\nu}^N}[\|\Phi_{\mathrm{Sig}}(\widehat{\mathrm{pp}}^{(1)}(\mathbf{X})) - \Phi_{\mathrm{Sig}}(\widehat{\mathrm{pp}}^{(1)}(\mathbf{Y}))\|^2_{H^{(2)}}]$$

and inserting the expressions of the empirical measures $\hat{\mu}^M = \frac{1}{M}\sum_{i=1}^M \delta_{\hat{x}^i}$ and $\hat{\nu}^N = \frac{1}{N}\sum_{j=1}^N \delta_{\hat{y}^j}$ into the above expectation, we obtain that

$$\mathbb{E}_{\hat{\mu}^M \otimes \hat{\nu}^N}[\|\Phi_{\text{Sig}}(\widehat{\text{pp}}^{(1)}(\mathbf{X})) - \Phi_{\text{Sig}}(\widehat{\text{pp}}^{(1)}(\mathbf{Y}))\|^2_{H^{(2)}}]$$
$$= \frac{1}{NM} \sum_{i=1}^{M} \sum_{j=1}^{N} \|\Phi_{\text{Sig}}(\widehat{\text{pp}}^{(1)}(\mathbf{X}))(\hat{x}^i) - \Phi_{\text{Sig}}(\widehat{\text{pp}}^{(1)}(\mathbf{Y}))(\hat{y}^j)\|^2_{H^{(2)}}. \tag{17}$$

Since

$$\begin{aligned}\|\Phi_{\text{Sig}}(\widehat{\text{pp}}^{(1)}(\mathbf{X}))(\hat{x}^i) - \Phi_{\text{Sig}}(\widehat{\text{pp}}^{(1)}(\mathbf{Y}))(\hat{y}^j)\|^2_{H^{(2)}} \\ = \langle \Phi_{\text{Sig}}(\widehat{\text{pp}}^{(1)}(\mathbf{X}))(\hat{x}^i) - \Phi_{\text{Sig}}(\widehat{\text{pp}}^{(1)}(\mathbf{Y}))(\hat{y}^j), \\ \Phi_{\text{Sig}}(\widehat{\text{pp}}^{(1)}(\mathbf{X}))(\hat{x}^i) - \Phi_{\text{Sig}}(\widehat{\text{pp}}^{(1)}(\mathbf{Y}))(\hat{y}^j)\rangle_{H^{(2)}}\end{aligned} \tag{18}$$

holds for every pair of indices (i, j), the task of computing this empirical estimator is in turn reduced to evaluating the empirical version of the rank-1 signature kernels

$$\boldsymbol{k}^{(1)}_{\text{Sig}}(\widehat{\text{pp}}^{(1)}(\mathbf{X})(\hat{x}^i), \widehat{\text{pp}}^{(1)}(\mathbf{Y})(\hat{y}^j)) = \langle \Phi_{\text{Sig}}(\widehat{\text{pp}}^{(1)}(\mathbf{X}))(\hat{x}^i), \Phi_{\text{Sig}}(\widehat{\text{pp}}^{(1)}(\mathbf{Y}))(\hat{y}^j)\rangle_{H^{(2)}},$$

for all i, j. By [30, Theorem 2.5] or [57, Section 7.2, PDE Solver], we know that the evaluation of the signature kernel

$$\langle \Phi_{\text{Sig}}(\widehat{\text{pp}}^{(1)}(\mathbf{X}))(\hat{x}^i), \Phi_{\text{Sig}}(\widehat{\text{pp}}^{(1)}(\mathbf{Y}))(\hat{y}^j)\rangle^2_{H^{(2)}}$$

of H-valued paths $\widehat{\text{pp}}^{(1)}(\mathbf{X})(\hat{x}^i)$ and $\widehat{\text{pp}^{(1)}}(\mathbf{Y})(\hat{y}^j)$ is the unique solution $u^{i,j}(T, T)$ of the following hyperbolic PDE, which is also called the *Goursat problem*:

$$\begin{aligned}\frac{\partial^2 u^{i,j}(s,t)}{\partial s \partial t} &= \Big\langle \frac{\partial}{\partial t}\widehat{\text{pp}}^{(1)}(\mathbf{X})_t(\hat{x}^i), \frac{\partial}{\partial s}\widehat{\text{pp}}^{(1)}(\mathbf{Y})_s(\hat{y}^j)\Big\rangle_H u^{i,j}(s,t) \\ &= \Big\langle \widehat{\text{pp}}^{(1)}(\mathbf{X})_t(\hat{x}^i) - \widehat{\text{pp}}^{(1)}(\mathbf{X})_{t-1}(\hat{x}^i), \\ &\qquad \widehat{\text{pp}}^{(1)}(\mathbf{Y})_s(\hat{y}^j) - \widehat{\text{pp}}^{(1)}(\mathbf{Y})_{s-1}(\hat{y}^j)\Big\rangle_H u^{i,j}(s,t),\end{aligned} \tag{19}$$

where we consider $\widehat{\text{pp}}^{(1)}(\mathbf{X})(\hat{x}^i)$ and $\widehat{\text{pp}}^{(1)}(\mathbf{Y})(\hat{y}^j)$ as continuous H-valued paths via linear interpolation. Furthermore, in view of (16) and the fact that $\boldsymbol{k}_{\text{Sig}}(\hat{x}^k, \hat{y}^\ell) = \langle \Phi_{\text{Sig}}(\hat{x}^k), \Phi_{\text{Sig}}(\hat{y}^\ell)\rangle_H$, using some linear algebra calculations we can readily check for $\widehat{\text{pp}}^{(1)}(\mathbf{X})_t(\hat{x}^i) = \hat{F}_t^{(\hat{x}^k)_{k\le M}}(\hat{x}^i)$ and $\widehat{\text{pp}}^{(1)}(\mathbf{Y})_s(\hat{y}^j) = \hat{F}_s^{(\hat{y}^\ell)_{\ell\le N}}(\hat{y}^j)$ that

$$\begin{aligned}\Big\langle \widehat{\text{pp}}^{(1)}(\mathbf{X})_t(\hat{x}^i), \widehat{\text{pp}}^{(1)}(\mathbf{Y})_s(\hat{y}^j)\Big\rangle_H &= \langle \hat{F}_t^{(\hat{x}^k)_{k\le M}}(\hat{x}^i), \hat{F}_s^{(\hat{y}^\ell)_{\ell\le N}}(\hat{y}^j)\rangle_H \\ = \langle \boldsymbol{k}^{X,t}_{\text{Sig}}(\hat{x}^i)(\mathbf{K}^X_{t,t} + M\lambda_M \mathrm{I}_M)^{-1}\Phi^{X,T}_{\text{Sig}}&, \boldsymbol{k}^{Y,s}_{\text{Sig}}(\hat{y}^j)(\mathbf{K}^Y_{s,s} + N\lambda_N \mathrm{I}_N)^{-1}\Phi^{Y,T}_{\text{Sig}}\rangle_H \\ &= \boldsymbol{k}^{\hat{x}^i}_s(\mathbf{K}^X_{t,t} + M\lambda_M \mathrm{I}_M)^{-1}\mathbf{K}^{X,Y}_{T,T}(\mathbf{K}^Y_{s,s} + N\lambda_N \mathrm{I}_N)^{-1}\boldsymbol{k}^{\hat{y}_j}_s,\end{aligned} \tag{20}$$

where $\boldsymbol{k}_t^{\hat{x}^i} \in \mathbb{R}^M$ and $\boldsymbol{k}_s^{\hat{y}^j} \in \mathbb{R}^N$ are the vectors

$$[\boldsymbol{k}_t^{\hat{x}^i}]_k = \boldsymbol{k}_{\mathrm{Sig}}(\hat{x}^k|_{I_t}, \hat{x}^i|_{I_t}), \quad [\boldsymbol{k}_s^{y}]_\ell = \boldsymbol{k}_{\mathrm{Sig}}(\bar{y}^\ell|_{I_s}, \hat{y}^j|_{I_s}), \quad 1 \le k \le M, 1 \le \ell \le N, \tag{21}$$

and $\mathbf{K}_{s,s}^X \in \mathbb{R}^{M\times M}$, $\mathbf{K}_{T,T}^{X,Y} \in \mathbb{R}^{M\times N}$ and $\mathbf{K}_{t,t}^Y \in \mathbb{R}^{N\times N}$ are the matrices

$$\begin{aligned}
&[\mathbf{K}_{t,t}^X]_{k,\ell} = \boldsymbol{k}_{\mathrm{Sig}}(\hat{x}^k|_{I_t}, \hat{x}^\ell|_{I_t}), \quad [\mathbf{K}_{T,T}^{X,Y}]_{k,\ell} = \boldsymbol{k}_{\mathrm{Sig}}(\hat{x}^k, \hat{y}^\ell), \\
&[\mathbf{K}_{s,s}^Y]_{k,\ell} = \boldsymbol{k}_{\mathrm{Sig}}(\hat{y}^k|_{I_s}, \hat{y}^\ell|_{I_s}).
\end{aligned} \tag{22}$$

Now, applying again the signature kernel PDE trick to the signature kernels above, we see that $\boldsymbol{k}_{\mathrm{Sig}}(\hat{x}^k, \hat{y}^\ell)$ is the unique solution $\tilde{u}^{k,\ell}(T,T)$ of the Goursat problem

$$\frac{\partial^2 \tilde{u}_{k,\ell}(s,t)}{\partial s \partial t} = \left\langle \frac{\partial}{\partial t}\hat{x}_t^k, \frac{\partial}{\partial s}\hat{y}_s^\ell \right\rangle_{\mathbb{R}^d} \tilde{u}^{k,\ell}(s,t), \quad (s,t) \in [0,T]\times[0,T].$$

Similarly, the quantity $\boldsymbol{k}_{\mathrm{Sig}}(\hat{x}^k|_{I_t}, \hat{x}^i|_{I_t})$ can be computed by finding the terminal value of $v_{k,i}(t,t)$ with $v_{k,i}$ being the unique solution to the Goursat problem

$$\frac{\partial^2 v_{k,i}(s,r)}{\partial s \partial r} = \left\langle \frac{\partial}{\partial s}\hat{x}_s^k, \frac{\partial}{\partial r}\hat{x}_r^i \right\rangle_{\mathbb{R}^d} v^{k,i}(s,r), \quad (s,r) \in [0,t]\times[0,t].$$

As a result, by solving a family of Goursat PDEs, We can compute the above vectors and matrices, and then insert them back into (20) we can get an estimation of the inner product $\langle \widehat{\mathrm{pp}}^{(1)}(\mathbf{X})_t(\hat{x}^i), \widehat{\mathrm{pp}}^{(1)}(\mathbf{Y})_s(\hat{y}^j)\rangle_H$; then we can obtain the coefficients of PDE (19) and are able to solve this equation to get the rank-1 signature kernel $\boldsymbol{k}_{\mathrm{Sig}}^{(1)}(\widehat{\mathrm{pp}}^{(1)}(\mathbf{X})(\hat{x}^i), \widehat{\mathrm{pp}}^{(1)}(\mathbf{Y})(\hat{y}^j))$ for each pair (i,j). Finally, in view of the polarization formula (18), we can use $\boldsymbol{k}_{\mathrm{Sig}}^{(1)}(\widehat{\mathrm{pp}}^{(1)}(\mathbf{X})(\hat{x}^i), \widehat{\mathrm{pp}}^{(1)}(\mathbf{Y})(\hat{y}^j))$ to compute the square of Hilbert norms $\|\Phi_{\mathrm{Sig}}(\widehat{\mathrm{pp}}^{(1)}(\mathbf{X}))(\hat{x}^i) - \Phi_{\mathrm{Sig}}(\widehat{\mathrm{pp}}^{(1)}(\mathbf{Y}))(\hat{y}^j)\|_{H^{(2)}}^2$ for each pair (i,j), which in turn allows us to obtain the sum

$$\frac{1}{NM}\sum_{i=1}^{M}\sum_{j=1}^{N} \|\Phi_{\mathrm{Sig}}(\widehat{\mathrm{pp}}^{(1)}(\mathbf{X}))(\hat{x}^i) - \Phi_{\mathrm{Sig}}(\widehat{\mathrm{pp}}^{(1)}(\mathbf{Y}))(\hat{y}^j)\|_{H^{(2)}}^2,$$

and the latter is exactly the empirical rank-1 signature kernel distance

$$\mathbb{E}_{\hat{\mu}^M \otimes \hat{\nu}^N}[\|\Phi_{\mathrm{Sig}}(\widehat{\mathrm{pp}}^{(1)}(\mathbf{X})) - \Phi_{\mathrm{Sig}}(\widehat{\mathrm{pp}}^{(1)}(\mathbf{Y}))\|_{H^{(2)}}^2] = \mathcal{D}_{\mathrm{Sig}}^{(1)}(\mathbf{X},\mathbf{Y})^2.$$

Here, we remark that one can use the finite difference method from [30, Section 3.1] to solve the above Goursat problem, see Algorithm 3 below. Note that therein f denotes the solution operator in this scheme which returns the outcome obtained in each step $s_p \times t_q \to s_{p+1} \times t_{q+1}$, where $(s_q)_q$ and $(t_p)_p$ are discretizations of $[0,T]$.

The whole algorithm can be summarized as follows via pseudo-code. Here we remark that in this algorithm, the distance $\mathcal{D}^{(1)}_{\text{Sig}}(\mathbf{X}, \mathbf{Y})$ in an infinite-dimensional Hilbert space can be efficiently estimated by solving a (possibly large) system of low-dimensional Hyperbolic PDEs (i.e., the Goursat problem), whose coefficients rely on the chosen sample paths $\hat{x}^i \sim \text{law}(X)$ and $\hat{y}^j \sim \text{law}(Y)$.

Algorithm 3 PDESolve $O(T^2)$

1: **Input:** matrix $\mathbf{M} \in \mathbb{R}^{T\times T}$ where $[\mathbf{M}]_{p,q} = \langle \hat{x}^i_{s_{p+1}} - \hat{x}^i_{s_p}, \hat{y}^j_{t_{q+1}} - \hat{y}^j_{t_q} \rangle_{\mathbb{R}^d}$, full $\in$ {True, False}
2: **Output:** full solution $u \in \mathbb{R}^{N\times N}$ with $u[p,q] = \boldsymbol{k}_{\text{Sig}}(x|_{I_{t_p}}, y|_{I_{t_q}})$ or $u[-1,-1] = \boldsymbol{k}_{\text{Sig}}(x, y)$
3: $u[1,:] \leftarrow 1$
4: $u[:,1] \leftarrow 1$
5: **for** p from 1 to $T-1$ **do**
6:     **for** q from 1 to $T-1$ **do**
7:         $u[p+1, q+1] \leftarrow f(u[p, q+1], u[p+1, q], u[p, q], M[p, q])$
8:     **end for**
9: **end for**
10: **if** full **then return** u **else return** $u[-1,-1]$

Algorithm 4 ZerothRankGram $O(dM^2T^2)$

1: **Input:** sample paths $\{\hat{x}^i\}_{i=1}^M \sim \text{law}(X)$ and $\{\hat{y}^j\}_{j=1}^N \sim \text{law}(Y)$, full $\in$ {True, False}
2: **Output:** $\mathbf{G} \in \mathbb{R}^{M\times N\times P\times Q}$ where $\mathbf{G}[i,j,p,q] = \boldsymbol{k}_{\text{Sig}}(\hat{x}^i|_{I_{s_p}}, \hat{y}^j|_{I_{t_q}})$ or $\mathbf{G}[:,:,-1,-1]$
3: $M[i,j,p,q] \leftarrow \langle \hat{x}^i_{s_p}, \hat{y}^j_{t_q}\rangle$ $\forall i \in \{1,\dots,m\}$, $j \in \{1,\dots,n\}$, $p \in \{1,\dots,T\}$, $q \in \{1,\dots,T\}$
4: $\mathbf{M} \leftarrow \mathbf{M}[:,:,1:,1:] + \mathbf{M}[:,:,:-1,:-1] - \mathbf{M}[:,:,1:,:-1] - \mathbf{M}[:,:,:-1,1:]$
5: $\mathbf{G}[i,j] \leftarrow$ PDESolve($\mathbf{M}[i,j]$), $\forall i \in \{1,\dots,M\}$, $j \in \{1,\dots,N\}$
6: **if** full **then return G else return** $\mathbf{G}[:,:,-1,-1]$

Algorithm 5 InnerProdPredCondKME $O(T^2M^3)$

1: **Input:** three Gram matrices $\mathbf{G}_{XX}, \mathbf{G}_{XY}, \mathbf{G}_{YY}$ and hyperparameter λ, where $\mathbf{G}_{XY}[i,j,p,q] = \boldsymbol{k}_{\text{Sig}}(\hat{x}^i|_{I_{s_p}}, \hat{y}^j|_{I_{t_q}})$ and similar expressions for $\mathbf{G}_{XX}$ and $\mathbf{G}_{YY}$.
2: **Output:** returns an empirical estimator of $\mathbf{M} \in \mathbb{R}^{M\times N\times T\times T}$ where $\mathbf{M}[i,j,p,q] = \langle \widehat{\text{pp}}^{(1)}(\mathbf{X})(\hat{x}^i)_{s_p}, \widehat{\text{pp}}^{(1)}(\mathbf{Y})(\hat{y}^j)_{t_q}\rangle$
3: $\mathbf{W}_X[:,:,p] \leftarrow (\mathbf{G}_{XX}[:,:,p,p] + M\lambda I)^{-1}$, $\forall p \in \{1,\dots,T\}$
4: $\mathbf{W}_Y[:,:,q] \leftarrow (\mathbf{G}_{YY}[:,:,q,q] + N\lambda I)^{-1}$, $\forall q \in \{1,\dots,T\}$
5: **for** p from 1 to T **do**
6:     **for** q from 1 to T **do**
7:         $\mathbf{M}[:,:,p,q] \leftarrow \mathbf{G}_{XX}[:,:,p,p]^T \mathbf{W}_X[:,:,p] \mathbf{G}_{XY}[:,:,-1,-1] \mathbf{W}_Y[:,:,q] \mathbf{G}_{YY}[:,:,q,q]$
8:     **end for**
9: **end for**

Algorithm 6 FirstRankGram $O(T^2M^3)$

1: **Input:** $\mathbf{G}_{XX}, \mathbf{G}_{XY}, \mathbf{G}_{YY}$ and hyperparameter λ.
2: **Output:** an empirical estimator of $\mathbf{G}^1_{X,Y} \subset \mathbb{R}^{M\times N}$, where $\mathbf{G}^1_{X,Y}[i,j] = k^{(1)}_{\text{Sig}}(\widehat{\text{pp}}^{(1)}(\mathbf{X})(\hat{x}^i), \widehat{\text{pp}}^{(1)}(\mathbf{X})(\hat{y}^j))$
3: $\mathbf{M} \leftarrow$ InnerProdPredCondKME$(\mathbf{G}_{XX}, \mathbf{G}_{XY}, \mathbf{G}_{YY}, \lambda)$
4: $\mathbf{M} \leftarrow \mathbf{M}[:,:,1:,1:] + \mathbf{M}[:,:,:-1,:-1] - \mathbf{M}[:,:,1:,:-1] - \mathbf{M}[:,:,:-1,1:]$
5: $\mathbf{G}^1_{XY}[i,j] \leftarrow$ PDESolve$(\mathbf{M}[i,j])$, $\forall i \in \{1,\dots,M\}, \forall j \in \{1,\dots,N\}$
6: **return** $\mathbf{G}^1_{XY}$

Algorithm 7 FirstRankMMD $O(dM^2T^2 + T^2M^3)$

1: **Input:** sample paths $\{\hat{x}^i\}_{i=1}^M \sim \text{law}(X)$ and $\{\hat{y}^j\}_{j=1}^N \sim \text{law}(Y)$, hyperparameter λ.
2: **Output:** an empirical estimator of the rank-1 signature kernel MMD $\hat{\mathcal{D}}^{(1)}_{\text{Sig}}(\mathbf{X},\mathbf{Y})$.
3: $\mathbf{G}_{XX} \leftarrow$ ZerothRankGram$(\{\hat{x}^i\}_{i=1}^M, \{\hat{x}^i\}_{i=1}^M$, full = True)
4: $\mathbf{G}_{XY} \leftarrow$ ZerothRankGram$(\{\hat{x}^i\}_{i=1}^M, \{\hat{y}^j\}_{j=1}^N$, full = True)
5: $\mathbf{G}_{YY} \leftarrow$ ZerothRankGram$(\{\hat{y}^j\}_{j=1}^N, \{\hat{y}^j\}_{j=1}^N$, full = True)
6: $\mathbf{G}^1_{XX} \leftarrow$ FirstRankGram$(\mathbf{G}_{XX}, \mathbf{G}_{XX}, \mathbf{G}_{XX}, \lambda)$
7: $\mathbf{G}^1_{XY} \leftarrow$ FirstRankGram$(\mathbf{G}_{XX}, \mathbf{G}_{XY}, \mathbf{G}_{YY}, \lambda)$
8: $\mathbf{G}^1_{YY} \leftarrow$ FirstRankGram$(\mathbf{G}_{YY}, \mathbf{G}_{YY}, \mathbf{G}_{YY}, \lambda)$
9: **return** $\text{avg}(\mathbf{G}^1_{XX}) - 2\text{avg}(\mathbf{G}^1_{XY}) + \text{avg}(\mathbf{G}^1_{YY})$

3 Applications

In this section, we will mention two concrete applications of higher-rank (expected) signatures to solve problems related to adapted weak topologies.

3.1 Classification of Filtered Processes in Adapted Weak Topologies

This subsection is mainly borrowed from [28, Section 5.1], which showed a significant application of the higher-rank expected signatures to distinguish stochastic processes in hypothesis testing. The Python code of the algorithm is available at: https://github.com/PatricBonnier/Higher-rank-signature-regression.

First recall that expected signatures (the expectations of Φ_{Sig} in our notation) are currently explored in machine learning applications since they provide a hierarchical description of the law of a stochastic process; in the terminology of statistical learning the signature map is a so-called "universal and characteristic" feature map. However, the expected signature can only be used to metrize the weak convergence and completely ignores the filtration. We now use the algorithms from the previous section to demonstrate that already in the arguably simplest classification task, the use of higher-rank expected signatures can be beneficial.

Define two processes $\mathbf{X}, \mathbf{Y} \in \mathrm{FP}$ with their natural filtrations as follows[7]

$$\mathbf{X}^c : X_0 = 0, X_1 = N_1, \qquad X_2 = c + N_2,$$

$$\mathbf{Y}^c : Y_0 = 0, Y_1 = \sqrt{1-\varepsilon^2} M_1 + \varepsilon c, Y_2 = c + M_2.$$

where c denotes a constant and N_1, N_2, M_1, M_2 are independent binomial random variables; $\varepsilon > 0$ fixed. Note that for a fixed c and small ε, the processes $\mathbf{X}^c$ and $\mathbf{Y}^c$ have similar laws; despite $\mathbf{X}^c$ and $\mathbf{Y}^c$ being essentially different in the rank-1 adapted weak topology, analogous to Example 1.3. We now endow FP with a probability measure μ as follows: a sample from μ consists of sampling a $C \sim N(0, 1)$ and then selecting with probability 0.5 the process $\mathbf{X}^C$ and with probability 0.5 the process $\mathbf{Y}^C$. Figure 2 shows for each sample from μ one sample trajectory from this process.

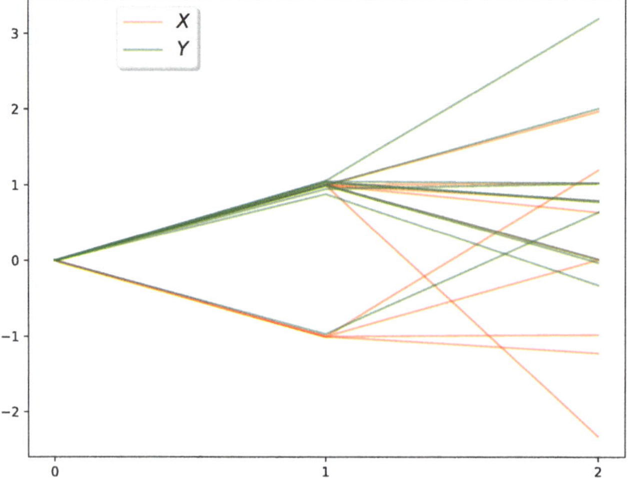

Fig. 2 The sample paths of μ are plotted above for $\varepsilon = 0.01$. They are marked as either $\mathbf{X}$ in red, or $\mathbf{Y}$ in green depending on if the sample comes from $\mathbf{X}^C$ or $\mathbf{Y}^C$

[7] This example is taken from [28, Section 5.1].

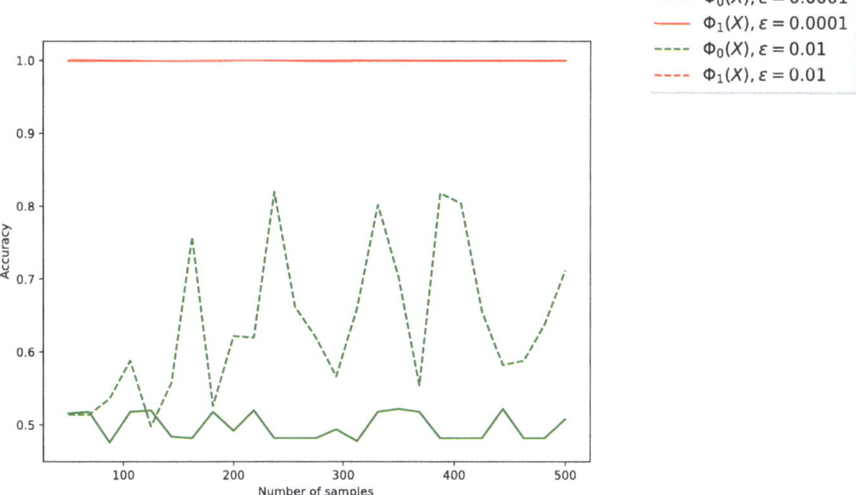

Fig. 3 The accuracies of the linear classifier trained on $\Phi_0(X)$ (stands for the classification based on the weak topology) and $\Phi_1(X)$ (stands for the classification based on the rank-1 adapted weak topology) is plotted against the number of samples used in green and red respectively. Solid lines are used for $\varepsilon = 10^{-4}$ and dashed lines for $\varepsilon = 10^{-2}$

We ran the following experiment: we sampled 1000 processes from μ and labelled the processes corresponding to whether $\mathbf{X}^c$ or $\mathbf{Y}^c$ was sampled. We then compare their distances in the rank-0 signature metric d_{Sig} (based on the usual signature) and in the rank-1 signature distance $d^{(1)}_{\text{Sig}}$ (based on the rank-1 signature) for each sample truncated at level 6 and level 3 respectively, by using Algorithm 1. This was then split into a training set and test set—both of size 500-and for $50 \leq m \leq 500$ a support vector machine classifier was trained on m data points from the training set with signature and rank-1 signature as features.

Figure 3 shows the accuracies of the resulting classifiers on the test set. Observe that for small values of ε, the classifier based on the usual signature feature (i.e., the criterion is the weak topology) is essentially guessing, and even for larger values it does not converge well; the classifier based on the rank-1 signature feature (i.e., the criterion is the rank-1 adapted weak topology) converges immediately however, which is to be expected as the rank-1 signature is able to separate $\mathbf{X}^c$ and $\mathbf{Y}^c$ independently of ε.

We emphasize that although this is a toy example, it demonstrate how the expected signature can fail to pick up essential properties of a model and that the higher-rank (expected) signature provide additional features that linearize complex dependencies between law and filtration.

3.2 Optimal Stopping via Adapted Distribution Regression

In this subsection, which is a shortened version of [47, Section 3], we will briefly show how to employ the higher-rank signatures in a kernel regression approach to learn the value functions of optimal stopping problems. By Theorem 2.6 we know that the rank-r signature kernel MMD $\mathcal{D}^{(r)}_{\mathrm{Sig}}(\cdot,\cdot)$ metrizes the rank-r adapted weak topology, for $r \in \mathbb{N}_0$. This fact together with a classical result from kernel learning implies the following important and useful theorem, whose proof can be found in [47, Appendix, proof of Theorem 2.14]:

Theorem 3.1 *Let $\sigma > 0$ be fixed and assume all filtered processes take values in a compact subset $\mathcal{K} \subset \mathbb{R}^d$. For each $r \in \mathbb{N}_0$ we define the kernel $K^{(r)}_{Sig} : FP \times FP \to \mathbb{R}$,*

$$K^{(r)}_{Sig}(\mathbf{X}, \mathbf{Y}) := \exp(-\sigma^2 \mathcal{D}^{(r)}_{Sig}(\mathbf{X}, \mathbf{Y})^2)$$

Then the RKHS generated by $K^{(r)}_{Sig}$ is dense in the space $C_b((FP, \mathcal{T}^{(r)}))$ w.r.t. the uniform topology. In particular, if $v(\cdot) \in C_b((FP, \mathcal{T}^{(r)}))$, then for any $\varepsilon > 0$, there exist an $M \in \mathbb{N}$, $\mathbf{Y}^i \in FP$ and scalars $a_i \in \mathbb{R}$, $i = 1, \ldots, M$, such that

$$\sup_{\mathbf{X} \in FP} \left| v(\mathbf{X}) - \sum_{i=1}^{M} a_i \exp(-\sigma^2 \mathcal{D}^{(r)}_{Sig}(\mathbf{X}, \mathbf{Y}^i)^2) \right| \le \varepsilon.$$

As the construction of $K^{(r)}_{\mathrm{Sig}}$ is based on the rank-r signature kernel MMD which is in turn induced by the rank-r signature kernel $\boldsymbol{k}^{(r)}_{\mathrm{Sig}}$, by an abuse of notation we will also call $K^{(r)}_{\mathrm{Sig}}$ the rank-r signature kernel.

Since the rank-0 adapted weak topology is exactly the weak topology, the above theorem yields that any weakly continuous bounded function $v(\cdot)$ on FP can be approximated by the (linear) regression of the classical signature kernel $K^{(0)}_{\mathrm{Sig}}(\mathbf{X}, \mathbf{Y}) = \exp(-\sigma^2 \mathcal{D}_{\mathrm{Sig}}(\mathbf{X}, \mathbf{Y})^2)$. Such an approach is called the distribution regression (via signature kernel) in [58]. This regression scheme is of particular interest in the context of mathematical finance: Suppose that $v : \mathrm{FP} \to \mathbb{R}$ is the pricing functional of a bounded payoff function of some path—dependent financial derivative (e.g., the European options), which is certainly weakly continuous, and we have a set of pairs $(\mathbf{Y}^j, y^j) \in \mathrm{FP} \times \mathbb{R}$, $j = 1, \ldots, M$, where $y^j = v(\mathbf{Y}^j)$ is the spotted fair price of model $\mathbf{Y}^j$, as the input, then one can perform the above signature kernel regression on this data set $(\mathbf{Y}^j, y^j)_{j=1,\ldots,M}$ to determine the coefficients a_j, $j = 1, \ldots, M$ such that

$$\left| v(\cdot) - \sum_{j=1}^{M} a_j \exp(-\sigma^2 \mathcal{D}_{\mathrm{Sig}}(\cdot, \mathbf{Y}^j)^2) \right| \le \varepsilon.$$

Moreover, if $\mathbf{X} \in \mathrm{FP}$ is a new financial model, to compute the fair price $v(\mathbf{X})$ on the market $\mathbf{X}$, one just needs to compute the signature kernel MMD $\mathcal{D}_{\mathrm{Sig}}(\cdot, \mathbf{Y}^j)^2$ for all $j = 1, \ldots, M$ by the second algorithm in the last section, and then use $\sum_{j=1}^{M} a_j \exp(-\sigma^2 \mathcal{D}_{\mathrm{Sig}}(\cdot, \mathbf{Y}^j)^2)$ as an approximating price (with an error up to ε).

Now let us consider the value function in the optimal stopping problem, or, in the context of mathematical finance, the pricing functional of American-type options,

$$v_c(\mathbf{X}) := \sup_{\tau \in \mathrm{ST}(\mathbf{X})} \mathbb{E}[c(X, \tau)]$$

for a given (bounded and continuous) function $c : (I \to \mathbb{R}^d) \times I \to \mathbb{R}$. From Example 1.3 we know that in general it is not weakly continuous, whilst by Lemma 1.9 we know that $v_c(\cdot)$ is continuous for $\mathcal{T}^{(T-1)}$. In fact, if we restrict to filtered processes with natural filtration, then on $\mathrm{FP}_{\mathrm{plain}}$ this functional is continuous w.r.t. the rank-1 adapted weak topology $\mathcal{T}^{(1)}$ (see [14, Theorem 1.1]). The above observations tell us that in order to apply a regression for computing the fair price of American-type options, one should use the rank-1 signature kernel MMD $\mathcal{D}^{(1)}_{\mathrm{Sig}}$ instead of $\mathcal{D}_{\mathrm{Sig}}$ in the usual distribution regression. We call such a regression approach with higher-rank signature kernels the *adapted distribution regression*.

To illustrate the power of adapted distribution regression, here we empirically investigate the effectiveness of the proposed signature kernel algorithm for two optimal stopping problems:

1. the pricing of an American basket option with geometric put payoff;
2. the pricing of American options under the rough Bergomi model.

More precisely, treating these as distribution regression problems from probability measures on path space to $\mathbb{R}$, we endow the classical Kernel Ridge Regression (KRR) algorithm with the characteristic kernel $K^{(1)}_{\mathrm{Sig}}$ studied in previous sections and compare its performance against classical algorithms. The code is available at https://github.com/maudl3116/higherOrderKME.

3.2.1 American Basket Option Pricing

The numerical example in this subsection is taken from [47, Section 3.5.1]. It is well known that an American option can be approximated by a Bermudan option, which can be exercised only at specific dates $t_0 < t_1 < t_2 < \ldots < t_N$. If the time grid is chosen sufficiently small, the American option is well approximated by the Bermudan option. For $d \in \mathbb{N}$, we consider the Black-Scholes model by using a d-dimensional geometric Brownian motion $(X_t)_{t \geq 0}$ for the stock prices. At time $t \in [0, T]$ with initial stock prices $X_t = x \in \mathbb{R}^d$, the price of the American option $V(t, x)$ can be formulated as the following optimal stopping problem

$$V(t,x) := \sup_{\tau\in[t,T]} \mathbb{E}\left[e^{-r(\tau-t)}\Phi(X_\tau)\middle| X_t = x\right]$$

with payoff function $\Phi : \mathbb{R}^d \to \mathbb{R}_+$, where the supremum is taken over all stopping times. Here, we consider a geometric put payoff defined as follows

$$\Phi(x) := \left(v - \left(\prod_{i=1}^{d} x_i\right)^{1/d}\right)_+$$

with strike $v > 0$.

To perform KRR, we consider the classical supervised learning setting and we form a dataset $\{(\mathcal{X}_i, y_i)\}_{i=1}^N$ of $N = 100$ input-output pairs, where

- $\mathcal{X}_i = \{x^{i,j} : [0,T] \to \mathbb{R}^d\}_{j=1}^{n_i}$ is a set of n_i paths sampled from a d-dimensional Black-Scholes model, with fixed interest rate $r = 2\%$, $d = 20$ and volatility sampled uniformly at random $\sigma_i \in [0.1, 0.5]$ over the time grid $\{0, T/\ell, 2T/\ell, \dots, T\}$ with $\ell = 10$ as in [43]. For simplicity we set $n_1 = \dots = n_N = n$;
- y_i is the price of the American option obtained via a binomial-tree method using a tree of depth 10,000 with strike $v = 100$ and maturity $T = 1$.

The dataset is randomly split into a training set of size $N_{\text{train}} = 90$ and a test set of size $N_{\text{test}} = 10$. We fit the KRR on the training set and evaluate the model on the test set.

We count how many sample paths n are required by both methods (namely the kernel induced by rank-0 signatures and rank-1 signatures) to produce an approximation error ϵ from the target price obtained using the lattice method. We use the Mean Absolute Percent Error (MAPE), that is $\epsilon = N_{\text{test}}^{-1} \sum_{i=1}^{N_{\text{test}}} |(\hat{y}_i - y_i)/y_i|$ where $\{\hat{y}_i\}_{i=1}^{N_{\text{test}}}$ are the prices predicted by either algorithm. In this supervised learning we use the code provided in [43] to apply the LS algorithm with basis functions given by the Taylor polynomials up to degree two, as recommended in [43], to get the price $y_i, i = 1, \dots, N_{\text{test}}$.

As shown in Table 1, the higher-rank signature methods, that is KRR with the rank-1 signature kernels $K_{\text{Sig}}^{(1)}$ require significantly fewer sample paths from the models to produce a desired precision than the classical signature kernel $K_{\text{Sig}}^{(0)}$. We also observe that for the (higher-rank) signature methods, the regression is fitted on N_{train} base models and so depends on $n \times N_{\text{train}}$ sample paths. However, for any new prediction, the $n \times N_{\text{train}}$ training sample paths are fixed and do not need to be re-sampled; only n additional samples are required. So for a total of N_{test} new predictions we will have that $n(N_{\text{train}} + N_{\text{test}}) \approx nN_{\text{test}}$ when $N_{\text{train}} \ll N_{\text{test}}$.

Table 1 Geometric put option pricing under the Black-Scholes model. The columns show the number of samples n to produce a MAPE ϵ on the test set according to different numerical methods

	$\epsilon = 10\%$	$\epsilon = 5\%$	$\epsilon = 2.5\%$
Rank-0 signature kernel $K_{\text{Sig}}^{(0)}$	50	250	1000
Rank-1 signature kernel $K_{\text{Sig}}^{(1)}$	50	200	500

3.2.2 Pricing American Options Under the Rough Bergomi Model

In this subsection, which is taken from [73], we make use of Support Vector Regression (SVR) equipped with the rank-1 signature kernel $K_{\text{Sig}}^{(1)}$ to address the pricing of American call options.

$$V(t,x) = \sup_{\tau \in [t,T]} \mathbb{E}\left[e^{-r(\tau - t)} \Phi(S_\tau) \middle| S_t = x \right],$$

where $\Phi(S_t) = (S_t - v)_+$, and the supremum is taken over all stopping times. Here, we assume that market price dynamics follow a *rough Bergomi model*, introduced in [18], which is defined as follows

$$S_t = \exp\left\{ \int_0^t \sqrt{V_u} \mathrm{d}B_u - \frac{1}{2} \int_0^t V_u \mathrm{d}u \right\}, \qquad B_u = \rho W_u^1 + \sqrt{1-\rho^2} W_u^2,$$

$$V_t = \xi \, \exp\left\{ \eta Y_t^a - \frac{\eta^2}{2} t^{2a+1} \right\}, \qquad Y_t^a = \sqrt{2a+1} \int_0^t (t-u)^a \mathrm{d}W_u^1,$$

where W^1, W^2 are two independent Brownian motions.

We employ the *hybrid scheme* of [24] to efficiently simulate sample paths of the Volterra process Y^a, with time complexity $\mathcal{O}(n \log(n))$, where n is the number of time steps. Option prices are numerically obtained using the Monte Carlo algorithm proposed by Longstaff and Schwartz (LS) in [59] with 500,000 sample paths from the rough Bergomi model with $\rho = -0.9$, $\eta = 1.9$ and $\xi = 0.09$. The strike is $v = 1$, the maturity is $T = 1$, we use 10 time steps and the interest rate is $r = 0.05$. For the distribution regression task, we consider 100 values $\{a_i\}_{i=1}^{100}$ for the roughness index a chosen uniformly at random from $]0.5, -0.2]$. Each a_i yields a different price process S^a. We regress the corresponding option price $V^a(0, S_0 = 1)$ on $m \in \mathbb{N}$ sample paths from S^a and use an SVR endowed with the rank-1 signature kernel $K_{\text{Sig}}^{(1)}$ and benchmark the method against the rank-0 signature kernel regression algorithm for the same number m of sample paths, where $m = 100, 250, 500$.

As shown in Table 2, SVR with kernel $K_{\text{Sig}}^{(1)}$ systematically outperforms the classical signature kernel $K_{\text{Sig}}^{(0)}$.

Table 2 American call option pricing under the rough Bergomi model. Standard deviations (in parenthesis) are obtained by repeating the experiment 10 times with different train-test splits (m-sample is fixed)

	Predictive MSE $\times 10^{-6}$		
	$m = 100$	$m = 250$	$m = 500$
Rank 0-Signature kernel $K_{\text{Sig}}^{(0)}$	5.27 ($\pm$ 1.53)	1.36 ($\pm$ 0.37)	0.98 ($\pm$ 0.20)
Rank 1-Signature kernel $K_{\text{Sig}}^{(1)}$	3.86 ($\pm$ 1.41)	1.40 ($\pm$ 0.42)	0.83 ($\pm$ 0.15)

4 Bibliographical Discussion

In this section we give a short bibliographical overview on past and recent developments with regard to adapted topologies.

The prediction process was introduced for the first time by Knight [53] in 1975 to study measurable processes in continuous time. In this very general setting, the usefulness of the prediction process stems from the fact that, even though no additional regularity of the process itself is a-priori assumed than measurability, when working with the right-continuous filtration generated by the process, it is more regular. In fact, the prediction process has càdlàg paths and the strong Markov property.

In the unpublished seminal monograph titled "Weak convergence and the general theory of processes" from 1981, Aldous [6] establishes the notion of "extended weak convergence". While our discussion from Sect. 1.3 is restricted to a discrete-time framework, his work was concerned with continuous time and stochastic bases of filtered processes that satisfy the usual conditions. With Knight's prediction processes as the main device, Aldous aimed to connect the theory of weak convergence as laid down in Billingsley's influential book [25] with, at the time, new developments in the 'general theory of stochastic processes' of the Strasbourg school and the martingale approach to stochastic processes. His main motivation stems from theoretical as well as practical deliberations on certain unsatisfactory features of the usual weak topology in regard to stochastic processes: On the one hand, stochastic operations such as the Doob-Meyer decomposition are not solely path transformation, and on the other hand, stochastic optimization problems such as optimal stopping are discontinuous w.r.t. the weak topology, cf. Example 1.3. We also refer to his work [5, 7].

From a model-theoretic perspective, Keisler [51] introduces in a seminar in 1979 the 'adapted probability logic' for filtered probability spaces with stochastic processes, which aims to also formalize the probabilistic language of stochastic processes. Subsequently, influenced by the work of Aldous, Hoover–Keisler [46] establish notions of equivalence of filtered processes based on higher-rank prediction processes, see Sect. 1.4, which tries to capture all relevant information, in particular the higher-order information that is unseen by the rank 1-prediction process as demonstrated in Example 1.6. They do so based on the concept of adapted distribution, cf. (11), which they introduce and then further explore. Two filtered processes have the same adapted distribution if they share the same 'probabilistic properties', such as having the same law, the same Doob-Meyer decomposition, the same value for filtration dependent functionals, and being a martingale or predictable. We refer to [46] for a precise discussion of this topic.

In independent work, Hellwig [42] studies in 1996 the question of continuity of sequential decision making problems in the context of mathematical economics. We put his framework in perspective with our initial setup from Sect. 1. For each time $t \in I$, we have a state space $\mathcal{X}_t$ that models the set of possible states of the world and $\mathcal{A}_t$ (where $\mathcal{A}_t$ is a compact metric space and $\mathcal{A} := \prod_{t \in I} \mathcal{A}_t$), the set of actions at time t, that are available to the agent. Parallel to the introduction, we choose as a mathematical base to model the uncertain evolution of the states of the world as well as information by a filtered process $\mathbf{X}$. For simplicity, we assume here that the stochastic basis of $\mathbf{X}$ supports a uniform random variable independent of $\mathcal{F}_T$, which allows us to skip working with additional randomization. This assumption is without loss of generality since any equivalence class of filtered processes admits a representative with that property. Given a certain state of the world $x \in \mathcal{X}$ and a chosen sequence of actions $a \in \mathcal{A}$, the agents utility is described by the utility function $u(x, a)$. The goal of the agent is to maximize its utility $U(\mathbf{X})$ by sequentially choosing actions where

$$U(\mathbf{X}) := \max_{A} \mathbb{E}\left[u(X, A)\right] \tag{23}$$

and the maximum is taken over the set of all admissible plan of actions $A = (A_t)_{t \in I}$ that are $(\mathcal{F}_t^{\mathbf{X}})$-adapted stochastic processes with paths in $\mathcal{A}$. Hellwig observed that, even when restricting to $\mathrm{FP}_{\mathrm{plain}}$ (that are filtered processes equipped with the natural filtration of the underlying process), the weak topology is not strong enough to have continuity of (23). Thus, a stronger notion of convergence has to be specified, the *information topology*. In view of the already introduced prediction process, this topology is easiest described as the initial topology w.r.t. the family of maps of marginals of the prediction process

$$\mathrm{FP} \to \mathcal{P}(\mathcal{X}) : \mathbf{X} \mapsto \mathrm{law}(\mathrm{pp}_t(\mathbf{X})), \quad t \in I.$$

This topology yields, when $u \in C_b(\mathcal{X} \times \mathcal{A})$, continuity of the utility maximization problem (23) on $\mathrm{FP}_{\mathrm{plain}}$, see also the discussion "On all adapted topologies are equal" below.

Starting from Wasserstein distances, which are a powerful tool to compare probability distributions, see [74, 78], and have become a staple even beyond applied and theoretical probability theory, versions of Wasserstein distances have been introduced for stochastic processes. In this context we want to mention the works of Vershik on the classification of filtrations [76, 77], where an iterated Kantorovich distance is central to his criterion for standardness of filtrations, see also [37, Theorem 2.7], and Rüschendorf's modification of the Wasserstein distance running under the name of 'Markov constructions' [71]. The extension from two to multiple time steps was done by Pflug [66] using the name "nested distance" to study continuity and stability of multistage stochastic optimization problems. We refer to [67] for a comprehensive overview on this extensive body of works. Lassalle [56] introduces a 'causal optimal transport problem' with an associated distance, to study concentration inequalities on the Wiener space, see also the recent article of Föllmer [39] in similar vein. In the same spirit, adapted Wasserstein distances [11, 17] were introduced to study sensitivity in mathematical finance [13, 16]. Notably, in finite discrete time the adapted Wasserstein distances coincides with the nested distance. It was shown in [17] that the adapted Wasserstein distance naturally extends to the space of filtered processes FP and defines there a complete and geodesically complete metric, compatible with the adapted weak topology. This insight allows to define new ways of taking averages and barycenters of families of stochastic processes, that respect geometry of the paths as well as the underlying filtrations. For example, the barycenter of a family of martingales is again a martingale which is not the case for classical Wasserstein barycenters, [4, 55]. Optimal transport-based methods for the numerical resolution were explored in [38, 68] and the adapted empirical measure for the statistical estimation of laws of stochastic processes was studied, e.g., in [1, 9]. Closely related to these advances is the distance for solution of SDEs [15, 26, 70] and the Knothe–Rosenblatt distance [21], which were recently investigated.

Finally, we also report on the approach based on signature methods in [28], which opens up the framework of adapted topologies to computationally efficient algorithms based on kernel methods. For more details, we refer to the preceding Sects. 2 and 3.

On All Adapted Topologies Are Equal

Remarkably, in discrete time all of the approaches discussed above yield the same topology on $\mathrm{FP}_{\mathrm{plain}}$, see [14, Theorem 1.1] and [63, Theorem 11]. This is of importance, since the nested path spaces of the higher-rank prediction processes are quite complex. Consequently, as long as one considers only processes with their generated filtration, which is typically the case in practice, from a topological point of view it is not necessary to consider higher-rank prediction processes. In mathematical terms, this means that when $\mathbf{X}, \mathbf{X}^1, \mathbf{X}^2, \dots,$ is a sequence in $\mathrm{FP}_{\mathrm{plain}}$, then $\mathbf{X}^n \to \mathbf{X}$ in the adapted weak topology $\mathcal{T}$ if and only if this convergence holds in the rank 1-adapted weak topology $\mathcal{T}^{(1)}$. We want to stress that in view of Example 1.3 (and in general for sequential decision making problems) this result requires that the limit of the sequence is element of $\mathrm{FP}_{\mathrm{plain}}$. This result emphasizes what was already observed in Example 1.6 that the higher order prediction process captures information of higher order, which is not necessary when considering natural filtrations.

Further Applications of Adapted Topologies

Over the past 40 years, the adapted weak topology and related concepts found application in multiple fields such as economics, mathematical finance, stochastic control, stochastic analysis and even machine learning. We try to provide a representative cross section: The adapted topologies found application in the study of optimal stopping [33], game options [35], robust hedging [36] and utility maximisation [19] with proportional transaction costs, convergence of filtrations and stability of backward SDEs [34], stability of Doob-Meyer decomposition [61], martingale representation [64], weak transport theory [3, 10, 12, 49], enlargement of filtration [2], Skorokhod representation of sequences of stochastic processes [23, 45], stochastic exit time problems [69], continuity in stochastic finance [13], sensitivity analysis of stochastic control problems [16], and in a hypothesis test for testing martingality [27]. Moreover, we refer to [72] for a comprehensive survey on information structures and associated stochastic control topologies for decision making problems. Related ideas were also recently explored in the context of machine learning; we mention, for example, chain rule optimal transport [62], causal Wasserstein GANs [52, 79], conditional optimal transport [48]. Higher-rank expected signatures [47, 73] were employed in the context of quantitative finance for real-world calibration and optimal stopping.

Acknowledgments CL will acknowledge the support from the National Key R&D Program of China (No. 2023YFA1010900).

References

1. B. Acciaio, S. Hou, Convergence of adapted empirical measures on $\mathbb{R}^d$. Preprint. arXiv:2211.10162 (2022)
2. B. Acciaio, J. Backhoff-Veraguas, A. Zalashko, Causal optimal transport and its links to enlargement of filtrations and continuous-time stochastic optimization. Stoch. Process. Appl. **130**(5), 2918–2953 (2020)
3. B. Acciaio, M. Beiglböck, G. Pammer, Weak transport for non-convex costs and model-independence in a fixed-income market. Math. Finance **31**(4), 1423–1453 (2021)
4. B. Acciaio, D. Kršek, G. Pammer, Multicausal transport: barycenters and dynamic matching. Preprint. arXiv:2401.12748 (2024)
5. D. Aldous, Stopping times and tightness. Ann. Probab. **6**, 335–340 (1978)
6. D. Aldous, Weak convergence and general theory of processes. Unpublished monograph: Department of Statistics, University of California, Berkeley, July 1981
7. D. Aldous, Stopping times and tightness. II. Ann. Probab. **17**, 586–595 (1989)
8. M. Avellaneda, A. Levy, A. Parás, Pricing and hedging derivative securities in markets with uncertain volatilities. Appl. Math. Finance **2**(2), 73–88 (1995)
9. J. Backhoff, D. Bartl, M. Beiglböck, J. Wiesel, Estimating processes in adapted Wasserstein distance. Ann. Appl. Probab. **32**(1), 529–550 (2022)
10. J. Backhoff-Veraguas, G. Pammer, Stability of martingale optimal transport and weak optimal transport. Ann. Appl. Probab. **32**(1), 721–752 (2022)
11. J. Backhoff-Veraguas, M. Beiglböck, Y. Lin, A. Zalashko, Causal transport in discrete time and applications. SIAM J. Optim. **27**(4), 2528–2562 (2017)
12. J. Backhoff-Veraguas, M. Beiglböck, G. Pammer, Existence, duality, and cyclical monotonicity for weak transport costs. Calc. Variations Partial Differential Equations **58**(6), 203 (2019)
13. J. Backhoff-Veraguas, D. Bartl, M. Beiglböck, M. Eder, Adapted Wasserstein distances and stability in mathematical finance. Finance Stoch. **24**(3), 601–632 (2020)

14. J. Backhoff-Veraguas, D. Bartl, M. Beiglböck, M. Eder, All adapted topologies are equal. Probab. Theory Relat. Fields **178**, 1125–1172 (2020)
15. J. Backhoff-Veraguas, S. Källblad, B.A. Robinson, Adapted Wasserstein distance between the laws of SDEs. Preprint. arXiv:2209.03243 (2022)
16. D. Bartl, J. Wiesel, Sensitivity of multiperiod optimization problems with respect to the adapted Wasserstein distance. SIAM J. Financ. Math. **14**(2), 704–720 (2023)
17. D. Bartl, M. Beiglböck, G. Pammer, The Wasserstein space of stochastic processes. arXiv:2104.14245v1 (2021)
18. C. Bayer, P. Friz, J. Gatheral, Pricing under rough volatility. Quant. Finance **16**(6), 887–904 (2016)
19. E. Bayraktar, L. Dolinskyi, Y. Dolinsky, Extended weak convergence and utility maximisation with proportional transaction costs. Finance Stoch. **24**(4), 1013–1034 (2020)
20. M. Beiglböck, P. Henry-Labordère, F. Penkner, Model-independent bounds for option prices—a mass transport approach. Finance Stoch. **17**(3), 477–501 (2013)
21. M. Beiglböck, G. Pammer, A. Posch, The Knothe–Rosenblatt distance and its induced topology. Preprint. arXiv:2312.16515 (2023)
22. M. Beiglböck, G. Pammer, S. Schrott, X. Zhang, Representing general stochastic processes as martingale laws. Preprint. arXiv:2312.16725 (2023)
23. M. Beiglböck, S. Pflügl, S. Schrott, A probabilistic view on the adapted Wasserstein distance. Preprint. arXiv:2406.19810 (2024)
24. M. Bennedsen, A. Lunde, M.S. Pakkanen, Hybrid scheme for Brownian semistationary processes. Finance Stoch. **21**(4), 931–965 (2017)
25. P. Billingsley, *Convergence of Probability Measures*. Wiley Series in Probability and Statistics: Probability and Statistics, 2nd edn. (Wiley, New York, 1999), A Wiley-Interscience Publication
26. J. Bion–Nadal, D. Talay, On a Wasserstein-type distance between solutions to stochastic differential equations. Ann. Appl. Probab. **29**(3), 1609–1639 (2019)
27. J. Blanchet, J. Wiesel, E. Zhang, Z. Zhang, Empirical martingale projections via the adapted Wasserstein distance. Preprint. arXiv:2401.12197 (2024)
28. P. Bonnier, C. Liu, H. Oberhauser, Adapted topologies and higher rank signatures. Ann. Appl. Probab. **33**(3), 2136–2175 (2023)
29. B. Bouchard, M. Nutz, Arbitrage and duality in nondominated discrete-time models. Ann. Appl. Probab. **25**(2), 823–859 (2015)
30. T. Cass, T. Lyons, C. Salvi, W. Yang, The Signature kernel is the solution of a Goursat PDE. SIAM J. Math. Data Sci. **3**(3), 873–899 (2021)
31. I. Chevyrev, H. Oberhauser, Signature moments to characterize laws of stochastic processes. J. Mach. Learn. Res. **23**, 1–42 (2022)
32. R. Cont, Model uncertainty and its impact on the pricing of derivative instruments. Math. Finance **16**(3), 519–547 (2006)
33. F. Coquet, S. Toldo, Convergence of values in optimal stopping and convergence of optimal stopping times. Electron. J. Probab. **12**(none), 207–228 (2007)
34. F. Coquet, J. Mémin, L. Słominski, On weak convergence of filtrations, in *Séminaire de probabilités XXXV* (2001), pp. 306–328
35. Y. Dolinsky, Applications of weak convergence for hedging of game options. Ann. Appl. Probab. **20**(5), 1891–1906 (2010)
36. Y. Dolinsky, Hedging of game options under model uncertainty in discrete time. Electron. Commun. Probab. **19**(none), 1–11 (2014)
37. L. Dubins, J. Feldman, M. Smorodinsky, B. Tsirelson, Decreasing sequences of σ-fields and a measure change for Brownian motion. Ann. Probab. **24**(2), 882–904 (1996)
38. S. Eckstein, G. Pammer, Computational methods for adapted optimal transport. Ann. Appl. Probab. **34**(1A), 675–713 (2024)
39. H. Föllmer, Optimal couplings on Wiener space and an extension of Talagrand's transport inequality, in *Stochastic Analysis, Filtering, and Stochastic Optimization: A Commemorative Volume to Honor Mark HA Davis's Contributions* (Springer, 2022), pp. 147–175

40. H. Föllmer, A. Schied, *Stochastic Finance: An Introduction in Discrete Time* (Walter de Gruyter, Berlin, 2011)
41. A. Galichon, P. Henry-Labordère, N. Touzi, A stochastic control approach to no-arbitrage bounds given marginals, with an application to lookback options. Ann. Appl. Probab. **24**(1), 312–336 (2014)
42. M.F. Hellwig, Sequential decisions under uncertainty and the maximum theorem. J. Math. Econ. **25**(4), 443–464 (1996)
43. C. Herrera, F. Krach, P. Ruyssen, J. Teichmann, Optimal stopping via randomized neural networks. Front. Math. Finance **3**(1), 31–77 (2024)
44. D.G. Hobson, Robust hedging of the lookback option. Finance Stoch. **2**, 329–347 (1998)
45. D.N. Hoover, Convergence in distribution and Skorokhod convergence for the general theory of processes. Probab. Theory Relat. Fields **89**, 239–259 (1991)
46. D.N. Hoover, H.J. Keisler, Adapted probability distributions. Trans. Am. Math. Soc. **286**(1), 159–201 (1984)
47. B. Horvath, M. Lemercier, C. Liu, T. Lyons, C. Salvi, Optimal stopping via distribution regression: a higher rank signature approach. arXiv: 2304.01479 (2023)
48. B. Hosseini, A.W. Hsu, A. Taghvaei, Conditional optimal transport on function spaces. Preprint. arXiv:2311.05672 (2023)
49. B. Jourdain, G. Pammer, An extension of martingale transport and stability in robust finance. Electron. J. Probab. **29**, 1–30 (2024)
50. C. Kardaras, G. Žitković, Stability of the utility maximization problem with random endowment in incomplete markets. Math. Finance **21**(2), 313–333 (2011)
51. J. Keisler, Hyperfinite models of adapted probability logic. Ann. Pure Appl. Logic, **31**, 71–86 (1986)
52. K. Klemmer, T. Xu, B. Acciaio, D.B. Neill, SPATE-GAN: Improved generative modeling of dynamic spatio-temporal patterns with an autoregressive embedding loss, in *Proceedings of the AAAI Conference on Artificial Intelligence*, vol. 36, no. 4 (June 2022), pp. 4523–4531
53. F.B. Knight, A predictive view of continuous time processes. Ann. Probab. **3**(4), 573–596 (1975)
54. D. Kramkov, M. Sîrbu, Sensitivity analysis of utility-based prices and risk-tolerance wealth processes. Annals Appl. Probab. **16**(4), 2140–2194 (2006)
55. D. Kršek, G. Pammer, General duality and dual attainment for adapted transport. Preprint. arXiv:2401.11958 (2024)
56. R. Lassalle, Causal transport plans and their Monge–Kantorovich problems. Stoch. Anal. Appl. **36**(3), 452–484 (2018)
57. D. Lee, H. Oberhauser, The signature kernel. arXiv:2305.04625 (2023)
58. M. Lemercier, C. Salvi, T. Damoulas, E. Bonilla, T. Lyons, Distribution regression for sequential data, in *International Conference on Artificial Intelligence and Statistics* (2021)
59. F.A. Longstaff, E.S. Schwartz, Valuing American options by simulation: a simple least-squares approach. Rev. Financ. Stud. **14**(1), 113–147 (2001)
60. T. Lyons, Uncertain volatility and the risk-free synthesis of derivatives. Appl. Math. Finance **2**(2), 117–133 (1995)
61. J. Mémin, Stability of Doob-Meyer decomposition under extended convergence. Acta Math. Appl. Sin. **19**(2), 177–190 (2003)
62. F. Nielsen, K. Sun, Chain rule optimal transport, in *Progress in Information Geometry: Theory and Applications* (Springer, 2021), pp. 191–217
63. G. Pammer, A note on the adapted weak topology in discrete time. Electron. Commun. Probab. **29**, 1–13 (2024)
64. A. Papapantoleon, D. Possamaï, and A. Saplaouras, Stability results for martingale representations: the general case. Trans. Am. Math. Soc. **372**(8), 5891–5946 (2019)
65. J. Park, K. Muandet, A measure-theoretic approach to kernel conditional mean embeddings, in *Advances in Neural Information Processing Systems*, vol. 33 (2020), pp. 21247–21259
66. G. Pflug, Version-independence and nested distributions in multistage stochastic optimization. SIAM J. Optim. **20**(3), 1406–1420 (2010)

67. G. Pflug, A. Pichler, *Multistage Stochastic Optimization*. Springer Series in Operations Research and Financial Engineering (Springer, Cham, 2014)
68. A. Pichler, M. Weinhardt, The nested Sinkhorn divergence to learn the nested distance. Comput. Manag. Sci. **19**(2), 269–293 (2022)
69. C. Reisinger, Y. Zhang, Regularity and stability of feedback relaxed controls. SIAM J. Control Optim. **59**(5), 3118–3151 (2021)
70. B.A. Robinson, M. Szölgyenyi, Bicausal optimal transport for SDEs with irregular coefficients. Preprint. arXiv:2403.09941 (2024)
71. L. Rüschendorf, The Wasserstein distance and approximation theorems. Probab. Theory Relat. Fields **70**(1), 117–129 (1985)
72. N. Saldi, S. Yüksel, Geometry of information structures, strategic measures and associated stochastic control topologies. Probab. Surv. **19**, 450–532 (2022)
73. C. Salvi, M. Lemercier, C. Liu, B. Horvath, T. Damoulas, T. Lyons, Higher order kernel mean embeddings to capture filtrations of stochastic processes, in *Advances in Neural Information Processing Systems*, vol. 34 (2021), pp. 16635–16647
74. F. Santambrogio, *Optimal Transport for Applied Mathematicians*, vol. 55, no. 58–63 (Birkäuser, New York, 2015), p. 94
75. M. Soner, N. Touzi, J. Zhang, Quasi-sure stochastic analysis through aggregation. Electron. J. Probab. **16**(none), 1844–1879 (2011)
76. A.M. Vershik, Theory of decreasing sequences of measurable partitions. Algebra Anal. **6**(4), 1–68 (1994)
77. A.M. Vershik, The theory of filtrations of subalgebras, standardness, and independence. Russian Math. Surv. **72**(2), 257 (2017)
78. C. Villani et al., *Optimal Transport: Old and New*, vol. 338 (Springer, Cham, 2009)
79. T. Xu, L.K. Wenliang, M. Munn, B. Acciaio, COT-GAN: Generating sequential data via causal optimal transport, in *Advances in Neural Information Processing Systems*, vol. 33 (2020), pp. 8798–8809

On Expected Signatures and Signature Cumulants in Semimartingale Models

Peter K. Friz, Paul P. Hager, and Nikolas Tapia

Abstract The signature transform, a Cartan type development, translates paths into high-dimensional feature vectors, capturing their intrinsic characteristics. Under natural conditions, the expectation of the signature determines the law of the signature, providing a statistical summary of the data distribution. This property facilitates robust modeling and inference in machine learning and stochastic processes. Building on previous work by the present authors (Friz et al., Unified signature cumulants and generalized Magnus expansions. In Forum of Mathematics, Sigma, vol. 10, p. e42, 2022) we here revisit the actual computation of expected signatures, in a general semimartingale setting. Several new formulae are given. A log-transform of (expected) signatures leads to log-signatures (signature cumulants), offering a significant reduction in complexity.

1 Introduction

Signatures and expected signatures have become increasingly important in recent years, offering a top-down description of sequential data and its statistics, respectively. To formalize this recall the tensor series over $\mathbb{R}^d$, denoted $\mathcal{T} := T\big((\mathbb{R}^d)\big)$, defined as infinite direct product of tensor powers of $\mathbb{R}^d$, with elements of the form

P. K. Friz
Institut für Mathematik, Technische Universität Berlin, Berlin, Germany

Weierstrass Institute, Berlin, Germany
e-mail: friz@math.tu-berlin.de

P. P. Hager (✉)
Department of Statistics and Operations Research, University of Vienna, Wien, Austria
e-mail: paul.peter.hager@univie.ac.at

N. Tapia
Weierstrass Institute, Berlin, Germany

Department of Mathematics, Humboldt University Berlin, Berlin, Germany
e-mail: tapia@wias-berlin.de

C. Bayer et al. (eds.), *Signature Methods in Finance*, Springer Finance Lecture Notes, https://doi.org/10.1007/978-3-031-97239-3_11

$$\mathbf{x} = (\mathbf{x}^{(0)}, \mathbf{x}^{(1)}, \mathbf{x}^{(2)}, \dots) \equiv \mathbf{x}^{(0)} + \mathbf{x}^{(1)} + \mathbf{x}^{(2)} + \cdots,$$

$\mathbf{x}^{(k)} \in (\mathbb{R}^d)^{\otimes k}$, $k \in \mathbb{N}$, with both $(\mathbb{R}^d)^{\otimes 0} \cong \mathbb{R}$ and $(\mathbb{R}^d)^{\otimes 1} \cong \mathbb{R}^d$ embedded in $\mathcal{T}$. Tensor series can be multiplied, so that $\mathcal{T}$ carries a natural algebra structure, exp and log are then defined by their usual power series. Given a smooth $\mathbb{R}^d$-valued path X, its *signature* on $[0, T]$, is obtained from solving the universal linear differential equation

$$\dot{\mathbf{S}} = \mathbf{S}\dot{X}, \qquad \mathbf{S}_0 = 1 \equiv (1, 0, 0, \dots) \in \mathcal{T}_1, \tag{1}$$

which just rephrases the common definition in terms of iterated integrals. We then set $\mathrm{Sig}(X)_{0,T} := \mathbf{S}_T$. Here and below, write $\mathcal{T}_\lambda$ for series starting with (scalar) λ. It is known, essentially as consequence of the chain rule, that the log-signature $\log \mathbf{S}_T$ takes values in $\mathcal{L} \equiv \mathrm{Lie}\big((\mathbb{R}^d)\big)(\subset \mathcal{T}_0)$, the (generically infinite) Lie series over $\mathbb{R}^d$. cf. [24, 29, 55].

Equation (1) remains meaningful when X is a sufficiently smooth $\mathcal{T}_0$-valued path, including the important case of $\mathcal{L}$-valued path; in which case $\mathbf{S}$ is known as smooth geometric rough path, of recent importance in signature kernel methods [6, 44].

Now, many paths of interest arise as sample paths from stochastic processes, far from continuously differentiable, leave alone smooth. A natural class of stochastic processes, where (1) still makes sense - as (Stratonovich) stochastic differential equation - is given by semimartingales. Let accordingly $\mathscr{S}^c(\mathbb{R}^d)$ denote the class of continuous, d-dimensional semimartingales on some filtered probability space $(\Omega, (\mathcal{F}_t)_{t\geq 0}, \mathbb{P})$. We recall that this is the somewhat decisive class of stochastic processes that allows for a reasonable, stochastic integration theory, notably required by no-arbitrage based continuous-time finance. Classic texts include [37, 42, 54, 56], as a concise introduction we recommend Chapter 1 of [2]. Similar to the above one considers, for $X \in \mathscr{S}^c(\mathbb{R}^d)$,

$$\mathrm{d}\mathbf{S} = \mathbf{S} \circ \mathrm{d}X, \qquad \mathbf{S}_0 = 1, \tag{2}$$

the solution of which is expressed in terms of iterated (Stratonovich) integrals. As before, (2) remains meaningful in greater generality, with X replaced by a $\mathcal{T}_0$-valued semimartingale, say $\mathbf{X} \in \mathscr{S}^c(\mathcal{T}_0)$. We then set $\mathrm{Sig}(\mathbf{X})_{0,T}(\omega) := \mathbf{S}_T(\omega)$, a (random) element in $\mathcal{T}_1$. Since $\mathcal{T}_0$ (resp. $\mathcal{T}_1$) are simple examples of Lie algebras (resp. groups), the linear Stratonovich SDE (2) then falls into a classical setting [33, 52].

Whenever the (generalized) signature $\mathrm{Sig}(\mathbf{X})_{0,T}$ is component-wise integrable, we define the *expected signature* and *signature cumulants* by

$$\boldsymbol{\mu}(T) := \mathbb{E}(\mathrm{Sig}(\mathbf{X})_{0,T}) \in \mathcal{T}_1, \quad \boldsymbol{\kappa}(T) := \log \boldsymbol{\mu}(T) \in \mathcal{T}_0.$$

Already in the case where $\mathbf{X}$ is deterministic and sufficiently regular, this leads to an interesting (ordinary differential) equation for $\boldsymbol{\kappa}$ with accompanying (Magnus) expansion, well understood as effective computational tool [7, 36]. Lyons [47] is an excellent survey with a variety of applications, ranging from machine learning to numerical algorithms on Wiener space known as cubature [48]; signature cumulants were named and first studied in their own right in [9], providing, in particular, formulas for converting signature moments to signature cumulants and vice versa.

In the special case of $d = 1$ and $\mathbf{X} = (0, X, 0, \dots)$ where X is a scalar semimartingale, $\boldsymbol{\mu}(T)$ and $\boldsymbol{\kappa}(T)$ are nothing but the sequence of moments and cumulants of the real valued random variable $X_T - X_0$. When $d > 1$, signature moments/cumulants provide an effective way to describe the process X on $[0, T]$, see [15, 43, 47]. The question arises how to compute them. If one takes $\mathbf{X}$ as d-dimensional Brownian motion, the signature cumulant $\boldsymbol{\kappa}(T)$ equals $(T/2)\mathbf{I}_d$, where $\mathbf{I}_d$ is the identity 2-tensor over $\mathbb{R}^d$. This is known as *Fawcett's formula*, [21, 48].

Exposing and building on [28], loosely speaking, our main functional equations (Theorems 3.4 and 3.6) are a vast generalization of Fawcett's formula. Projecting these equations to tensor levels, we obtain recursions for signature moments and signature cumulants respectively (Theorems 4.1 and 4.4). These recursions represent $\boldsymbol{\mu}^{(n)}$ in terms of $\boldsymbol{\mu}^{(1)}, \dots, \boldsymbol{\mu}^{(n-1)}$ and $\boldsymbol{\kappa}^{(n)}$ in terms of $\boldsymbol{\kappa}^{(1)}, \dots, \boldsymbol{\kappa}^{(n-1)}$ using the underlying dynamic structure of the probability space.

This chapter is organized as follows: Sect. 2 introduces the mathematical preliminaries, including tensor algebra, continuous semimartingales taking values therein, generalized signatures, and their conditional expectations. In Sect. 3, we derive functional equations for the expected signature and signature cumulants, transitioning from discrete-time processes to continuous semimartingales. Section 4 develops the resulting recursive formulas for signature moments and cumulants and demonstrates how they relate to the so-called diamond products of semimartingales. In Sect. 5, we examine the implications of the general functional equations for multivariate moments and cumulants by projecting to the symmetric algebra. In Sect. 6, we apply the main results to time-inhomogeneous Lévy processes and Brownian rough paths. Finally, Sect. 7 summarizes all main results and corollaries and gives an outlook to potential applications and directions for future research.

2 Preliminaries

2.1 The Tensor Algebra and Tensor Series

Denote by $T(\mathbb{R}^d)$ the tensor algebra over $\mathbb{R}^d$, i.e.

$$T(\mathbb{R}^d) := \bigoplus_{k=0}^{\infty} (\mathbb{R}^d)^{\otimes k},$$

elements of which are *finite* sums (tensor polynomials) of the form

$$\mathbf{x} = \sum_{k\geq 0} \mathbf{x}^{(k)} = \sum_{w\in\mathcal{W}^d} \mathbf{x}^w e_w \tag{3}$$

with $\mathbf{x}^{(k)} \in (\mathbb{R}^d)^{\otimes k}$, $\mathbf{x}^w \in \mathbb{R}$ and linear basis vectors $e_w := e_{i_1}\cdots e_{i_k} \in (\mathbb{R}^d)^{\otimes k}$ where w ranges over all words $w = i_1\cdots i_k \in \mathcal{W}_d$ over the alphabet $\{1,\ldots,d\}$. Note $\mathbf{x}^{(k)} = \sum_{|w|=k} \mathbf{x}^w e_w$ where $|w|$ denotes the length a word w. The element $e_\emptyset = 1 \in (\mathbb{R}^d)^{\otimes 0} \cong \mathbb{R}$ is the neutral element of the concatenation (tensor) product, which is obtained by linear extension of $e_w e_{w'} = e_{ww'}$ where $ww' \in \mathcal{W}_d$ denotes concatenation of two words. We thus have, for $\mathbf{x},\mathbf{y} \in T(\mathbb{R}^d)$,

$$\mathbf{x}\mathbf{y} = \sum_{k\geq 0}\sum_{\ell=0}^{k} \mathbf{x}^{(\ell)}\mathbf{y}^{(k-\ell)} = \sum_{w\in\mathcal{W}^d}\left(\sum_{w_1w_2=w} \mathbf{x}^{w_1}\mathbf{y}^{w_2}\right) e_w \in T(\mathbb{R}^d). \tag{4}$$

Note that we reserve the usual product symbol $\otimes$ for another product that will be introduced below. This will also prevent an overflow of product symbols throughout the text.

This extends naturally to *infinite* sums, i.e., tensor series, elements of the "completed" tensor algebra

$$\mathcal{T} := T\big((\mathbb{R}^d)\big) := \prod_{k=0}^{\infty} (\mathbb{R}^d)^{\otimes k},$$

which are written as in (3), but now as formal infinite sums with identical notation and multiplication rules; the resulting algebra $\mathcal{T}$ obviously extends $T(\mathbb{R}^d)$. Denote by $\mathcal{T}_0$ and $\mathcal{T}_1$ the subspaces of tensor series starting with 0 and 1 respectively; that is, $\mathbf{x} \in \mathcal{T}_0$ (resp. $\mathcal{T}_1$) if and only if $\mathbf{x}^\emptyset = 0$ (resp. $\mathbf{x}^\emptyset = 1$). Restricted to $\mathcal{T}_0$ and $\mathcal{T}_1$ respectively, the exponential and logarithm in $\mathcal{T}$, defined by the power series,

$$\exp\colon \mathcal{T}_0 \to \mathcal{T}_1, \quad \mathbf{x} \mapsto \exp(\mathbf{x}) := 1 + \sum_{k=1}^{\infty} \frac{1}{k!}\mathbf{x}^k,$$

$$\log\colon \mathcal{T}_1 \to \mathcal{T}_0, \quad 1+\mathbf{x} \mapsto \log(1+\mathbf{x}) := \sum_{k=1}^{\infty} \frac{(-1)^{k+1}}{k}\mathbf{x}^k,$$

are globally defined and inverse to each other. We will usually abbreviate $e^{\mathbf{x}} = \exp(\mathbf{x})$. The vector space $\mathcal{T}_0$ becomes a *Lie algebra* with the commutator bracket $[-,-]\colon \mathcal{T}_0 \otimes \mathcal{T}_0 \to \mathcal{T}_0$ given by

$$[\mathbf{x},\mathbf{y}] := \mathbf{x}\mathbf{y} - \mathbf{y}\mathbf{x}, \quad \mathbf{x},\mathbf{y} \in \mathcal{T}_0.$$

Define the adjoint operator associated to a Lie-algebra element $\mathbf{y} \in \mathcal{T}_0$ by

$$\operatorname{ad}\mathbf{y}\colon \mathcal{T}_0 \to \mathcal{T}_0,\ \mathbf{x} \mapsto [\mathbf{y}, \mathbf{x}]\,.$$

The exponential image $\mathcal{T}_1 = \exp(\mathcal{T}_0)$ is a Lie group, at least formally so. We refrain from equipping the infinite-dimensional $\mathcal{T}_1$ with a differentiable structure, not necessary in view of the "locally finite" nature of the group law $(\mathbf{x}, \mathbf{y}) \mapsto \mathbf{x}\mathbf{y}$, by which we mean that the coefficients of the product $\mathbf{x}\mathbf{y}$ can be computed with a finite number of operations as is clear from Eq. (4).

For two tensor series $\mathbf{x}, \mathbf{y}$ we define their *outer tensor product*

$$\mathbf{x} \otimes \mathbf{y} := \sum_{n=0}^{\infty} \left(\sum_{i=0}^{n} \mathbf{x}^{(i)} \otimes \mathbf{y}^{(n-i)} \right).$$

We emphasize that here $\otimes$ does *not* denote the (inner tensor) product in $\mathcal{T}$, for which we did not reserve a symbol, but it denotes another (outer) tensor product. In particular, this expression is not the same as Eq. (4).

Given linear maps $g, f : \mathcal{T} \to \mathcal{T}$ we define

$$(g \odot f)(\mathbf{x} \otimes \mathbf{y}) = g(\mathbf{x}) f(\mathbf{y}) \in \mathcal{T}$$

whenever $\mathbf{x}, \mathbf{y} \in \mathcal{T}$, where the multiplication on the right-hand side is performed in $\mathcal{T}$ according to Eq. (4). It then extends by linearity to the linear span of outer tensors which we will denote by $\mathcal{T} \otimes \mathcal{T}$.

For $n \in \mathbb{N}$, the subspace

$$\mathcal{I}_n := \prod_{k=n+1}^{\infty} (\mathbb{R}^d)^{\otimes k}$$

is a two sided ideal of $\mathcal{T}$, meaning that $\mathcal{T}\mathcal{I}_n + \mathcal{I}_n\mathcal{T} \subseteq \mathcal{I}_n$, which turns out to be the right condition under which the quotient space $\mathcal{T}/\mathcal{I}_n$ has a natural algebra structure. We can identify $\mathcal{T}/\mathcal{I}_n$ with

$$\mathcal{T}^n := \bigoplus_{k=0}^{n} (\mathbb{R}^d)^{\otimes k},$$

equipped with the truncated tensor product

$$\mathbf{x}\mathbf{y} = \sum_{k=0}^{n} \sum_{\ell_1+\ell_2=k} \mathbf{x}^{(\ell_1)} \mathbf{y}^{(\ell_2)} = \sum_{w \in \mathcal{W}^d, |w| \le n} \left(\sum_{w_1 w_2 = w} \mathbf{x}^{w_1} \mathbf{y}^{w_2} \right) e_w \in \mathcal{T}^n,$$

which formally is the same as Eq. (4) after setting $\mathbf{x}^{(k)} = 0$ for all $k > n$ as the above definition suggests. We denote the canonical *projection map* by $\pi_{(0,n)}\colon \mathcal{T} \to \mathcal{T}^n$. The usual power series in $\mathcal{T}^n$ define $\exp_n\colon \mathcal{T}_0^n \to \mathcal{T}_1^n$ with inverse $\log_n\colon \mathcal{T}_1^n \to \mathcal{T}_0^n$, and we may again abuse notation and write exp and log when no confusion arises. As before, $\mathcal{T}_0^n$ has a natural Lie algebra structure, and $\mathcal{T}_1^n$ (now finite dimensional) is a *bona fide* Lie group.

We equip $T(\mathbb{R}^d)$ with the norm

$$|a|_{T(\mathbb{R}^d)} := \max_{k\in\mathbb{N}} |a^{(k)}|_{(\mathbb{R}^d)^{\otimes k}},$$

where $|\cdot|_{(\mathbb{R}^d)^{\otimes k}}$ is the Euclidean norm on $(\mathbb{R}^d)^{\otimes k} \cong \mathbb{R}^{d^k}$, which makes it a Banach space. The same norm makes sense in $\mathcal{T}^n$, and since the definition is consistent in the sense that $|a|_{\mathcal{T}^k} = |a|_{\mathcal{T}^n}$ for any $a \in \mathcal{T}^n$ and $k \geq n$, and $|a|_{\mathcal{T}^n} = |a|_{(\mathbb{R}^d)^{\otimes n}}$ for any $a \in (\mathbb{R}^d)^{\otimes n}$. We will drop the index whenever it is possible and write simply $|a|$. For a word $w \in \mathcal{W}_d$ with $|w| > 0$ we define the directional derivative for a function $f\colon \mathcal{T} \to \mathbb{R}$ by $(\partial_w f)(\mathbf{x}) := \partial_t f(\mathbf{x} + te_w)\big|_{t=0}$, whenever the derivative exists. In particular, for the exponential function we have

$$\begin{aligned}
\partial_w \exp(\mathbf{x}) &= G(\operatorname{ad}\mathbf{x})(e_w)\exp(\mathbf{x})\\
\partial_{w'}\partial_w \exp(\mathbf{x}) &= Q(\operatorname{ad}\mathbf{x})(e_w \otimes e_{w'})\exp(\mathbf{x}),
\end{aligned}$$

where the operators $G(\operatorname{ad}\mathbf{x})\colon \mathcal{T}_0 \to \mathcal{T}_0$ and $Q(\operatorname{ad}\mathbf{x})\colon \mathcal{T}_0 \otimes \mathcal{T}_0 \to \mathcal{T}_0$ are defined by the power series

$$\begin{aligned}
G(\operatorname{ad}\mathbf{x})(\mathbf{y}) &= \frac{e^{\operatorname{ad}\mathbf{x}} - 1}{\operatorname{ad}\mathbf{x}}(\mathbf{y}) = \sum_{k=0}^{\infty} \frac{1}{(k+1)!}(\operatorname{ad}\mathbf{x})^k(\mathbf{y})\\
&= \mathbf{y} + \frac{1}{2}[\mathbf{x},\mathbf{y}] + \frac{1}{6}[\mathbf{x},[\mathbf{x},\mathbf{y}]] + \cdots\\
Q(\operatorname{ad}\mathbf{x})(\mathbf{y}\otimes\mathbf{y}') &= 2\sum_{n,m=0}^{\infty} \frac{(\operatorname{ad}\mathbf{x})^n \odot (\operatorname{ad}\mathbf{x})^m}{(n+1)!m!(n+m+2)}(\mathbf{y}\otimes\mathbf{y}')\\
&= \mathbf{y}\mathbf{y}' + \frac{1}{3}[\mathbf{x},\mathbf{y}]\mathbf{y}' + \frac{1}{3}\mathbf{y}[\mathbf{x},\mathbf{y}'] + \frac{1}{6}[\mathbf{x},\mathbf{y}][\mathbf{x},\mathbf{y}'] + \cdots
\end{aligned} \tag{5}$$

The first identity is well known and can be traced back to Schur [57]. With this identities one also obtains the following formal Taylor expansion of the exponential map:

$$\exp(\mathbf{x})\exp(-\mathbf{y}) - 1 = G(\operatorname{ad}\mathbf{x})(\mathbf{x}-\mathbf{y}) + Q(\operatorname{ad}\mathbf{x})((\mathbf{x}-\mathbf{y})\otimes(\mathbf{x}-\mathbf{y})) + \cdots \tag{6}$$

where we recall that the symbol $\otimes$ is meant as an *outer* tensor product. Naturally, similar formulas hold in the truncated tensor algebra where exp is replaced by $\exp_n$.

2.2 Semimartingales

Let $\mathscr{C}$ be the space of continuous adapted processes $X\colon \Omega \times [0,T) \to \mathbb{R}$, with $T \in (0,\infty]$, defined on some filtered probability space $(\Omega, (\mathcal{F}_t)_{0\le t\le T}, \mathbb{P})$. The space of *continuous semimartingales* $\mathscr{S}^c$ is given by the processes $X \in \mathscr{C}$ that can uniquely be decomposed as

$$X_t = X_0 + M_t + A_t,$$

where $M \in \mathscr{M}_{\mathrm{loc}}$ is a continuous local martingale, and $A \in \mathscr{V}$ is an adapted process of locally bounded variation, both started at zero. The (predictable) quadratic variation process of X is denoted by $\langle X\rangle_t$. Covariation angle brackets $\langle X, Y\rangle$, for another real-valued semimartingale Y, are defined by polarization. For $q \in [1,\infty)$, write $\mathcal{L}^q = L^q(\Omega, \mathcal{F}, \mathbb{P})$, then a Banach space $\mathscr{H}^q \subset \mathscr{S}^c$ is given by those $X \in \mathscr{S}^c$ with $X_0 = 0$ and

$$\|X\|_{\mathscr{H}^q} := \left\| \langle M\rangle_T^{1/2} + \int_0^T |\mathrm{d}A_s| \right\|_{\mathcal{L}^q} < \infty.$$

For a process $X \in \mathscr{C}$ we define

$$\|X\|_{\mathscr{S}^q} := \left\| \sup_{0\le t\le T} |X_t| \right\|_{\mathcal{L}^q}$$

and define the space $\mathscr{S}^q \subset \mathscr{S}^c$ of semimartingales $X \in \mathscr{S}^c$ such that $\|X\|_{\mathscr{S}^q} < \infty$. Note that there exists a constant $c_q > 0$ depending on q such that (see [54, Ch. V, Theorem 2])

$$\|X\|_{\mathscr{S}^q} \le c_q \|X\|_{\mathscr{H}^q}. \tag{7}$$

We view d-dimensional semimartingales, $X = \sum_{i=1}^d X^i e_i \in \mathscr{S}^c(\mathbb{R}^d)$, as special cases of tensor series valued semimartingales $\mathscr{S}^c(\mathcal{T})$ of the form

$$\mathbf{X} = \sum_{w\in\mathcal{W}_d} \mathbf{X}^w e_w$$

with each component $\mathbf{X}^w$ a real-valued continuous semimartingale. This extends mutatis mutandis to the spaces of $\mathcal{T}$-valued adapted continuous processes $\mathscr{C}(\mathcal{T})$, martingales $\mathscr{M}(\mathcal{T})$ and processes of finite variation $\mathscr{V}(\mathcal{T})$. Note also that we typically deal with $\mathcal{T}_0$-valued semimartingales $\mathbf{X}$ which amounts to have only words with length $|w| \ge 1$. Standard notions such as continuous local martingale $\mathbf{X}^c$ and jump process $\Delta\mathbf{X}_t = \mathbf{X}_t - \mathbf{X}_{t-}$ are defined componentwise. To weaken integrability requirements later, we often deal semimartingales taking values in the truncated tensor algebra $\mathcal{T}^N$ and will omit specifically extending notations when it is straightforward.

Brackets Now let $\mathbf{X}$ and $\mathbf{Y}$ be $\mathcal{T}$-valued semimartingales. We define the (non-commutative) *outer quadratic covariation bracket* of $\mathbf{X}$ and $\mathbf{Y}$ by

$$\langle\!\langle \mathbf{X}, \mathbf{Y} \rangle\!\rangle_t := \sum_{w_1, w_2 \in \mathcal{W}_d} \langle \mathbf{X}^{w_1}, \mathbf{Y}^{w_2} \rangle_t e_{w_1} \otimes e_{w_2} \in \mathcal{T} \otimes \mathcal{T}.$$

Similarly, define the (non-commutative) *inner quadratic covariation bracket* by

$$\langle \mathbf{X}, \mathbf{Y} \rangle_t := \sum_{w \in \mathcal{W}_d} \left(\sum_{w_1 w_2 = w} \langle \mathbf{X}^{w_1}, \mathbf{Y}^{w_2} \rangle_t \right) e_w \in \mathcal{T}.$$

As usual, we may write $\langle\!\langle \mathbf{X} \rangle\!\rangle \equiv \langle\!\langle \mathbf{X}, \mathbf{X} \rangle\!\rangle$ and $\langle \mathbf{X} \rangle \equiv \langle \mathbf{X}, \mathbf{X} \rangle$.

$\mathscr{H}$-Spaces The definition of $\mathscr{H}^q$-norm naturally extends to tensor valued martingales. More precisely, for $\mathbf{X}^{(n)} \in \mathscr{S}^c((\mathbb{R}^d)^{\otimes n})$ with $n \in \mathbb{N}_{\geq 1}$ and $q \in [1, \infty)$ we define

$$\|\mathbf{X}^{(n)}\|_{\mathscr{H}^q} := \|\mathbf{X}^{(n)}\|_{\mathscr{H}^q((\mathbb{R}^d)^{\otimes n})} := \left\| |\langle \mathbf{M} \rangle|_T^{1/2} + |\mathbf{A}|_{1\text{-var};[0;T]} \right\|_{\mathcal{L}^q},$$

where $\mathbf{X}^{(n)} = \mathbf{M} + \mathbf{A}$ with $\mathbf{M} \in \mathscr{M}_{\mathrm{loc}}((\mathbb{R}^d)^{\otimes n})$ and $\mathbf{A} \in \mathscr{V}((\mathbb{R}^d)^{\otimes n})$, with

$$|\mathbf{A}|_{1\text{-var};[0;T]} := \sup_{0 \leq t_1 \leq \cdots \leq t_k \leq T} \sum_{t_i} \left| \mathbf{A}_{t_{i+1}} - \mathbf{A}_{t_i} \right| \leq \sum_{w \in \mathcal{W}_d, |w| = n} \int_0^T \left| \mathrm{d}A_s^w \right|,$$

with the supremum taken over all partitions of the interval $[0, T]$. One may readily check that

$$\|\mathbf{X}^{(n)}\|_{\mathscr{H}^q} \leq \sum_{w \in \mathcal{W}_d, |w| = n} \|\mathbf{X}^w\|_{\mathscr{H}^q}$$

and for $\mathbf{X}^{(n)} \in \mathscr{M}_{\mathrm{loc}}((\mathbb{R}^d)^{\otimes n})$:

$$\|\mathbf{X}^{(n)}\|_{\mathscr{H}^q} = \left\| |\langle \mathbf{X}^{(n)} \rangle|_T^{1/2} \right\|_{\mathcal{L}^q}.$$

Further define the following subspace $\mathscr{H}^{q,N} \subset \mathscr{S}^c(\mathcal{T}_0^N)$ of homogeneously q-integrable semimartingales

$$\mathscr{H}^{q,N} := \left\{ \mathbf{X} \in \mathscr{S}^c(\mathcal{T}_0^N) \;\middle|\; \mathbf{X}_0 = 0,\ \|\!|\mathbf{X}|\!\|_{\mathscr{H}^{q,N}} < \infty \right\},$$

where for any $\mathbf{X} \in \mathscr{S}^c(\mathcal{T}^N)$ we define

$$\|\!|\mathbf{X}\|\!|_{\mathscr{H}^{q,N}} := \sum_{n=1}^{N} \left(\|\mathbf{X}^{(n)}\|_{\mathscr{H}^{qN/n}} \right)^{1/n}.$$

Note that $\|\!|\cdot\|\!|_{\mathscr{H}^{q,N}}$ is sub-additive and positive definite on $\mathscr{H}^{q,N}$ and it is *homogeneous* under dilation in the sense that

$$\|\!|\delta_\lambda \mathbf{X}\|\!|_{\mathscr{H}^{q,N}} = |\lambda| \; \|\!|\mathbf{X}\|\!|_{\mathscr{H}^{q,N}}, \quad \delta_\lambda \mathbf{X} := (\mathbf{X}^{(0)}, \lambda \mathbf{X}^{(1)}, \dots, \lambda^N \mathbf{X}^{(N)}), \quad \lambda \in \mathbb{R}.$$

We also introduce the following subspace of $\mathscr{S}^c(\mathcal{T})$

$$\mathscr{H}^{\infty-}(\mathcal{T}) := \left\{ \mathbf{X} \in \mathscr{S}^c(\mathcal{T}) : \; \mathbf{X}^w \in \mathscr{H}^q, \; \forall\, 1 \le q < \infty, \; w \in \mathcal{W}_d \right\}.$$

Note that if $\mathbf{X} \in \mathscr{S}^c(\mathcal{T})$ such that $\|\!|\mathbf{X}^{(0,N)}\|\!|_{\mathscr{H}^{1,N}} < \infty$ for all $N \in \mathbb{N}_{\ge 1}$ then it also holds $\mathbf{X} \in \mathscr{H}^{\infty-}(\mathcal{T})$.

Stochastic Integrals We are now going to introduce a notation for the stochastic integration with respect to tensor valued semimartingales. Denote by $\mathcal{L}(\mathcal{T};\mathcal{T}) = \{f : \mathcal{T} \to \mathcal{T} \mid f \text{ is linear}\}$ the space of endomorphisms on $\mathcal{T}$ and let

$$\mathbf{F} : \Omega \times [0, T] \to \mathcal{L}(\mathcal{T};\mathcal{T}), \quad (t, \omega) \mapsto \mathbf{F}_t(\omega; \cdot)$$

such that

$$(\mathbf{F}_t(\mathbf{x}))_{0 \le t \le T} \in \mathscr{C}(\mathcal{T}), \quad \text{for all } \mathbf{x} \in \mathcal{T} \tag{8}$$

$$\text{and} \quad \mathbf{F}_t(\omega; \mathcal{I}_n) \subset \mathcal{I}_n, \quad \text{for all } n \in \mathbb{N}, \; (\omega, t) \in \Omega \times [0, T], \tag{9}$$

where $\mathcal{I}_n \subset \mathcal{T}$ was introduced in Sect. 2.1, consisting of series with tensors of level $n+1$ and higher. In this case, we can define the stochastic Itô integral of $\mathbf{F}$ with respect to $\mathbf{X} \in \mathscr{S}^c(\mathcal{T})$ by

$$\int_{(0,\cdot]} \mathbf{F}_t \, \mathrm{d}\mathbf{X}_t := \sum_{w \in \mathcal{W}_d} \; \sum_{v \in \mathcal{W}_d,\, |v| \le |w|} \int_{(0,\cdot]} \mathbf{F}_t(e_v)^w \mathrm{d}\mathbf{X}_t^v e_w \in \mathscr{S}^c(\mathcal{T}). \tag{10}$$

For example, let $\mathbf{Y}, \mathbf{Z} \in \mathscr{C}(\mathcal{T})$ and define $\mathbf{F} := \mathbf{Y} \,\mathrm{Id}\, \mathbf{Z}$, i.e. $\mathbf{F}_t(\mathbf{x}) = \mathbf{Y}_t \,\mathbf{x}\, \mathbf{Z}_t$, the concatenation product from the left and right, for all $\mathbf{x} \in \mathcal{T}$. Then we see that $\mathbf{F}$ indeed satisfies the conditions (8) and (9) and we have

$$\int_{(0,\cdot]} (\mathbf{Y}_t \,\mathrm{Id}\, \mathbf{Z}_t)(\mathrm{d}\mathbf{X}_t) = \int_{(0,\cdot]} \mathbf{Y}_t \mathrm{d}\mathbf{X}_t \mathbf{Z}_t = \sum_{w \in \mathcal{W}_d} \left(\sum_{w_1 w_2 w_3 = w} \int_{(0,\cdot]} \mathbf{Z}_t^{w_1} \mathbf{Y}_t^{w_3} \, \mathrm{d}\mathbf{X}_t^{w_2} \right) e_w. \tag{11}$$

Another important example is given by $\mathbf{F} = (\mathrm{ad}\, \mathbf{Y})^k := \mathrm{ad}\, \mathbf{Y} \circ \cdots \circ \mathrm{ad}\, \mathbf{Y}$ (k-times) for any $\mathbf{Y} \in \mathscr{C}(\mathcal{T}_0)$ and $k \in \mathbb{N}$. Indeed, we immediately see $\mathbf{F}$ satisfies the condition (9) and since the iteration of adjoint operations can be expanded in terms of left- and right-multiplication, we also see that $\mathbf{F}$ satisfies (8). More generally, integrals with respect to power series of adjoint operators are well defined. Indeed, introducing for $\ell \in (\mathbb{N}_{\geq 1})^k$ the notation $\ell = (l_1, \dots, l_k)$, $|\ell| := k$ and $||\ell|| := l_1 + \dots + l_k$ and $(\mathrm{ad}\,\mathbf{x}^{(\ell)}) = (\mathrm{ad}\,\mathbf{x}^{(l_1)} \cdots \mathrm{ad}\,\mathbf{x}^{(l_k)})$ for any $\mathbf{x} \in \mathcal{T}_0$. Then for any $(a_k)_{k=0}^{\infty} \subset \mathbb{R}$ and $\mathbf{X} \in \mathscr{S}^c(\mathcal{T}_0)$ the integral

$$\int_{(0,\cdot]} \left[\sum_{k=0}^{\infty} a_k (\mathrm{ad}\,\mathbf{Y}_t)^k \right] (\mathrm{d}\mathbf{X}_t) = \sum_{n=1}^{\infty} \sum_{k=0}^{n-1} \sum_{\|\ell\|=n,\, |\ell|=k+1} a_k \int_{(0,\cdot]} \mathrm{ad}\,\mathbf{Y}_t^{(\ell)} (\mathrm{d}\mathbf{X}_t^{(l_1)}) \tag{12}$$

is well defined in the sense of (11). The definition of the integral with integrands of the form

$$\mathbf{F} : \Omega \times [0, T] \to \mathcal{L}(\mathcal{T} \otimes \mathcal{T}; \mathcal{T})$$

with respect to processes $\mathbf{X} \in \mathscr{S}^c(\mathcal{T} \otimes \mathcal{T})$ is completely analogous.

2.3 Generalized Signatures

We now give precise meaning to $\mathrm{d}\mathbf{S} = \mathbf{S} \circ \mathrm{d}\mathbf{X}$, or component-wise, for every word $w \in \mathcal{W}_d$,

$$\mathrm{d}\mathbf{S}^w = \sum_{w_1 w_2 = w} \mathbf{S}^{w_1} \circ \mathrm{d}\mathbf{X}^{w_2},$$

where the driving noise $\mathbf{X}$ is a $\mathcal{T}_0$-valued continuous semimartingale, i.e. $\mathbf{X} \in \mathscr{S}^c(\mathcal{T}_0)$. The Itô integral meaning of this equation, when started at time s from $\mathbf{s} \in \mathcal{T}_1$, for times $t \geq s$, is given by

$$\mathbf{S}_t = \mathbf{s} + \int_{(s,t]} \mathbf{S}_u \,\mathrm{d}\mathbf{X}_u + \frac{1}{2} \int_s^t \mathbf{S}_u \,\mathrm{d}\,\langle \mathbf{X} \rangle_u\,, \tag{13}$$

leaving the component-wise version to the reader. We have

Proposition 2.1 *Suppose $\mathbf{X} \in \mathscr{S}^c(\mathcal{T}_0)$. For every $s \in [0, T]$ and $\mathbf{s} \in \mathcal{T}_1$, equation* (13) *has a unique global solution on $\mathcal{T}_1$ starting from $\mathbf{S}_s = \mathbf{s}$.*

Proof Note that $\mathbf{S}$ solves (13) iff $\mathbf{s}^{-1}\mathbf{S}$ solves the same equation started from $1 \in \mathcal{T}_1$. We may thus take $\mathbf{s} = 1$ without loss of generality. The graded structure of our problem, and more precisely that $\mathbf{X} = (0, X, \mathbb{X}, \dots)$ in (13) has no scalar component, shows that the (necessarily) unique solution is given explicitly by iterated integration, as may be seen explicitly when writing out $\mathbf{S}^{(0)} \equiv 1$, $\mathbf{S}_t^{(1)} = \int_s^t \mathrm{d}X = X_{s,t} \in \mathbb{R}^d$,

$$\mathbf{S}_t^{(2)} = \int_{(s,t]} \mathbf{S}_u^{(1)} \, \mathrm{d}X_u + \mathbb{X}_t - \mathbb{X}_s + \frac{1}{2} \langle X \rangle_{s,t} \in (\mathbb{R}^d)^{\otimes 2},$$

and so on. (In particular, we do not need to rely on abstract existence, uniqueness results for SDEs [40] or Lie group stochastic exponentials [33].) □

Definition 2.2 For $\mathbf{X} \in \mathscr{S}^c(\mathcal{T}_0)$ and $s \in [0, T]$ we define $\mathrm{Sig}(\mathbf{X}|_{[s,\cdot]}) \equiv \mathrm{Sig}(\mathbf{X})_{s,\cdot}$ as the unique solution to (13) such that $\mathrm{Sig}(\mathbf{X})_{s,s} = 1$. We call $\mathrm{Sig}(\mathbf{X})_{s,t}$ the generalized signature (short: signature) of $\mathbf{X}$ on $[s, t]$.

Remark 2.3 In rough path theory, one defines signatures for geometric rough paths. This is consistent with Definition 2.2 when $\mathbf{X} \in \mathscr{S}^c(\mathcal{L})$ (recall $\mathcal{L} = \mathrm{Lie}\big((\mathbb{R}^d)\big)$), including the special case of a continuous semimartingale with values in $\mathbb{R}^d$.

The following can be seen as a (generalized) Chen relation.

Lemma 2.4 *Let $\mathbf{X} \in \mathscr{S}^c(\mathcal{T}_0)$ and $0 \le s \le t \le u \le T$. Then the following identity holds with probability one, for all such s, t, u,*

$$\mathrm{Sig}(\mathbf{X})_{s,t}\mathrm{Sig}(\mathbf{X})_{t,u} = \mathrm{Sig}(\mathbf{X})_{s,u}. \tag{14}$$

Proof Call $\Phi_{t\leftarrow s}\mathbf{s} := \mathbf{S}_t$ the solution to (13) at time $t \ge s$, started from $\mathbf{S}_s = \mathbf{s}$. By uniqueness of the solution flow, we have $\Phi_{u\leftarrow t} \circ \Phi_{t\leftarrow s} = \Phi_{u\leftarrow s}$. It now suffices to remark that, thanks to the multiplicative structure of (13) we have $\Phi_{t\leftarrow s}\mathbf{s} = \mathbf{s}\mathrm{Sig}(\mathbf{X})_{s,t}$. □

As usual, when $\mathbf{X} \in \mathscr{S}^c(\mathcal{T}_0^N)$ for some $N \ge 1$ the development (13) is understood within the truncated tensor algebra $\mathcal{T}_1^N$ and therefore $\mathrm{Sig}(\mathbf{X})_{s,t}$ is an element in $\mathcal{T}_1^N$.

2.4 Expected Signatures and Signature Cumulants

Throughout this section let $\mathbf{X} \in \mathscr{S}^c(\mathcal{T}_0)$ be defined on a filtered probability space $(\Omega, \mathcal{F}, (\mathcal{F}_t)_{0\le t\le T}, \mathbb{P})$ satisfying the usual conditions and with the property that every martingale has a continuous version. Recall that $\mathbb{E}_t$ denotes the conditional expectation with respect to the sigma algebra $\mathcal{F}_t$. When $\mathbb{E}(|\mathrm{Sig}(\mathbf{X})_{0,t}^w|) < \infty$ for all $0 \le t \le T$ and all words $w \in \mathcal{W}_d$, then the *(conditional) expected signature*

$$\boldsymbol{\mu}_t(T) := \mathbb{E}_t\left(\mathrm{Sig}(\mathbf{X})_{t,T}\right) = \sum_{w\in\mathcal{W}_d} \mathbb{E}_t(\mathrm{Sig}(\mathbf{X})_{t,T}^w)e_w \in \mathcal{T}_1, \quad 0 \le t \le T,$$

is well defined. In this case, we can also define the *(conditional) signature cumulant* of $\mathbf{X}$ by

$$\boldsymbol{\kappa}_t(T) := \log\left(\boldsymbol{\mu}_t(T)\right) \in \mathcal{T}_0, \quad 0 \le t \le T.$$

An important observation is the following

Lemma 2.5 *Given* $\mathbb{E}(|\mathrm{Sig}(\mathbf{X})^w_{0,t}|) < \infty$ *for all* $0 \le t \le T$ *and words* $w \in \mathcal{W}_d$, *then* $\boldsymbol{\mu}(T) \in \mathscr{S}^c(\mathcal{T}_1)$ *and* $\boldsymbol{\kappa}(T) \in \mathscr{S}^c(\mathcal{T}_0)$.

Proof It follows from the relation (14) that

$$\boldsymbol{\mu}_t(T) = \mathbb{E}_t\left(\mathrm{Sig}(\mathbf{X})_{t,T}\right) = \mathbb{E}_t\left(\mathrm{Sig}(\mathbf{X})^{-1}_{0,t}\mathrm{Sig}(\mathbf{X})_{0,T}\right) = \mathrm{Sig}(\mathbf{X})^{-1}_{0,t}\mathbb{E}_t\left(\mathrm{Sig}(\mathbf{X})_{0,T}\right).$$

Therefore projecting to the tensor components we have

$$\boldsymbol{\mu}_t(T)^w = \sum_{w_1 w_2 = w} (-1)^{|w_1|}\mathrm{Sig}(\mathbf{X})^{\overline{w_1}}_{0,t}\mathbb{E}_t\left(\mathrm{Sig}(\mathbf{X})^{w_2}_{0,T}\right), \quad 0 \le t \le T, \quad w \in \mathcal{W}_d.$$

where $\overline{w_1}$ denotes word reversal. Since $(\mathrm{Sig}(\mathbf{X})^w_{0,t})_{0\le t\le T}$ and $(\mathbb{E}_t(\mathrm{Sig}(\mathbf{X})^w_{0,T})_{0\le t\le T}$ are semimartingales (the latter in fact a martingale), it follows from Itô's product rule that $\boldsymbol{\mu}^w(T)$ is also a semimartingale for all words $w \in \mathcal{W}_d$, hence $\boldsymbol{\mu}(T) \in \mathscr{S}^c(\mathcal{T}_1)$. Further, recall that $\boldsymbol{\kappa}(T) = \log(\boldsymbol{\mu}(T))$ and therefore it follows from the definition of the logarithm on $\mathcal{T}_1$ that each component $\boldsymbol{\kappa}(T)^w$ with $w \in \mathcal{W}_d$ is a polynomial of $(\boldsymbol{\mu}(T)^v)_{v\in\mathcal{W}_d, |v|\le|w|}$. Hence it follows again by Itô's product rule that $\boldsymbol{\kappa}(T) \in \mathscr{S}^c(\mathcal{T}_0)$. □

It is of strong interest to have a more explicit necessary condition for the existence of the expected signature. The following theorem below, the proof of which can be found in [28, Section 7.2], yields such a criterion.

Theorem 2.6 *Let* $q \in [1, \infty)$ *and* $N \in \mathbb{N}_{\ge 1}$, *then there exist two constants* $c, C > 0$ *depending only on* d, N *and* q, *such that for all* $\mathbf{X} = \mathbf{X}^{(0,N)} \in \mathscr{H}^{q,N}$

$$c\|\!|\mathbf{X}|\!\|_{\mathscr{H}^{q,N}} \le \|\!|\mathrm{Sig}(\mathbf{X})_{0,\cdot}|\!\|_{\mathscr{H}^{q,N}} \le C\|\!|\mathbf{X}|\!\|_{\mathscr{H}^{q,N}}.$$

In particular, if $\mathbf{X} \in \mathscr{H}^{\infty-}(\mathcal{T}_0)$ *then* $\mathrm{Sig}(\mathbf{X})_{0,\cdot} \in \mathscr{H}^{\infty-}(\mathcal{T}_1)$ *and the expected signature exists.*

Remark 2.7 Let $\mathbf{X} = (0, M, 0, \ldots, 0)$ where $M \in \mathscr{M}(\mathbb{R}^d)$ is a martingale, then

$$\|\!|\mathbf{X}|\!\|_{\mathscr{H}^{q,N}} = \|M\|_{\mathscr{H}^{qN}} = \||\langle M\rangle_T|^{1/2}\|_{\mathscr{L}^{qN}},$$

and we see that the above estimate implies that

$$\max_{n=1,\ldots,N} \|\mathrm{Sig}(\mathbf{X})^{(n)}_{0,\cdot}\|^{1/n}_{\mathscr{S}^{qN/n}} \le C\|M\|_{\mathscr{H}^{qN}}.$$

This estimate is already known and follows from the Burkholder-Davis-Gundy inequality for enhanced martingales, which was first proved in [23].

3 Functional Equations for the Expected Signature

In this section we present central functional equations for the expected signature and signature cumulants of semimartingales. For better accessibility we will first present the formulas for discrete processes, which then will inform the results for purely continuous processes via a limiting procedure. We will however prove the results in the continuous case directly by a stochastic calculus approach.

3.1 Discrete Processes

Let $(\Omega, \mathcal{G}, \mathbb{P})$ be a probability space with filtration $(\mathcal{G}_j)_{j=0,\dots,J}$. Further, let $(\mathbf{X}_j)$ be a $\mathcal{T}_0$-valued adapted process. We can define a discrete version of the signature by multiplying exponential increments

$$\mathbf{S}_0 = 1, \qquad \mathbf{S}_j = \mathbf{S}_{j-1}e^{\mathbf{X}_j - \mathbf{X}_{j-1}}, \qquad \text{for } j = 1, \dots, J.$$

This is precisely the signature of the paths obtained from linearly interpolating the points $\mathbf{X}_0, \mathbf{X}_1, \dots, \mathbf{X}_J$ in $\mathcal{T}_0$. Assume for simplicity that $\mathbf{X}$ has moments of all orders, i.e., $\mathbb{E}[|\mathbf{X}_j^{(n)}|^p] < \infty$ for all $n, p \in \mathbb{N}_{\geq 1}$, for otherwise we need to truncate tensor levels. We then correspondingly define the conditional expected signature of $(\mathbf{X}_j)$ by

$$\boldsymbol{\mu}_j := \mathbb{E}\big[\mathbf{S}_j^{-1}\mathbf{S}_J \,|\mathcal{G}_j\big], \qquad j = 0, \dots, J.$$

A simple computation yields for $j = J-1, \dots, 0$:

$$\begin{aligned}\boldsymbol{\mu}_j &= \mathbb{E}\big[e^{\mathbf{X}_{j+1}-\mathbf{X}_j} \cdots e^{\mathbf{X}_J - \mathbf{X}_{J-1}} \,|\mathcal{G}_j\big] \\ &= \mathbb{E}\big[e^{\mathbf{X}_{j+1}-\mathbf{X}_j}\mathbb{E}\big[e^{\mathbf{X}_{j+2}-\mathbf{X}_{j+1}} \cdots e^{\mathbf{X}_J-\mathbf{X}_{J-1}} \,|\mathcal{G}_{j+1}\big] \,|\mathcal{G}_j\big] \\ &= \mathbb{E}\big[e^{\mathbf{X}_{j+1}-\mathbf{X}_j}\boldsymbol{\mu}_{j+1} \,|\mathcal{G}_j\big].\end{aligned}$$

These identities permit to calculate $\boldsymbol{\mu}$ in a backwards induction. What is not revealed, however, is that to compute $\boldsymbol{\mu}^{(n)}$ only strictly lower tensor levels $\boldsymbol{\mu}^{(1)}, \dots \boldsymbol{\mu}^{(n-1)}$ are needed. To see that this is really the case, we first bring the identity into the following difference form

$$0 = \mathbb{E}\big[(e^{\mathbf{X}_{j+1}-\mathbf{X}_j} - 1)\boldsymbol{\mu}_{j+1} + (\boldsymbol{\mu}_{j+1} - \boldsymbol{\mu}_j) \,|\mathcal{G}_j\big]. \tag{15}$$

Summing over $\{j, j+1, \dots, J-1\}$ and conditioning to $\mathcal{G}_j$ we obtain the following

Theorem 3.1 *Given that the adapted $\mathcal{T}_0$-valued process $(\mathbf{X}_j)$ satisfies the moment condition $\mathbb{E}[|\mathbf{X}_j^{(n)}|^p] < \infty$ for all $n, p \in \mathbb{N}_{\geq 1}$, its conditional expected signature*

$(\boldsymbol{\mu}_j)$ is uniquely characterized by the equation

$$\boldsymbol{\mu}_j = 1 + \mathbb{E}\Bigg[\sum_{i=j}^{J-1}(e^{\mathbf{X}_{i+1}-\mathbf{X}_i} - 1)\boldsymbol{\mu}_{i+1}\,\Bigg|\mathcal{G}_j\Bigg], \qquad j = 0, \ldots, J.$$

In particular, projecting to tensor levels yields the following recursion over $n \in \{1, 2, \ldots\}$ and $j = \{J, J-1, \ldots, 1\}$

$$\boldsymbol{\mu}_j^{(n)} = \sum_{k=0}^{n-1}\sum_{i=j}^{J-1}\sum_{\|\ell\|=n-k}\frac{1}{|\ell|!}\mathbb{E}\Big[(\mathbf{X}_{i+1}-\mathbf{X}_i)^{(\ell)}\boldsymbol{\mu}_{i+1}^{(k)}\,\Big|\mathcal{G}_j\Big],$$

where for $\ell \in \mathbb{N}_{\geq 1}^k$ and $\mathbf{x} \in \mathcal{T}_0$ we set $\mathbf{x}^{(\ell)} = \mathbf{x}^{(l_1)} \cdots \mathbf{x}^{(l_k)}$.

Example 3.2 (Markov Chains; See also [10, Section 5]) Consider a d-dimensional Markov chain (X_j) and set $(\mathbf{X}_j^{(1)}) = (X_j)$ and $\mathbf{X}_j^{(n)} \equiv 0$ for $n \neq 1$. Denote by $p_j^{(i)}(x, \mathrm{d}y)$ the step-i transition probability kernel of X starting in $x \in \mathbb{R}^d$ at time j with $p_j^{(0)}(x, \mathrm{d}y) = \delta_x(\mathrm{d}y)$. We see inductively that almost surely $\boldsymbol{\mu}_j^{(n)} = f_j^n(X_j)$ where the functions $f_j^n : \mathbb{R}^d \to (\mathbb{R}^d)^{\otimes n}$ satisfy the recursive scheme

$$f_j^n(x) = \sum_{k=1}^{n-1}\sum_{i=j}^{J-1}\frac{1}{n!}\int_{\mathbb{R}^d}\int_{\mathbb{R}^d}(z-y)^{\otimes(n-k)}f_{i+1}^k(z)p_i^{(1)}(x, \mathrm{d}y)p_j^{(i-j)}(y, \mathrm{d}z)$$

for $n = 1, 2 \ldots$ and $j = J-1, \ldots, 0$ with and $f_j^0 = 1$, $f_J^n = 0$ for $n \geq 1$.

Example 3.3 (Random Walks in the Group [11, 14]) Let $(g_j)_{j=1,\ldots,J}$ be an IID sequence with values in the Lie algebra $\mathcal{T}_0^N$ and exponential image $G_j = e^{g_j} \in \mathcal{T}_1$. Consider the random walk $\mathbf{X}_j = \sum_{k=1}^j g_k$ and its signature

$$\mathbf{S}_j = \prod_{k=1}^{j} G_k, \quad j = 1, \ldots, J.$$

Then either directly, or following the computations above, we obtain the conditional expected signature

$$\boldsymbol{\mu}_j = \mathbb{E}\big(\mathbf{S}_j^{-1}\mathbf{S}_J\big|\mathcal{G}_j\big) = M^{J-j}, \quad j = 0, \ldots, J$$

assuming that $M = \mathbb{E}(G_1)$ is (componentwise) well defined.

For instance, the case of a planar lattice random walk is covered by $N = 1, d = 2$, and equal weighted point masses at $\pm e_1 \pm e_2$, that is, $\mathbf{X}$ is a simple random walk in $\mathbb{R}^2$. A direct computation shows that

$$M = \frac{1}{4}\left(e^{e_1} + e^{-e_1} + e^{e_2} + e^{-e_2}\right) = \frac{1}{2}\left(\cosh(e_1) + \cosh(e_2)\right),$$

where $\cosh\colon \mathcal{T} \to \mathcal{T}$ is defined by its power series (see Sect. 2.2). In particular, for any tensor series $\mathbf{x} \in \mathcal{T}_0$,

$$\cosh(\mathbf{x}) = 1 + \frac{1}{2}\mathbf{x}^2 + \cdots$$

For $\lambda > 0$, let $\delta_\lambda : \mathcal{T} \to \mathcal{T}$ denote the dilation operator. Diffusive rescaling of moments yields

$$\delta_{J^{-1/2}}\boldsymbol{\mu}_0 = (\delta_{J^{-1/2}}M)^J = \left(\mathbf{1} + \frac{1}{2J}(e_1^2 + e_2^2) + \mathrm{o}(J^{-1})\right)^J.$$

Upon taking the limit as $J \to \infty$ we recover Fawcett's formula for planar Brownian motion.

For an example with $N = 2$ consider equal weighted point masses at $\{\pm[e_i, e_j] : 1 \le i < j \le d\}$. A similar argument leads to a Fawcett-type formula

$$\lim_{J\to\infty} \delta_{J^{-1/4}}\boldsymbol{\mu}_0(J) = e^{\frac{2}{d(d-1)}\sum_{i<j}([e_i,e_j])^2}.$$

for the level-4 Brownian rough paths that recently arose in [32] as limits of fractional Brownian motion with $H \to \frac{1}{4}$ (see also Sect. 6.2).

We close the discussion on discrete processes by arguing how to extend these results to the continuous case. Expressing the functional equation from Theorem 3.1 in the alternative form

$$\boldsymbol{\mu}_j = 1 + \mathbb{E}\left[\sum_{i=j}^{J-1}(e^{\mathbf{X}_{i+1}-\mathbf{X}_i} - 1)(\boldsymbol{\mu}_i + (\boldsymbol{\mu}_{i+1} - \boldsymbol{\mu}_i))\,\middle|\,\mathcal{G}_j\right],$$

we see that in the diffusive limit, setting $\delta\mathbf{X}_i = \mathbf{X}_{i+1} - \mathbf{X}_i$, we can expand

$$\sum_i (e^{\mathbf{X}_{i+1}-\mathbf{X}_i} - 1)\boldsymbol{\mu}_i \approx \sum_i (\delta\mathbf{X}_i + \frac{1}{2}(\delta\mathbf{X}_i)^2 + \cdots)\boldsymbol{\mu}_i,$$

and similarly, with $\delta\boldsymbol{\mu}_i = \boldsymbol{\mu}_{i+1} - \boldsymbol{\mu}_i$,

$$\sum_i (e^{\mathbf{X}_{i+1}-\mathbf{X}_i} - 1)(\boldsymbol{\mu}_{i+1} - \boldsymbol{\mu}_i) \approx \sum_i (\delta\mathbf{X}_i + \cdots)\delta\boldsymbol{\mu}_i,$$

to anticipate the functional relation of Theorem 3.4, derived by stochastic calculus.

For the conditional signature cumulants $(\boldsymbol{\kappa}_j) = (\log \boldsymbol{\mu}_j)$ we obtain a first identity by multiplying (15) with $e^{-\boldsymbol{\kappa}_j}$ from the right:

$$0 = \mathbb{E}\big[(e^{\mathbf{X}_{j+1}-\mathbf{X}_j} - 1)e^{\boldsymbol{\kappa}_{j+1}}e^{-\boldsymbol{\kappa}_j} + (e^{\boldsymbol{\kappa}_{j+1}}e^{-\boldsymbol{\kappa}_j} - 1) \,\big|\mathcal{G}_j\big].$$

Even in the commutative (one-dimensional) case the above identity gives non-trivial relations for conditional cumulants (see Sect. 5.3.2). In contrast to the continuous case, transforming the above formula entirely to logarithmic coordinates under the use of the Baker-Campbell-Hausdorff (BCH) formula is in general not possible, since after all the logarithm is non-linear and cannot be exchanged with the conditional expectation. Nevertheless, a martingale transform yields the identity

$$0 = \mathbb{E}\left[\sum_{i=j}^{J-1} H(\operatorname{ad}\boldsymbol{\kappa}_i)\Big((e^{\mathbf{X}_{i+1}-\mathbf{X}_i} - 1)e^{\boldsymbol{\kappa}_{i+1}}e^{-\boldsymbol{\kappa}_i} + (e^{\boldsymbol{\kappa}_{i+1}}e^{-\boldsymbol{\kappa}_i} - 1)\Big) \,\Big|\mathcal{G}_j\right],$$

where following Eq. (6) a second order expansion in the terms $\delta\mathbf{X}$ and $\delta\boldsymbol{\kappa}_i = \boldsymbol{\kappa}_{i+1} - \boldsymbol{\kappa}_i$ yields

$$e^{\boldsymbol{\kappa}_{i+1}}e^{-\boldsymbol{\kappa}_i} - 1 = G(\operatorname{ad}\boldsymbol{\kappa}_i)(\delta\boldsymbol{\kappa}_i) + Q(\operatorname{ad}\boldsymbol{\kappa}_i)(\delta\boldsymbol{\kappa}_i \otimes \delta\boldsymbol{\kappa}_i) + \dots$$

and

$$(e^{\mathbf{X}_{i+1}-\mathbf{X}_i} - 1)e^{\boldsymbol{\kappa}_{i+1}}e^{-\boldsymbol{\kappa}_i} = \Big(\delta\mathbf{X}_i + \frac{1}{2}(\delta\mathbf{X}_i)^2 + \dots\Big)\Big(1 + G(\operatorname{ad}\boldsymbol{\kappa}_i)(\delta\boldsymbol{\kappa}_i) + \dots\Big),$$

allows us to anticipate that the diffusive limit leads to the functional equation given in Theorem 3.6 below.

3.2 The Continuous Case

Throughout this section we will assume that $(\Omega, \mathcal{F}, (\mathcal{F}_t), \mathbb{P})$ is a filtered probability space satisfying the usual conditions and with the property that every martingale has a continuous version. This is for instance the case if the filtration is the natural and completed filtration of a Brownian motion.

To streamline the derivation, we will start from first principles and introduce the key concepts from Itô-calculus in the non-commutative setting when first needed, not focusing on integrability considerations which are fully resolved in [28]. For $\mathbf{X} \in \mathscr{H}^{\infty-}(\mathcal{T}_0)$ the conditional expected signature and the signature cumulants

$$\boldsymbol{\mu}(T) = (\mathbb{E}_t(\operatorname{Sig}(\mathbf{X})_{t,T}))_{0\le t\le T} \in \mathscr{S}^c(\mathcal{T}_1), \quad \boldsymbol{\kappa}(T) = (\log \boldsymbol{\mu}_t(T))_{0\le t\le T} \in \mathscr{S}^c(\mathcal{T}_0),$$

are well defined. The derivation starts by defining the martingale

$$\mathbf{M} := (\mathbb{E}_t[\mathrm{Sig}(\mathbf{X})_{0,T}])_{0\le t\le T} \in \mathscr{M}(\mathcal{T}_1),$$

which is continuous under the given assumptions on the filtration. It factors into

$$\mathbf{M}_t \;=\; \mathrm{Sig}(\mathbf{X})_{0,t}\mathbb{E}_t[\mathrm{Sig}(\mathbf{X})_{t,T}] \;=\; \mathrm{Sig}(\mathbf{X})_{0,t}\boldsymbol{\mu}_t(T).$$

One is now inclined to apply Itô's rule to the product on the right-hand side as this should reveal non-trivial cancellations implied by the martingality of $\mathbf{M}$. For ease of notation we denote $\mathbf{S}_t := \mathrm{Sig}(\mathbf{X})_{0,t}$ and $\boldsymbol{\mu}_t := \boldsymbol{\mu}_t(T)$. Then the definitions put forward in Sect. 2.2 directly yield the product rule

$$\begin{aligned}\mathbf{M}_t &= 1 + \int_0^t \mathrm{d}\mathbf{S}_u\,\boldsymbol{\mu}_u + \int_0^t \mathbf{S}_u\,\mathrm{d}\boldsymbol{\mu}_u + \langle \mathbf{S}, \boldsymbol{\mu}\rangle_t \\ &= 1 + \int_0^t \mathbf{S}_u\left(\left(\mathrm{d}\mathbf{X}_u + \frac{1}{2}\mathrm{d}\langle \mathbf{X}\rangle_u\right)\boldsymbol{\mu}_u + \mathrm{d}\boldsymbol{\mu}_u + \mathrm{d}\langle \mathbf{X}, \boldsymbol{\mu}\rangle_u\right) \qquad (16)\end{aligned}$$

where in the second line we have used the Itô-integral form of $\mathbf{S}$ in (13). The martingality of $\mathbf{M}$ and invertability of the tensor $\mathbf{S}_t$ for all times $t \in [0, T]$ imply that the term in the bracket is the differential of a local martingale. By Theorem 2.6 we have $\mathbf{S}, \boldsymbol{\mu} \in \mathscr{H}^{\infty-}(\mathcal{T}_0)$ and it is therefore not difficult to conclude that this local martingale is also a true martingale. Taking conditional expectations of its integral then yields the following functional equation $\boldsymbol{\mu}$:

Theorem 3.4 *The conditional expected signature $\boldsymbol{\mu} = \boldsymbol{\mu}(T)$ of $\mathbf{X} \in \mathscr{H}^{\infty-}(\mathcal{T}_0)$ is the unique solution (up to indistinguishably) of the following functional equation*

$$\boldsymbol{\mu}_t = 1 + \mathbb{E}_t\left\{\int_t^T \left(\mathrm{d}\mathbf{X}_u + \frac{1}{2}\mathrm{d}\langle \mathbf{X}\rangle_u\right)\boldsymbol{\mu}_u + \langle \mathbf{X}, \boldsymbol{\mu}\rangle_{t,T}\right\}, \qquad 0 \le t \le T. \qquad (17)$$

Furthermore, if $\mathbf{X} \in \mathscr{H}^{1,N}$ for some $N \in \mathbb{N}_{\ge 1}$, then the identity (17) still holds true for the truncated expected signature $\boldsymbol{\mu} := (\mathbb{E}_t(\mathrm{Sig}(\mathbf{X}^{(0,N)})_{t,T}))_{0\le t\le T}$.

It is crucial to understand that the uniqueness part of the above statement follows by projecting the above equation to a tensor level, say $n \ge 1$, and then noting that the above right-hand side only depends on tensor levels $\boldsymbol{\mu}^{(k)}$ for $k \le n-1$. In other words, Eq. (17) directly leads to a recursive scheme for calculating $\boldsymbol{\mu}$ which is explicitly spelled out in Corollary 4.1 of Sect. 4.

Moving forward to the signature cumulants $\boldsymbol{\kappa} := \boldsymbol{\kappa}.(T)$ one needs to further resolve the differential of $\boldsymbol{\mu} = e^{\boldsymbol{\kappa}}$ in (16). The key is following Itô-formula for the tensor exponential map.

Lemma 3.5 *For $\mathbf{Z} \in \mathscr{S}^c(\mathcal{T}_0)$ it holds*

$$e^{\mathbf{Z}_t} - e^{\mathbf{Z}_0} = \int_0^t G(\mathrm{ad}\,\mathbf{Z}_u)(\mathrm{d}\mathbf{Z}_u)e^{\mathbf{Z}_u} + \int_0^t Q(\mathrm{ad}\,\mathbf{Z}_u)(\mathrm{d}\langle\!\langle \mathbf{Z}, \mathbf{Z}\rangle\!\rangle_u)e^{\mathbf{Z}_u},$$

for all $0 \le t \le T$.

A version of the above lemma in a matrix setting previously appeared in [38]; the full proof can be found in [28, Lemma 7.8]. Loosely speaking, the form of the right-hand side follows immediately from the order-2 Taylor expansion for the tensor exponential given in Eq. (6).

Applying Theorem 3.5 to $\boldsymbol{\kappa}$ in (16) and collecting terms, we arrive at

$$\mathbf{M}_t = 1 + \int_0^t \mathbf{S}_u \mathrm{d}\mathbf{L}_u e^{\boldsymbol{\kappa}_u(T)},$$

where $\mathbf{L} \in \mathscr{S}^c(\mathcal{T}_0)$ is given by

$$\mathbf{L}_t = \mathbf{X}_t + \frac{1}{2}\langle \mathbf{X} \rangle_t + \int_0^t G(\operatorname{ad}\boldsymbol{\kappa}_{u-})(\mathrm{d}\boldsymbol{\kappa}_u) + \frac{1}{2}\int_t^T Q(\operatorname{ad}\boldsymbol{\kappa}_{u-})(\mathrm{d}\langle\!\langle \boldsymbol{\kappa}, \boldsymbol{\kappa} \rangle\!\rangle_u)$$
$$+ \int_t^T (\mathrm{Id} \odot G(\operatorname{ad}\boldsymbol{\kappa}_{u-}))(\mathrm{d}\langle\!\langle \mathbf{X}, \boldsymbol{\kappa} \rangle\!\rangle_u).$$

Again the martingality of $\mathbf{M}$ and invertability of $\mathbf{S}_t, e^{-\boldsymbol{\kappa}_t} \in \mathscr{S}^c(\mathcal{T}_1)$ imply that $\mathbf{L}$ is a local martingale. The main work done in the proof of [28, Theorem 4.1] is to show that under the given assumptions on $\mathbf{X}$ it also holds that $\mathbf{L}$ is also a true martingale. Taking expectations then yields a first functional equation for $\boldsymbol{\kappa}$:

$$\mathbb{E}_t[\mathbf{L}_{t,T}] = 0.$$

Similarly to the functional equation (17) for $\boldsymbol{\mu}$, the above equation leads to a recursive scheme over tensor levels which uniquely determines $\boldsymbol{\kappa}$. However, there is still a degree of implicitness that can be removed.

To this end note that the operator $G(\operatorname{ad}\mathbf{x})$ has the following inverse[1]

$$H(\operatorname{ad}\mathbf{x}) = \frac{\operatorname{ad}\mathbf{x}}{\exp(\operatorname{ad}\mathbf{x}) - 1} := \sum_{k=0}^{\infty} \frac{B_k}{k!}(\operatorname{ad}\mathbf{x})^k, \tag{18}$$

with Bernoulli numbers $(B_k)_{k \ge 0} = (1, -\frac{1}{2}, \frac{1}{6} \dots)$. Integrating $H(\operatorname{ad}\boldsymbol{\kappa})$ against $\mathrm{d}\mathbf{L}$ and verifying that the resulting process is in $\mathscr{M}(\mathcal{T}_0)$ we arrive at a second functional equation for $\boldsymbol{\kappa}$:

Theorem 3.6 *The signature cumulant* $\boldsymbol{\kappa}$ *of* $\mathbf{X} \in \mathscr{H}^{\infty-}(\mathcal{T}_0)$ *is the unique solution (up to indistinguishably) of the following functional equation*

[1] This follows from the fact that for linear operators, power series calculus is multiplicative in the sense that the composition $G(L)H(L)$ equals $GH(L)$ where $GH(z)$ is the Cauchy product of G and H. Moreover, $G(z)$ and $H(z)$ are multiplicative inverses as power series so that $GH(z) = 1$.

$$
\begin{aligned}
\kappa_t = \mathbb{E}_t \Bigg\{ & \int_t^T H(\operatorname{ad}\kappa_u)(\mathrm{d}\mathbf{X}_u) + \frac{1}{2}\int_t^T H(\operatorname{ad}\kappa_u)(\mathrm{d}\langle \mathbf{X}\rangle_u) \\
& + \frac{1}{2}\int_t^T H(\operatorname{ad}\kappa_u)\, Q(\operatorname{ad}\kappa_u)(\mathrm{d}\langle\!\langle \kappa, \kappa\rangle\!\rangle_u) \\
& + \int_t^T H(\operatorname{ad}\kappa_u)\, (\mathrm{Id}\odot G(\operatorname{ad}\kappa_u))(\mathrm{d}\langle\!\langle \mathbf{X}, \kappa\rangle\!\rangle_u) \Bigg\},
\end{aligned}
\tag{19}
$$

for all $0 \leq t \leq T$. *Furthermore, if* $\mathbf{X} \in \mathscr{H}^{1,N}$ *for some* $N \in \mathbb{N}_{\geq 1}$, *then the identity* (19) *still holds true for the truncated signature cumulant* $\kappa := (\log \mathbb{E}_t(\mathrm{Sig}(\mathbf{X}^{(0,N)})_{t,T}))_{0\leq t\leq T}$.

4 Recursive Formulas and Diamond Products

Theorems 3.4 and 3.6 allow for an iterative computation of the expected signature and signature cumulants, by projecting the functional equations to tensor levels. In the following subsections we will explicitly spell out these recursive schemes and then rephrase them in terms of so called *diamond products* as they were introduced for scalar semimartingales in [3].

While all recursions can be stated for the general càdlàg case [28, Section 4.2], we only present the continuous case for simplicity. Therefore we shall presume the assumptions on the filtration from the continuous setting in Sect. 3.2 throughout this section.

4.1 Recursive Formula for Signature Moments

This recursive scheme trivially starts with $\mu^{(0)} \equiv 0$ and the first level of signature moments

$$
\mu_t^{(1)} = \mathbb{E}_t\left(\mathbf{X}_{t,T}^{(1)}\right).
$$

and continuous with a first recursion in the second level

$$
\mu_t^{(2)} = \mathbb{E}_t\left[\mathbf{X}_{t,T}^{(2)} + \frac{1}{2}\left\langle \mathbf{X}^{(1)}\right\rangle_{t,T} + \int_t^T \mathrm{d}\mathbf{X}_u^{(1)}\,\mu_u^{(1)} + \left\langle \mathbf{X}^{(1)}, \mu^{(1)}\right\rangle_{t,T}\right]. \tag{20}
$$

Projecting the functional equation (17) for μ to higher tensor levels immediately gives the following

Corollary 4.1 *Let* $\mathbf{X} \in \mathscr{H}^{1,N}$ *for some* $N \in \mathbb{N}_{\geq 1}$. *Then for* $n = 1, \ldots, N$ *we have*

$$\boldsymbol{\mu}_t^{(n)} = \sum_{k=0}^{n-1} \mathbb{E}_t \left\{ \int_t^T \mathrm{d}\mathbf{X}_u^{(n-k)} \, \boldsymbol{\mu}_u^{(k)} + \frac{1}{2} \sum_{i=1}^{n-k} \int_t^T \mathrm{d}\left\langle \mathbf{X}^{(i)}, \mathbf{X}^{(n-k-i)} \right\rangle_u \boldsymbol{\mu}_u^{(k)} + \left\langle \mathbf{X}^{(n-k)}, \boldsymbol{\mu}^{(k)} \right\rangle_{t,T} \right\}. \tag{21}$$

Example 4.2 Consider the special case with vanishing higher order components, $\mathbf{X}^{(i)} \equiv 0$, for $i \neq 1$, and $\mathbf{X} = \mathbf{X}^{(1)} \equiv M$, a d-dimensional continuous square-integrable martingale. We then have $\boldsymbol{\mu}^{(1)} \equiv 0$ and it directly follows from Stratonovich-Itô correction that

$$\boldsymbol{\mu}_t^{(2)} = \mathbb{E}_t \int_t^T (M_u - M_t) \circ \mathrm{d}M_u = \frac{1}{2} \mathbb{E}_t \left\langle M \right\rangle_{t,T} = \frac{1}{2} \mathbb{E}_t \left\langle \mathbf{X}^{(1)} \right\rangle_{t,T} .$$

which is indeed a (very) special case of the general expression for $\boldsymbol{\mu}^{(2)}$.

Assuming that $M \in \mathscr{H}^N$ the recursion proceeds for levels $n \in \{2, 3, \ldots, N\}$ and simplifies due to the martingality to

$$\boldsymbol{\mu}_t^{(n)} = \mathbb{E}_t \left\{ \frac{1}{2} \int_t^T \mathrm{d} \left\langle M, M \right\rangle_u \boldsymbol{\mu}_u^{(n-2)} + \left\langle M, \boldsymbol{\mu}^{(n-1)} \right\rangle_{t,T} \right\}.$$

In case M is a Gaussian martingale of the form $M_t = \int_0^t \sigma(s).\mathrm{d}B_s \in \mathscr{H}^{\infty-}$ for an m-dimensional Brownian motion B and $\sigma \in L^2([0, T]; \mathbb{R}^{d \times m})$, an induction immediately shows that $\boldsymbol{\mu}$ is deterministic. Whence brackets with $\boldsymbol{\mu}$ vanish and we obtain $\boldsymbol{\mu}^{(n)} \equiv 0$ for odd n and

$$\boldsymbol{\mu}_t^{(n)} = \frac{1}{2} \int_t^T \sigma.\sigma^T(u) \, \boldsymbol{\mu}_u^{(n-2)} \, \mathrm{d}u, \qquad \text{for even } n. \tag{22}$$

When M is a standard Brownian motion, i.e., $\sigma \equiv \mathbf{I}_d \in \mathbb{R}^{d \times d}$ is the identity matrix, we now readily verify Fawcett's formula [19, 21]

$$\boldsymbol{\mu}_t = \exp\left(\frac{T-t}{2} \mathbf{I}_d\right), \qquad t \in [0, T].$$

Remark 4.3 Fawcett's formula is reminiscent of the case of real-valued Gaussian vectors, which are in fact characterized by having non-vanishing cumulants only at degree 2, and one might wonder whether that is also the case for expected signatures as they play a role analogue to characteristic functions for paths. In that case, the formula is valid *only* for Brownian motion. Other Gaussian processes may have arbitrarily complicated cumulants as can already be seen from Theorem 4.2 above. For an even more elementary example, let $d > 1$ and consider a centered Gaussian vector $Z = (Z^1, \ldots, Z^d) \in \mathbb{R}^d$ with covariance matrix Σ, and $X_t := tZ$. Clearly, the law of X_t is Gaussian for every fixed $t > 0$. Moreover,

$$\mathrm{Sig}(X)_{t,T} = \exp\left((T-t)Z\right)$$

so that the first few terms of the expected signature are $\boldsymbol{\mu}_t^{(1)}(T) = \boldsymbol{\mu}_t^{(3)}(T) = 0$ (in fact all odd degrees vanish), and

$$\boldsymbol{\mu}_t^{(2)}(T) = \frac{(T-t)^2}{2} \sum_{i_1,i_2=1}^{d} \Sigma^{i_1 i_2} e_{i_1 i_2}$$

$$\boldsymbol{\mu}_t^{(4)}(T) = \frac{(T-t)^4}{4!} \sum_{i_1,i_2,i_3,i_4=1}^{d} (\Sigma^{i_1 i_2}\Sigma^{i_3 i_4} + \Sigma^{i_1 i_3}\Sigma^{i_2 i_4} + \Sigma^{i_1 i_4}\Sigma^{i_2 i_3}) e_{i_1 i_2 i_3 i_4}$$

where in the last term we have used Isserlis's theorem to express mixed moments as products of covariances. Using the power series expansion of log we see that the first four signature cumulants are $\boldsymbol{\kappa}_t^{(1)}(T) = \boldsymbol{\kappa}_t^{(3)}(T) = 0$, and

$$\begin{aligned} \boldsymbol{\kappa}_t^{(2)}(T) &= \boldsymbol{\mu}_t^{(2)}(T) \\ \boldsymbol{\kappa}_t^{(4)}(T) &= \boldsymbol{\mu}_t^{(4)}(T) - \frac{1}{2}\boldsymbol{\mu}_t^{(2)}(T)\boldsymbol{\mu}_t^{(2)}(T) \\ &= \frac{(T-t)^4}{4!} \sum_{i_1,i_2,i_3,i_4=1}^{d} (\Sigma^{i_1 i_3}\Sigma^{i_2 i_4} + \Sigma^{i_1 i_4}\Sigma^{i_2 i_3} - 2\Sigma^{i_1 i_2}\Sigma^{i_3 i_4}) e_{i_1 i_2 i_3 i_4}. \end{aligned}$$

In particular, the fourth cumulant does not vanish. A general diagrammatic expansion, analog to Isserlis' theorem, for the expected signature of a Gaussian processes in terms of its correlation function has been recently obtained by Cass and Ferrucci [13].

4.2 *Recursive Formula for Signature Cumulants*

The first level of signature cumulants are identical to the signature moments

$$\boldsymbol{\kappa}_t^{(1)} = \boldsymbol{\mu}_t^{(1)} = \mathbb{E}_t\left(\mathbf{X}_{t,T}^{(1)}\right).$$

The formula we obtain from projecting (19) to the second tensor level reads

$$\boldsymbol{\kappa}_t^{(2)} = \mathbb{E}_t\left[\mathbf{X}_{t,T}^{(2)} + \frac{1}{2}\left\langle \mathbf{X}^{(1)}\right\rangle_{t,T} + \frac{1}{2}\int_t^T \left[\boldsymbol{\kappa}_u^{(1)}, \mathrm{d}\mathbf{X}_u^{(1)}\right] + \frac{1}{2}\left\langle \boldsymbol{\kappa}^{(1)}\right\rangle_{t,T} + \left\langle \mathbf{X}^{(1)}, \boldsymbol{\kappa}^{(1)}\right\rangle_{t,T}\right].$$

From the very definition of the logarithm we must have $\boldsymbol{\kappa}^{(2)} = \boldsymbol{\mu}^{(2)} - \frac{1}{2}\boldsymbol{\mu}^{(1)}\boldsymbol{\mu}^{(1)}$, which is in line with the above expression for $\boldsymbol{\kappa}^{(2)}$ and the previous formula (20) for

$\boldsymbol{\mu}^{(2)}$, but already requires a little work to verify. Indeed, using the martingality of $\mathbf{X}^{(1)} + \boldsymbol{\kappa}^{(1)} = \mathbb{E}_{\cdot}[\mathbf{X}^{(1)}_T]$ we resolve both expressions using the Itô-product rule in the following calculation

$$\mathbb{E}_t\left[\frac{1}{2}\int_t^T \mathrm{d}\mathbf{X}^{(1)}_u \boldsymbol{\kappa}^{(1)}_u\right] = \mathbb{E}_t\left[-\frac{1}{2}\int_t^T \mathrm{d}\boldsymbol{\kappa}^{(1)}_u \boldsymbol{\kappa}^{(1)}_u\right]$$
$$= \frac{1}{2}\mathbb{E}_t\left[-\boldsymbol{\kappa}^{(1)}_t\boldsymbol{\kappa}^{(1)}_t - \frac{1}{2}\int_t^T \boldsymbol{\kappa}^{(1)}_u \mathrm{d}\mathbf{X}^{(1)}_u + \frac{1}{2}\left\langle \boldsymbol{\kappa}^{(1)}\right\rangle_{t,T}\right].$$

The general recursion for $\boldsymbol{\kappa}$ is combinatorially more involved. To present it as concise as possible, we recall the following notation from Sect. 2.2: For $\ell \in (\mathbb{N}_{\geq 1})^k$ we write $\ell = (l_1, \ldots, l_k)$, $|\ell| := k$ and $||\ell|| := l_1 + \cdots + l_k$. Furthermore, for $0 \leq i, j \leq k$ we define $\ell_{i:j} = (l_{i+1}, \ldots, l_j)$ if $i < j$ and $\ell_{i:j} = ()$ otherwise. Moreover $(\operatorname{ad}\mathbf{x}^{(\ell)}) = (\operatorname{ad}\mathbf{x}^{(l_1)} \cdots \operatorname{ad}\mathbf{x}^{(l_k)})$ for any $\mathbf{x} \in \mathcal{T}_0$ and $(\operatorname{ad}\mathbf{x}^{()}) = \operatorname{Id}$.

Corollary 4.4 *Let $\mathbf{X} \in \mathscr{H}^{1,N}$ for some $N \in \mathbb{N}_{\geq 1}$, then we have*

$$\boldsymbol{\kappa}^{(n)}_t = \mathbb{E}_t\left(\mathbf{X}^{(n)}_{t,T}\right) + \sum_{|\ell|\geq 2,\ ||\ell||=n} \mathbb{E}_t\Bigg(\operatorname{HMag}^1(\mathbf{X}, \boldsymbol{\kappa}; \ell)_{t,T} + \frac{1}{2}\operatorname{HMag}^2(\mathbf{X}, \boldsymbol{\kappa}; \ell)_{t,T}$$
$$+ \operatorname{HQua}(\boldsymbol{\kappa}; \ell)_{t,T} + \operatorname{HCov}(\mathbf{X}, \boldsymbol{\kappa}; \ell)_{t,T}\Bigg) \tag{23}$$

where the summation is over $\ell \in \mathbb{N}^k_{\geq 1}$, $k \in \mathbb{N}_{\geq 1}$ and

$$\operatorname{HMag}^1(\mathbf{X}, \boldsymbol{\kappa}; \ell)_{t,T} = \frac{B_{k-1}}{(k-1)!}\int_t^T \operatorname{ad}\boldsymbol{\kappa}^{(\ell_{1:k})}_u \left(\mathrm{d}\mathbf{X}^{(l_1)}_u\right)$$

$$\operatorname{HMag}^2(\mathbf{X}, \boldsymbol{\kappa}; \ell)_{t,T} = \frac{B_{k-2}}{(k-2)!}\int_t^T \operatorname{ad}\boldsymbol{\kappa}^{(\ell_{2:k})}_u \left(\mathrm{d}\left\langle \mathbf{X}^{(l_1)}, \mathbf{X}^{(l_2)}\right\rangle_u\right)$$

$$\operatorname{Qua}(\boldsymbol{\kappa}; \ell)_{t,T} = \frac{1}{k!}\sum_{m=2}^{k}\binom{k-1}{m-1}\int_t^T \left(\operatorname{ad}\boldsymbol{\kappa}^{(\ell_{2:m})}_{u-} \odot \operatorname{ad}\boldsymbol{\kappa}^{(\ell_{m:k})}_u\right)$$
$$\times \left(\mathrm{d}\langle\!\langle \boldsymbol{\kappa}^{(l_1)}, \boldsymbol{\kappa}^{(l_2)}\rangle\!\rangle_u\right)$$

$$\operatorname{Cov}(\mathbf{X}, \boldsymbol{\kappa}; \ell)_{t,T} = 1_{\{k\geq 2\}}\frac{1}{(k-1)!}\int_t^T \left(\operatorname{Id} \odot \operatorname{ad}\boldsymbol{\kappa}^{(l_{2:k})}_{u-}\right)\left(\mathrm{d}\langle\!\langle \mathbf{X}^{(l_1)}, \boldsymbol{\kappa}^{(l_2)}\rangle\!\rangle_u\right)$$

$$\operatorname{HQua}(\boldsymbol{\kappa}; \ell)_{t,T} = \int_t^T \sum_{j=2}^{k}\frac{B_{k-j}}{(k-j)!}\operatorname{ad}\boldsymbol{\kappa}^{(\ell_{j:k})}_{u-}\left(\mathrm{dQua}(\boldsymbol{\kappa}; \ell_{0:j})_u\right)$$

$$\mathrm{HCov}(\mathbf{X}, \boldsymbol{\kappa}; \ell)_{t,T} = \int_t^T \sum_{j=1}^{k} \frac{B_{k-j}}{(k-j)!} \operatorname{ad} \boldsymbol{\kappa}_{u-}^{(\ell_{j:k})} \big(\mathrm{dCov}(\mathbf{X}, \boldsymbol{\kappa}; \ell_{0:j})_u\big)$$

Example 4.5 In the Gaussian martingale case from Theorem 4.2, i.e., where $\mathbf{X} = (0, \int_0^\cdot \sigma(s).\mathrm{d}B_s, 0, \dots) \in \mathscr{H}^{-\infty}(\mathcal{T}_0)$, an induction yields that $\boldsymbol{\kappa}^{(n)}$ is deterministic for all $n \in \mathbb{N}$. Indeed, firstly the HMag^1-terms vanish due to martingality of $\mathbf{X}$. Secondly, from the an induction hypothesis $\boldsymbol{\kappa}^{(0,n-1)}$ is deterministic, hence, of finite variation and all cross-variation terms vanish. Finally, the remaining HMag^2-terms are deterministic due $\langle \mathbf{X} \rangle_t = \sigma.\sigma^T(t)$. The recursion thus dramatically simplifies to $\boldsymbol{\kappa}^{(n)} \equiv 0$ for odd n and

$$\boldsymbol{\kappa}_t^{(n)} = \sum_{\|\ell\|=n-2} \frac{B_{|\ell|}}{|\ell|!} \int_t^T \operatorname{ad} \boldsymbol{\kappa}_u^{(\ell)} \big(\sigma.\sigma^T(u)\big) \mathrm{d}u, \qquad \text{for even } n.$$

This is precisely the (deterministic) Magnus expansion of the logarithm of the solution of Eq. (22). We will revisit this connection in Sect. 6.1 in the more general setting of time dependent Lévy-processes.

Note that $\boldsymbol{\kappa}_t$ is a Lie-series over symmetric 2-tensors $\mathrm{Sym}(\mathbb{R}^d \otimes \mathbb{R}^d) \subset \mathcal{T}_0$. In the standard Brownian case $\sigma \equiv \mathbf{I}_d$, all commutators vanish and we obtain Fawcett's formula in logarithmic form $\boldsymbol{\kappa}_t = \frac{1}{2}(T-t)\mathbf{I}_d$.

4.3 *Diamond Products*

We extend the notion of the *diamond product* introduced in [3], which we recall, for continuous scalar semimartingales to our setting. Denote by $\mathbb{E}_t$ the conditional expectation with respect to the sigma algebra $\mathcal{F}_t$.

Definition 4.6 For $X, Y \in \mathscr{S}^c(\mathbb{R})$ define

$$(X \diamond Y)_t(T) := \mathbb{E}_t\left(\langle X, Y \rangle_{t,T}\right).$$

For $\mathcal{T}$-valued continuous semimartingales we have the following

Definition 4.7 For $\mathbf{X}$ and $\mathbf{Y}$ in $\mathscr{S}^c(\mathcal{T})$ define

$$(\mathbf{X} \diamond \mathbf{Y})_t(T) := \mathbb{E}_t\big(\langle \mathbf{X}, \mathbf{Y} \rangle_{t,T}\big) = \sum_{w \in \mathcal{W}_d} \left(\sum_{w_1 w_2 = w} (\mathbf{X}^{w_1} \diamond \mathbf{Y}^{w_2})_t(T) \right) e_w \in \mathcal{T}$$

whenever the $\mathcal{T}$-valued quadratic covariation which appears on the right-hand side is integrable. We also define an *outer diamond* by

$$(\mathbf{X} \bullet \mathbf{Y})_t(T) := \mathbb{E}_t(\langle\!\langle \mathbf{X}, \mathbf{Y} \rangle\!\rangle_{t,T}) = \sum_{w_1, w_2 \in \mathcal{W}_d} (\mathbf{X}^{w_1} \diamond \mathbf{Y}^{w_2})_t(T) e_{w_1} \otimes e_{w_2} \in \mathcal{T} \otimes \mathcal{T}.$$

The following lemma allows us to rewrite all recursion from the previous sections in terms of diamond products.

Lemma 4.8 ([28, Lemma 2.2]) *Let $p, q, r \in [1, \infty)$ such that $1/p+1/q+1/r < 1$ and let $X \in \mathscr{M}^c_{\text{loc}}((\mathbb{R}^d)^{\otimes l})$, $Y \in \mathscr{M}^c_{\text{loc}}((\mathbb{R}^d)^{\otimes m})$, and $Z \in \mathscr{C}((\mathbb{R}^d)^{\otimes n})$ with $l, m, n \in \mathbb{N}$, such that $\|X\|_{\mathscr{H}^p}$, $\|Y\|_{\mathscr{H}^q}$, $\|Z\|_{\mathscr{S}^r} < \infty$. Then it holds for all $0 \le t \le T$*

$$\mathbb{E}_t\left(\int_t^T Z_u \mathrm{d}(X \diamond Y)_u(T)\right) = -\mathbb{E}_t\left(\int_t^T Z_u \mathrm{d}\,\langle X, Y \rangle_u\right).$$

The proof relies on standard approximation results for continuous semimartingales as it is fairly clear from the definition of the diamond product that the identity holds for simple processes Z, since $(X \diamond Y)_T(T) = 0$. The Kunita-Watanabe inequality ensures that the expectation on the right-hand side is well defined.

For clarity we focus on the case where $\mathbf{X} = \mathbf{M} \in \mathscr{H}^{\infty-}$ is a martingale, for otherwise one needs to carry along conditional expectations of the finite variation parts. For the conditional expected signature, the recursion from Theorem 4.1 is then conveniently rewritten into

$$\boldsymbol{\mu}_t^{(n)} = \sum_{k=0}^{n-1} \mathbb{E}_t\left\{\frac{1}{2}\sum_{i=1}^{n-k} \int_t^T \mathrm{d}\big(\mathbf{M}^{(i)} \diamond \mathbf{M}^{(n-k-i)}\big)_u(T)\, \boldsymbol{\mu}_u^{(k)}\right\} + (\mathbf{M}^{(n-k)} \diamond \boldsymbol{\mu}^{(k)})_t(T).$$

Similarly, for the signature cumulants we obtain

$$\mathbb{E}_t\big(\mathrm{HMag}^2(\mathbf{X}, \boldsymbol{\kappa}; \ell)_{t,T}\big) = \frac{B_{k-2}}{(k-2)!} \int_t^T \operatorname{ad} \boldsymbol{\kappa}_u^{(\ell_{2:k})}\Big(\mathrm{d}(\mathbf{X}^{(l_1)} \diamond \mathbf{X}^{(l_2)})_u(T)\Big),$$

$$\mathbb{E}_t\big(\mathrm{Qua}(\boldsymbol{\kappa}; \ell)_{t,T}\big)$$

$$= -\mathbb{E}_t\left\{\frac{1}{k!}\sum_{m=2}^{k}\binom{k-1}{m-1}\int_t^T \left(\operatorname{ad} \boldsymbol{\kappa}_{u-}^{(\ell_{2:n})} \odot \operatorname{ad} \boldsymbol{\kappa}_{u-}^{(\ell_{m:k})}\right)\left(\mathrm{d}(\boldsymbol{\kappa}^{(\ell_1)} \bullet \boldsymbol{\kappa}^{(\ell_2)})_u(T)\right)\right\}.$$

and

$$\mathbb{E}_t\big(\mathrm{Cov}(\mathbf{X}, \boldsymbol{\kappa}; \ell)_{t,T}\big) = -\mathbb{E}_t\left\{\frac{1}{(k-1)!}\int_t^T \left(\mathrm{Id} \odot \operatorname{ad} \boldsymbol{\kappa}_{u-}^{(\ell_{3:k})}\right)\left(\mathrm{d}(\mathbf{X}^{(\ell_1)} \bullet \boldsymbol{\kappa}^{(\ell_2)})_u(T)\right)\right\}.$$

When $d = 1$ (or in the projection onto the symmetric algebra, cf. Sect. 5) the cumulant recursion takes a particularly simple form, since $\operatorname{ad} \mathbf{x} \equiv 0$ for all $\mathbf{x} \in \mathcal{T}_0$. Equation (23) then becomes

$$\kappa_t^{(n)}(T) = \mathbb{E}_t\left(\mathbf{X}_{t,T}^{(n)}\right) + \frac{1}{2}\sum_{k=1}^{n-1}((\mathbf{X}^{(k)} + \kappa^{(k)}) \diamond (\mathbf{X}^{(n-k)} + \kappa^{(n-k)}))_t(T).$$

We shall revisit this in a multivariate setting and comment on related works in Sect. 5.

4.4 *Remark on Tree Representation*

As illustrated in the previous section, in the case where $d = 1$ (or when projecting onto the symmetric algebra cf. Sect. 5), the cumulant recursion takes a particularly simple form. The algebraic perspective in the setting of Friz et al. [27] gives a tree series expansion of cumulants using binary trees. This representation follows from the fact that the diamond product of semimartingales is commutative but not associative. As an example (with notations taken from Sect. 5.3), in case of a one-dimensional continuous martingale, the first terms are

$$\mathbb{K}_t(T) = \bullet + \frac{1}{2}\,\cdot + \frac{1}{2}\,\cdot + \frac{1}{2}\,\cdot + \frac{1}{8}\,\cdot + \cdots$$

This expansion is organized (graded) in terms of the number of leaves in each tree, and each leaf represents the underlying martingale.

In the deterministic case, tree expansions are also known for the Magnus expansion [35] and the BCH formula [12]. These expansions, also in terms of binary trees, are *different* from the ones above as the trees are required to be *planar* to account for the non-commutativity of the Lie algebra. As an example, consider a matrix Lie group G with Lie algebra $\mathfrak{g}$. Let Y solve the matrix ODE $\dot{Y}_t = A_t Y_t$ for some $\mathfrak{g}$-valued path A, and set $\Omega_t(T) = \log(e^{Y_t} e^{-Y_T})$. We have

$$\Omega_t(T) = \bullet + \frac{1}{2}\,\cdot + \frac{1}{12}\,\cdot + \frac{1}{4}\,\cdot + \cdots$$

In this expansion, the nodes represent the underlying vector field and edges represent integration (with respect to time) and application of the Lie bracket, coming from the ad operator. It is an interesting open question to find a unified tree representation that accounts for the unified functional recursion of our Corollary 4.4.

5 Multivariate Moments and Cumulants

We saw that $\mathcal{T} := T((\mathbb{R}^d))$ is the natural state space for signatures, expected signatures and their logarithms. When $d = 1$, the signature of a path $X : [0, T]$

is nothing more than the sequence

$$\left(1, X_{0,T}, \frac{(X_{0,T})^2}{2}, \frac{(X_{0,T})^3}{3!}, \dots\right) \equiv 1 + X_{0,T} + \frac{(X_{0,T})^2}{2} + \frac{(X_{0,T})^3}{3!} + \dots$$

The expected signatures is then exactly the sequence of moments of the random variable $X_{0,T}$, to the extent of being well-defined and up to factorial constants. Similarly, the signatures cumulants correspond to the sequence of classical cumulants of $X_{0,T}$. Since $T\big((\mathbb{R}^d)\big)$ is a commutative algebra if (and only if) $d = 1$, our previous expression for expected signatures and signatures cumulants simplify dramatically, without becoming trivial (as pointed out in several works [3, 27, 30, 41]). We can capture multivariate moments and cumulants, by working with the "commutative shadow" [4] of $\mathcal{T}$ which we now introduce.

5.1 The Symmetric Algebra

The *symmetric algebra* over $\mathbb{R}^d$, denoted by $S(\mathbb{R}^d)$ is the quotient of $T(\mathbb{R}^d)$ by the two-sided ideal I generated by $\{xy - yx : x, y \in \mathbb{R}^d\}$. A linear basis of $S(\mathbb{R}^d)$ is then given by $\{\hat{e}_v\}$ over non-decreasing words, $v = (i_1, \dots, i_n) \in \widehat{\mathcal{W}}_d$, with $1 \le i_1 \le \dots \le i_n \le d, n \ge 0$. Every $\hat{\mathbf{x}} \in S(\mathbb{R}^d)$ can be written as finite sum,

$$\hat{\mathbf{x}} = \sum_{v \in \widehat{\mathcal{W}}_d} \hat{\mathbf{x}}^v \hat{e}_v,$$

and we have an immediate identification with polynomials in d commuting indeterminates. The canonical projection

$$\pi_{\mathrm{Sym}} : T(\mathbb{R}^d) \twoheadrightarrow S(\mathbb{R}^d), \quad \mathbf{x} \mapsto \sum_{w \in \mathcal{W}_d} \mathbf{x}^w \hat{e}_{\hat{w}}, \tag{24}$$

where $\hat{w} \in \hat{\mathcal{W}}_d$ denotes the non-decreasing reordering of the letters of the word $w \in \mathcal{W}_d$, is an algebra epimorphism, which extends to an epimorphism $\pi_{\mathrm{Sym}} : \mathcal{T} \twoheadrightarrow \mathcal{S}$ where $\mathcal{S} = S\big((\mathbb{R}^d)\big)$ is the algebra completion, identifiable as formal series in d commuting indeterminates. As a vector space, $\mathcal{S}$ can be identified with *symmetric* formal tensor series. Denote by $\mathcal{S}_0$ and $\mathcal{S}_1$ the affine space given by those $\hat{\mathbf{x}} \in \mathcal{S}$ with $\hat{\mathbf{x}}^{\emptyset} = 0$ and $\hat{\mathbf{x}}^{\emptyset} = 1$ respectively. The usual power series in $\mathcal{S}$ define $\widehat{\exp}: \mathcal{S}_0 \to \mathcal{S}_1$ with inverse $\widehat{\log}: \mathcal{S}_1 \to \mathcal{S}_0$ and we have

$$\pi_{\mathrm{Sym}} \exp(\mathbf{x} + \mathbf{y}) = \widehat{\exp}(\hat{\mathbf{x}})\widehat{\exp}(\hat{\mathbf{y}}), \quad \mathbf{x}, \mathbf{y} \in \mathcal{T}_0$$
$$\pi_{\mathrm{Sym}} \log(\mathbf{x}\mathbf{y}) = \widehat{\log}(\hat{\mathbf{x}}) + \widehat{\log}(\hat{\mathbf{y}}), \quad \mathbf{x}, \mathbf{y} \in \mathcal{T}_1.$$

We shall abuse notation in what follows and write $e^{(\cdot)}$ (resp. log), instead of $\widehat{\exp}$ (resp. $\widehat{\log}$).

All definitions for tensor valued continuous (and càdlàg) semimartingales have a straightforward extension to $\mathcal{S}$-valued process. In particular, given $\mathbf{X}$ and $\mathbf{Y}$ in $\mathscr{S}^c(\mathcal{S})$, the *inner* quadratic covariation is given by

$$\langle \mathbf{X}, \mathbf{Y} \rangle = \sum_{w_1, w_2 \in \widehat{\mathcal{W}}_d} \langle \mathbf{X}^{w_1}, \mathbf{Y}^{w_2} \rangle \hat{e}_{w_1} \hat{e}_{w_2}.$$

Write $\mathcal{S}^N$ for the truncated symmetric algebra, linearly spanned by $\{\hat{e}_w : w \in \widehat{\mathcal{W}}_d, |w| \leq N\}$ and $\mathcal{S}_0^N$ for those elements with zero scalar entry. In complete analogy with non-commutative setting discussed above, we then write $\widehat{\mathscr{H}}^{q,N} \subset \mathscr{S}^c(\mathcal{S}_0^N)$ for the corresponding space of homogeneously q-integrable semimartingales.

Finally, also the definition of diamond products from Sect. 4.3 extends immediately to $\mathcal{S}$-valued semimartingales. In particular, given $\hat{\mathbf{X}}$ and $\hat{\mathbf{Y}}$ in $\mathscr{S}^c(\mathcal{S})$, we have

$$(\hat{\mathbf{X}} \diamond \hat{\mathbf{Y}})_t(T) := \mathbb{E}_t\left(\langle \hat{\mathbf{X}}, \hat{\mathbf{Y}} \rangle_{t,T}\right) = \sum_{w_1, w_2 \in \widehat{\mathcal{W}}_d} (\hat{\mathbf{X}}^{w_1} \diamond \hat{\mathbf{Y}}^{w_2})_t(T) \hat{e}_{w_1} \hat{e}_{w_2} \in \mathcal{S},$$

where the last expression is given in terms of diamond products of scalar semimartingales.

5.2 *Moments and Cumulants*

We quickly discuss the development of a symmetric algebra valued semimartingale, more precisely $\hat{\mathbf{X}} \in \mathscr{S}^c(\mathcal{S}_0)$, in the group $\mathcal{S}_1$. That is, we consider

$$\mathrm{d}\hat{\mathbf{S}} = \hat{\mathbf{S}} \circ \mathrm{d}\hat{\mathbf{X}}. \tag{25}$$

It is immediate (validity of chain rule) that the unique solution to this equation, at time $t \geq s$, started at $\hat{\mathbf{S}}_s = \hat{\mathbf{s}} \in \mathcal{S}_1$ is given by

$$\hat{\mathbf{S}}_t := e^{\hat{\mathbf{X}}_t - \hat{\mathbf{X}}_s} \hat{\mathbf{s}} \in \mathcal{S}_1$$

and we also write $\hat{\mathbf{S}}_{s,t} = e^{\hat{\mathbf{X}}_t - \hat{\mathbf{X}}_s}$ for this solution started at time s from $1 \in \mathcal{S}_1$. The relation to signatures is as follows.

Proposition 5.1 *Let* $\mathbf{X}$ *and* $\mathbf{Y}$ *be* $\mathcal{T}$*-valued semimartingales and define* $\hat{\mathbf{X}} := \pi_{\mathrm{Sym}}(\mathbf{X})$ *and* $\hat{\mathbf{Y}} := \pi_{\mathrm{Sym}}(\mathbf{Y})$. *Then for all* $0 \leq s \leq t \leq T$ *it holds almost surely*

$$\pi_{\mathrm{Sym}} \int_s^t \mathbf{X} \mathrm{d} \mathbf{Y} = \int_s^t \hat{\mathbf{X}} \, \mathrm{d} \hat{\mathbf{Y}}, \tag{26}$$

and

$$\pi_{\mathrm{Sym}} \mathrm{Sig}(\mathbf{X})_{s,t} = \hat{\mathbf{S}}_{s,t} = e^{\hat{\mathbf{X}}_t - \hat{\mathbf{X}}_s}.$$

Proof (i) That the projections $\hat{\mathbf{X}}, \hat{\mathbf{Y}}$ define $\mathcal{S}$-valued semimartingales follows from the componentwise definition and the fact that the canonical projection is linear. In particular, the right-hand side of Eq. (26) is well defined. To show Eq. (26) we apply the canonical projection π_{Sym} to both sides of Eq. (11) after choosing $Z_t \equiv \mathbf{1}$, and using the explicit action of π_{Sym} on basis tensors we obtain the identity

$$\pi_{\mathrm{Sym}} \int \mathbf{X} \, \mathrm{d} \mathbf{Y} = \sum_{w \in \mathcal{W}^d} \left(\sum_{uv = w} \int \mathbf{X}^u \, \mathrm{d} \mathbf{Y}^v \right) \hat{e}_{\hat{w}} = \int \hat{\mathbf{X}} \, \mathrm{d} \hat{\mathbf{Y}}.$$

by Eq. (24). Part (ii) is then immediate. □

Assuming componentwise integrability, we then define the conditional *symmetric moments* and *cumulants* of the $\mathcal{S}$-valued semimartingale $\hat{\mathbf{X}}$ by

$$\hat{\boldsymbol{\mu}}_t(T) := \mathbb{E}_t \left(e^{\hat{\mathbf{X}}_T - \hat{\mathbf{X}}_t} \right) \in \mathcal{S}_1,$$
$$\hat{\boldsymbol{\kappa}}_t(T) := \log \left(\hat{\boldsymbol{\mu}}_t(T) \right) \in \mathcal{S}_0,$$

for $0 \leq t \leq T$. If $\hat{\mathbf{X}} = \pi_{\mathrm{Sym}}(\mathbf{X})$ for a $\mathcal{T}$-valued semimartingale $\mathbf{X}$, with expected signature and signature cumulants $\boldsymbol{\mu}(T)$ and $\boldsymbol{\kappa}(T)$, it is then clear that the symmetric moments and cumulants of $\hat{\mathbf{X}}$ are obtained by projection,

$$\hat{\boldsymbol{\mu}}(T) = \pi_{\mathrm{Sym}}(\boldsymbol{\mu}(T)), \quad \hat{\boldsymbol{\kappa}}(T) = \pi_{\mathrm{Sym}}(\boldsymbol{\kappa}(T)).$$

Example 5.2 Let X be an $\mathbb{R}^d$-valued martingale in $\mathscr{H}^{\infty-}$, and $\hat{\mathbf{X}}_t := \sum_{i=1}^d X_t^i \hat{e}_i$. Then

$$\begin{aligned} \hat{\boldsymbol{\mu}}_t(T) &= \sum_{n=0}^{\infty} \frac{1}{n!} \mathbb{E}_t[(X_T - X_t)^n] \\ &= 1 + \sum_{n=1}^{\infty} \frac{1}{n!} \sum_{i_1, \ldots, i_n = 1}^{d} \mathbb{E}_t \left[(X_T^{i_1} - X_t^{i_1}) \cdots (X_T^{i_n} - X_t^{i_n}) \right] \hat{e}_{\widehat{i_1 \cdots i_n}}, \end{aligned}$$

consists of the (time-t conditional) multivariate moments of $X_T - X_t \in \mathbb{R}^d$. Here, the series on the right hand side is understood in the formal sense. It readily follows, also noted in [9, Example 3.3], that $\hat{\boldsymbol{\kappa}}_t(T) = \log(\hat{\boldsymbol{\mu}}_t(T))$ consists precisely of the multivariate cumulants of $X_T - X_t$. Note that the symmetric moments and cumulants of the scaled process aX, $a \in \mathbb{R}$, is precisely given by $\delta_a \hat{\boldsymbol{\mu}}$ and $\delta_a \hat{\boldsymbol{\kappa}}$ where the linear dilation map is defined by $\delta_a : \hat{e}_w \mapsto a^{|w|} \hat{e}_w$. The situation is similar for $a \cdot X = (a_1 X^1, \ldots, a_d X^d)$, $a \in \mathbb{R}^d$, but now with $\delta_a : \hat{e}_w \mapsto a^w \hat{e}_1^{|w|}$ with $a^w = a_1^{n_1} \cdots a_d^{n_d}$ where n_i denotes the multiplicity of the letter $i \in \{1, \ldots, d\}$ in the word w.

We next consider linear combinations, $\hat{\mathbf{X}} = aX + b\langle X \rangle$, for general pairs $a, b \in \mathbb{R}$, having already dealt with $b = 0$. The special case $b = -a^2/2$, by scaling there is no loss in generality to take $(a, b) = (1, -1/2)$, yields a (at least formally) familiar exponential martingale identity.

Example 5.3 Let X be an $\mathbb{R}^d$-valued martingale in $\mathscr{H}^{\infty-}$, and define

$$\hat{\mathbf{X}}_t := \sum_{i=1}^{d} X_t^i \hat{e}_i - \frac{1}{2} \sum_{1 \le i \le j \le d} \langle X^i, X^j \rangle_t \hat{e}_{ij}.$$

In this case we have trivial symmetric cumulants, $\hat{\boldsymbol{\kappa}}_t(T) = 0$ for all $0 \le t \le T$. Indeed, Itô's formula shows that $t \mapsto \exp(\hat{\mathbf{X}}_t)$ is an $\mathcal{S}_1$-valued martingale, so that

$$\hat{\boldsymbol{\mu}}_t(T) = \mathbb{E}_t[e^{\hat{\mathbf{X}}_T - \hat{\mathbf{X}}_t}] = e^{-\hat{\mathbf{X}}_t} \mathbb{E}_t[e^{\hat{\mathbf{X}}_T}] = 1.$$

While the symmetric cumulants of the last example carries no information, it suffices to work with

$$\hat{\mathbf{X}} = \sum_{i=1}^{d} a^i X^i \hat{e}_i + \sum_{1 \le i \le j \le d} b_{jk} \langle X^j, X^k \rangle \hat{e}_{ij}$$

in which case $\hat{\boldsymbol{\mu}} = \hat{\boldsymbol{\mu}}(a, b), \hat{\boldsymbol{\kappa}} = \hat{\boldsymbol{\kappa}}(a, b)$ contains full information of the joint moments of X and its quadratic variation process. A recursion of these was constructed as diamond expansion in [27].

5.3 Diamond Relations for Multivariate Cumulants

We will demonstrate how a symmetrization of the functional equations from Sect. 3 and the recursions from Sect. 4 lead to generalized view on the cumulant recursions from [3, 27, 41]. As before we will first divide the focus between the continuous case and the discrete setting. Referring to [28, Section 5.2] for a unification of both settings into a general càdlàg form.

5.3.1 The Continuous Case

We assume that the usual assumptions from the continuous setting in Sect. 3.2 on the filtration are in place. Then following Theorem 4.7 the diamond product of $\hat{\mathbf{X}}, \hat{\mathbf{Y}} \in \mathscr{S}^c(\mathcal{S}_0)$ is another continuous $\mathcal{S}_0$-valued semimartingale given by

$$(\hat{\mathbf{X}} \diamond \hat{\mathbf{Y}})_t(T) = \mathbb{E}_t\big(\langle \hat{\mathbf{X}}, \hat{\mathbf{Y}}\rangle_{t,T}\big) = \sum (\mathbb{E}_t \langle \hat{\mathbf{X}}^{w_1}, \hat{\mathbf{Y}}^{w_2}\rangle_{t,T}) \hat{e}_{w_1} \hat{e}_{w_2},$$

with summation over all $w_1, w_2 \in \widehat{\mathcal{W}}_d$, provided all brackets are integrable. This trivially adapts to $\mathcal{S}^N$-valued semimartingales, $N \in \mathbb{N}_{\geq 1}$, in which case all words have length at most N, the summation is restricted accordingly to $|w_1| + |w_2| \leq N$.

Theorem 5.4 *Let $\Xi = (0, \Xi^{(1)}, \Xi^{(2)}, \dots)$ be an $\mathcal{F}_T$-measurable random variable with values in $\mathcal{S}_0$, componentwise in $\mathcal{L}^{\infty-}$. Then*

$$\mathbb{K}_t(T) := \log \mathbb{E}_t\left(e^{\Xi}\right)$$

satisfies the following functional equation, for all $0 \leq t \leq T$,

$$\mathbb{K}_t(T) = \mathbb{E}_t \Xi + \frac{1}{2}(\mathbb{K} \diamond \mathbb{K})_t(T). \tag{27}$$

Furthermore, if $N \in \mathbb{N}_{\geq 1}$, and $\Xi = (\Xi^{(1)}, \dots, \Xi^{(N)})$ is $\mathcal{F}_T$-measurable with graded integrability condition

$$\|\Xi^{(n)}\|_{\mathcal{L}^{N/n}} < \infty, \qquad n = 1, \dots, N, \tag{28}$$

then the identity (27) holds for the cumulants upto level N, i.e. for $\mathbb{K}^{(0,N)} := \log(\mathbb{E}_t(e^{\Xi^{(0,N)}})$ with values in $\mathcal{S}_0^N$.

Remark 5.5 Identity (27) is reminiscent to the quadratic form of the generalized Riccati equations for affine diffusions. The relation can be presented more explicitly when the involved processes are assumed to have a Markov structure and the functional signature cumulant equation reduces to a PDE system (see [28, Section 6.2.2]). The framework described here, however, requires neither Markov nor affine structure. In [28, Section 6.3] it is exemplified for affine Volterra processes that such computations are also possible in the fully non-commutative setting.

Proof We first observe that since $\Xi \in \mathcal{L}^{\infty-}$, by Doob's maximal inequality and the BDG inequality, we have that $\hat{\mathbf{X}}_t := \mathbb{E}_t \Xi$ is a continuous martingale in $\mathscr{H}^{\infty-}(\mathcal{S}_0)$. In particular, thanks to Theorem 2.6, the signature moments are well defined. According to Sect. 5.2, the signature is then given by

$$\mathrm{Sig}(\hat{\mathbf{X}})_{t,T} = e^{\Xi - \mathbb{E}_t \Xi},$$

hence $\hat{\kappa}_t(T) = \mathbb{K}_t(T) - \hat{\mathbf{X}}_t$.

Projecting Eq. (19) onto the symmetric algebra yields

$$\begin{aligned}\hat{\kappa}_t(T) &= \mathbb{E}_t\left\{\hat{\mathbf{X}}_{t,T} + \frac{1}{2}\langle\hat{\mathbf{X}}\rangle_{t,T} + \frac{1}{2}\langle\kappa(T)\rangle_{t,T} + \langle\hat{\mathbf{X}}, \kappa(T)\rangle_{t,T}\right\} \\ &= \mathbb{E}_t\left\{\Xi + \frac{1}{2}\langle\mathbb{K}(T)\rangle_{t,T}\right\} - \hat{\mathbf{X}}_t,\end{aligned}$$

and Eq. (27) follows upon recalling that $(\mathbb{K} \diamond \mathbb{K})_t(T) = \mathbb{E}_t\langle\mathbb{K}(T)\rangle_{t,T}$. The proof of the truncated version is left to the reader. □

As a corollary, we provide a general view on recent results of [3, 27, 41].

Corollary 5.6 *The conditional multivariate cumulants $(\mathbb{K}_t)_{0\le t\le T}$ of a random variable Ξ with values in $\mathcal{S}_0(\mathbb{R}^d)$, componentwise in $\mathcal{L}^{\infty-}$ satisfy the recursion*

$$\mathbb{K}_t^{(1)} = \mathbb{E}_t(\Xi^{(1)}) \quad \textit{and} \quad \mathbb{K}_t^{(n)} = \mathbb{E}_t(\Xi^{(n)}) + \frac{1}{2}\sum_{k=1}^{n}\left(\mathbb{K}^{(k)} \diamond \mathbb{K}^{(n-k)}\right)_t(T) \quad \textit{for } n \ge 2, \tag{29}$$

The analogous statement holds true in the N-truncated setting, i.e. as recursion for $n = 1, .., N$ under the condition (28).

Remark 5.7 In absence of higher order information, i.e., $\Xi^{(2)} = \Xi^{(3)} = \ldots \equiv 0$, this type of cumulant recursion appears in [41]; and under optimal integrability conditions on $\Xi^{(1)}$ with finite Nth moments in [27]. (The latter requires a localization argument which is avoided here by directly working in the correct algebraic structure.)

5.3.2 The Discrete Case

Consider a probability space $(\Omega, \mathcal{G}, \mathbb{P})$ with discrete filtration $(\mathcal{G}_j)_{j=0,\ldots,J}$ and a $\mathcal{G}_T$-measurable $\mathcal{S}_0$-valued random variable Ξ. Assuming that Ξ is componentwise in $\mathcal{L}^{\infty-}$, we are again interested in calculating $(\mathbb{K}_j) := (\log \mathbb{E}(e^{\Xi}\,|\mathcal{G}_j))$. Instead of projecting the identities from the non-commutative setting in Sect. 3.1 to the symmetric algebra, we present a direct derivation of the corresponding identities.

Using the multiplicative property of the exponential map in the symmetric algebra, the martingale property of $e^{\mathbb{K}}$ can be rewritten to

$$0 = \mathbb{E}\big(e^{\mathbb{K}_{j+1}-\mathbb{K}_j} - 1\,\big|\mathcal{G}_j\big).$$

A summation of the above identity paired with

$$\mathbb{K}_j = \Xi - (\mathbb{K}_J - \mathbb{K}_j) = \Xi - \sum_{i=j}^{J-1}(\mathbb{K}_{i+1} - \mathbb{K}_i)$$

and conditioning to $\mathcal{G}_j$ then yields the following.

Theorem 5.8 *Given that the $\mathcal{G}_J$-measurable $\mathcal{S}_0$-valued random variable is componentwise in $\mathcal{L}^{\infty-}$. It follows that $(\mathbb{K}_j) := (\log \mathbb{E}(e^{\Xi} | \mathcal{G}_j))$ satisfies for $j = J, J-1, \ldots, 0$:*

$$\mathbb{K}_j = \mathbb{E}(\Xi | \mathcal{G}_j) + \mathbb{E}\bigg(\sum_{i=j}^{J-1} e^{\mathbb{K}_{i+1} - \mathbb{K}_i} - 1 - (\mathbb{K}_{i+1} - \mathbb{K}_i) \bigg| \mathcal{G}_j \bigg). \tag{30}$$

In particular, projecting to symmetric tensor levels we see that the conditional cumulants of Ξ satisfy for all $n = 1, 2, \ldots$ and $j = J, J-1, \ldots, 1$:

$$\mathbb{K}_j^{(n)} = \mathbb{E}(\Xi^{(n)} | \mathcal{G}_j) + \mathbb{E}\bigg(\sum_{\|\ell\|=n, |\ell|=k} \frac{1}{k!} (\mathbb{K}_{i+1} - \mathbb{K}_i)^{(\ell)} \bigg| \mathcal{G}_j \bigg),$$

This is really the result we get from projecting the Theorem 3.1 to the symmetric algebra (upon using that the symmetric signature cumulants are given by $\hat{\boldsymbol{\kappa}}_j = \mathbb{K}_j - \mathbb{E}(\Xi | \mathcal{G}_j)$). Nevertheless, due the ad hoc derivation presented here, it might seem surprising that the resulting expansion leads to any non-trivial relations for conditional cumulants at all.

While on the first level one still trivially has $(\mathbb{K}_j^{(1)}) = (\mathbb{E}(\Xi^{(1)} | \mathcal{G}_j))$, on the second level we have

$$\mathbb{K}_j^{(2)} = \mathbb{E}(\Xi^{(2)} | \mathcal{G}_j) + \frac{1}{2} \mathbb{E}\bigg(\sum_{i=j}^{J-1} (\mathbb{K}_{j+1}^{(1)} - \mathbb{K}_j^{(1)})^2 \bigg| \mathcal{G}_j \bigg)$$

which one recognizes, in case $\Xi^{(2)} = 0$, as the energy identity for the discrete square-integrable martingale $(\mathbb{K}_j^{(1)}) = (\mathbb{E}(\Xi^{(1)} | \mathcal{G}_j))$. Going further in the recursion yields increasingly non-obvious relations. Taking $\Xi^{(2)} = \Xi^{(3)} = \ldots \equiv 0$ for notational simplicity gives

$$\mathbb{K}_j^{(3)} = \mathbb{E}\bigg(\sum_{i=j}^{J-1} \frac{1}{6} (\mathbb{K}_{i+1}^{(1)} - \mathbb{K}_i^{(1)})^3 + (\mathbb{K}_{i+1}^{(1)} - \mathbb{K}_i^{(1)})(\mathbb{K}_{i+1}^{(2)} - \mathbb{K}_i^{(2)}) \bigg| \mathcal{G}_j \bigg). \tag{31}$$

It is interesting to note that inductively the identity for $\mathbb{K}^{(n)}$ can by expressed in terms of variations (of variations) of the martingale $\mathbb{K}^{(1)}$, which relates to the *Bartlett identities* for martingales that have appeared in the statistics literature, cf. Mykland [53] and the references therein.

Remark 5.9 The second formula in Theorem 5.8, in terms of summation over partitions ℓ, is sub-optimal in the sense that due to commutativity there are many repeated terms. For example, the factor $\frac{1}{2}$ in front of the second term in Eq. (31) suggested by that formula is actually just 1, since both compositions $\ell = (1, 2)$ and

$\ell = (2, 1)$ give the same product. In general one can take care of this algebraically by using (rescaled) Bell polynomials, which in the context of conditional cumulants was realized in [30].

6 Applications

6.1 Time-Inhomogeneous Lévy Processes

We now wish to derive a formula for the expected signature of time-inhomogeneous Lévy processes. This class of processes models trajectories having jumps, possibly infinitely many. To avoid the technical difficulties entailed by general càdlàg[2] stochastic integration, we will only consider time-inhomogeneous Lévy process that have finitely many jumps. Specifically, for a filtered probability space $(\Omega, \mathcal{F}, \mathbb{P}, (\mathcal{F}_t))$ satisfying the usual conditions we consider a process of the form

$$X = \int_0^\cdot b(u)\mathrm{d}u + \sum_{k=1}^{m} \int_0^\cdot \sigma_k(u)\mathrm{d}B_u^k + N, \tag{32}$$

where $b : [0, T] \to \mathbb{R}^d$ is integrable, $\sigma_k : [0, T] \to \mathbb{R}^d$ is square-integrable, $B = (B^1, \cdots, B^m)$ is a Brownian motion with respect to $(\mathcal{F}_t)$ and N is a time-inhomogenous *compound Poisson process* with respect to $(\mathcal{F}_t)$, i.e., N is a process with piecewise constant right-continuous sample paths in $\mathbb{R}^d$ and is fully characterized by the existence of a finite *intensity measure* K so that

$$\mathbb{E}_s\Big(\sum_{s<u\leq t} f(\Delta N_u)\Big) = \int_s^t \int_{\mathbb{R}^d} f(x)K_u(\mathrm{d}x)\mathrm{d}u, \tag{33}$$

for all $0 \leq s \leq t \leq T$ and bounded measurable functions $f : \mathbb{R}^d \to \mathbb{R}$, where

$$\Delta N_t := N_t - N_{t-} := N_t - \lim_{\varepsilon \to 0} N_{t-\varepsilon}.$$

The interpretation of signature as the "geometric" development of a $\mathcal{T}_0$-value semimartingale $\mathbf{X}$ in the group $\mathcal{T}_1$ extents from the continuous [33] to the càdlàg case [18] and is consistent with the Marcus [5, 22, 40, 50, 51] interpretation of

$$\mathrm{d}\mathbf{S}_t = \mathbf{S}_t \circ \mathrm{d}\mathbf{X}_t, \qquad t \geq s, \quad \mathbf{S}_s = 1 \in \mathcal{T}_1. \tag{34}$$

[2] From French "continu(e) à droit, limite a gauche" (right-continuous with left limits).

The definition of the signature $\mathrm{Sig}(\mathbf{X})_{s,\cdot}$ as the solution to (34) results in the desired property that, whenever the process jumps, the signature is multiplied by the exponential displacement, i.e.,

$$\mathrm{Sig}(\mathbf{X})_{s,t} = \mathrm{Sig}(\mathbf{X})_{s,t-}e^{\Delta \mathbf{X}_t}.$$

For the time-inhomogeneous Lévy processes X of the form (32) we can resolve the above Marcus definition of the signature $\mathbf{S} = \mathrm{Sig}(X)_{s,\cdot}$ by simply integrating between the finite number of jump times. This finally leads to the Itô-integral from

$$\mathbf{S}_t = 1 + \int_s^t \mathbf{S}_{u-}\left(b(u) + \frac{1}{2}a(u)\right)\mathrm{d}u + \sum_{k=1}^m \int_s^t \mathbf{S}_{u-}\,\sigma_k(u)\mathrm{d}B_u^k + \sum_{s<u\le t} \mathbf{S}_{u-}\left(e^{\Delta N_u} - 1\right), \tag{35}$$

where $a(t) := \sum_{k=1}^m (\sigma_k(t))^{\otimes 2} \in \mathrm{Sym}(\mathbb{R}^{d\times d})$.

We then have the following result for the expected signature and signature-cumulants of time-inhomogeneous Lévy processes.

Proposition 6.1 *Let X be a process of the from* (32) *with an intensity measure K satisfying the moment condition*

$$\int_0^T \int_{\mathbb{R}^d} |x|^q K_t(\mathrm{d}x)\mathrm{d}t < \infty, \qquad q \ge 1, \tag{36}$$

Then its conditional expected signature $\boldsymbol{\mu} = (\mathbb{E}_t(\mathrm{Sig}(X)_{t,T})_{0\le t\le T}$ is the solution of the following backwards differential equation in $\mathcal{T}_1$:

$$\boldsymbol{\mu}_t = 1 + \int_t^T \mathfrak{y}(u)\,\boldsymbol{\mu}_u \mathrm{d}u, \qquad 0 \le t \le T, \tag{37}$$

where

$$\mathfrak{y}(t) := b(t) + \frac{1}{2}a(t) + \int_{\mathbb{R}^d} (e^{\mathbf{x}} - 1)K_t(\mathrm{d}\mathbf{x}) \in \mathcal{T}_0. \tag{38}$$

Moreover, the conditional signature cumulants $\boldsymbol{\kappa} = \log(\boldsymbol{\mu})$ are the solution of

$$\boldsymbol{\kappa}_t = \int_t^T H(\mathrm{ad}\,\boldsymbol{\kappa}_u)(\mathfrak{y}(u))\,\mathrm{d}u, \qquad 0 \le t \le T.$$

Remark 6.2 The case of general time-homogeneous Lévy process, i.e., semi-martingales with independent increments that are continuous in probability, is treated in [28, Corollary 6.5]. Allowing for infinitely many jumps by extending K

to a Lévy-measure, i.e., by requiring K to only be a σ-finite measure satisfying $\int_0^T \int_{\mathbb{R}^d} 1 \wedge |x|^2 K_t(\mathrm{d}x)\mathrm{d}t < \infty$, a martingale compensation of the small jumps is required and results in subtracted indicator function $\mathbf{1}_{|x|>1}$ appearing in (36) and (38).

Proof To avoid using general stochastic integration, we present a direct proof following [22]. This prove bases on the identity

$$\mathbb{E}_s\left[\sum_{s<u\leq T} H_{u-} f(\Delta N_u)\right] = \mathbb{E}_s\left[\int_s^t \int_{\mathbb{R}^d} H_{u-} f(\Delta N_u) K_u(\mathrm{d}x)\mathrm{d}u\right], \tag{39}$$

which is proven using a measure theoretic induction starting with (33) for $H \equiv 1$ and all bounded measurable functions $f : \mathbb{R}^d \to \mathbb{R}$, and then extending over simple processes H, i.e. piecewise constant adapted bounded processes, to all càdlàg adapted process H with $\|H\|_{\mathscr{S}^1} < \infty$. We refer to [22, Lemma 43] for a more detailed proof in the time-homogeneous case.

As before we will omit proving the integrability properties of $\mathrm{Sig}(X)_{0,T}$. As proven in [28, Corollary 6.5] the moment condition (36) suffices to show that $\|\mathrm{Sig}(X)^w_{0,\cdot}\|_{\mathscr{S}^q} < \infty$ for all $w \in \mathcal{W}_d$ and $q > 1$. Hence, the expected signature is well defined and it immediately follows from the independence of increments of X with respect to the underlying filtration that

$$\boldsymbol{\mu}_s(t) = \mathbb{E}_s[\mathrm{Sig}(X)_{s,t}] = \mathbb{E}_s[\mathrm{Sig}(X)_{s,t}], \qquad 0 \leq s \leq t.$$

Taking expectation of each of the terms in Itô-integral form (35) of $\mathbf{S}_t = \mathrm{Sig}(X)_{s,t}$ we obtain using the integrability of $\mathbf{S}$ and the Fubini's theorem that firstly

$$\mathbb{E}_s\left[\int_s^t \mathbf{S}_{u-}\left(b(u) + \frac{1}{2}a(u)\right)\mathrm{d}u\right] = \int_s^t \boldsymbol{\mu}_t(u-)\left(b(u) + \frac{1}{2}a(u)\right)\mathrm{d}u,$$

secondly using the martingality of the stochastic integral with respect to Brownian motion

$$\mathbb{E}_s\left[\sum_{k=1}^m \int_s^t \mathbf{S}_{u-}\,\sigma_k(u)\mathrm{d}B^k_u\right] = 0$$

and finally using (39) and once more Fubini's theorem

$$\mathbb{E}_s\left[\sum_{s<u\leq T} \mathbf{S}_{u-}(e^{\Delta N_u} - 1)\right] = \int_s^t \int_{\mathbb{R}^d} \boldsymbol{\mu}_s(u-)(e^{\mathbf{x}} - 1)K_u(\mathrm{d}x)\mathrm{d}u.$$

Summarizing, this show that $\boldsymbol{\mu}_s$ satisfies the forward equation

$$\mu_s(t) = 1 + \int_s^t \mu_s(u-)\mathfrak{y}(u)\mathrm{d}u, \qquad s \le t \le T.$$

This implies in particular that $t \mapsto \mu_s(t)$ is absolutely continuous. Furthermore, by the uniqueness of the solution flow we also have the identity $\mu_0(t)\mu_t(T) = \mu_0(T)$. Hence, $t \mapsto \mu_t(T) = \mu_0(t)^{-1}\mu_0(T)$ is absolutely continuous as well, and differentiation of the previous identity yields

$$\mu_0(t)\mathfrak{y}(t)\mu_t(T) + \mu_0(t)\frac{\mathrm{d}}{\mathrm{d}t}\mu_t(T) = 0,$$

for almost every $t \in [0, T]$. Multiplication with $\mu_0(t)^{-1}$ and integration over $[t, T]$ then yields the backwards equation.

The second part of the statement follows by noting that from the first part of the proof we have

$$\mathbb{E}_t\big(\mathrm{Sig}(X)_{t,T}\big) = \mathrm{Sig}\Big(\int_0^\cdot \mathfrak{y}(u)\mathrm{d}u\Big)_{t,T}.$$

Hence, μ is really the signature of the continuous deterministic $\mathcal{T}_0$-valued path $\mathbf{X} := \int_0^\cdot \mathfrak{y}(u)\mathrm{d}u$. An application of Theorem 3.6 then directly leads to the stated expansion for the log-signature $\log(\mathrm{Sig}(\mathbf{X})_{t,T}) = \log(\mu_t(T))$. □

We conclude this section with an important consequence that concerns the *convergence radius* of the expected signature, which for a general tensor series $\mathbf{x} \in \mathcal{T}$ is defined as the maximal $r \in [0, \infty)$ such that

$$\sum_{n=0}^{\infty} \lambda^n |\mathbf{x}^{(n)}| < \infty, \qquad \text{for all } \lambda \in (0, r). \tag{40}$$

In [15] it was proven that if the expected signature has infinite convergence radius it characterizes the law of the signature, hence, the law of the underlying process when the signature itself is characterizing. Thus, rendering the question of infinite convergence radius a central but in general difficult question. For Brownian motion it is a consequence of Fawcett's formula [19], for Lévy-process a consequence of the signature Lévy-Khinchin formula in [22]. Apart from that there is a negative answer in [8, 45] for the case of Brownian motion stopped at the exit of a domain. The following results gives necessary condition for time-inhomogeneous $\mathfrak{g}^N$-valued Lévy processes.

Corollary 6.3 *Let X, μ and $\mathfrak{y}$ be as in Theorem 6.1. Then for any $N \in \mathbb{N}$ and $\lambda > 0$ it holds for all $0 \le t \le T$:*

$$\sum_{n=0}^{N} \lambda^n |\mu_t^{(n)}| \le \exp\left(\sum_{n=1}^{N} \lambda^n \int_t^T |\mathfrak{y}^{(n)}(s)|\mathrm{d}s\right).$$

In particular, $\boldsymbol{\mu}_0$ *has infinite convergence radius if* $\int_0^T |\mathfrak{y}(u)|\mathrm{d}u$ *has.*

Proof Projecting (37) to tensor level $n \in \mathbb{N}$ and using the compatibility of the norm on $(\mathbb{R}^d)^{\otimes n}$ we have for any $\lambda > 0$ and all $t \in [0, T]$:

$$\lambda^n |\boldsymbol{\mu}_t^{(n)}| \leq \sum_{k=0}^{n} \int_t^T \lambda^k |\boldsymbol{\mu}_s^{(k)}| \lambda^{n-k} |\mathfrak{y}^{(n-k)}(s)| \mathrm{d}s$$

Summation of the above inequality yields for all $t \in [0, T]$:

$$\begin{aligned}\sum_{n=0}^{N} \lambda^n |\boldsymbol{\mu}_t^{(n)}| &\leq 1 + \sum_{n=0}^{N} \sum_{k=0}^{n} \int_t^T \lambda^k |\boldsymbol{\mu}_s^{(k)}| \lambda^{n-k} |\mathfrak{y}^{(n-k)}(s)| \mathrm{d}s \\ &\leq 1 + \int_t^T \left(\sum_{n=0}^{N} \lambda^n |\boldsymbol{\mu}_s^{(n)}| \right) \left(\sum_{n=0}^{N} \lambda^n |\mathfrak{y}^{(n)}(s)| \right) \mathrm{d}s.\end{aligned}$$

The stated estimate then follows from Gronwall's lemma and the second claim by talking taking limits for $N \to \infty$. □

6.2 Brownian Rough Paths

This section is framed within rough path terminology and is directed to a reader with special interest into the topic. The unfamiliar reader may use [21] as a starting point. We denote by $\mathcal{G}^M \subset \mathcal{T}_1^M$ the free step-M nilpotent free Lie group and recall that the space of weakly geometric rough paths consists of continuous finite p-variation paths $\mathcal{W}\Omega_T^{c,p} = C^p([0, T]; (\mathcal{G}^M, |||\cdot|||))$ with $M \leq p < M + 1$ and where $|||\cdot|||$ is a homogeneous norm on $\mathcal{G}^M$. In accordance with [14, 22] we call a random variable $\mathbf{Y}$ with values in $\mathcal{W}\Omega_T^{c,p}$ a *Brownian rough path with drift* if it has stationary and independent increments $\mathbf{Y}_{s,t} := \mathbf{Y}_s^{-1}\mathbf{Y}_t$. For the means of this section we denote the signature of $\mathbf{Y}$, i.e., the full Lyons lift, by $\mathbf{Y}^{<\infty}$.

Here, we will specifically consider the case where $M = 2N$ and $\mathbf{Y} = \mathbf{B}$ is the Stratonovich development of Brownian motion in the free step-N nilpotent Lie algebra $\mathfrak{g}^N = \log \mathcal{G}^N$. More precisely, let $\mathbb{B} = \sum_{i \in I} B^i \mathfrak{u}_i$, where $\{\mathfrak{u}_i\}_{i \in I}$ is a orthonormal system in $\mathfrak{g}^N \subset \mathcal{T}_0^N$ and $(B^i)_{i \in I}$ is a $|I|$-dimensional Brownian motion with possibly mutually correlated coefficients. Define the correlation matrix of $\mathbb{B}$ by

$$\Sigma := \sum_{i,j \in I} \mathbb{E}(B_1^i B_1^j) \mathfrak{u}_i \mathfrak{u}_j \quad \in \mathrm{Sym}(\mathfrak{g}^N \otimes \mathfrak{g}^N) \subset \mathcal{T}_0^{2N}.$$

The truncated Stratonovich development $\mathbf{B} := \pi_{(0,2N)}\mathrm{Sig}(\mathbb{B})_{0,\cdot}$ takes values in the group $\mathcal{G}^{2N}$ and following standard arguments from [21, Section 3.2] we see that the samples of $\mathbf{B}$ are (weakly) geometric rough path for any $p \in (2N, 2N+1)$. Clearly, this is still the case after adding a drift $\mathbb{B} \rightsquigarrow \mathbb{B}(\mathfrak{y}) := (\mathbb{B} + t\mathfrak{y})_{t\geq 0}$ for any constant vector $\mathfrak{y} \in \mathfrak{g}^{2N}$. The resulting development $\mathbf{B}(\mathfrak{y})$ is a Brownian rough-path with drift, as independence and stationarity of the increments follow directly from the definition. Its signature is given by the full development $\mathbf{B}(\mathfrak{y})^{<\infty} = \mathrm{Sig}(\mathbb{B}(\mathfrak{y}))_{0,\cdot}$.

Theorem 3.4 implies the following generalization of Fawcett's formula.

Corollary 6.4 *The expected signature of a Brownian rough path* $\mathbf{B}(\mathfrak{y})$ *with correlation* Σ *and drift* $\mathfrak{y}$ *is given by*

$$\mathbb{E}\big(\mathbf{B}(\mathfrak{y})_T^{<\infty}\big) = \exp\left(T\mathfrak{y} + \frac{T}{2}\Sigma\right).$$

Remark 6.5 The above suggest to define a "standard" Brownian rough path by requiring additionally that $\log \mathbb{E}(\mathbf{Y}_{s,t}) = \frac{1}{2}(t-s)\sum_{i\in I}(\mathfrak{u}_i)^2$ and that $\{\mathfrak{u}_i\}$ spans $\mathfrak{g}^N$. However, this is only a appropriate definition as far as it is appropriate to call the Stratonovich lift the "standard" lift of a Brownian motion.

Remark 6.6 The results from the previous section, can be extended to time-inhomogeneous *Lévy rough paths* in the sense of [14, 22], without any additional effort. Indeed, instead of starting with the differential characteristics (b, a, K) in $\mathbb{R}^d$ we could have likewise used characteristics in $\mathfrak{g}^N$, resulting in a $\mathfrak{g}^N$-valued semimartingale $\mathbf{X}$ with independent increments. The truncated Marcus lift $\mathbf{S} = \pi_{(0,2N)}\mathrm{Sig}(\mathbb{X})$ then has independent group increments $\mathbf{S}_{s,t} = \mathbf{S}_s^{-1}\mathbf{S}_t$ and the sample paths are càdlàg (weakly) geometric rough paths in the sense of [22, 25].

Proof From the definition we then have $\mathbf{B}(\mathfrak{y}) := \pi_{(0,2N)}\mathrm{Sig}(\mathbb{B}(\mathfrak{y}))_{0,\cdot}$ where $\mathbb{B}$ is Brownian motion in $\mathfrak{g}^N$. For any $w \in \mathcal{W}_d$ and $q \geq 1$ we have

$$\|\mathbb{B}(\mathfrak{y})^w\|_{\mathscr{H}^q} = \left\| \langle \sum_i B^i \mathfrak{u}_i^w \rangle_T^{1/2} + T|\mathfrak{y}^w| \right\|_{\mathcal{L}^q} = \sqrt{T}\left| \sum_{i,j\in I} \Sigma^{ij}\mathfrak{u}_i^w \mathfrak{u}_j^w \right|^{1/2} + T|\mathfrak{y}^w| < \infty.$$

Hence, $\mathbb{B}(\mathfrak{y}) \in \mathscr{H}^{\infty-}$ and thus Theorems 3.4 and 4.1 apply to the conditional expected signature $\boldsymbol{\mu}_t := \mathbb{E}_t(\mathrm{Sig}(\mathbb{B}(\mathfrak{y}))_{t,T})$. Due to martingality of $\mathbb{B}$ we have

$$\boldsymbol{\mu}_t^{(1)} = \mathbb{E}\Big(\mathbb{B}_{t,T}^{(1)} + (T-t)\mathfrak{y}^{(1)}\Big) = (T-t)\mathfrak{y}^{(1)}$$

Further, due to the deterministic quadratic variation $\langle \mathbb{B} \rangle_t = t\Sigma$ one sees inductively by following the recursion in Theorem 4.1 that $\boldsymbol{\mu}^{(n)}$ is deterministic and hence by Theorem 3.4 satisfies

$$\boldsymbol{\mu}_t = 1 + \int_t^T \Big(\mathfrak{y} + \frac{1}{2}\Sigma\Big)\boldsymbol{\mu}_u du.$$

We readily conclude by noting that the unique solutions to the above ordinary differential equation in $\mathcal{T}_1$ is given by $\boldsymbol{\mu}_t = e^{(T-t)(\mathfrak{y}+\frac{1}{2}\Sigma)}$. □

We mention two important examples: Firstly, the zero mass limit of a *physical Brownian motion in a magnetic field* ([26], [28, Section 3.4], [46]), which is a Brownian rough path with $N = 1$ and drift $\mathfrak{y} = (0, 0, A)$ where $A \in \mathfrak{so}(\mathbb{R}^d \otimes \mathbb{R}^d)$ is an antisymmetric matrix that depends on the physical environment. In line with [46] the above result immediately gives that the expected signature of the small mass limit is given by $e^{T(A+\frac{1}{2}\Sigma)}$.

As a second example we mention the recent finding by Hairer [32] that the $H \to \frac{1}{4}$ limit of a suitably renormalized fractional Brownian motion converges to a pure area rough path with coefficients given by standard Brownian motion. This limit is naturally included in our framework as a Brownian rough path of $\mathbb{B} = \sum_{i \in I} \mathfrak{u}_i B^i$ where the corresponding orthonormal system of $\mathfrak{g}^2$ is given by $\{\mathfrak{u}_i : i \in I\} = \{[e_i, e_j] : 1 \leq i < j \leq d\}$. Theorem 6.4 yields the corresponding generalization of Fawcett's formula. We note that it might be of interest to obtain this formula as the rescaled limits for $H \to \frac{1}{4}$ in [13].

7 Conclusion

We summarize the results of this chapter and comment on their practical relevance. An overview of the different formulas and corresponding recursions with the relation to recent and classical results from the literature is displayed in Figs. 1 and 2.

- Theorem 3.4 and Theorem 3.6, more precisely Eqs. (17) and (19), respectively characterize expected signatures and signature cumulants of semimartingales

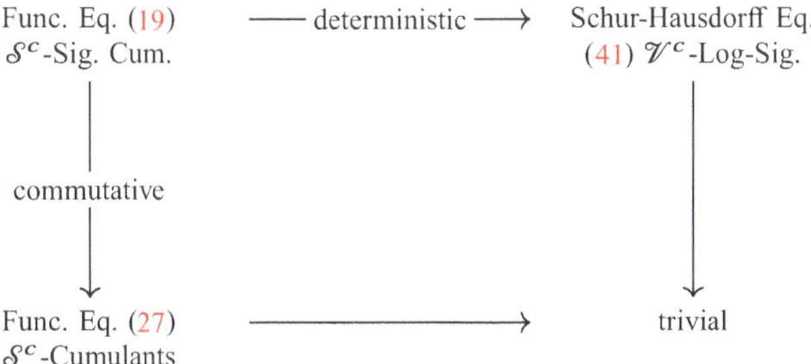

Fig. 1 An overview of the different formulas and corresponding recursions (in Fig. 2) with the relation to recent and classical results from the literature. Recall that $\mathcal{S}^c$ stands for continuous semimartingales and $\mathcal{V}^c$ for continuous processes of finite variation. Here, "trivial" refers to the linear equation $\mathrm{d}\hat{\mathbf{S}} = \hat{\mathbf{S}}\mathrm{d}\hat{\mathbf{X}}$ with $\hat{\mathbf{S}}_0 = 1 \in S$ in the commutative case $\hat{\mathbf{X}} \in \mathcal{V}^c(S)$, which is solved by $e^{\hat{\mathbf{X}}-\hat{\mathbf{X}}_0}$ with logarithm $\hat{\mathbf{X}} - \hat{\mathbf{X}}_0$

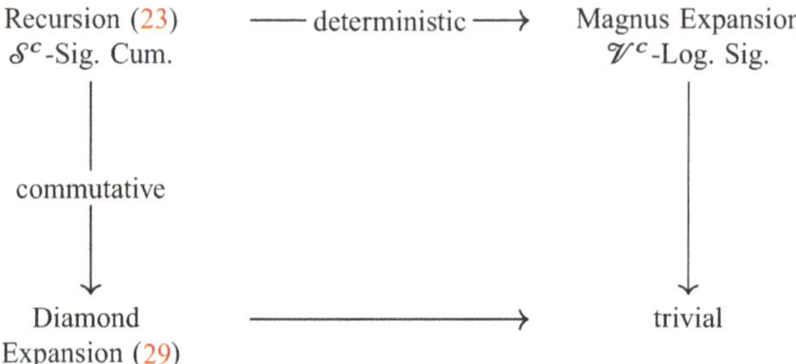

Fig. 2 Accompanying recursions to Fig. 1

by dynamic formulas, with respect to the underlying filtration. The functional equation (19) for signature cumulants is represented in the top-left corner of Fig. 1.

- Projecting (17) and (19) to tensor levels we obtain recursions for signature moments and signature cumulants respectively which are presented in Theorems 4.1 and 4.4. Specifically, these recursions represent $\boldsymbol{\mu}^{(n)}$ in terms of $\boldsymbol{\mu}^{(1)}, \ldots, \boldsymbol{\mu}^{(n-1)}$ (and analogously for $\boldsymbol{\kappa}$). The recursion for signature cumulants is represented by the top-left corner in Fig. 2.
- Even when an explicit form of $\boldsymbol{\mu}_t$ or $\boldsymbol{\kappa}_t$ is obtainable, the recursive nature of the expressions for each homogeneous component can be useful in some numerical scenarios. This provides a means of computation which can in principle be more efficient than the naive Monte Carlo approach. This is most evidently the case in discretized Markovian situations (compare with Theorem 3.2), where the recursion from Theorem 3.1 directly turns into a inductive backwards scheme, where conditional expectations can numerically be approximated using functional linear regression procedures.
- The most classical consequence of Theorem 3.6 appears when $\mathbf{X}$ is a deterministic continuous semimartingale, i.e., in particular the components of $\mathbf{X}$ are continuous paths of finite variation. The signature cumulants are then just the log-signature $\boldsymbol{\kappa}_t(T) = \log \mathrm{Sig}(\mathbf{X})_{t,T}$ and what remains of (19) is a classical differential equation due to [34, 57], here in backward form

$$-\mathrm{d}\boldsymbol{\kappa}_t(T) = H(\mathrm{ad}\,\boldsymbol{\kappa}_t)\mathrm{d}\mathbf{X}_t. \tag{41}$$

The accompanying recursion is then precisely Magnus expansion [7, 35, 36, 49]. This transition to the deterministic case is represented in Figs. 1 and 2 by going from the top-left to the top-right corner.

- Projecting to the symmetric tensor algebra $\mathcal{S}$ (Sects. 5.1 and 5.2), i.e., enforcing commutativity, signature moments and signature cumulants turn into the sequence of classical multivariate moments and cumulants of random variable $\mathbf{X}_{0,T}$. Projecting the functional equation (19) to $\mathcal{S}$ and recasting conditional quadratic variation increments as diamond products (Theorem 4.7) we obtain the cumulant equation (27) originally obtain in [27]. This consequence is depicted in Fig. 1 by going from the upper-left to the lower-left corner.

- Equation (27) should be compared with Riccati's ordinary differential equation from affine process theory [16, 17, 39]. Of course, our results, in particular Eqs. (19) and (27), are not restricted to affine semimartingales. In turn, expected signatures and cumulants—and subsequently all our statements above these—require moments, which is not required for the Riccati evolution of the characteristic function of affine processes. Of recent interest, explicit diamond expansions have been obtained for "rough affine" processes, non-Markov by nature, with cumulant generating function characterized by Riccati Volterra equations, see [1, 27, 31]. It is remarkable that analytic tractability remains intact when one passes to path space and considers signature cumulants (see in particular [28, Section 6.3]).
- Projecting (19) to tensor levels, we obtained a *diamond expansion* for multivariate cumulants (29). Such expansions where previously obtained in [3, 27, 30, 41], together with a range of applications, from quantitative finance (including rough volatility models [1, 31]), to statistical physics: in [41] the authors rely on such formulas to compute the cumulant-generating function of log-correlated Gaussian fields, more precisely approximations thereof, that underlies the Sine-Gordon model, which is a key ingredient in their renormalization procedure.
- In mathematical finance, a case of particular interest is the $\mathcal{S}$-valued semimartingale $\hat{\mathbf{X}} = (0, aX, b\langle X\rangle, 0, \dots)$, for a d-dimensional continuous martingale X. The resulting diamond expansion from [27], which improves and unifies previous results [3, 41], allowing to calculate joint cumulants of log-price and integrated volatility [20].
- We end this summary with a comment on questions of *convergence*: Basic facts of analytic functions show that *classical* moment- and cumulant-generating functions, for random variables with finite moments of all orders, have a radius of convergence $\rho \geq 0$, directly related to growth of the corresponding sequence. Convergence, in the sense $\rho > 0$, implies that the moment problem is well-posed. That is, the moments (equivalently: cumulants) determine the law of the underlying random variable. (Note that this might not be the case if $\rho = 0$. See also [27] for a related discussion in the context of diamond expansions.) Similarly, understanding the growth of expected signatures or signature cumulants is important task: in a celebrated paper [15] it was shown that under infinite convergence radius of the expected signature, in the sense of (40), the "expected signature problem" is well-posed; that is, the expected signature (equivalently: signature cumulants) determines the law of the random signature. In Sect. 6.1 we show for the case of time-inhomogeneous Lévy processes how our functional equation lead to a verifiable sufficient condition for expected signature to have an infinite convergence radius.

Acknowledgments PKF and NT acknowledge support from DFG CRC/TRR 388 "Rough Analysis, Stochastic Dynamics and Related Fields", projects A02, A05 (PF) and B01 (NT). PKF is also supported by the DFG Excellence Cluster MATH+ via a MATH+ Distinguished Fellowship.

References

1. E. Abi Jaber, M. Larsson, S. Pulido, Affine Volterra processes. Ann. Appl. Probab. **29**(5), 3155–3200 (2019)
2. Y. Aït-Sahalia, J. Jacod, *High-frequency Financial Econometrics* (Princeton University Press, Princeton, 2014)
3. E. Alòs, J. Gatheral, R. Radoičić, Exponentiation of conditional expectations under stochastic volatility. Quant. Finance **20**(1), 13–27 (2020)
4. C. Améndola, P. Friz, B. Sturmfels, Varieties of signature tensors, in *Forum of Mathematics, Sigma*, vol. 7 (Cambridge University Press, Cambridge, 2019)
5. D. Applebaum, *Lévy Processes and Stochastic Calculus*, volume 116 of Camb. Stud. Adv. Math., 2nd edn. (Cambridge University Press, Cambridge, 2009)
6. C. Bellingeri, P.K. Friz, S. Paycha, R. Preiß, Smooth rough paths, their geometry and algebraic renormalization. Vietnam J. Math. **50**(3), 719–761 (2022)
7. S. Blanes, F. Casas, J. Oteo, J. Ros, The Magnus expansion and some of its applications. Phys. Rep. **470**(5–6), 151–238 (2009)
8. H. Boedihardjo, J. Diehl, M. Mezzarobba, H. Ni, The expected signature of Brownian motion stopped on the boundary of a circle has finite radius of convergence. Bull. Lond. Math. Soc. **53**(1), 285–299 (2021)
9. P. Bonnier, H. Oberhauser, Signature cumulants, ordered partitions, and independence of stochastic processes. Bernoulli **26**(4), 2727–2757 (2020)
10. P. Bonnier, C. Liu, H. Oberhauser, Adapted topologies and higher rank signatures. Ann. Appl. Probab. **33**(3), 2136–2175 (2023)
11. E. Breuillard, P. Friz, M. Huesmann, From random walks to rough paths. Proc. Am. Math. Soc. **137**(10), 3487–3496 (2009)
12. F. Casas, A. Murua, An efficient algorithm for computing the Baker-Campbell-Hausdorff series and some of its applications. J. Math. Phys. **50**(3), 033513, 23 (2009)
13. T. Cass, E. Ferrucci, On the Wiener chaos expansion of the signature of a Gaussian process. Probab. Theory Relat. Fields **189**, 909–947 (2024)
14. I. Chevyrev, Random walks and Lévy processes as rough paths. Probab. Theory Relat. Fields **170**(3–4), 891–932 (2018)
15. I. Chevyrev, T. Lyons, Characteristic functions of measures on geometric rough paths. Ann. Probab. **44**(6), 4049–4082 (2016)
16. C. Cuchiero, D. Filipović, E. Mayerhofer, J. Teichmann, Affine processes on positive semidefinite matrices. Ann. Appl. Probab. **21**(2), 397–463 (2011)
17. D. Duffie, D. Filipović, W. Schachermayer, Affine processes and applications in finance. Ann. Appl. Probab. **13**(3), 984–1053 (2003)
18. A. Estrade, Exponentielle stochastique et intégrale multiplicative discontinues. Ann. Inst. Henri Poincaré, Probab. Stat. **28**(1), 107–129 (1992)
19. T. Fawcett, Problems in stochastic analysis: connections between rough paths and non-commutative harmonic analysis. PhD thesis, University of Oxford, 2002
20. P. Friz, J. Gatheral, Diamonds and forward variance models. Preprint, arXiv:2205.03741 [q-fin.MF] (2022)
21. P.K. Friz, M. Hairer, *A Course on Rough Paths*. Universitext, 2nd edn. (Springer International Publishing, Berlin, 2020)
22. P.K. Friz, A. Shekhar, General rough integration, Lévy rough paths and a Lévy–Kintchine-type formula. Ann. Probab. **45**(4), 2707–2765 (2017)

23. P.K. Friz, N.B. Victoir, *The Burkholder-Davis-Gundy Inequality for Enhanced Martingales*. Lecture Notes in Mathematics, vol. 1934 (Springer, Berlin, 2006)
24. P.K. Friz, N.B. Victoir, *Multidimensional Stochastic Processes as Rough Paths: Theory and Applications*. Cambridge Studies in Advanced Mathematics (Cambridge University Press, 2010)
25. P.K. Friz, H. Zhang, Differential equations driven by rough paths with jumps. J. Differ. Equations **264**(10), 6226–6301 (2018)
26. P. Friz, P. Gassiat, T. Lyons, Physical Brownian motion in a magnetic field as a rough path. Trans. Am. Math. Soc. **367**(11), 7939–7955 (2015)
27. P.K. Friz, J. Gatheral, R. Radoičić, Forests, cumulants, martingales. Ann. Probab. **50**(4), 1418–1445 (2022)
28. P.K. Friz, P.P. Hager, N. Tapia, Unified signature cumulants and generalized Magnus expansions, in *Forum of Mathematics, Sigma*, vol. 10 (Cambridge University Press, Cambridge, 2022), p. e42
29. P.K. Friz, T. Lyons, A. Seigal, Rectifiable paths with polynomial log-signature are straight lines. Bull. Lond. Math. Soc. **56**, 2922 (2024)
30. M. Fukasawa, K. Matsushita, Realized cumulants for martingales. Electron. Commun. Probab. **26**, 10 (2021), Id/No 12
31. J. Gatheral, M. Keller-Ressel, Affine forward variance models. Finance Stoch. **23**(3), 501–533 (2019)
32. M. Hairer, Renormalisation in the presence of variance blowup. Preprint, arXiv:2401.10868 [math.PR] (2024)
33. M. Hakim-Dowek, D. Lépingle, L'exponentielle stochastique des groupes de Lie, in *Séminaire de Probabilités XX 1984/85* (Springer, 1986), pp. 352–374
34. F. Hausdorff, Die symbolische Exponentialformel in der Gruppentheorie. Ber. Verh. Kgl. Sächs. Ges. Wiss. Leipzig. Math. Phys. Kl. **58**, 19–48 (1906)
35. A. Iserles, S.P. Nørsett, On the solution of linear differential equations in Lie groups. Philos. Trans. R. Soc. A **357**(1754), 983–1019 (1999)
36. A. Iserles, H. Munthe-Kaas, S. Nørsett, A. Zanna, Lie-group methods. Acta Numer. **9**, 1 (2005)
37. J. Jacod, A.N. Shiryaev, *Limit Theorems for Stochastic Processes*. Number 488 in Grundlehren der Mathematischen Wissenschaften [Fundamental Principles of Mathematical Sciences] (Springer, Berlin, 2003)
38. K. Kamm, S. Pagliarani, A. Pascucci, On the stochastic Magnus expansion and its application to SPDEs. J. Sci. Comput. **89**(3), 31 (2021), Id/No 56
39. M. Keller-Ressel, W. Schachermayer, J. Teichmann, Affine processes are regular. Probab. Theory Related Fields **151**(3–4), 591–611 (2011)
40. T.G. Kurtz, E. Pardoux, P. Protter, Stratonovich stochastic differential equations driven by general semimartingales. Ann. Inst. Henri Poincaré Probab. Stat. **31**(2), 351–377 (1995)
41. H. Lacoin, R. Rhodes, V. Vargas, A probabilistic approach of ultraviolet renormalization in the boundary sine-Gordon model. Probab. Theory Relat. Fields **185**(1–2), 1–40 (2023)
42. J.-F. Le Gall, *Brownian Motion, Martingales, and Stochastic Calculus* (Springer, 2016)
43. Y. LeJan, Z. Qian, Stratonovich's signatures of Brownian motion determine Brownian sample paths. Probab. Theory Related Fields **157**, 209 (2011)
44. M. Lemercier, T. Lyons, C. Salvi, Log-PDE Methods for Rough Signature Kernels, eprint arXiv:2404.02926 (2025). https://doi.org/10.48550/arXiv.2404.02926
45. S. Li, H. Ni, Expected signature of stopped Brownian motion on d-dimensional $C^{2,\alpha}$-domains has finite radius of convergence everywhere: $2 \leq d \leq 8$. J. Funct. Anal. **282**(12), 45 (2022), Id/No 109447
46. S. Li, H. Ni, Q. Zhu, Small mass limit of expected signature for physical Brownian motion. Preprint, arXiv:2305.00343 [math.PR] (2023) (2023)
47. T. Lyons, Rough paths, signatures and the modelling of functions on streams, in *Proceedings of the International Congress of Mathematicians—Seoul 2014*, vol. IV (Kyung Moon Sa, Seoul, 2014), pp. 163–184

48. T. Lyons, N. Victoir, Cubature on Wiener space. Proc. R. Soc. Lond. Ser. A Math. Phys. Eng. Sci. **460**(2041), 169–198 (2004)
49. W. Magnus, On the exponential solution of differential equations for a linear operator. Commun. Pure Appl. Math. **7**(4), 649–673 (1954)
50. S.I. Marcus, Modeling and analysis of stochastic differential equations driven by point processes. IEEE Trans. Inform. Theory **24**(2), 164–172 (1978)
51. S.I. Marcus, Modeling and approximation of stochastic differential equations driven by semimartingales. Stochastics **4**(3), 223–245 (1981)
52. H.P. McKean, *Stochastic Integrals*. Number 353 in AMS Chelsea Publishing Series (American Mathematical Society, Piscataway, 1969)
53. P.A. Mykland, Bartlett type identities for martingales. Ann. Stat. **22**(1), 21–38 (1994)
54. P.E. Protter, *Stochastic Integration and Differential Equations*. Stochastic Modelling and Applied Probability, 2nd edn. (Springer, Berlin Heidelberg, 2005)
55. C. Reutenauer, Free Lie algebras, in *Handbook of Algebra*, vol. 3 (Elsevier, Amsterdam, 2003), pp. 887–903
56. D. Revuz, M. Yor, *Continuous Martingales and Brownian Motion*. Grundlehren der mathematischen Wissenschaften (Springer, Berlin Heidelberg, 2004)
57. F. Schur, Zur Theorie der endlichen Transformationsgruppen. Math. Ann. **38**, 263–286 (1891)

The manufacturer's authorised representative in the EU is Springer Nature Customer Service Centre GmbH, Europaplatz 3, 69115 Heidelberg, Germany. If you have any concerns regarding our products, please contact ProductSafety@springernature.com

Printed and bound by CPI Group (UK) Ltd, Croydon, CR0 4YY
07/07/2026
02160920-0004